Nalini Anantharaman • Ashkan Nikeghbali
Michael Th. Rassias
Editors

Frontiers in Analysis and Probability

In the Spirit of the Strasbourg-Zürich Meetings

Springer

Editors
Nalini Anantharaman
IRMA
University of Strasbourg
Strasbourg, France

Ashkan Nikeghbali
Institute of Mathematics
University of Zürich
Zürich, Switzerland

Michael Th. Rassias
Institute of Mathematics
University of Zürich
Zürich, Switzerland

ISBN 978-3-030-56411-7 ISBN 978-3-030-56409-4 (eBook)
https://doi.org/10.1007/978-3-030-56409-4

Mathematics Subject Classification: 32-XX, 34-XX, 35-XX, 37-XX, 39-XX, 42-XX, 43-XX, 46-XX, 47-XX, 60-XX, 65-XX, 81-XX

This Springer imprint is published by the registered company Springer Nature Switzerland AG.
The registered company address is: Gewerbestrasse 11, 6330 Cham, Switzerland

Preface

This volume presents papers devoted to a broad spectrum of areas of Mathematical Analysis and Probability Theory, in the spirit of the topics treated in the so-called Strasbourg–Zürich Meetings. These meetings have been organized twice a year since 2015, taking place once in Zürich and once in Strasbourg each year, and constitute a place of vibrant mathematical communication that gathers experts from all over the world.

Topics treated within the scope of this volume include the study of monochromatic random waves defined for general Riemannian manifolds, notions of entropy related to a compact manifold of negative curvature, interacting electrons in a random background, l^p-cohomology (in degree one) of graph and its connections with other topics, limit operators for circular ensembles, polyharmonic functions for finite graphs and Markov chains, the ETH approach to quantum mechanics, two-dimensional quantum Yang–Mills theory, Gibbs measures of nonlinear Schrödinger equations, interfaces in spectral asymptotics, and nodal sets.

The papers published in this volume have been contributed by experts from the international community, who have presented the state-of-the-art research in the corresponding problems treated. The effort has been made for the present volume to be a valuable source for both graduate students and research mathematicians working in analysis, probability as well as their interconnections and applications.

We express our warmest thanks to all the contributing authors of this volume, who have participated in this collective effort. Last but not least, we would like to extend our appreciation to the Springer staff for their valuable help throughout the publication process of this work.

Strasbourg, France — Nalini Anantharaman

Zürich, Switzerland — Ashkan Nikeghbali

Zürich, Switzerland — Michael Th. Rassias

Contents

Monochromatic Random Waves for General Riemannian Manifolds

Yaiza Canzani

1 Introduction

This is a survey article on the recent developments on monochromatic random waves for general Riemannian manifolds obtained in [7–10, 30]. Let (M, g) be a compact, smooth, Riemannian manifold without boundary of dimension $n \geq 2$, and write Δ_g for the corresponding positive definite Laplace–Beltrami operator. Consider an orthonormal basis $\{\varphi_{\lambda_j}\}_{j=1}^{\infty}$ of $L^2(M, g)$ consisting of real-valued eigenfunctions

$$\Delta_g \varphi_{\lambda_j} = \lambda_j^2 \varphi_{\lambda_j},$$

with eigenvalues $0 = \lambda_0 < \lambda_1 \leq \lambda_2 \leq \cdots \nearrow \infty$, normalized so that $\|\varphi_{\lambda_j}\|_{L^2} = 1$.

Laplace eigenfunctions have been a common object of study for the mathematical physics community since they encode how heat and waves propagate across M. From a quantum mechanics point of view, $|\varphi_{\lambda_j}(x)|^2$ is the probability density for finding a quantum particle of energy λ_j^2 at the point x. It is therefore a natural problem to try to understand how φ_{λ_j} behaves. For example, one would like to understand how many minimums and maximums φ_{λ_j} has, as they are the most likely places for the quantum particles to be found at. See Figure 1. Similarly, understanding the geometry of the zero set of φ_{λ_j} would yield information on the structure of the least likely places for the quantum particles. See Figure 2.

However, it is often the case that studying such questions for φ_{λ_j} defined on a general manifold is quite hard, as eigenfunctions cannot be computed explicitly.

Y. Canzani (✉)
Department of Mathematics, University of North Carolina, Chapel Hill, Chapel Hill, NC, USA
e-mail: canzani@email.unc.edu

N. Anantharaman et al. (eds.), *Frontiers in Analysis and Probability*,
https://doi.org/10.1007/978-3-030-56409-4_1

Fig. 1 Zeros and critical points for an eigenfunction on a torus. The function takes positive values on the shaded black areas, and negative values on the white areas. The maximums for the function are attained at the red points, while the minimums occur at the blue points. This picture was created by E. Vouga

Fig. 2 Nodal domains of a monochromatic random wave on the round sphere. Picture created by D. Beliaev

Even more so, for high energies, numerical methods cannot approximate the eigenfunctions in an efficient way that would keep track of features such as the structure of their zero set. It is then natural to randomize the problem and to study how the eigenfunctions behave on average.

It is then natural to work with *monochromatic random waves* of frequency λ. These are random fields on M defined by

$$\phi_\lambda := \frac{1}{\sqrt{\dim H_{\eta,\lambda}}} \sum_{\lambda_j \in [\lambda, \lambda+\eta(\lambda)]} a_j \varphi_{\lambda_j}, \tag{1}$$

where the coefficients $a_j \sim N(0,1)$ are real valued, i.i.d, standard Gaussian random variables, $\eta = \eta_\lambda = \eta(\lambda)$ is a non-negative function satisfying $\eta(\lambda) = o(\lambda)$ as $\lambda \to \infty$, and

$$H_{\eta,\lambda} := \bigoplus_{\lambda_j \in [\lambda, \lambda+\eta_\lambda]} \ker(\Delta_g - \lambda_j^2 \mathrm{Id}).$$

We write

$$\phi_\lambda \in \mathrm{RW}_\lambda(M, g, \eta)$$

for short. The ensembles ϕ_λ are Gaussian models for eigenfunctions of the Laplacian with eigenvalue approximately equal to λ^2 on a compact Riemannian manifold (M, g). In the setting of a general smooth manifold, the ensembles RW_λ were first defined by Zelditch in [36]. Zelditch was inspired in large part by the influential work of Berry [3], which proposes that random planar waves on Euclidean space and flat tori are good semiclassical models for high frequency eigenfunctions in quantum systems whose classical dynamics are chaotic. As we will see in Section 2, when properly scaled, the waves $\phi_\lambda \in \mathrm{RW}_\lambda(M, g, \eta)$ behave like random planar waves. Random planar waves are Laplace eigenfunctions with eigenvalue 1, and since their frequency is fixed to be 1 they are said to be monochromatic. The fact that the scaled ϕ_λ behave like random planar waves as $\lambda \to \infty$ is the reason why waves in $\mathrm{RW}_\lambda(M, g, \eta)$ are said to be monochromatic.

On round spheres and flat tori the Laplace eigenvalues occur with large multiplicity. Indeed, $\dim H_{0,\lambda}$ grows like λ^{n-1} when λ is an eigenvalue. Therefore, in these cases, one typically takes $\eta \equiv 0$ so that $\phi_\lambda \in \mathrm{RW}_\lambda(M, g, 0)$ is an exact eigenfunction and $\lambda \in \{\lambda_j\}$. However, for a generic metric on any smooth compact manifold M, the eigenvalues λ_j^2 are simple. It is then natural to take η so that $\dim H_{\eta,\lambda}$ has the same rate of growth in powers of λ as the dimension of the eigenspaces for a round sphere. In particular, it is known [7] that if (M, g) has at least one non self-focal point (that is, there exists $x \in M$ so that $|\mathcal{L}_{x,x}| = 0$, see (11)), then for every $c > 0$ there exists $C > 0$ such that $\dim H_{c,\lambda}$ grows like $C\lambda^{n-1}$ as $\lambda \to \infty$. Since the existence of a non self-focal point is a very weak condition, it is customary to work with random waves in $\mathrm{RW}_\lambda(M, g, c)$ for some $c > 0$.

This survey article focuses on the results of [7–10, 30]. The results in [7, 9] were the first ones to allow for the treatment of monochromatic random waves to take place on general manifolds by establishing that, when properly rescaled, the waves have a universal behavior. Prior to these results, monochromatic random waves had only been studied for the torus or the sphere. The article [8] is the first one in the literature pertaining statistics of the size of zero set and of the numbers of critical points for monochromatic random waves on general Riemannian manifolds. The results in [10, 30] deal with the study of the diffeomorphism types of the components of the zero sets of the monochromatic random waves, and of the nesting configurations of the components. These results build on the ground breaking work of Nazarov–Sodin [31].

In this article we discuss the following aspects of $\phi_\lambda \in \mathrm{RW}_\lambda(M, g, \eta)$.

- **Section 2:** Universal behavior of ϕ_λ.
- **Section 3:** Number of critical points and size of the zero set of ϕ_λ.
- **Section 4:** Structure of the zero set of ϕ_λ.

The literature about random waves is extensive and rapidly evolving. This survey by no means attempts to give an overall account of every known result. There are numerous works directly related to the topics of this survey, including [4–6, 11, 14–16, 19–22, 25–29, 31, 33, 34].

2 Universal Behavior of ϕ_λ

By the Kolmogorov Consistency Theorem, the law of $\phi_\lambda \in \mathrm{RW}_\lambda(M, g, \eta)$, which is a centered smooth Gaussian field, is completely characterized by its covariance kernel

$$\Pi_{\eta,\lambda}(x, y) := \mathrm{Cov}\,(\phi_\lambda(x), \phi_\lambda(y)) = \frac{1}{\dim H_{\eta,\lambda}} \sum_{\lambda_j \in [\lambda, \lambda+\eta\lambda]} \varphi_{\lambda_j}(x)\varphi_{\lambda_j}(y),$$

where $x, y \in M$. The function $\Pi_{\eta,\lambda}(x, y)$ is the Schwartz kernel for the orthogonal projection operator $\Pi_{\eta,\lambda} : L^2(M, g) \to H_{\eta,\lambda}$, normalized to have unit trace. The study of local quantities, such as the size of the zero set of ϕ_λ, or the number of critical points of ϕ_λ, hinges on understanding the statistics of ϕ_λ, as $\lambda \to \infty$, restricted to "wavelength balls" of radius $\approx \lambda^{-1}$ around a fixed point $x \in M$. After rescaling by $1/\lambda$, the function ϕ_λ has frequency approximately equal to 1 on such balls in the sense that it solves the approximate local eigenvalue equation

$$\Delta_{T_x M} \phi_\lambda(x + \tfrac{u}{\lambda}) \approx \phi_\lambda(x + \tfrac{u}{\lambda}), \tag{2}$$

where $\Delta_{T_x M}$ denotes the *flat* Laplacian on the tangent space at x, $T_x M$. One could therefore expect, after the scaling, for the Gaussian random wave $u \mapsto \phi_\lambda(x + \frac{u}{\lambda})$ to behave like a Gaussian random wave ϕ_∞ on $\mathbb{R}^n \cong T_x M$ satisfying

$$\Delta_{\mathbb{R}^n} \phi_\infty = \phi_\infty.$$

The latter is called a random planar wave, and we discuss them in Section 2.1. Moreover, we shall see in Section 2.2, that for a generic Riemannian metric on M, the rescaled covariance kernel $\Pi_{\eta,\lambda}$ of $\phi_\lambda \in \mathrm{RW}_\lambda(M, g, \eta)$ converges in the C^∞ topology to that of a random planar wave ϕ_∞ on $\mathbb{R}^n \cong T_x M$.

2.1 Random Planar Waves

Let $\sigma_{S^{n-1}}$ be the Haar measure on the round sphere S^{n-1}, normalized so that $\sigma_{S^{n-1}}(S^{n-1}) = 1$. Using that the transformation $\xi \mapsto -\xi$ preserves S^{n-1}, choose a real-valued orthonormal basis $\{\psi_j\}_{j=1}^\infty$ of $L^2(S^{n-1}, \sigma_{S^{n-1}})$ satisfying

$$\psi_j(-\xi) = (-1)^{\epsilon_j} \psi_j(\xi), \qquad \epsilon_j \in \{0, 1\}. \tag{3}$$

A *random planar wave* is defined to be the random real-valued function ϕ_∞ on $\mathbb{R}^n$ given by

$$\phi_\infty(u) = \sum_{j=1}^{\infty} b_j \, i^{\eta_j} \, \widehat{\psi_j}(u), \tag{4}$$

where

$$\widehat{\psi_j}(u) = \int_{\mathbb{R}^n} \psi_j(\xi) e^{-i\langle u,\xi\rangle} d\sigma_{S^{n-1}}(\xi), \tag{5}$$

and the b_j's are i.i.d, real valued, standard Gaussian random variables. We write

$$\phi_\infty \in \mathrm{RW}_1(\mathbb{R}^n, g_{\mathbb{R}^n}),$$

for short, where $g_{\mathbb{R}^n}$ is the Euclidean metric. We note that the fields in $\mathrm{RW}_1(\mathbb{R}^n, g_{\mathbb{R}^n})$ do not depend on the choice of the orthonormal basis $\{\psi_j\}$. In addition, since the Euclidean Laplacian is $\Delta_{g_{\mathbb{R}^n}} = -\sum_{k=1}^n \partial_{u_k}^2$, and $\Delta_{g_{\mathbb{R}^n}} e^{-i\langle u,\xi\rangle} = e^{-i\langle u,\xi\rangle} \sum_{k=1}^n \xi_k^2$, it is immediate that

$$\Delta_{g_{\mathbb{R}^n}} \phi_\infty = \phi_\infty.$$

As explained in the introduction, random planar waves are often called monochromatic random waves because their frequency (the square root of their eigenvalue) is *equal* to 1.

Next, note that the distributional identity $\sum_{j=1}^\infty \phi_j(\xi)\phi_j(\eta) = \delta(\xi - \eta)$ on S^{n-1} together with (3) lead to the explicit expression for the covariance function:

$$\Pi_\infty(u, v) := \mathrm{Cov}(\phi_\infty(u), \phi_\infty(v)) = \int_{\mathbb{R}^n} e^{i\langle u-v,\xi\rangle} d\sigma_{S^{n-1}}(\xi), \tag{6}$$

where $u, v \in \mathbb{R}^n$. From (4) it follows that almost all ϕ_∞'s are analytic in u [1]. It is also known that

$$\Pi_\infty(u, v) = \frac{1}{(2\pi)^{\frac{n}{2}}} \frac{J_\nu(|u - v|)}{|u - v|^\nu}, \tag{7}$$

where J_ν is the Bessel function of index $\nu := \frac{n-2}{2}$.

There is a natural choice of a basis for $L^2(S^{n-1}, d\sigma_{S^{n-1}})$ given by spherical harmonics. Let $\{Y_m^\ell\}_{m=1}^{d_{\ell,n}}$ be a real-valued basis for the space of spherical harmonics $\mathcal{E}_\ell(S^{n-1})$ of eigenvalue $\ell(\ell + n - 2)$, where $d_{\ell,n} = \dim \mathcal{E}_\ell(S^{n-1})$. In [10, Corollary 2.2] we prove that the monochromatic Gaussian ensembles ϕ_∞'s take the form

$$\phi_\infty(u) = (2\pi)^{\frac{n}{2}} \sum_{\ell=0}^{\infty} \sum_{m=1}^{d_{\ell,n}} b_{\ell,m}\, Y_m^\ell\left(\frac{u}{|u|}\right) \frac{J_{\ell+\nu}(|u|)}{|u|^\nu},$$

where the $b_{\ell,m}$'s are i.i.d standard Gaussian variables.

2.2 Points of Isotropic Scaling

The discussion around (2) shows that it is natural to study ϕ_λ by fixing $x \in M$ and considering the rescaled pullback of ϕ_λ to the tangent space $T_x M$. We denote this pullback by

$$\phi_\lambda^x(u) := \phi_\lambda\Big(\exp_x\Big(\frac{u}{\lambda}\Big)\Big), \tag{8}$$

where $\exp_x : T_x M \to M$ is the exponential map. The dilated functions ϕ_λ^x are centered Gaussian fields on $T_x M$, and we denote their scaled covariance kernel by

$$\Pi_{\eta,\lambda}^x(u, v) := \mathrm{Cov}(\phi_\lambda^x(u), \phi_\lambda^x(v)) = \Pi_{\eta,\lambda}\Big(\exp_x\Big(\frac{u}{\lambda}\Big),\ \exp_x\Big(\frac{v}{\lambda}\Big)\Big).$$

When x is a point of isotropic scaling (see Definition 1 below), we shall see that the kernels $\Pi_{\eta,\lambda}^x$ converge to the covariance kernel of a random planar wave

$$\phi_\infty^x \in \mathrm{RW}_1(T_x M, g_x).$$

Here, g_x denotes the constant coefficient metric obtained by freezing g at x. By the Kolmogorov Extension Theorem, together with (6), the random wave ϕ_∞^x is completely characterized by its two point correlation function kernel

$$\Pi_\infty^x(u, v) = (2\pi)^{\frac{n}{2}} \frac{J_{\frac{n-2}{2}}\big(\|u - v\|_{g_x}\big)}{\|u - v\|_{g_x}^{\frac{n-2}{2}}} = \int_{S_x M} e^{i\langle u-v,\xi\rangle_{g_x}}\, d\sigma_{S_x M}(\xi). \tag{9}$$

Here J_ν denotes a Bessel function of the first kind with index ν, $S_x M$ is the unit sphere in $T_x M$ with respect to g_x, and $d\sigma_{S_x M}$ is the hypersurface measure on $S_x M$.

Definition 1 A point $x \in M$ is a *point of isotropic scaling*, denoted $x \in \mathcal{IS}(M, g, \eta)$, if for every non-negative function r_λ satisfying $r_\lambda = o(\lambda)$ as $\lambda \to \infty$, and all $\alpha, \beta \in \mathbb{N}^n$, we have

$$\sup_{u,v \in B_{r_\lambda}} \left| \partial_u^\alpha \partial_v^\beta \left[\Pi_{\eta,\lambda}^x(u, v) - \Pi_\infty^x(u, v)\right]\right| = o_{\alpha,\beta}(1) \tag{10}$$

as $\lambda \to \infty$, where the rate of convergence depends on α, β and B_r denotes a ball of radius r centered at $0 \in T_x M$. We also say that M is a *manifold of isotropic scaling* if

$$M = \mathcal{IS}(M, g, \eta)$$

and if the convergence in (10) is uniform over $x \in M$ for each $\alpha, \beta \in \mathbb{N}^n$.

Verifying that $x \in M$ belongs to $\mathcal{IS}(M, g, \eta)$ is difficult to do directly, except on simple examples such as the flat torus. We briefly recall several settings in which $\mathcal{IS}(M, g, \eta)$ is known to be large.

- Let S^n be the n-sphere equipped with the round metric g_{S^n}. The Mehler–Heine asymptotics [24] imply that

$$\mathcal{IS}(S^n, g_{S^n}, 0) = S^n,$$

  when the limit in (10) is taken along the sequence of eigenvalues $\lambda_j \to \infty$ for the sphere. In this case, the ϕ_λ's are known as random spherical harmonics.
- Let $\mathbb{T}^n$ be the n-dimensional torus equipped with the flat metric $g_{\mathbb{T}^n}$. When $n \geq 5$ we have that $\mathcal{IS}(\mathbb{T}^n, g_{\mathbb{T}^n}, 0) = \mathbb{T}^n$. For $2 \leq n \leq 4$, the asymptotics (10) hold at every $x \in \mathbb{T}^n$ but only for a density one subsequence of eigenvalues [13]. In this case, the ϕ_λ's are known as random trigonometric polynomials.
- The pointwise Weyl law [17] implies that if $\lim_{\lambda\to\infty} \eta_\lambda = \infty$, then $\mathcal{IS}(M, g, \eta) = M$.

In addition, it is very likely that if (M, g) has no conjugate points, then the condition

$$\lim_{\lambda\to\infty} \log(\lambda) \cdot \eta_\lambda = \infty$$

implies $\mathcal{IS}(M, g, \eta) = M$. This was proved by B. Keeler in [18], but with the convergence in (10) only holding for $\alpha = \beta = 0$. Note that if (M, g) has negative sectional curvature everywhere, then it has no conjugate points and all points are non self-focal. In contrast, there exist smooth perturbations of the round metric on S^2 for which $\mathcal{IS}(S^2, g, 1) \subsetneq S^2$ (see [23, 35]).

For $x, y \in M$ let

$$\mathcal{L}_{x,y} = \{\xi \in S_x M : \exists t > 0 \text{ s.t. } \exp_x(t\xi) = y\} \tag{11}$$

be the set of directions that generate geodesic arcs from x to y. The set $\mathcal{L}_{x,y}$ is contained in $S_x M$ and $S_x M$ is endowed with the Liouville measure. The corresponding volume of $\mathcal{L}_{x,y}$ is denoted by $|\mathcal{L}_{x,y}|$.

The main result of this section is the following, and it was proved in [7, 9].

Theorem 1 *Let (M, g) be a compact, smooth, Riemannian manifold, with no boundary. Let η be a non-negative function with* $\liminf_{\lambda\to\infty} \eta_\lambda > 0$. *Let $x \in M$ be so that $|\mathcal{L}_{x,x}| = 0$. Then,*

$$x \in \mathcal{IS}(M, g, \eta). \tag{12}$$

By [32, Lem 6.1], the condition that $|\mathcal{L}_{x,x}| = 0$ for all $x \in M$ is generic in the space of Riemannian metrics on a fixed compact smooth manifold M.

Definition 1 gives that if $x \in \mathcal{IS}(M, g, \eta)$, then the scaling limit of waves in $\mathrm{RW}_\lambda(M, g, \eta)$ around x is universal in the sense that it depends only on the dimension of M. In the language of Nazarov–Sodin [31] the asymptotics (10) imply that if $M = \mathcal{IS}(M, g, \eta)$, then the ensembles $\mathrm{RW}_\lambda(M, g, \eta)$ have translation invariant local limits.

3 Number of Critical Points and Size of the Zero Set

Define the measures of integration over the zero set $\{\phi_\lambda = 0\}$ and the set of critical points $\{d\phi_\lambda = 0\}$ by

$$\mathrm{Z}_\lambda(\psi) := \int_{\phi_\lambda^{-1}(0)} \psi(x) d\sigma_{Z_\lambda}(x) \qquad \text{and} \qquad \mathrm{Crit}_\lambda(\psi) := \sum_{d\phi_\lambda(x)=0} \psi(x),$$

where $\psi : M \to \mathbb{R}$ and σ_{Z_λ} is the $(n-1)$-dimensional Hausdorff measure over $\{\phi_\lambda = 0\}$. This section is divided into two parts. In Section 3.1 we give asymptotics for $\mathbb{E}[Z_\lambda]$ and $\mathbb{E}[\mathrm{Crit}_\lambda]$, and bounds for their variances. The results in Section 3.1 rely heavily on a careful analysis of what happens for the scaled random waves ϕ_λ^x. The results for the localized waves are discussed in Section 3.2.

Previous results on the Hausdorff measure of the zero sets focus primarily on exactly solvable examples. On round spheres, for instance, Bérard [2] proved (14) (example (1) on p.3). Later, in the same setting, Neuheisel [25] and Wigman [33] obtained upper bounds for the variance that are of polynomial order in λ. Further, on S^2, Wigman [34] found that the variance actually grows like $\lambda^{-2} \log \lambda$ as $\lambda \to \infty$. On flat tori $\mathbb{T}^n$ (for exact eigenfunctions) Rudnick and Wigman [29] computed the expected value of the total Hausdorff measure of the zero set and gave an upper bound of the form $\lambda^2 (\dim(H_{0,\lambda}))^{-1/2}$ on its variance. Subsequently, on $\mathbb{T}^2$, Krishnapur, Kurlberg, and Wigman [19] found that the variance is asymptotic to a constant, while Marinucci, Peccati, Rossi, and Wigman proved that the size of the zero set converges to a limiting distribution that is not Gaussian and depends on the angular distribution of lattice points on circles [22].

The behavior of the number of critical points has been studied in detail on S^2. Nicolaescu [26] studied the expected value of the number of critical points, obtaining (15). The variance was studied by Cammarota, Marinucci, and Wigman [6].

They obtain a polynomial upper bound. This upper bound was later improved by Cammarota and Wigman [5] who proved that the variance grows like $\lambda^2 \log \lambda$ (as opposed to our $\lambda^{7/2}$ estimate) as $\lambda \to \infty$. Finally, for smooth domain in $\mathbb{R}^2$, Nourdin–Peccati–Rossi [27] prove that both for real and complex random waves, the Hausdorff measure of the nodal set is asymptotically normal in the high frequency limit.

3.1 Global Statistics

The main result in this section gives asymptotics for the expected value, and estimates for the variance, of the linear statistics of Z_λ, Crit_λ that are valid for generic Riemannian metrics on M. For the estimates about the means of $Z_\lambda(\psi)$, $\text{Crit}_\lambda(\psi)$ one needs to ask that (M, g) be a manifold of isotropic scaling (see Definition 1). This is true for any manifold with negative curvature, or with no conjugate points. The variance estimates are more delicate, so one needs to ask in addition that the restrictions of ϕ_λ to small balls centered at different points become asymptotically uncorrelated. This is the following definition.

Definition 2 The random waves $\phi_\lambda \in \text{RW}_\lambda(M, g, \eta)$ are said to have *short-range correlations* if for each $\varepsilon > 0$ and every $\alpha, \beta \in \mathbb{N}$

$$\sup_{\{x,y:\, d_g(x,y) \geq \lambda^{-1+\varepsilon}\}} \left| \nabla_x^\alpha \nabla_y^\beta \Pi_{\eta,\lambda}(x, y) \right| = o_\varepsilon(\lambda^{\alpha+\beta}), \tag{13}$$

as $\lambda \to \infty$, where ∇_x, ∇_y are covariant derivatives.

This condition is again generic in the space of Riemannian metrics on (M, g) and is satisfied for example if for any pair of points $x, y \in M$ the measure of geodesic arcs joining them is zero. That is, if $\left|\mathcal{L}_{x,y}\right| = 0$ for all $x, y \in M$, then the random waves in $\text{RW}_\lambda(M, g, \eta)$ have short-range correlations.

The condition that $\left|\mathcal{L}_{x,y}\right| = 0$ for all $x, y \in M$ is known to happen on manifolds of negative curvature, or more generally, with no conjugate points (see [8, Section 1.5]). It is likely that a similar argument would show that $\left|\mathcal{L}_{x,y}\right| = 0$ for all $x, y \in M$ is also generic but have not checked the details. It is known, however, that $\left|\mathcal{L}_{x,y}\right| = 0$ holds for all $x, y \in M$ if (M, g) is negatively curved or, more generally, has no conjugate points.

We are ready to state the main theorem of this section. This result was proved in [8].

Theorem 2 *Let (M, g) be a smooth, compact, Riemannian manifold of dimension $n \geq 2$ with no boundary. Let $\eta = \eta(\lambda)$ be a non-negative function satisfying $\eta(\lambda) = o(\lambda)$ as $\lambda \to \infty$. Let $\phi_\lambda \in \text{RW}_\lambda(M, g, \eta)$ and suppose that M is a manifold of isotropic scaling (Definition 1). Then, for any bounded measurable function $\psi : M \to \mathbb{R}$,*

$$\lim_{\lambda\to\infty} \mathbb{E}\left[\lambda^{-1} Z_\lambda(\psi)\right] = \frac{1}{\sqrt{\pi n}} \frac{\Gamma\left(\frac{n+1}{2}\right)}{\Gamma\left(\frac{n}{2}\right)} \int_M \psi(x) dv_g(x) \tag{14}$$

and

$$\lim_{\lambda\to\infty} \mathbb{E}\left[\lambda^{-n} \operatorname{Crit}_\lambda(\psi)\right] = C_n \int_M \psi(x) dv_g(x), \tag{15}$$

where C_n is a positive constant that depends only on n. Suppose further that ϕ_λ has short-range correlations in the sense of (13). *Then,*

$$\operatorname{Var}\left[\lambda^{-1} Z_\lambda(\psi)\right] = O(\lambda^{-\frac{n-1}{2}}) \tag{16}$$

and

$$\operatorname{Var}\left[\lambda^{-n} \operatorname{Crit}_\lambda(\psi)\right] = O\left(\lambda^{-\frac{n-1}{2}}\right), \tag{17}$$

as $\lambda\to\infty$.

Theorem 2 is the first result with a non-trivial variance estimate for the Hausdorff measure of the nodal set of random waves for a generic smooth Riemannian manifold (for real analytic (M, g) a weaker estimate was given in [36, Cor. 2]). A version of (14) was also stated, with a heuristic proof, in [36, Prop. 2.3] for both Zoll and aperiodic manifolds.

We also note that the test function ψ in Theorem 2 can be replaced by a function $\psi(x) = \psi(x, \phi_\lambda(x), D^2\phi_\lambda(x), \ldots)$ depending on the jets of ϕ_λ provided $\psi : \mathbb{R}^n \times C^0(\mathbb{R}^n, \mathbb{R}^k) \to \mathbb{R}$ is bounded and continuous when $C^0(\mathbb{R}^n, \mathbb{R}^k)$ is equipped with the topology of uniform convergence on compact sets. Hence, for example, we could study the distribution of critical values by taking $\psi(u, \phi_\lambda) = \mathbf{1}_{\{\phi_\lambda^x \geq \alpha\}}(u)$, for $\alpha \in \mathbb{R}$.

In addition, the proof of Theorem 2 actually shows that (14) holds as soon as almost every point is a point of isotropic scaling. That is, it holds provided

$$\operatorname{vol}_g(M \backslash \mathcal{IS}(M, g, \eta)) = 0,$$

(see Definition 1).

Furthermore, by the Borel–Cantelli Lemma, if $n \geq 4$ and ϕ_j are independent frequency $j \in \mathbb{N}$ random waves on (M, g), then (16) shows that the total nodal set measure $j^{-1} Z_j(\psi) - \mathbb{E}\left[j^{-1} Z_j(\psi)\right]$ converges almost surely to 0.

Finally, when $n = 2$ we have $C_2 = \mathbb{E}\left[\operatorname{Crit}_{\infty,1}\right] = \frac{1}{4\pi\sqrt{6}}$ where C_2 is the dimensional constant in (15), see (25).

Theorem 2 hinges on a careful study of the statistics of ϕ_λ when restricted to "wavelength balls" of radius $\approx \lambda^{-1}$ around a fixed point $x \in M$ of isotropic scaling. The results that describe the behavior of Z_λ or $\operatorname{Crit}_\lambda$ restricted to these shrinking balls are described in Sections 3.2.1 and 3.2.2, respectively. The results are "glued" to obtain Theorem 2. Glueing variance estimates is a delicate matter. It is instrumental to the proof that the waves have short-range correlations.

3.2 *Local Statistics*

In this section we discuss the behavior of the zero sets and of the critical points for the scaled waves ϕ_λ^x. When x is a point of isotropic scaling the behavior of the scaled random wave ϕ_λ^x converges to that of the random planar wave $\phi_\infty^x \in \mathrm{RW}_1(T_xM, g_x)$. One can therefore prove much stronger results on statistics for ϕ_λ^x than ϕ_λ.

3.2.1 Local Universality of Zeros

Consider the rescaled random wave ϕ_λ^x for $x \in \mathcal{IS}(M, g, \eta)$ and denote by Z_λ^x its Riemannian hypersurface (i.e. Hausdorff) measure:

$$Z_\lambda^x(A) := \sigma_{Z_\lambda}\Big(\big(\phi_\lambda^x\big)^{-1}(0) \cap A\Big), \qquad \forall A \subseteq T_xM \text{ measurable.}$$

The main result concerns the restriction of Z_λ^x to various balls B_r of radius r centered at $0 \in T_xM$. We set

$$Z_{\lambda,r}^x := \frac{\mathbf{1}_{B_r} \cdot Z_\lambda^x}{\mathrm{vol}(B_r)} \quad \text{and} \quad Z_{\infty,r}^x := \frac{\mathbf{1}_{B_r} \cdot Z_\infty^x}{\mathrm{vol}(B_r)}. \tag{18}$$

We have denoted by $\mathbf{1}_{B_r}$ the characteristic function of the ball B_r and by Z_∞^x the hypersurface measure on $(\phi_\infty^x)^{-1}(0)$ for $\phi_\infty^x \in \mathrm{RW}_1(T_xM, g_x)$. Again, for various measures μ, we write $\mu(\psi)$ for integration of a measurable function ψ against μ. In particular,

$$Z_{\lambda,r}^x(1) = \frac{\mathcal{H}^{n-1}\Big(\big(\phi_\lambda^x\big)^{-1}(0) \cap B_r\Big)}{\mathrm{vol}(B_r)}.$$

The following result is proved in [8]. See Figure 3 for a depiction of the statement.

Theorem 3 (Weak Convergence of Zero Set Measures) *Let (M, g) be a smooth, compact, Riemannian manifold of dimension $n \geq 2$ with no boundary. Let $\eta = \eta(\lambda)$ be a non-negative function satisfying $\eta(\lambda) = o(\lambda)$ as $\lambda \to \infty$. Fix a non-negative function r_λ that satisfies $r_\lambda = o(\lambda)$ as $\lambda \to \infty$. Let $\phi_\lambda \in \mathrm{RW}_\lambda(M, g, \eta)$ and $x \in \mathcal{IS}(M, g, \eta)$. Suppose $\lim_{\lambda\to\infty} r_\lambda$ exists and equals $r_\infty \in (0, \infty]$.*

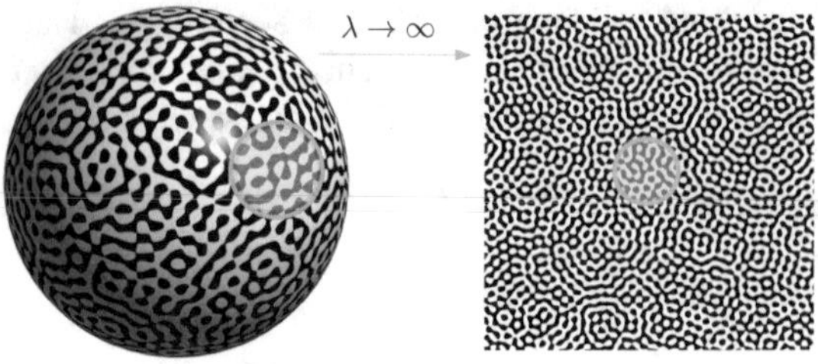

Fig. 3 Depiction of the universal behavior displayed by monochromatic random waves. The zero set measure for the monochromatic random wave on the sphere (left) converges to the zero set measure for the random planar wave (right)

Case 1 ($r_\infty < \infty$): *The measures Z^x_{λ,r_λ} converge to Z^x_{∞,r_∞} weakly in distribution. That is, for any bounded, measurable function $\psi : T_xM \to \mathbb{R}$*

$$Z^x_{\lambda,r_\lambda}(\psi) \quad \xrightarrow{d} \quad Z^x_{\infty,r_\infty}(\psi) \tag{19}$$

as $\lambda \to \infty$, where $\xrightarrow{d}$ denotes convergence in distribution.

Case 2 ($r_\infty = \infty$): *We have the following convergence in probability to a constant:*

$$Z^x_{\lambda,r_\lambda}(1) \quad \xrightarrow{p} \quad \frac{1}{\sqrt{\pi n}} \frac{\Gamma\left(\frac{n+1}{2}\right)}{\Gamma\left(\frac{n}{2}\right)}, \tag{20}$$

as $\lambda \to \infty$. In particular,

$$\lim_{\lambda\to\infty} \mathrm{Var}\left[Z^x_{\lambda,r_\lambda}(1)\right] = 0. \tag{21}$$

The function ψ in (19) can be allowed to depend on the jets $D^j\phi_\lambda$, $j \geq 1$. More precisely, $\psi(u)$ can be replaced by $\psi(u, W(u))$, where W is a random field so that $u \mapsto (\phi^x_\lambda(u), W(u))$ is a continuous Gaussian field with values in $\mathbb{R}^{1+k}$ and $\psi : \mathbb{R}^n \times C^0(\mathbb{R}^n, \mathbb{R}^k) \to \mathbb{R}$ is bounded and continuous when $C^0(\mathbb{R}^n, \mathbb{R}^k)$ is equipped with the topology of uniform convergence on compact sets. Since $(\phi^x_\lambda(u), D\phi^x_\lambda(u), D^2\phi^x_\lambda(u), \ldots)$ is a smooth Gaussian field, we may take $W(u) = \left(D^j\phi_\lambda(u),\ j \geq 1\right)$. Similarly, in (20) and (21), the function $1 = 1(u)$ can be replaced by $\psi(W(u))$ where again $\psi : C^0(\mathbb{R}^n, \mathbb{R}^k) \to \mathbb{R}$ is bounded and continuous in the topology of uniform convergence on compact sets. The only difference is that (20) then reads

$$Z^x_{\lambda,r_\lambda}(\psi) - \mathbb{E}\left[Z^x_{\infty,r_\lambda}(\psi)\right] \quad \xrightarrow{p} \quad 0.$$

The relations (20) and (21) hold even if the balls B_{r_λ} in the definition of Z^x_{λ,r_λ} are replaced by any $\lambda-$dependent sets A_{λ,r_λ} for which the diameter is bounded above and below by constant times r_λ, and whose volume tends to infinity when $r_\lambda \to \infty$.

The rates of convergence in (19)–(21) - even after the generalizations indicated above- are uniform as x varies over a compact set $S \subset \mathcal{IS}(M, g, \eta)$ as long as the convergence in (10) is uniform over S.

3.2.2 Local Universality of Critical Points

Let $x \in M$ and for each $r > 0$ define the normalized counting measure

$$\mathrm{Crit}^x_{\lambda,r} := \frac{1}{\mathrm{vol}(B_r)} \sum_{\substack{d\phi^x_\lambda(u)=0 \\ u \in B_r}} \delta_u \tag{22}$$

of critical points in a ball of radius r. We define $\mathrm{Crit}^x_{\infty,r}$ in the same way as $\mathrm{Crit}^x_{\lambda,r}$ but with ϕ^x_λ replaced by $\phi^x_\infty \in \mathrm{RW}_1(T_xM, g_x)$, and continue to write $\mu(\psi)$ for the pairing of a measure μ with a function ψ. For example,

$$\mathrm{Crit}^x_{\lambda,r}(1) = \frac{\#\{u \in B_r : d\phi^x_\lambda(u) = 0\}}{\mathrm{vol}(B_r)}.$$

Theorem 4 *Let (M, g) be a smooth, compact, Riemannian manifold of dimension $n \geq 2$ with no boundary. Let $\eta = \eta(\lambda)$ be a non-negative function satisfying $\eta(\lambda) = o(\lambda)$ as $\lambda \to \infty$. Fix a non-negative function r_λ that satisfies $r_\lambda = o(\lambda)$ as $\lambda \to \infty$. Let $\phi_\lambda \in \mathrm{RW}_\lambda(M, g, \eta)$ and $x \in \mathcal{IS}(M, g, \eta)$. Suppose that $\lim_{\lambda\to\infty} r_\lambda$ exists and equals $r_\infty \in (0, \infty]$.*

Case 1. ($r_\infty < \infty$): *For $k = 1, 2$ and each bounded measurable function $\psi : T_xM \to \mathbb{R}$*

$$\lim_{\lambda\to\infty} \mathbb{E}\left[\mathrm{Crit}^x_{\lambda,r_\lambda}(\psi)^k\right] = \mathbb{E}\left[\mathrm{Crit}^x_{\infty,r_\infty}(\psi)^k\right]. \tag{23}$$

Case 2. ($r_\infty = \infty$): *We have*

$$\lim_{\lambda\to\infty} \mathrm{Var}[\mathrm{Crit}^x_{\lambda,r_\lambda}(1)] = \mathbb{E}\left[\mathrm{Crit}^x_{\infty,1}(1)\right]. \tag{24}$$

This limit is the expected number of critical points in a ball of radius 1 for frequency 1 random waves on $\mathbb{R}^n$, which is independent of x.

The moments $\mathbb{E}\left[(\mathrm{Crit}^x_{\infty,r_\infty}(\psi))^k\right]$ are finite for $k = 1, 2$. In particular, if $\dim(M) = 2$, then for every $x \in M$

$$\mathbb{E}\left[\mathrm{Crit}^x_{\infty,1}(1)\right] = \frac{1}{4\pi\sqrt{6}}. \tag{25}$$

The balls B_{r_λ} in (24) can be replaced by any λ−dependent sets A_{λ,r_λ} for which the diameter is bounded above and below by a constant times r_λ and whose volume tends to infinity with r_λ.

Both ψ in (23) and the function 1 being integrated against $\mathrm{Crit}^x_{\lambda,r_\lambda}$ in (24) can be replaced by a bounded continuous function of the jets of ϕ_λ, giving information for instance about critical points filtered by critical value.

Also, the rates of convergence in (23) and (24) are uniform over $x \in S \subset \mathcal{IS}(M, g, \eta)$ if (10) is uniform over S.

On the n-dimensional flat torus, Nicolaescu [26] obtained several results related to Theorem 4 in the $r_\infty < \infty$ case.

4 Structure of the Zero Set

Let (M, g) be a Riemannian manifold, and let ϕ be an eigenfunction for the Laplace operator. The zero set $\phi^{-1}(0) = \{x \in M : \phi(x) = 0\}$ decomposes into a collection of connected components which we denote by $\mathcal{C}(\phi)$. See Figure 4. Our interest is in the diffeomorphism types of the components in $\mathcal{C}(\phi)$. For generic ϕ the components of $\mathcal{C}(\phi)$ are smooth $(n-1)$-dimensional manifolds. The connected components of $M\backslash\phi^{-1}(0)$ are the nodal domains of ϕ and our interest is in their nesting properties, again for generic ϕ.

The results presented in this section build on the ground breaking work of Nazarov–Sodin [31]. They studied the number of nodal domains for monochromatic random waves on manifolds with isotropic scaling. They proved that there exists a positive constant C so that the mean number of nodal domains for ϕ_λ grows like $C\lambda^n$. The approach of [10, 30] to study the diffeomorphism types of the zero set components is very similar in spirit to the work [31] as the rationale is that one is counting components of the zero set with a given diffeomorphism type. A similar argument is carried to deal with the nesting configurations.

The argument developed by Nazarov–Sodin hinges on the fact that most zero set components lie within a ball of radius R/λ for $R > 0$ large enough. One can therefore count the number of components of ϕ_λ^x within the ball $B(0, R) \subset T_xM$. The latter is done using the universal behavior of ϕ_λ^x guaranteed by the fact that M is a manifold of isotropic scaling.

The works of Gayet–Welshinger [14–16] are also very related to the results described in this section, only that they are not applicable to monochromatic random waves.

4.1 *Diffeomorphism Types*

Let $\mathcal{D}_{n-1}$ denote the (countable and discrete) set of diffeomorphism classes of compact connected smooth $(n-1)$-dimensional manifolds that can be embedded in $\mathbb{R}^n$. The compact components c in $\mathcal{C}(\phi)$ give rise to elements $D(c)$ in $\mathcal{D}_{n-1}$ (here we are assuming that ϕ is generic with respect to a Gaussian measure so that $\phi^{-1}(0)$ is smooth).

Let $\phi_\lambda \in \mathrm{RW}(M, g, \eta)$. The diffeomorphism types exhibited by the components of $\phi_\lambda^{-1}(0)$ are described by the probability measure $\mu_{\mathcal{D}(\phi_\lambda)}$ on $\mathcal{D}_{n-1}$ given by

$$\mu_{\mathcal{D}(\phi_\lambda)} := \frac{1}{|\mathcal{C}(\phi_\lambda)|} \sum_{c \in \mathcal{C}(\phi_\lambda)} \delta_{D(c)},$$

where δ_D is a point mass at $D \in \mathcal{D}_{n-1}$. The following is part of the main theorem in [30, Theorem 1.1].

Theorem 5 *There exists a probability measure $\mu_{\mathcal{D}}$ supported on $\mathcal{D}_{n-1}$ such that the following holds. Let (M, g) be a smooth, compact, Riemannian manifold of dimension $n \geq 2$ with no boundary. Let $\eta = \eta(\lambda)$ be a non-negative function satisfying $\eta(\lambda) = o(\lambda)$ as $\lambda \to \infty$. Suppose that M is a manifold of isotropic scaling. Then, for any given $D \in \mathcal{D}_{n-1}$ and $\varepsilon > 0$,*

$$\lim_{\lambda \to \infty} \mathbb{P}\Big(\phi_\lambda \in \mathrm{RW}_\lambda(M, g, \eta) : |\mu_{\mathcal{D}(\phi_\lambda)}(D) - \mu_{\mathcal{D}}(D)| > \varepsilon\Big) = 0.$$

The theorem asserts that there exists a probability measure $\mu_{\mathcal{D}}$ on $\mathcal{D}_{n-1}$ to which $\mu_{\mathcal{D}(\phi)}$ approaches as $\lambda \to \infty$, for almost all ϕ. The probability measure $\mu_{\mathcal{D}}$ is universal in that it only depends on the dimension n of M.

For $n \geq 4$, little is known about the space $\mathcal{D}_{n-1}$. In particular, there is no classification for the diffeomorphism types of $(n-1)$-dimensional smooth

Fig. 4 Zero set of a random planar wave in $\mathbb{R}^3$. Picture created by A. Barnett

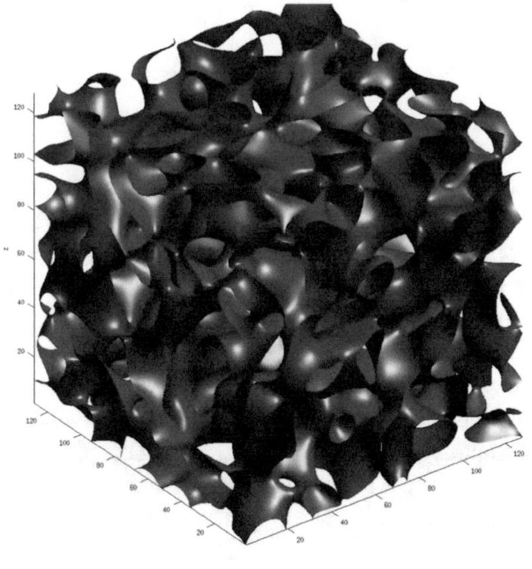

manifolds. This makes it difficult to study the support of $\mu_{\mathcal{D}}$. Remarkably, it is possible to prove that the support of $\mu_{\mathcal{D}}$ is all of $\mathcal{D}_{n-1}$. This result is proved in [10].

Theorem 6 *Every atom of $\mathcal{D}_{n-1}$ is positively charged by $\mu_{\mathcal{D}}$. That is,*

$$\text{supp}(\mu_{\mathcal{D}}) = \mathcal{D}_{n-1}.$$

Theorem 6 asserts that every diffeomorphism type that can occur will do so with a positive probability for the universal distribution of topological types of random monochromatic waves in [30].

The proof of Theorem 6 relies on the fact that for a manifold of isotropic scaling the statistics of ϕ_λ^x converge to those of ϕ_∞^x for every $x \in M$. Indeed, the proof reduces to establishing the following result.

Theorem 7 *Given $D \in \mathcal{D}_{n-1}$ there exists $\phi \in \ker(\Delta_{\mathbb{R}^n} - Id)$ and $c \in \mathcal{C}(\phi)$ for which $D(c) = D$.*

Theorem 7 is of basic interest in the understanding of the possible shapes of nodal sets and domains of eigenfunctions in $\mathbb{R}^n$ (it applies equally well to any eigenfunction with eigenvalue $\lambda^2 > 0$ instead of 1). To prove Theorem 7 one applies Whitney's approximation Theorem to realize c as an embedded real analytic submanifold of $\mathbb{R}^n$. Then, following some techniques in [12] one can find suitable approximations of $\phi \in \ker(\Delta_{g_{\mathbb{R}^n}} - 1)$ and whose zero set contains a diffeomorphic copy of c. The construction of ϕ hinges on the Lax–Malgrange Theorem and Thom's Isotopy Theorem.

The reduction from Theorem 7 to Theorem 6 is abstract and is based on the "soft" techniques in [30, 31]. In particular, it offers us no lower bounds for these probabilities. Developing such lower bounds is an interesting problem.

4.2 Nesting Configurations

Let ϕ be a Laplace eigenfunction for a Riemannian manifold (M, g). The connected components of $M \backslash \phi^{-1}(0)$ are the nodal domains of ϕ and our interest is in their nesting properties, again for generic ϕ. Let $\mathcal{U}$ be a coordinate patch for M. The components of $\mathcal{C}(\phi)$ that are contained in $\mathcal{U}$ are denoted by $\mathcal{C}_{\mathcal{U}}(\phi)$. To each compact $c \in \mathcal{C}_{\mathcal{U}}(\phi)$ we associate a finite connected rooted tree as follows. By the Jordan–Brouwer separation Theorem each component $c \in \mathcal{C}(\phi)$ has an exterior and interior. We choose the interior to be the end that is contained within $\mathcal{U}$. The nodal domains of ϕ, which are in the interior of c, are taken to be the vertices of a graph. Two vertices share an edge if the respective nodal domains have a common boundary component (unique if there is one). This gives a finite connected rooted tree denoted $T(c)$; the root being the domain adjacent to c (see Figure 5).

The reason for working in a coordinate patch $\mathcal{U}$ for M is that for general (M, g) there is no global way to define a tree that describes the nesting configuration of the

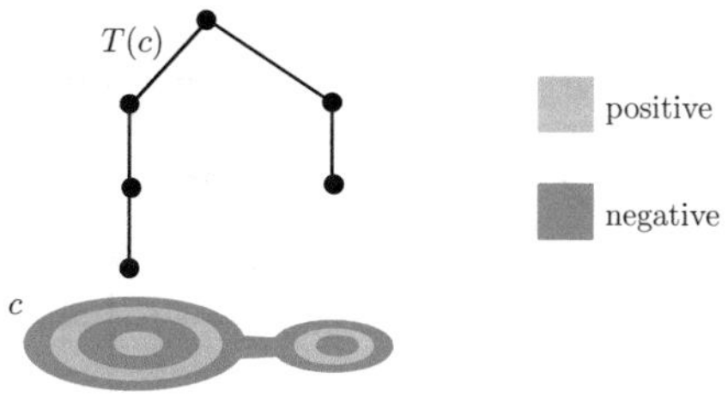

Fig. 5 This picture shows a nodal domain configuration, where positive nodal domains are depicted in orange and negative nodal domains are green. The corresponding rooted tree is shown

zero set in all of M, for all $c \in C(\phi)$. The reason is that a zero set component may not divide M into two different regions. It is important to note that in a coordinate patch this is always the case. However, according to [31] almost all c's localize to small coordinate patches. This inconvenience is the reason why [10] is written for $M = S^n$ the round sphere. By the Jordan–Brouwer separation Theorem, on S^n every component of the zero set separates S^n into two distinct components. This gives that the nesting graph for the zero sets is a rooted tree well defined without the need for a coordinate patch.

Let $\mathcal{T}$ be the collection (countable and discrete) of finite connected rooted trees. The distribution of nested ends of nodal domains of ϕ that lie within $\mathcal{U}$ is described by the measure $\mu_{\mathcal{T}(\phi),\mathcal{U}}$ on $\mathcal{T}$ given by

$$\mu_{\mathcal{T}(\phi),\mathcal{U}} := \frac{1}{|\mathcal{C}_{\mathcal{U}}(\phi)|} \sum_{c \in \mathcal{C}_{\mathcal{U}}(\phi)} \delta_{T(\phi)},$$

where δ_T is the point mass at $T \in \mathcal{T}$.

The following is part of the main theorem in [30, Theorem 1.1]. Also, see [10, Remark 2].

Theorem 8 *There exists a probability measure $\mu_{\mathcal{T}}$ supported on $\mathcal{T}$ such that the following holds. Let (M, g) be a smooth, compact, Riemannian manifold of dimension $n \geq 2$ with no boundary. Let $\eta = \eta(\lambda)$ be a non-negative function satisfying $\eta(\lambda) = o(\lambda)$ as $\lambda \to \infty$. Suppose that M is a manifold of isotropic scaling and let $\mathcal{U}$ be a coordinate patch for M. Then, for any given $T \in \mathcal{T}$ and $\varepsilon > 0$,*

$$\lim_{\lambda \to \infty} \mathbb{P}\Big(\phi_\lambda \in \mathrm{RW}_\lambda(M, g, \eta) : |\mu_{\mathcal{T}(\phi),\mathcal{U}}(T) - \mu_{\mathcal{T}}(T)| > \varepsilon\Big) = 0.$$

Theorem in [30] asserts that there exists a probability measure $\mu_{\mathcal{T}}$ on $\mathcal{T}$ to which $\mu_{\mathcal{T}(\phi),\mathcal{U}}$ approaches as $\lambda \to \infty$, for almost all ϕ provided M is a manifold of isotropic scaling.

The probability measure $\mu_{\mathcal{T}}$ is universal in that it only depends on the dimension n of M. The following result is part of theorem in [10] and deals with the support of $\mu_{\mathcal{T}}$.

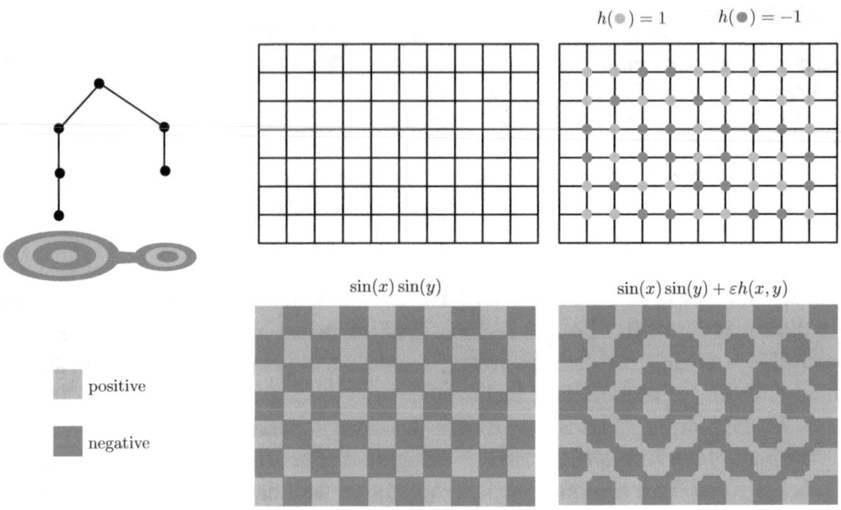

Fig. 6 This picture shows how to perturb the zero set of $\psi(x, y) = \sin(x)\sin(y)$ by adding $h \in \ker(\Delta_{\mathbb{R}^n} - I)$ that we prescribe on the singularities of ψ that lie in its zero set so that the zero set of $\phi = \psi + \varepsilon h$, for $\varepsilon > 0$ small, has the correct nesting configuration

Theorem 9 *Every atom of* $\mathcal{T}$ *is positively charged by* $\mu_{\mathcal{T}}$. *That is,*

$$\mathrm{supp}(\mu_{\mathcal{T}}) = \mathcal{T}.$$

The proof of Theorem 9 hinges on the fact that any rooted tree can be realized by elements of $\ker(\Delta_{\mathbb{R}^n} - I)$ as described by the following result.

Theorem 10 *Given* $T \in \mathcal{T}$ *there exists* $\phi \in \ker(\Delta_{\mathbb{R}^n} - Id)$ *and* $c \in \mathcal{C}(\phi)$ *for which* $T(c) = T$.

As far as Theorem 10, the case $n = 2$ is resolved in [30] using a deformation of $\sin(\pi x)\sin(\pi y)$ and a combinatorial chess board type argument. This is described in Figure 6. In higher dimensions, for example $n = 3$, one proceeds by deforming

$$\psi(x, y, z) = \sin(\pi x)\sin(\pi y)\sin(\pi z).$$

This ψ has enough complexity to produce all elements in $\mathcal{T}$ after deformation. However, it is much more difficult to study than the 2-dimensional case. Unlike $\sin(\pi x)\sin(\pi y)$, the zero set $\psi^{-1}(0)$ has point and 1-dimensional edge singularities. The analysis of its resolution under deformation requires a lot of care, especially as far as engineering elements of $\mathcal{T}$. The pay off as we noted is that it is rich enough to prove Theorem 10.

Acknowledgments The author is very grateful to her collaborators B. Hanin and P. Sarnak. The author would also like to thank the Alfred P. Sloan Foundation for their support.

References

1. R. Adler, J. Taylor, *Random Fields and Geometry*. Springer Monographs in Mathematics, vol. 115 (Springer, New York, 2009)
2. P. Bérard, Volume des ensembles nodaux des fonctions propres du laplacien. Semin. Theor. Spectr. Geom. **3**, 1–9 (1984)
3. M. Berry, Regular and irregular semiclassical wavefunctions. J. Phys. A Math. Gen. **10**(12), 2083 (1977)
4. V. Cammarota, D. Marinucci, A quantitative central limit theorem for the Euler-Poincaré characteristic of random spherical eigenfunctions. Ann. Probab. **46**(6), 3188–3228 (2018)
5. V. Cammarota, I. Wigman, Fluctuations of the total number of critical points of random spherical harmonics. Stoch. Processes Their Appl. **127**(12), 3825–3869 (2017)
6. V. Cammarota, D. Marinucci, I. Wigman, On the distribution of the critical values of random spherical harmonics. J. Geom. Anal. **26**(4), 3252–3324 (2016)
7. Y. Canzani, B. Hanin, Scaling limit for the Kernel of the spectral projector and remainder estimates in the pointwise Weyl law. Anal. Partial Differ. Equ. **8**(7), 1707–1731 (2015)
8. Y. Canzani, B. Hanin, Local Universality for zeros and critical points of monochromatic random waves (2016). Preprint, arXiv:1610.09438
9. Y. Canzani, B. Hanin, C^∞ scaling asymptotics for the spectral projector of the Laplacian. J. Geom. Anal. **28**(1), 111–122 (2018)
10. Y. Canzani, P. Sarnak, Topology and nesting of the zero set components of monochromatic random waves. Commun. Pure Appl. Math. **72**(2), 343–374 (2019)
11. F. Dalmao, I. Nourdin, G. Peccati, M. Rossi, Phase singularities in complex arithmetic random waves. Electron. J. Probab. 24, 1–45 (2019)
12. A. Enciso, D. Peralta-Salas, Submanifolds that are level sets of solutions to a second-order elliptic PDE. Adv. Math. **249**, 204–249 (2013)
13. P. Erdös, R.R. Hall, On the angular distribution of Gaussian integers with fixed norm. Discrete Math. **200**, 87–94 (1999) (Paul Erdös memorial collection)
14. D. Gayet, J. Welschinger, Betti numbers of random nodal sets of elliptic pseudo-differential operators (2014). Preprint, arXiv:1406.0934
15. D. Gayet, J. Welschinger, Expected topology of random real algebraic submanifolds. J. Inst. Math. Jussieu **14**(04), 673–702 (2015)
16. D. Gayet, J. Welschinger, Universal components of random nodal sets. Commun. Math. Phys. 1–21 (2015). arXiv:1503.01582
17. L. Hörmander, The spectral function of an elliptic operator. Acta Math. **121**(1), 193–218 (1968)
18. B. Keeler, A logarithmic improvement in the two point Weyl Law for manifolds without conjugate points (2019). Preprint, arXiv:1905.05136
19. M. Krishnapur, P. Kurlberg, I. Wigman, Nodal length fluctuations for arithmetic random waves. Ann. Math. **177**, 699–737 (2013)
20. D. Marinucci, I. Wigman, The defect variance of random spherical harmonics. J. Phys. A Math. Theor. **44**(35), 355206 (2011)
21. D. Marinucci, I. Wigman, On nonlinear functionals of random spherical eigenfunctions. Commun. Math. Phys. **327**(3), 849–872 (2014)
22. D. Marinucci, G. Peccati, M. Rossi, I. Wigman, Non-universality of nodal length distribution for arithmetic random waves. Geom. Funct. Anal. **26**(3), 926–960 (2016)
23. J. Marklof, S. O'Keefe, Weyl law and quantum ergodicity for maps with divided phase space. Nonlinearity **18**, 277–304 (2005)
24. F. Mehler, Ueber die Vertheilung der statischen Elektricität in einem von zwei Kugelkalotten begrenzten Körper. J. Reine Angew. Math. **68**, 134–150 (1868)
25. J. Neuheisel, The asymptotic distribution of nodal sets on spheres. Diss. Johns Hopkins University, 2010

26. L. Nicolaescu, Critical sets of random smooth functions on products of spheres (2010). Preprint, arXiv:1008.5085
27. I. Nourdin, G. Pecatti, M. Rossi, Nodal statistics of planar random waves (2017). Preprint, arXiv:1708.02281
28. M. Rossi, Random nodal lengths and Wiener chaos (2018). Preprint, arXiv:1803.09716
29. Z. Rudnick, I. Wigman, On the volume of nodal sets for eigenfunctions of the Laplacian on the torus. Ann. Henri Poincare **9**(1), 109–130 (2008)
30. P. Sarnak, I. Wigman, Topologies of nodal sets of random band-limited functions. Commun. Pure Appl. Math. **72**(2), 275–342 (2019)
31. M. Sodin, F. Nazarov, Asymptotic laws for the spatial distribution and the number of connected components of zero sets of Gaussian random functions (2015). Preprint, arXiv:1507.02017
32. C. Sogge, S. Zelditch, Riemannian manifolds with maximal eigenfunction growth. Duke Math. J. **114**(3), 387–437 (2002)
33. I. Wigman, On the distribution of the nodal sets of random spherical harmonics. J. Math. Phys. **50**(1), 013521 (2009)
34. I. Wigman, Fluctuations of the nodal length of random spherical harmonics. Commun. Math. Phys. **298**(3), 787–831 (2010)
35. S. Zelditch, On the rate of quantum ergodicity. II. Lower bounds. Commun. Partial Differ. Equ. **19**(9–10), 1565–1579 (1994)
36. S. Zelditch, Real and complex zeros of Riemannian random waves. Contemp. Math. **14**, 321 (2009)

A Brief Review of the "ETH-Approach to Quantum Mechanics"

Jürg Fröhlich

1 Introduction—Comments on the Foundations of Quantum Mechanics and Purpose of Paper

Let me start with a few general remarks: I consider it to be an intellectual scandal that, nearly one hundred years after the discovery of matrix mechanics by *Heisenberg, Born, Jordan* and *Dirac*, many or most professional physicists—experimentalists and theorists alike— admit to being confused about the deeper meaning of Quantum Mechanics (QM), or are trying to evade taking a clear standpoint by resorting to agnosticism or to overly abstract formulations of QM that often only add to the confusion. Attempts to replace QM by some alternative deterministic theory, one that does not have a "measurement problem," yet reproduces important predictions of QM, do not appear to have been very successful, so far. Unfortunately, most physicists have prejudices preventing them from taking a fresh, unbiased look at the subject, and discussions of the foundations of QM tend to be surprisingly emotional. *I feel it is time to change this situation!*

My own interests in the foundations of Quantum Mechanics were aroused in courses on QM taught by *Klaus Hepp* and *Markus Fierz* in the late sixties of the past century, which I took as an undergraduate student. I suppose that most serious students of Physics develop such interests during their first courses on QM. But I felt that the subject had better remain a hobby until later in my career. Not least because of the appearance of partly contradictory novel *"interpretations of QM"*, all of which left me unsatisfied, (see, e.g., [1, 2], and [3] for a brief survey), my views of the foundations of QM actually remained quite confused until a little more than ten years ago (which did not prevent me from giving talks about the subject—some

J. Fröhlich (✉)
Department of Physics, ETH Zurich, 8093 Zurich, Switzerland
e-mail: juerg@phys.ethz.ch

N. Anantharaman et al. (eds.), *Frontiers in Analysis and Probability*,
https://doi.org/10.1007/978-3-030-56409-4_2

with modest impact—in numerous places). But when I was approaching mandatory retirement I felt an urge to clarify my understanding of some of the subjects I had to teach to my students for thirty years—thermodynamics, effective dynamics (in particular Brownian motion), and, foremost, the foundations of QM; see [4–7] and references given there, the last two papers having some relevance for the foundations of QM.[1] At the beginning of 2012, my interests in this subject became more serious, and I pursued them in joint efforts with my last PhD student, *Baptiste Schubnel*. Later, some further colleagues got interested in our efforts, including *M. Ballesteros, Ph. Blanchard, N. Crawford, J. Faupin*, and *M. Fraas*, who collaborated with us in changing configurations. At this point, I wish to thank my collaborators for their support in this endeavor, as well as quite a few colleagues—too many to mention all of them—who were willing to listen to me and discuss ideas on basic questions concerning the foundations of QM with me. *D. Dürr* and *S. Goldstein* deserve my thanks for the encouragement and understanding they have provided.

In this paper, I present a sketch of the "ETH-Approach to Quantum Mechanics" [8–10]. The ETH-Approach is supposed to lay the foundations of a logically coherent quantum theory of *"events"* [11] and of *direct* or *projective measurements* of physical quantities (serving to record "events") that does not require invoking any "deos ex machina," such as "observers"; (see also [2]). I have given quite a few talks about this new approach. Technical details have been presented in a short course taught at Les Diablerets, in January of 2017 [12], and in [13, 14]. Our work has profited from ideas proposed by the late *Rudolf Haag* [11], from a paper of *D. Buchholz* and the late *J. E. Roberts* [15], and from discussions with Buchholz. In completing this paper I enjoyed receiving feedback from a very careful referee who found many typos and pointed out various unclear statements. A form of the ETH-Approach compatible with Einstein causality and Relativity Theory is sketched in [16]. But a comprehensive review of our work has not been written, yet.

Wide-spread recent interest in foundational problems surrounding QM has been triggered by problems in quantum information theory and by the 2012 Nobel Prize in Physics awarded to *S. Haroche* [17] and *D. Wineland*. Their discoveries, as well as results described in [18, 19], and references given there have influenced some of our own work on the theory of indirect measurements in QM, which has appeared in [20–22] and is briefly sketched at the end of this paper. The theory of indirect ("non-demolition-" and "weak-") measurements is quite well developed and clear, *assuming* one understands what "events" and "direct measurements and observations" are, specifically *direct* observations of "probes" used to *indirectly* retrieve information on physical systems. The theory of "events" and of "direct (projective) measurements" actually constitutes the deep and controversial part of the foundations of QM, and it is a novel approach to this theory that I intend to outline in this paper.

---

[1] I think it is more appropriate to speak of the "foundations of QM," rather than "interpretations of QM." We have to understand what QM tells us about Nature, *what it means* - once this is accomplished, the correct interpretation of the theory will come almost automatically.

2 Standard Formulation of Quantum Mechanics and Its Shortcomings

In our courses on Quantum Mechanics, physical systems, S, are often described as pairs, $(\mathcal{H}, U)$, of a Hilbert space, $\mathcal{H}$, of pure state vectors and a propagator, U, consisting of unitary operators $\big(U(t,t')\big)_{t,t'\in\mathbb{R}}$, acting on $\mathcal{H}$ *seemingly* describing the time evolution of state vectors in $\mathcal{H}$ from time t' to time t. The state space $\mathcal{H}$ of physically realistic systems tends to be infinite-dimensional (but separable). Alas, all infinite-dimensional separable Hilbert spaces are isomorphic, and the data invariantly encoded in the pair $(\mathcal{H}, U)$ do not tell us anything interesting about the physics of S, beyond spectral properties of the operators $U(t,t')$, (i.e., "energy levels"); and they lead one to the mistaken impression that QM might be a *linear* and *deterministic* theory—alas, one that is entirely inadequate to describe events and the outcome of observations and measurements.

We must therefore clarify what should be added to the formalism of QM in order to capture its fundamentally probabilistic nature and to arrive at a mathematical structure that enables one to describe **physical phenomena** ("events") in *isolated open* systems S, without a need to appeal to the intervention of "observers" with "free will"—as is done in the conventional *"Copenhagen Interpretation of QM"*— or to assume that other "ghosts" not intrinsic to the theory come to our rescue.

Isolated open systems: An *isolated system* S is one that, for all practical purposes, does not have any interactions with its complement, i.e., with the rest of the Universe; meaning that, for periods of time much longer than the time of monitoring it, interactions between the degrees of freedom of S and those of its complement can be neglected in the description of the Heisenberg-picture time evolution of operators. This does, however, *not* exclude that the *state* of S may be entangled with the state of its complement. The special role played by isolated systems in discussions of the foundations of QM stems from the fact that, *only for an isolated system, S*, the time evolution in the **Heisenberg picture** of arbitrary operators acting on $\mathcal{H}$ is given by conjugation with the unitary propagator, U, of S (determined by its *Hamiltonian*). An isolated system S is called *open* if it can emit modes to the outside world (the complement of S) that eventually cannot be recorded, anymore, by any devices belonging to S, yet can be in a state entangled with the state of S after emission. The reader may think of photons or gravitons emitted by an isolated system S that escape from detection by any devices in S. (See also Definition 1, below.) □

Physical quantities characteristic of a system S are described by certain self-adjoint linear operators, $X = X^*$, acting on $\mathcal{H}$. This feature is common to *all physical theories* used at present.[2] The Copenhagen Interpretation of Quantum Mechanics then stipulates that there are *"observers"* with "free will" who can

[2]In classical theories, these operators generate an *abelian* (C^*-) algebra, and time evolution is given by a *-automorphism group of this algebra generated by a vector field on its spectrum; while, in QM, the algebra generated by operators representing physical quantities (and events) is

decide to measure such physical quantities arbitrarily quickly, at arbitrary times, and at an arbitrary rate. It is argued that the time evolution of physical states of S is determined by its unitary propagator U, which solves a (deterministic) *Schrödinger equation*, *except* when a measurement of a physical quantity represented by an operator $X = X^*$ is made: Immediately after the measurement of X the state of S, according to the Copenhagen Interpretation, is in an eigenstate of X corresponding to the measured value of X. If this value is not recorded, one is advised to use a density matrix describing an *incoherent* superposition of eigenstates of X, chosen in accordance with *Born's Rule*, to describe the future evolution of S.

For a variety of reasons, this is not a satisfactory recipe for how to apply QM to describe physical phenomena! One might want to view the evolution of states in the presence of measurements, as described in the Copenhagen Interpretation of QM, as some kind of stochastic process. But the problem is that one is dealing with a stochastic process that does *not* have a classical state space, and that it is *transition amplitudes*, rather than *transition probabilities*, that are given by matrix elements of a family of operators (the propagator U) satisfying a group composition law, i.e., a kind of Chapman–Kolmogorov equation.[3] According to the Copenhagen Interpretation, predicting/determining the transition probabilities describing the stochastic time evolution of states of S in the presence of repeated measurements would apparently require knowing what kind of physical quantities are measured by the intervention of "observers," and at what times these measurements are made. For, any intermediate intervention of an "observer" destroys "interference effects"; and hence it seemingly affects the value of the transition probability between an initial state of S in the past and a target state in the future, *even* if a sum over *all possible* outcomes of the intermediate intervention is taken.[4] Without complete information on all intermediate measurements performed on S, which, in the Copenhagen Interpretation, is *not* provided by the theory, reliable predictions of future states of the system and of future expectation values of physical quantities become impossible. As a result, the Copenhagen Interpretation renders QM nearly "unpredictive"—even though, by experience, it is a heuristic framework supplementing QM that works well for many or most "practical purposes," because, much of the time (in particular when using a scattering matrix), one is interested in predicting the outcome of only *a single* measurement. The situation is hardly improved in a definitive way by resorting to concepts such as "decoherence" and interpretations such as "consistent histories" [1], "many worlds," etc.. (See [2, 23] for further information.)

---

non-commutative, and time evolution is given by a *-automorphism group of such an algebra *only* if the system is *isolated*.

[3]It is advocated by certain groups of people that the problem arising from this fact can be remedied by invoking the phenomenon of "decoherence" and appealing to the "consistency" of histories of events [1]. But I find the arguments supporting this point of view unconvincing.

[4]This is the case unless perfect "decoherence" holds.

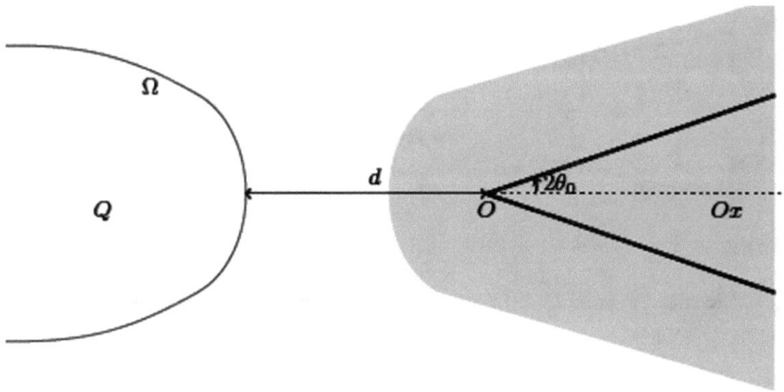

Fig. 1 ↑ Q = sub-system "confined" to Ω ↑ Particle P propagating into shaded cone

Before proceeding to describe the "*ETH*-Approach," I recall an argument, presented in detail in [13], that shows that the Schrödinger equation does *not* describe the time evolution of *states* of systems in the presence of "events" or "measurements," *assuming* that the usual correlations between the outcomes of Bell-type measurements, claimed to be confirmed in many experiments, hold.

We consider the following Gedanken-Experiment [13] (see Figures 1 and 2), which, ultimately, will show that *time evolution of states in QM is intrinsically stochastic, in spite of the deterministic nature of the Schrödinger equation.*

We prepare the system $Q \vee P$ in a state with the property that particle P propagates into the shaded cone opening to the right, as indicated in Figure 1, except for tiny tails leaking beyond this region, while the degrees of freedom of Q remain confined to a vicinity of the region Ω in the complement of the shaded cone, except for tiny tails. Thanks to cluster properties, expectation values of the Heisenberg-picture time evolution of physical quantities, such as spin, momentum, etc. referring to P in this state then turn out to be essentially *independent* of the time evolution of the degrees of freedom of Q. In other words, interaction terms in the Hamiltonian of the system coupling P to Q can be neglected. This is discussed in much detail in [13].

More concretely, we study the following system sketched in Figure 2.

Temporary assumptions (leading to a contradiction):

- P and P': Two spin-$\frac{1}{2}$ particles prepared in a spin-singlet initial state, $\psi_{L/R}$, localized, initially, in the central region shown in Figure 2; the orbital wave function of P is chosen such that P propagates into the cone opening to the right (except for very tiny tails) and that it will eventually undergo a Stern–Gerlach spin measurement, while the orbital wave function of P', an electron, is chosen such that this particle propagates into the cone opening to the left, with only very tiny tails leaking beyond this cone into the half-space to the right of the spin filter. (One may assume, for simplicity, that there are no terms in the total Hamiltonian

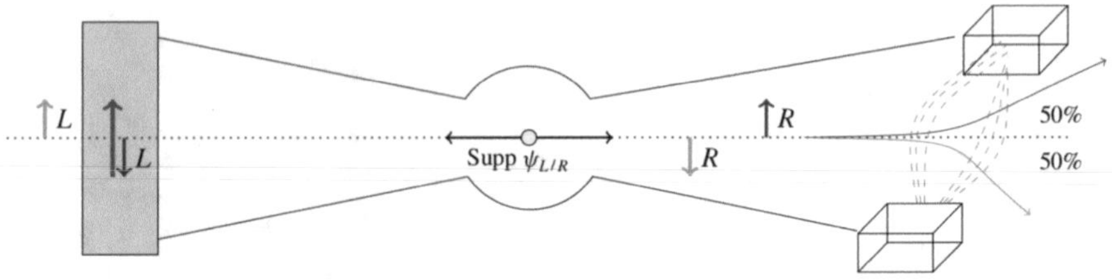

Fig. 2 Q:={spin filter ∨ particle P'} cone opening to right:= ess. supp of orbital wave function of P

of the system describing direct interactions between P and P'.) The spin filter (e.g., a spontaneously magnetized metallic film) is prepared in a poorly known initial state.

- *The dynamics of the* state *of the* total *system is assumed to be fully determined by a Schrödinger equation given by a concrete self-adjoint Hamiltonian containing only short-range interaction terms.* In particular, the initial state of the total system (consisting of the spin filter, the two particles and possibly some Stern–Gerlach equipment serving to measure a component of the spin of particle P) is assumed to determine whether particle P' will pass through the spin filter, or not, (given that the initial state of $P' \vee P$ is a spin-singlet state, with P' and P moving into opposite cones). Since it is assumed that a Schrödinger equation determines the evolution of *states* of this system, the Schrödinger picture and the Heisenberg picture are equivalent.
- *Correlations between the outcomes of spin measurements of P' and of P are assumed to be those predicted by standard quantum mechanics*, (relying on the "Copenhagen interpretation" and apparently confirmed in many experiments): We first note that if P' passes through the spin filter, then its spin is "up," (i.e., aligned with the majority spin of electrons in the spin filter), if it does not pass through the filter, (i.e., if it hops into a vacant state localized inside the spin filter), its spin is "down." The second assumption stated above then says that, whether P' passes through the filter, or not, is determined by the *initial state* of the total system and by solving a *deterministic Schrödinger equation*. In addition to the two assumptions already stated, we also assume that if the spin of P' is measured to be "up," the spin of P is measured to be "down" (for example, in a Stern–Gerlach experiment involving a magnetic field parallel to the majority spin of the spin filter), and if the spin of P' is "down," then the spin of P is "up."

Next, we recall the

Fact: Expectation values of observables (such as spin, momentum, etc.) referring to particle P in the state of the system described above are *independent* of the degrees of freedom of $Q := \{P' \vee \text{ spin filter}\}$, for arbitrarily long times, up to very tiny corrections. Thus, to a very good approximation, their evolution can be assumed to be given by free-particle dynamics. This is a consequence of our choice of an initial state (propagation properties of the orbital wave functions of P and P') and of cluster properties of the time evolution—as shown in [13].

It follows that, to a very good approximation, the *spin of P is conserved before it is measured* $\Rightarrow$

$$\underline{\text{Expectation value of spin of } P \approx 0, \forall \text{ times before measurement time,}}$$
$$\underline{\textit{independently} \text{ of the evolution of } Q = \{P' \vee \text{ spin filter}\}!}$$

But *this* **contradicts** *the third (last) assumption stated above:* The first two assumptions imply that the values of the z-component of the spin of P' measured with the help of the spin filter do apparently *not* introduce any bias in the outcomes of measurements of the z-component of the spin of P. In other words, the second assumption stated above is incompatible with the *Bell-type "non-locality"* of Quantum Mechanics, as expressed in the third assumption stated above.

This argument is robust, in the sense that it suffices to assume that correlations between measurements of a component of the spin of P' and a component of the spin of P are fairly close to those predicted by the Bell-type non-locality described in the third assumption.

Conclusion: If the third assumption holds true, then the quantum-mechanical time evolution of *states* of physical systems in the presence of measurements (or "events") is *not* given by a deterministic Schrödinger equation, and the equivalence of the Heisenberg picture and the Schrödinger picture apparently fails. Quantum Mechanics appears to be intrinsically probabilistic (and "non-local," in the sense of Bell-type correlations—which does, however, *not* invalidate locality in the sense of "Einstein causality")! These conclusions agree with ones reached by studying Gedanken-experiments such as *"Wigner's friend"* and other related ones, e.g., one recently proposed in [24].

Our task is thus to find out what one has to add to a minimal formulation of Quantum Mechanics in order to be able to describe the *stochastic dynamics of states* of physical systems in the presence of "events" and their recordings (in projective measurements), in such a way that correlations between the outcomes of measurements agree with the Bell-type "non-locality" of Quantum Mechanics—without the need to assume that "observers" intervene. The results reviewed in the next section are intended to report on some progress in this direction.

3 Summary of the "*ETH*-Approach"

In this section I briefly describe the so-called *ETH-Approach to Quantum Mechanics* [8–10, 12–14], which is designed to retain attractive features of the Copenhagen Interpretation but eliminates its fatal weaknesses; and I note that "*E*" stands for "Events," "*T*" for "Trees," and "*H*" for "Histories." In the following, I attempt to explain what these terms mean, and why the concepts underlying the "*ETH*-Approach" are important for an understanding of the foundations of Quantum

Mechanics (QM). The basic premises and contentions of this approach are as follows:

I. Potential Events. In the ETH-Approach to QM, *Time*, denoted by t, is taken as an irreducible concept. It is described by the real line, $\mathbb{R}$, with its usual order relation.[5] But in order to make the following discussion mathematically watertight it is advisable to sometimes assume that time is *discretized*, $t \in \mathbb{Z}$. An important idea underlying the ETH-Approach is that time is not merely a parameter, but that it can be monitored by recording "events" happening in an isolated open system. (The precise meaning of this idea will become clearer later on.)

Let $t_0 \in \mathbb{R}$ be the time of the present. We consider an isolated open physical system S and we denote by $\mathcal{H}$ the Hilbert space of pure state vectors of S. Our first task is to clarify what is meant by *"potential events"* in S that may happen at some future time $t > t_0$, or later: Potential events are described by families, $\{\pi_\xi, \xi \in \mathcal{X}\}$ of orthogonal projections acting on $\mathcal{H}$, with the properties that

$$\pi_\xi \cdot \pi_\eta = \delta_{\xi\eta}\, \pi_\xi, \ \ \forall \xi, \eta \text{ in } \mathcal{X}, \quad \text{(disjointeness)}$$
$$\sum_{\xi \in \mathcal{X}} \pi_\xi = \mathbb{1}, \quad \text{(partition of unity)}. \tag{1}$$

For simplicity we henceforth assume that the sets $\mathcal{X}$ labelling the projections that represent potential events are countable, discrete sets. (This merely serves to avoid technical complications in our exposition; of course, continuous spectra occur, too.) In the Heisenberg picture, which we will use henceforth, the concrete projection operators acting on the Hilbert space $\mathcal{H}$ of S representing a *specific* potential event, e.g., the click of a detector belonging to S when it is hit by a certain type of particle in S, depend on the time $t > t_0$ in the future when the event might happen. In an autonomous system, the concrete projection operators representing a *specific* potential event that may happen at a time $t > t_0$ or at another time $t' > t_0$ are unitarily conjugated to one another by the propagator $U(t, t')$ of the system; (Heisenberg-picture evolution of operators). All projection operators representing potential events that may happen at some time $t > t_0$, or later, generate a *-algebra denoted by $\mathcal{E}_{\geq t}$. It immediately follows from this definition that

$$\mathcal{E}_{\geq t'} \subseteq \mathcal{E}_{\geq t}, \quad \text{if } t' > t.$$

Remark The concrete projection operators representing some potential event that may happen in system S (see Equations (1)) depend on the time t when the potential event would start to happen and on the time-interval during which it would happen. More concretely, if $\hat{A}_i$, $i = 1, 2, \ldots,$ are abstract operators

[5]The role of *space*-time in a relativistic version of the "ETH-Approach" is discussed in [16].

representing physical quantities of S, (e.g., a component of the spin of a certain species of particles localized in a certain region of physical space and measured in a Stern–Gerlach experiment), and if $A_i(t)$ denotes the Heisenberg-picture operator on $\mathscr{H}$ representing $\hat{A}_i$ at time t, then a potential event arising from monitoring the quantities $\hat{A}_i$, $i = 1, 2, \ldots$, which starts to happen at time t, consists of a family of projections satisfying Equations (1) that are functionals of the operators

$$\{A_i(t') \,|\, i = 1, 2, \ldots;\ t' \in [t, T), \text{ for some } T \text{ with } t < T \leq \infty\} \qquad \square.$$

This remark is inspired by general wisdom from local quantum field theory.

For simplicity we assume that *all physically relevant states* of S can be described by density matrices acting on $\mathscr{H}$, and that the algebras $\mathscr{E}_{\geq t}$ are closed in the weak topology of the algebra, $B(\mathscr{H})$, of all bounded operators acting on $\mathscr{H}$. Typically, all the algebras $\mathscr{E}_{\geq t}$ are then isomorphic to one *universal* (von Neumann) algebra[6] $\mathscr{N}$, i.e.,

$$\mathscr{E}_{\geq t} \simeq \mathscr{N}, \quad \forall t \in \mathbb{R}. \tag{2}$$

The algebra, $\mathscr{E}$, of all potential events that may happen in the course of history is defined by

$$B(\mathscr{H}) \supseteq \mathscr{E} := \overline{\bigvee_{t \in \mathbb{R}} \mathscr{E}_{\geq t}}, \tag{3}$$

(where the closure is taken in the operator norm of $B(\mathscr{H})$).

II. The Principle of Diminishing Potentialities. In the quantum theory of (autonomous) systems with finitely many degrees of freedom—as treated in our introductory courses on QM—the algebras $\mathscr{E}_{\geq t}$ turn out to be *independent* of time t; and usually $\mathscr{E}_{\geq t} = B(\mathscr{H})$. For such systems, one *cannot* develop a sensible quantum theory of events, and it is impossible to come up with a logically coherent, intrinsically quantum-mechanical description of the retrieval of information on such systems, i.e., of measurements, *without* adding further quantum systems with infinitely many degrees of freedom that serve to "measure" the former systems (or without resorting to something like "Copenhagen"). In this respect, quantum systems with finitely many degrees of freedom are as "interesting" as the space-time region outside the event horizon of a black hole: no information can be extracted! In order to encounter non-trivial dependence of the algebras $\mathscr{E}_{\geq t}$ on time t, we must consider *isolated (open) systems with infinitely many degrees of freedom* and with the property

[6] In local relativistic quantum theories with massless particles, the algebra $\mathscr{N}$ tends to be a von Neumann algebra of type III; see [15].

that the propagator U of S is generated by a Hamiltonian whose spectrum does *not* have any isolated eigenvalues, and (if time is continuous) the spectrum is *unbounded above and below*, or, in relativistic quantum theory, it is *semi-bounded, but without any spectral gaps*; i.e., we must assume that there exist massless modes.

Our contention is that a basic property of a quantum theory of isolated open systems, S, enabling one to describe *events* and their *recording* in projective measurements of physical quantities is captured in the following **"Principle of Diminishing Potentialities"** (PDP):

$$\boxed{\mathcal{E}_{\geq t'} \subsetneqq \mathcal{E}_{\geq t} \subsetneqq \mathcal{E}, \quad \text{whenever} \quad t' > t.} \tag{4}$$

To be more precise, one expects that if time is continuous the relative commutant

$$\left(\mathcal{E}_{\geq t'}\right)' \cap \mathcal{E}_{\geq t}, \quad \text{with } t' > t,$$

is an infinite-dimensional, non-commutative algebra. (If time is discrete this relative commutant can, however, be a finite-dimensional algebra.) Examples of non-relativistic and relativistic systems satisfying property (4) will be discussed elsewhere, (see also [12]).[7] Here I just mention that (PDP), in the sense of a relativistic variant of Equation (4), is a *theorem* in local relativistic quantum field theories with massless particles in four space-time dimensions.[8] This follows from important results in [15] and is used in [16].

Definition 1 *Isolated open systems* S (featuring events) are henceforth *defined* in terms of a filtration, $\{\mathcal{E}_{\geq t}\}_{t\in\mathbb{R}}$ (or, for the sake of simplicity and precision, $\{\mathcal{E}_{\geq t}\}_{t\in\mathbb{Z}}$), of (von Neumann) algebras satisfying the *"Principle of Diminishing Potentialities"* (4), all represented on a common Hilbert space $\mathcal{H}$, whose projections describe potential events. □

If Ω denotes the density matrix on $\mathcal{H}$ representing the actual state of a system S, we use the notation

$$\omega(X) := tr(\Omega\, X), \qquad \forall X \in B(\mathcal{H}),$$

to denote the expectation value of the operator X in the state ω determined by Ω. We define

$$\omega_t(X) := \omega(X), \qquad \forall X \in \mathcal{E}_{\geq t}, \tag{5}$$

[7] I sometimes fear that unrealistically simple examples advanced with the intention to clarify aspects of the foundations of QM have had the opposite effect: They have contributed to clouding our views.

[8] and the algebras $\mathcal{E}_{\geq t}$, $t \in \mathbb{R}$, are von Neumann algebras of type III.

i.e., ω_t is the *restriction* of the state ω to the algebra $\mathcal{E}_{\geq t}$.

Note that, as a consequence of (PDP) and of *entanglement*, the restriction, ω_t, of a state ω on the algebra $\mathcal{E}$ to a subalgebra $\mathcal{E}_{\geq t} \subset \mathcal{E}$ will usually be **mixed** *even* if ω is a **pure** state on $\mathcal{E}$.

III. Actual Events. Henceforth we only study isolated open systems S for which (PDP), in the form of Equation (4), holds. Let $\{\pi_\xi, \xi \in \mathcal{X}\} \subset \mathcal{E}_{\geq t}$ be a potential event that might start to happen at some time t, with $\{\pi_\xi, \xi \in \mathcal{X}\}$ *not* contained in $\mathcal{E}_{\geq t'}$, for $t' > t$. Tentatively, we say that this potential event **actually starts to happen** at time t iff

$$\omega_t(X) = \sum_{\xi \in \mathcal{X}} \omega_t\big(\pi_\xi \, X \, \pi_\xi\big), \qquad \forall X \in \mathcal{E}_{\geq t}, \tag{6}$$

meaning that ω_t is an incoherent superposition of states labelled by the points $\xi \in \mathcal{X}$; in other words, off-diagonal expectations, $\omega_t\big(\pi_\xi \, X \, \pi_\eta\big)$, $\xi \neq \eta$, do *not* contribute to the right side of (6). Equation (6) is equivalent to saying that the projections $\pi_\xi, \xi \in \mathcal{X}$, belong to the *centralizer* of the state ω_t.

Given a *-algebra $\mathcal{M}$ and a state ω on $\mathcal{M}$, the centralizer, $\mathcal{C}_\omega(\mathcal{M})$, of the state ω is defined to be the subalgebra of $\mathcal{M}$ spanned by all operators, Y, in $\mathcal{M}$ with the property that

$$\omega([Y, X]) = 0, \qquad \forall X \in \mathcal{M}.$$

The *center* of the centralizer, denoted by $\mathcal{Z}_\omega(\mathcal{M})$, is the abelian subalgebra of the centralizer consisting of all operators in $\mathcal{C}_\omega(\mathcal{M})$ commuting with all other operators in $\mathcal{C}_\omega(\mathcal{M})$.

We note that the center, $\mathcal{Z}(\mathcal{M})$, of the algebra $\mathcal{M}$ is contained in $\mathcal{Z}_\omega(\mathcal{M})$, *for all* states ω.

Definition 2 A potential event $\{\pi_\xi, \xi \in \mathcal{X}\} \subset \mathcal{E}_{\geq t}$, with $\{\pi_\xi, \xi \in \mathcal{X}\}$ not contained in $\mathcal{E}_{\geq t'}$, for $t' > t$, *actually starts to happen* at time t iff $\mathcal{Z}_{\omega_t}(\mathcal{E}_{\geq t})$ is *non-trivial*,

$$\{\pi_\xi, \xi \in \mathcal{X}\} \text{ generates } \mathcal{Z}_{\omega_t}\big(\mathcal{E}_{\geq t}\big), \tag{7}$$

and

$$\omega_t(\pi_{\xi_j}) \text{ is } \textit{strictly positive}, \;\; \xi_j \in \mathcal{X}, \;\; j = 1, 2, \ldots, n, \tag{8}$$

for some $n \geq 2$. □

IV. The fundamental Axiom. We are now in a position to describe the evolution of states in the ETH-Approach to QM. Let ω_t be the state of an isolated system S right before time t. Let us suppose that an event $\{\pi_\xi, \xi \in \mathcal{X}\}$ generating $\mathcal{Z}_{\omega_t}(\mathcal{E}_{\geq t})$ starts to happen at time t, in the sense of Definition 2.

<u>**Axiom**</u> The actual state of the system S right *after* time t when the event $\{\pi_\xi, \xi \in \mathscr{X}\}$ has started to happen is given by one of the states

$$\omega_{t,\xi_*}(\cdot) := [\omega_t(\pi_{\xi_*})]^{-1}\, \omega_t\big(\pi_{\xi_*}(\cdot)\pi_{\xi_*}\big), \tag{9}$$

for some $\xi_* \in \mathscr{X}$ with $\omega_t(\pi_{\xi_*}) > 0$, ("**state-collapse** postulate"[9]. The probability for the system S to be found in the state ω_{t,ξ_*} right after time t when the event $\{\pi_\xi, \xi \in \mathscr{X}\}$ has started to happen is given by **Born's Rule**, i.e., by

$$prob\{\xi_*, t\} = \omega_t(\pi_{\xi_*}). \tag{10}$$

□

Remarks

(1) The projection π_{ξ_*} selecting the actual state ω_{t,ξ_*} of S (and sometimes also the point $\xi_* \in \mathscr{X}$) is called the *"actual event"* happening at time t.
(2) The contents and meaning of this Axiom are clear and mathematically watertight as long as time is discrete. (If time is continuous further precision ought to be provided.)

This Axiom, Equations (9) and (10), conveys the following picture of quantum dynamics: In Quantum Mechanics, the <u>evolution of states</u> of an isolated open system S featuring events, in the sense of Definitions 1 and 2 proposed above, is given by a (rather unusual novel type of) <u>stochastic branching process</u>, whose state space is what I call the *"non-commutative spectrum"*, $\mathfrak{Z}_S$, of S. Assuming that Equation (2) holds, the non-commutative spectrum of S is defined by

$$\mathfrak{Z}_S := \bigcup_{\omega} \mathscr{Z}_\omega(\mathscr{N}), \quad \text{with} \quad \mathfrak{X}_S := \bigcup_{\omega} \text{spec}\Big(\mathscr{Z}_\omega(\mathscr{N})\Big), \tag{11}$$

where the union over ω is a disjoint union, and ω ranges over *all* physical states of S.[10] Equation (7) and **Born's Rule**, Equation (10), specify the *branching probabilities* of the process.

The above picture of the stochastic time evolution of states of an isolated open system S is illustrated, metaphorically (for discrete time), in Figure 3. It differs substantially from and supercedes the *"decoherence mumbo-jumbo."*

Let us suppose, for the sake of simplicity and mathematical precision, that time is discrete, ($t \in \mathbb{Z}$). It is important to note that, in general, the events (described by orthogonal projections in $\mathscr{E}_{\geq t'}$) predicted to happen at a later time $t' > t$ on

[9] a rather unfortunate name!

[10] The set $\mathfrak{X}_S$ can also be defined in terms of a certain "flag manifold" associated with the Hilbert space $\mathscr{H}$.

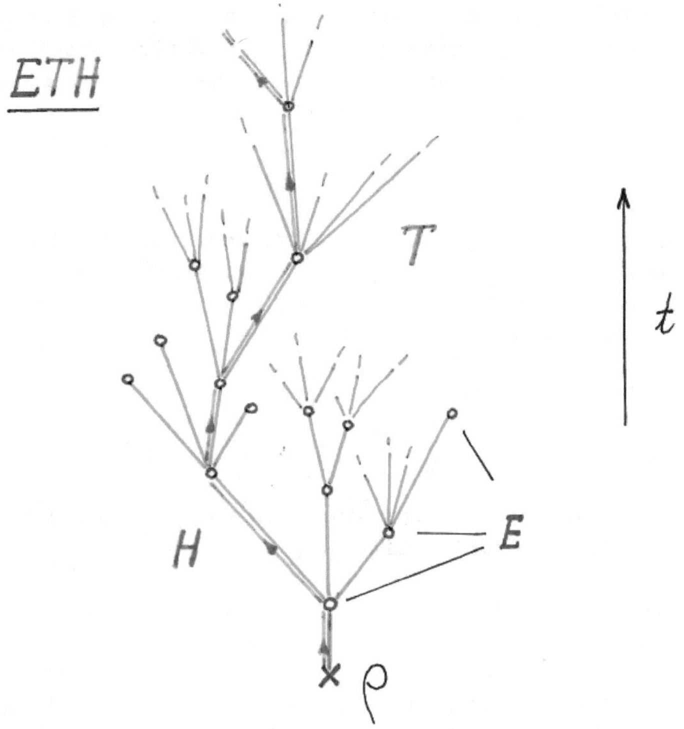

Fig. 3 Time evolution of a state of S with initial condition $\omega := \rho$
E: "Events," T: "Tree" of possible future states, H: "History" of actual events/states

the basis of the states $\omega_{t,\xi}, \xi \in \mathscr{X}$, where $\{\pi_\xi, \xi \in \mathscr{X}\}$ generates $\mathscr{Z}_{\omega_t}(\mathscr{E}_{\geq t})$, are *different* from the events one would predict to happen at time t' on the basis of the state $\omega_t|_{\mathscr{E}_{\geq t'}}$, used when the actual event happening at time t is not known (i.e., has not been recorded); and the projections representing these different sets of events usually do *not* commute with one another. Furthermore, for $t' > t$, the operators in $\mathscr{Z}_{\omega_{t,\xi}}(\mathscr{E}_{\geq t'})$ and in $\mathscr{Z}_{\omega_{t,\eta}}(\mathscr{E}_{\geq t'}), \xi, \eta \in \mathscr{X}$, (with $\omega_t(\pi_\xi)$, $\omega_t(\pi_\eta)$ strictly positive), but $\xi \neq \eta$, do *not* in general commute with each other. This is a *fundamental difference* between the *"non-commutative branching processes,"* described here, and classical stochastic branching processes.

The discussion above is mathematically sound if time is discrete, but requires more precision if time is taken to be continuous.

To be on the safe side, we temporarily choose time to be discrete ($t \in \mathbb{Z}$). Let H be the Hamiltonian of an isolated open system, and suppose that

$$\|e^{iH} - \mathbf{1}\| \ll 1\,. \tag{12}$$

Let us suppose that $\{\pi_{t,\xi}, \xi \in \mathscr{X}_t\}$ is an event that starts to happen at time t, provided the state of S at time t is given by ω_t; (i.e., $\{\pi_{t,\xi}, \xi \in \mathscr{X}_t\}$ generates $\mathscr{Z}_{\omega_t}(\mathscr{E}_{\geq t})$). Let ξ_* be the element of $\mathscr{X}_t$ with the property that, in accordance with the Axiom stated in IV., above, the state of S right after time t is given by

$$\omega_{t,\xi_*}(\cdot) := [\omega_t(\pi_{t,\xi_*})]^{-1}\, \omega_t\big(\pi_{t,\xi_*}(\cdot)\pi_{t,\xi_*}\big)\,,$$

with $\omega_t\big(\pi_{t,\xi_*}\big) > 0$; i.e., π_{t,ξ_*} is the "actual event" happening at time t. Let $t' = t+1$ be the time following t, and let $\{\pi_{t',\xi}, \xi \in \mathscr{X}_{t'}\}$ be the event that starts to happen at time t', *provided* that the state of S at time t' is given by ω_{t,ξ_*}. Then assumption (12) suggests that there exists an element $\xi_\natural \in \mathscr{X}_{t'}$ with the property that

$$\begin{aligned} &\omega_{t,\xi_*}\big(\pi_{t',\xi_\natural}\big) \approx 1, \quad \text{but} \\ &\omega_{t,\xi_*}\big(\pi_{t',\xi}\big) \ll 1, \;\; \forall\, \xi \neq \xi_\natural\,,\; \xi \in \mathscr{X}_{t'}\,. \end{aligned} \tag{13}$$

According to the Axiom in IV., in particular **Born's Rule**, the actual state of S right after time t' is then very likely given by

$$\omega_{t,\xi_*,t',\xi_\natural}(\cdot) := [\omega_{t,\xi_*}(\pi_{t',\xi_\natural})]^{-1}\omega_{t,\xi_*}\big(\pi_{t',\xi_\natural}(\cdot)\pi_{t',\xi_\natural}\big) \approx \omega_{t,\xi_*}(\cdot)\,.$$

The state $\omega_{t,\xi_*,t',\xi_\natural}$ is close to the one that would commonly be used in the Heisenberg picture of quantum mechanics in the absence of any "measurements" or "events" after time t, namely the state $\omega_{t,\xi_*}(\cdot)$.

However, for purely statistical (*entropic*!) reasons, every once in a while, i.e., at rare times t', an event $\pi_{t',\xi}$ is realized that has a *very small Born probability*, $\omega_{t'}(\pi_{t',\xi}) \ll 1, \quad \xi \in \mathscr{X}_{t'}$.

Digression on "Missing Information" associated with events:[11]

Given the event $\{\pi_{t,\xi}, \xi \in \mathscr{X}_t\}$ happening at time t, assuming that ω_t is the actual state of S right before time t, we define the *"missing information"* (or *"entropy production"*), $\sigma(\omega_t, \mathscr{X}_t)$, associated with this event by

$$\sigma(\omega_t, \mathscr{X}_t) := -\sum_{\xi \in \mathscr{X}_t} \omega_t(\pi_{t,\xi}) \cdot \ell n\big(\omega_t(\pi_{t,\xi})\big) \tag{14}$$

Assuming that (12) holds, the "missing information" associated with most events that ever happen is very small. If the "missing information" associated with *all* events were tiny, then taking the state of S in the Heisenberg picture to be constant in time would be a good approximation to its stochastic evolution. However, every once in a while, events corresponding to a *large* "missing information" (entropy production) may be encountered, and these are the events that will most likely be

[11] This digression can be omitted at first reading, and the reader is invited to proceed to point V., below.

noticed and recorded, because they trigger a substantial change of the state of S. (Some people will want to call them "measurements.")

Let t_0 be the time at which the system S has been prepared in a state ω, (as discussed in [14]), and $t_j := t_0 + j \in \mathbb{Z}$; further, let π_{t_j,ξ_j} be the *actual event* happening at time t_j, given the initial state ω of S and earlier actual events π_{t_ℓ,ξ_ℓ}, $\ell < j$, $j = 1, 2, \ldots, n$; (see Definition 2 and Axiom). We define

$$\mu_\omega\big(\xi_1, \xi_2, \ldots, \xi_n|X\big) := \omega\Big(\prod_{j=1}^{n} \pi_{t_j,\xi_j} \cdot X \cdot X^* \cdot \big(\prod_{j=1}^{n} \pi_{t_j,\xi_j}\big)^*\Big), \tag{15}$$

where the product is ordered according to $\prod_{j=1}^{n} a_j = a_1 \cdot a_2 \cdots a_n$, and X is an arbitrary non-zero operator in $\mathcal{E}_{\geq t}$, for some $t > t_n$, with $\omega\big(X \cdot X^*\big) > 0$. Then $\mu_\omega(\ldots|X)$ is a positive measure on the Cartesian product $\bigtimes_{j=1}^{n} \mathcal{X}_{t_j}$. *Note that the space $\mathcal{X}_{t_{k+1}}$ depends on the choice of ω and on* **all** *the actual events $\pi_{t_1,\xi_1}, \ldots, \pi_{t_k,\xi_k}$ that happened at times $t_1 < \cdots < t_k$,* **before** *t_{k+1}; with $k = 1, 2, \ldots, n-1$.* For any m, with $0 < m < n$, we set

$$X(\underline{\xi}^{(m,n)}) := \prod_{j=m+1}^{n} \pi_{t_j,\xi_j} \cdot X,$$

and $X(\underline{\xi}^{(n,n)}) := X$. Then

$$\mu_\omega\big(\xi_1, \ldots, \xi_n|X\big) = \mu_\omega\big(\xi_1, \ldots, \xi_m|X(\underline{\xi}^{(m,n)})\big).$$

The measure $\mu_\omega\big(\ldots|X\big)$ has the (possibly somewhat perplexing) property that

$$\sum_{\xi_{k+1},\ldots,\xi_m} \mu_\omega\big(\xi_1, \ldots, \xi_k, \xi_{k+1}, \ldots, \xi_m|X(\underline{\xi}^{(m,n)})\big) =$$

$$= \mu_\omega(\xi_1, \ldots, \xi_k|X(\underline{\xi}^{(m,n)})\big), \tag{16}$$

for arbitrary k, with $1 \leq k \leq m \leq n$, as one easily verifies. (Identity (16) may look familiar to the reader from a similar one satisfied by the "Lüders–Schwinger–Wigner formula" [25] for the probability of a sequence of outcomes of measurements, assuming perfect *decoherence*. However, it actually has quite a different origin!) It is sometimes convenient to define $\mu_\omega\big(\ldots|X\big)$ as a measure on the space

$$\mathfrak{X}_n := \big(\mathfrak{X}_S\big)^{\times n},$$

where $\mathfrak{X}_S$ has been defined in Equation (11), with the convention that

$$\pi_{t_k,\xi} = 0, \quad \text{unless } \xi \in \mathcal{X}_{t_k} \subset \mathfrak{X}_S.$$

For $X = \mathbb{1}$, $\mu_\omega(\ldots|\mathbb{1})$ is a *probability measure* on $\mathfrak{X}_n$. If arbitrarily long sequences of events are considered, it is useful to introduce the "path space"

$$\mathfrak{X}_\infty := \varinjlim_{n\to\infty} \mathfrak{X}_n \, .$$

Thanks to property (16), the measures $\mu_\omega(\ldots|\mathbb{1})$ determine a unique probability measure on $\mathfrak{X}_\infty$. This follows from a well-known lemma due to *Kolmogorov.*

Next, we define the *"missing information per event"* of a sequence of events, as follows:

$$\sigma_n(\mu_\omega) := -\frac{1}{n} \sum_{\xi_1,\ldots,\xi_n} \mu_\omega(\xi_1,\ldots,\xi_n|\mathbb{1}) \cdot \ell n\big(\mu_\omega(\xi_1,\ldots,\xi_n|\mathbb{1})\big)\, ,$$

and

$$\sigma(\mu_\omega) := \text{limsup}_{n\to\infty}\sigma_n(\mu_\omega). \tag{17}$$

If events happening at times $t_1, \ldots, t_n$ are not recorded, then $\sigma_n(\mu_\omega)$ is a measure of how much the state of the system at time $t > t_n$ deviates from the (initial) state ω used in the Heisenberg picture of standard QM.

Of particular interest is the so-called *relative entropy*

$$S_n\big(\mu_\omega\|\mu_\omega^{opp}\big) := \sum_{\xi_1,\ldots,\xi_n} \mu_\omega(\xi_1,\ldots,\xi_n|\mathbb{1}) \times \\ \times \Big(\ell n\, \mu_\omega(\xi_1,\ldots,\xi_n|\mathbf{1}) - \ell n\, \mu_\omega^{opp}(\xi_1,\ldots,\xi_n|\mathbb{1})\Big)\, , \tag{18}$$

where

$$\mu_\omega^{opp}(\xi_1,\ldots,\xi_n|\mathbb{1}) := \omega\Big(\big(\prod_{j=1}^{n}\pi_{t_j,\xi_j}\big)^* \cdot \prod_{j=1}^{n}\pi_{t_j,\xi_j}\Big)$$

is the measure obtained when the order of the events is (time-)reversed. The relative entropy $S_n\big(\mu_\omega\|\mu_\omega^{opp}\big)$ is *non-negative*, and its growth in n, as $n \to \infty$, is a measure of the *irreversibility* of histories of events featured by the system and reflects the "arrow of time."

End of Digression.

V. Recording events by "projective measurements" of physical quantities. We consider an isolated open system S described in terms of a filtration $\{\mathcal{E}_{\geq t}\}_{t\in\mathbb{R}}$ of algebras represented on its Hilbert space $\mathcal{H}$ of pure state vectors, as described in Definition 1, (paragraph I.). We propose to clarify how events happening in S can be recorded by projectively (directly) measuring *"physical quantities"* characteristic of S. (Time may be taken to be continuous; but, for the sake of simplicity and mathematical precision, the reader is invited to continue to assume that $t \in \mathbb{Z}$.)

Definition 3 A *"physical quantity"* characteristic of S is an abelian (C^*-) algebra, $\mathcal{Q}$, with the property that, for each time t, there exists a representation, $\sigma_t^{\mathcal{Q}}$, of $\mathcal{Q}$ on $\mathcal{H}$ as a subalgebra of $\mathcal{E}_{\geq t}$. □

For autonomous systems, the representations $\sigma_t^{\mathcal{Q}}$ and $\sigma_{t'}^{\mathcal{Q}}$ are unitarily equivalent, with

$$\sigma_t^{\mathcal{Q}}(A) = U(t', t)\, \sigma_{t'}^{\mathcal{Q}}(A)\, U(t, t'), \quad \forall A \in \mathcal{Q},$$

where $U(t', t) = \exp\big(i(t - t')H\big)$ is the propagator of S, with t, t' arbitrary times; (Heisenberg-picture dynamics).

For simplicity, we assume that the physical quantities $\mathcal{Q}$ available to identify properties of S or record events all have discrete spectrum; i.e.,

$$\mathcal{Q} = \langle \Pi_\eta^{\mathcal{Q}} | \eta \in \mathcal{Y}^{\mathcal{Q}} \rangle, \tag{19}$$

where $\mathcal{Y}^{\mathcal{Q}} \equiv \mathrm{spec}(\mathcal{Q})$ is a discrete set, which we view as a subset of the real line, and the operators $\Pi_\eta^{\mathcal{Q}}$ are disjoint orthogonal projections. (Of course, continuous spectra can arise, too. But in order to avoid technical complications, we ignore them here.) We can then describe $\mathcal{Q}$ as the algebra given by all functions of a single self-adjoint operator, $\widehat{Y}$, with discrete spectrum, $\mathrm{spec}(\widehat{Y}) \simeq \mathcal{Y}^{\mathcal{Q}}$, and spectral projections $\Pi_\eta^{\mathcal{Q}}$. For every time t, there exists a self-adjoint operator, $Y(t) = \sigma_t^{\mathcal{Q}}(\widehat{Y})$, acting on $\mathcal{H}$ that represents $\widehat{Y}$ at time t.

It is interesting to ask whether physical quantities can serve to detect or record events happening in S. For a discrete set

$$\mathcal{O}_S = \{\mathcal{Q}_j\}_{j \in \mathfrak{J}}$$

of physical quantities characteristic of S, it is arbitrarily unlikely that one of the algebras $\sigma_t^{\mathcal{Q}_j}(\mathcal{Q}_j)$, $j \in \mathfrak{J}$, has a non-trivial intersection with (e.g., contains or is contained in) an algebra $\mathcal{Z}_{\omega_t}(\mathcal{E}_{\geq t})$ describing the event happening at time t, for some state ω_t. To cope with this problem, we have to understand how well $\mathcal{Z}_{\omega_t}(\mathcal{E}_{\geq t})$ can be approximated by an algebra generated by a family, $\{Q_\alpha(t)\}_{\alpha=0}^N$, of disjoint orthogonal projections contained in (or equal to) an algebra $\sigma_t^{\mathcal{Q}}(\mathcal{Q})$, for some $\mathcal{Q} \in \mathcal{O}_S$.

There are different ways of quantifying how well the algebra generated by $\{Q_\alpha(t)\}_{\alpha=0}^N$ approximates the event described by $\mathcal{Z}_{\omega_t}(\mathcal{E}_{\geq t})$. To keep our discussion brief, it is convenient to introduce *"conditional expectations"* of algebras:

Definition 4 Let $\mathcal{N}$ be a (von Neumann) subalgebra of a (von Neumann) algebra $\mathcal{M}$. A linear map

$$\epsilon_\omega : \mathscr{M} \underset{\text{onto}}{\rightarrow} \mathscr{N} \tag{20}$$

is a *conditional expectation* from $\mathscr{M}$ onto $\mathscr{N}$ with respect to a normal state ω on $\mathscr{M}$ iff

(i) $\|\epsilon_\omega(X)\| \leq \|X\|, \quad \forall X \in \mathscr{M}$
(ii) $\epsilon_\omega(X) = X, \quad \forall X \in \mathscr{N}$
(iii) $\omega \circ \epsilon_\omega = \omega$
(iv) $\epsilon_\omega(AXB) = A\epsilon_\omega(X)B, \quad \forall\, A, B, \in \mathscr{N}, \forall\, X \in \mathscr{M}$ □

Conditional expectations have the following properties:

(v) $\epsilon_\omega(X^*X) \geq 0, \quad \forall\, X \in \mathscr{M}$
(vi) $\epsilon_\omega : \mathscr{M} \rightarrow \mathscr{N}$ is completely positive, and $\epsilon_\omega(\mathbb{1}_\mathscr{M}) = \mathbb{1}_\mathscr{N}$

See, e.g., [26] for an exposition of the theory of conditional expectations. Under very general assumptions, there exist conditional expectations

$$\epsilon_{\omega_t} : \mathscr{E}_{\geq t} \rightarrow \mathscr{Z}_{\omega_t}\big(\mathscr{E}_{\geq t}\big), \tag{21}$$

for arbitrary times t.

Let ω_t be the state of a system S right before an event $\{\pi_\xi, \xi \in \mathscr{X}_t\}$ generating $\mathscr{Z}_{\omega_t}(\mathscr{E}_{\geq t})$ starts to happen. I propose to clarify in which way a physical quantity $\mathscr{Q} \in \mathscr{O}_S$ can be used to record this event, and how precisely the value of this quantity identifies the *actual* event, $\xi_* \in \mathscr{X}_t$, happening at time t.

We assume that there exists a physical quantity $\mathscr{Q}$ and a family of disjoint orthogonal projections $\{\widehat{Q}_\alpha\}_{\alpha=0}^N \subset \mathscr{Q}$, $N \geq 2$, with the following properties:

(a) $\sum_{\alpha=0}^N Q_\alpha(t) = \mathbb{1}$, where $Q_\alpha(t) = \sigma_t^{\mathscr{Q}}(\widehat{Q}_\alpha)$, $\alpha = 1, \ldots, N$, $\forall t$;
(b) there exists a positive number $\delta \ll 1$ such that

$$\omega_t\Big(\sum_{\alpha=1}^N Q_\alpha(t)\Big) \geq 1 - \delta \quad \text{(or, equivalently, } \omega_t\big(Q_0(t)\big) \leq \delta\,\text{)};$$

(c) Given an operator $X \in \mathscr{E}_{\geq t}$, we define

$$\text{dist}\big(X, \mathscr{Z}_{\omega_t}(\mathscr{E}_{\geq t})\big) := \|X - \epsilon_{\omega_t}(X)\|.$$

We assume that

$$\text{dist}\big(Q_\alpha(t), \mathscr{Z}_{\omega_t}(\mathscr{E}_{\geq t})\big) < \delta\,, \quad \text{for } \alpha = 1, \ldots, N\,. \tag{22}$$

In the following, we use the notation $\mathscr{O}(\varepsilon)$ to denote any real number whose absolute value is bounded above by $const.\,\varepsilon$, where $const.$ is a *uniformly bounded* positive constant. Properties (a) through (c) of $\{\widehat{Q}_\alpha\}_{\alpha=0}^N$ can be used to derive the following equations:

For an arbitrary operator $X \in \mathcal{E}_{\geq t}$,

$$
\begin{aligned}
\omega_t(X) &= \sum_{\alpha=1}^{N} \omega_t\big(Q_\alpha(t)\, X\big) + \mathcal{O}(\delta \|X\|) \\
&= \sum_{\alpha=1}^{N} \omega_t\big(Q_\alpha(t)[Q_\alpha(t)X]\big) + \mathcal{O}(\delta \|X\|) \\
&= \sum_{\alpha=1}^{N} \omega_t\big(\epsilon_{\omega_t}(Q_\alpha(t))[Q_\alpha(t)X]\big) + \mathcal{O}(\delta\, N \|X\|) \\
&= \sum_{\alpha=1}^{N} \omega_t\big(Q_\alpha(t)X\, \epsilon_{\omega_t}(Q_\alpha(t))\big) + \mathcal{O}(\delta\, N \|X\|) \\
&= \sum_{\alpha=1}^{N} \omega_t\big(Q_\alpha(t)X\, Q_\alpha(t)\big) + \mathcal{O}(\delta\, N \|X\|).
\end{aligned}
\tag{23}
$$

Apparently, if $\delta N \ll 1$, then, to a good approximation, the state ω_t is an incoherent superposition of eigenstates of the disjoint projections $Q_\alpha(t)$, $\alpha = 1, \ldots, N$. We then say that, at approximately time t, "a projective (direct) measurement of $\mathcal{Q}$ takes place."

Definition 5 (Resolution of $\mathcal{Q}$ in Recording an Event) Assuming that $\mathcal{X}_t$ is a countable set, then, for any $\delta \in (0, 1)$, there exists a subset $\mathcal{X}_t^{(M)} \subseteq \mathcal{X}_t$ whose cardinality is given by a finite integer M such that

$$
\omega_t\Big(\sum_{\xi \in \mathcal{X}_t^{(M)}} \pi_{t,\xi} \Big) \geq 1 - \delta\,.
$$

Then, for an arbitrary operator $X \in \mathcal{E}_{\geq t}$,

$$
\omega_t(X) = \sum_{\xi \in \mathcal{X}_t^{(M)}} \omega_t\big(\pi_{t,\xi}\, X\, \pi_{t,\xi}\big) + \mathcal{O}(\delta\, \|X\|)\,.
$$

The *"resolution"* of $\{Q_\alpha(t)\}_{\alpha=0}^N \subset \mathcal{Q}$ in recording the event $\{\pi_{t,\xi}, \xi \in \mathcal{X}_t\}$ starting to happen at time t is defined by

$$
\mathfrak{R} := \frac{N}{M} \cdot (1 - \delta)\,, \quad \text{for } 2 \leq N \leq M\,, \quad (\mathfrak{R} = 0, \text{ for } N = 1)\,. \tag{24}
$$

□

It turns out that property (c), Equation (22), above, implies that, given an orthogonal projection $Q_\alpha(t) \in \sigma_t^{\mathcal{Q}}(\mathcal{Q})$, there exists an orthogonal projection $P_\alpha \in \mathcal{Z}_{\omega_t}(\mathcal{E}_{\geq t})$ such that

$$\| Q_\alpha(t) - P_\alpha \| < \mathcal{O}(\delta) . \tag{25}$$

A proof of this simple lemma can be found in the appendix of [3].

Since the projections $\pi_{t,\xi}, \xi \in \mathcal{X}_t$ generate the abelian algebra $\mathcal{Z}_{\omega_t}(\mathcal{E}_{\geq t})$, we have that

$$\pi_{t,\xi} \cdot P = \pi_{t,\xi}, \quad \text{or } \pi_{t,\xi} \cdot P = 0, \qquad \forall \xi \in \mathcal{X}_t , \tag{26}$$

for any orthogonal projection $P \in \mathcal{Z}_{\omega_t}(\mathcal{E}_{\geq t})$. Equations (25) and (26) then imply the

Result For any $\alpha = 1, \ldots, N$, and for all $\xi \in \mathcal{X}_t$,

$$\boxed{\|\pi_{t,\xi}\, Q_\alpha(t) - \pi_{t,\xi}\| < \mathcal{O}(\delta) , \quad \text{or} \quad \|\pi_{t,\xi}\, Q_\alpha(t)\| < \mathcal{O}(\delta) .}$$

Suppose that the physical quantity $\mathcal{Q}$ is generated by all functions of a single self-adjoint operator $\hat{Y}$. Then the best estimate for the value of $\hat{Y}$ right after time t when the event $\{\pi_{t,\xi} | \xi \in \mathcal{X}_t\}$ has started to happen is an eigenvalue of $\hat{Y}$ corresponding to an eigenstate of the operator $Y(t) \equiv \sigma_t^{\mathcal{Q}}(\hat{Y})$ in the range of the projection $Q_\alpha(t)$. The state of S right after time t is then given by

$$[\omega_t(\pi_{t,\xi_\flat})]^{-1} \omega_t\big(\pi_{t,\xi_\flat}(\cdot)\pi_{t,\xi_\flat}\big) ,$$

for some $\xi_\flat \in \mathcal{X}_t$ for which

$$\|\pi_{t,\xi_\flat} Q_\alpha(t) - \pi_{t,\xi_\flat}\| < \mathcal{O}(\delta) . \tag{27}$$

Furthermore: The higher the resolution, $\mathfrak{R}$, of $\mathcal{Q}$ in recording the event $\{\pi_{t,\xi}, \xi \in \mathcal{X}_t\}$, the more precise the information provided by a measurement of $\mathcal{Q}$ is; if $N = M$ and δ is sufficiently small, then every $\widehat{Q}_\alpha$ determines a unique point $\xi_\flat \in \mathcal{X}_t$ with the property that $\| Q_\alpha(t) - \pi_{t,\xi_\flat} \| < \mathcal{O}(\delta)$. (In the limit where $\delta \to 0$ the information on the event that starts to happen at time t becomes totally accurate.)

Remarks

(1) The main results of this paragraph are Equation (23), the **Result** stated above, and Equation (27).

(2) The concepts presented in paragraph V. and results closely related to the ones described above can be obtained without ever using the theory of conditional expectations. However, their use renders the presentation more elegant.

This completes our review of the "*ETH*-Approach to Quantum Mechanics" in a non-relativistic setting. Some idealized models fitting into this framework are discussed elsewhere, [12]. A relativistic form of this approach will be presented in [16]. The material in [16] leads one to speculate that a logically coherent quantum theory of events, measurements, and observations in *realistic* autonomous isolated (open) systems—not involving the intervention of "observers"—can only be developed in the realm of *local relativistic quantum theories* with *massless* particles, and for even-dimensional space-times.

4 Scattered Remarks About Indirect Measurements, Conclusions

I start this section with a few comments on "indirect measurements" (see [19, 27] for important early results) and then sketch some conclusions.

Let S be an isolated open system, as discussed in Sections 2 and 3. I assume that the system has been prepared in such a way that there is a specific physical quantity, $\mathcal{Q}$, characteristic of S that repeatedly records events featured by S (i.e., is "measured projectively"), at times $t_1 < t_2 < \cdots < t_n, n \in \mathbb{N}$, as discussed in paragraph V. of Section 4, Equations (23) and (27). Let us assume that the spectrum of $\mathcal{Q}$ is a finite set $\mathscr{Y}^{\mathcal{Q}} = \{0, 1, \ldots, k\}$, so that $\mathcal{Q}$ is generated by a single self-adjoint operator, $\widehat{Y}$, with eigenvalues $0, 1, 2, \ldots, k$. Let

$$\underline{\eta}^{(n)} := \{\eta_1, \eta_2, \ldots, \eta_n\}, \quad \eta_j \in \mathscr{Y}^{\mathcal{Q}}, \; j = 1, 2, \ldots, n, \tag{28}$$

be the sequence of values of $\widehat{Y}$ measured at times $t_1, t_2, \ldots, t_n$, as explained in paragraph V. of Section 4. This means that the state of S right after time t_j is in an approximate eigenstate corresponding to the eigenvalue η_j of the operator $Y(t_j)$ representing $\widehat{Y}$ at time t_j, for $j = 1, 2, \ldots, n$, as expressed in Equation (23). The sequence $\underline{\eta}^{(n)}$ is called a "measurement protocol" of length n. As an example, $\widehat{Y}$ may describe the functioning of k different detectors that click when a certain type of particle (e.g., a photon or an atom), called a *"probe,"* belonging to S impacts them, with the following meaning of its eigenvalues:

$$\eta = 0 \leftrightarrow \text{ none of the detectors clicks}, \; \eta = \ell \leftrightarrow \text{ detector } \ell \text{ has clicked}, \quad \ell = 1, \ldots, k.$$

Given a measurement protocol $\underline{\eta}^{(n)}$ of length n, we define the frequency (of occurrence) of the value $\eta \in \mathscr{Y}^{\mathcal{Q}}$ by

$$f_\eta\big(\underline{\eta}^{(n)}\big) := \frac{1}{n}\Big(\sum_{j=1}^{n} \delta_{\eta\,\eta_j}\Big). \tag{29}$$

Note that

$$f_\eta\big(\underline{\eta}^{(n)}\big) \geq 0, \quad \text{and} \sum_{\eta=1}^{k} f_\eta\big(\underline{\eta}^{(n)}\big) = 1\,.$$

Of particular interest is the asymptotics of $f_\eta\big(\underline{\eta}^{(n)}\big)$, as $n \to \infty$. Let us temporarily assume that, $\forall \eta = 0, 1, \ldots, k$, the limit of $f_\eta\big(\underline{\eta}^{(n)}\big)$, as $n \to \infty$, exists whenever a copy of S prepared in a fixed initial state is subjected to very many repeated measurements of $\widehat{Y}$, with

$$\lim_{n\to\infty} f_\eta\big(\underline{\eta}^{(n)}\big) \in \{p(\eta|\alpha)\}_{\alpha=1}^{N}\,, \tag{30}$$

for some $N < \infty$; (this is a *"Law of Large Numbers,"* see [20]). In (30),

$$p(\eta|\alpha) \geq 0, \;\text{ and} \sum_{\eta=1}^{k} p(\eta|\alpha) = 1\,, \tag{31}$$

for all $\alpha = 1, \ldots, N$, for some $N < \infty$. Apparently, the probability measures $p(\cdot|\alpha)$, $\alpha = 1, \ldots, N$, describe all possible limiting values the frequencies $f_{(\cdot)}(\underline{\eta}^{(n)})$ may converge to. We propose to interpret the parameter α as follows: α characterizes a *time-independent* property of S, i.e., it is an eigenvalue of a self-adjoint operator, A, on $\mathscr{H}$ representing a physical quantity of S that commutes with the operators $Y(t_j)$, $j = 1, 2, \ldots$, and is a *conservation law*, meaning that A is time-independent (under the Heisenberg time evolution of operators on $\mathscr{H}$). Such an indirect measurement of A is called a *"non-demolition measurement."* One expects that conservation laws are elements of

$$\mathscr{E}_\infty := \bigwedge_{t\in\mathbb{R}} \mathscr{E}_{\geq t}\,,$$

where $\mathscr{E}_\infty$ is an algebra in the center of the algebra $\mathscr{E}$ defined in (3) ("asymptotic abelianness" in time). Under suitable hypotheses this expectation can actually be proven.

Thus, if the frequencies $f_\eta\big(\underline{\eta}^{(n)}\big)$ are seen to converge to the value $p(\eta|\alpha_*)$, as $n \to \infty$, $\eta \in \mathscr{Y}^{\mathscr{Q}}$, for some $\alpha_* \in \text{spec}(A)$, and if the measures $p(\cdot|\alpha)$ separate points in the spectrum, spec(A), of A, then we *know* that, asymptotically, as $t \to \infty$, the value of the conservation law A approaches α_*. (The fact that the measures $p(\cdot|\alpha)$ may depend on α in a non-trivial way, at all, is a consequence of *"entanglement"*; see [18–20].)

Evidently, one would like to prove (30) and to predict the probability of indirectly measuring a value α_* for A, assuming one knows the initial state of S. However, this can only be done if the events encoded by the values $\eta_1, \eta_2, \ldots,$ of the physical quantity $\hat{Y}$, which is measured at times $t_1, t_2, \ldots,$ are the *only* events happening in

S. For a limited class of systems (see [18, 20]), one can prove that if this is the case then (30) holds, the state of S approaches an eigenstate of A corresponding to some eigenvalue $\alpha_* \in \text{spec}(A)$, as time $t \to \infty$, (*"purification"*), and the probability of measuring the value α_* is given by *Born's Rule* applied to the initial state of S and the operator A, see [20].

Usually, operators on $\mathscr{H}$ representing physical quantities of S are *not* time-independent. If the rate of change in time of a physical quantity, A, of S that one attempts to measure *indirectly*, as described above, is very *small*, as compared to the rate of repeated projective measurements of the physical quantity $\hat{Y}$ used to determine the value of A,[12] then it turns out that, to good accuracy, the dynamics of the state of the system S is described by a *Markov jump process* on the set of eigenspaces of the operator A. The sample paths of this process describe **"quantum jumps"** of (the state of) S from one approximate eigenstate of A to another one. This picture has been given a precise meaning in [20, 22], in the framework of some simple models.

Concluding Remarks:

(1) The *ETH*-Approach to *QM* sketched in this paper is a "Quantum Mechanics without observers." It introduces a precise notion of "events" into the quantum formalism; and it furnishes quantum theory with a clear "ontology."

(2) The *ETH*-Approach establishes a precise formalism to describe the *stochastic time evolution of states* of isolated (open) systems featuring events. As I have tried to explain, while, for an *isolated* system, the Heisenberg-picture time evolution of operators, in particular of physical quantities characteristic of such a system, determined by the unitary propagator of the system is perfectly adequate, the time evolution of its *states* is described by a novel kind of *stochastic branching process* with a "non-commutative state space." This is described in some detail in paragraph IV. of Section 3. The analysis presented there shows that it is simply *not true*—in any naive sense—that the "Heisenberg picture" and the "Schrödinger picture" are equivalent.

(3) It is explained in paragraph V. of Section 3 what a "physical quantity" characteristic of an isolated open system is, what it means to measure such a quantity "projectively," and how "projective measurements" of physical quantities can be used to record events. This also lays a basis for a precise *theory of indirect measurements*.

(4) It is important to note that, in the *ETH*-Approach to *QM*, the expected value of a *conservation law* represented by a self-adjoint operator A in the actual state of an isolated open system featuring events is **not** constant in time (as it would be if states evolved according to the Schrödinger equation).

(5) A *"passive state"* of an isolated open system S prepared at some time t_0 is a state ω for which $\mathscr{Z}_{\omega_t}(\mathscr{E}_{\geq t}) = \{\mathbb{C}\,\mathbb{1}\}$, for all times $t > t_0$. We expect that it often happens that states of S approach "passive states" asymptotically, as

[12] One speaks of a "weak measurement" of A.

$t \to \infty$, (with $\sigma(\mu_\omega) = 0$, see (17)). Thermal equilibrium states are "passive states."

(6) Clearly, the ETH-Approach to QM is so general that, for the time being, it is very hard to use it to carry out explicit calculations for *realistic* model systems and to show in which way its predictions *differ*—usually (hopefully) only ever so slightly—from those made on the basis of, for example, the Copenhagen Interpretation of QM, or *Bohmian Mechanics.* I emphasize, however, that *differences in the predictions of the ETH-Approach and other versions of QM—however small they may be—really exist!*

(7) After completion of this work *Bernard Kay* has pointed out to me that in two of his papers—see [28]—ideas somewhat related to some of the ideas proposed in the present paper have been described. I thank Bernard for valuable discussions.

References

1. R.B. Griffiths, Consistent histories and the interpretation of quantum mechanics. J. Stat. Phys. **36**, 219–272 (1984); M. Gell-Mann, J.B. Hartle, Classical equations for quantum systems. Phys. Rev. D **47**, 3345–3382 (1993)
2. D. Dürr, S. Teufel, *Bohmian Mechanics – The Physics and Mathematics of Quantum Theory* (Springer, Berlin, 2009)
3. J. Fröhlich, B. Schubnel, Do we understand quantum mechanics – finally?, in *Proceedings of Conference in Memory of Erwin Schrödinger*, Vienna, January 2011 (2012)
4. W.K. Abou Salem, J. Fröhlich, Status of the fundamental laws of thermodynamics. J. Stat. Phys. **126**, 1045–1068 (2007)
5. W. De Roeck, J. Fröhlich, Diffusion of a massive quantum particle coupled to a quasi-free thermal medium. Commun. Math. Phys. **303**, 613–707 (2011)
6. J. Fröhlich, Z. Gang, A. Soffer, Friction in a model of Hamiltonian dynamics. Commun. Math. Phys. **315**, 401–444 (2012); J. Fröhlich, Z. Gang, Emission of Cherenkov radiation as a mechanism for Hamiltonian friction. Adv. Math. **264**, 183–235 (2014)
7. R. Bauerschmidt, W. De Roeck, J. Fröhlich, Fluctuations in a kinetic transport model for quantum friction. J. Phys. A Math. Theor. **47**, 275003 (2014). arXiv:1403.5790
8. J. Fröhlich, B. Schubnel, Quantum probability theory and the foundations of quantum mechanics, in *The Message of Quantum Science – Attempts Towards a Synthesis*, ed. by Ph. Blanchard, J. Fröhlich (Springer, Berlin, 2015). arXiv:1310.1484
9. Ph. Blanchard, J. Fröhlich, B. Schubnel, A 'Garden of Forking Paths' – the quantum mechanics of histories of events. Nucl. Phys. B **912**, 463–484 (2016)
10. B. Schubnel, Mathematical results on the foundations of quantum mechanics. PhD thesis, 2014. https://doi.org/10.3929/ethz-a-010428944
11. R. Haag, Fundamental irreversibility and the concept of events. Commun. Math. Phys. **132**, 245–251 (1990); R. Haag, Events, histories, irreversibility, in *Quantum Control and Measurement.* Proc. ISQM, ARL Hitachi, ed. by H. Ezawa, Y. Murayama (North Holland, Amsterdam, 1993); Ph. Blanchard, A. Jadczyk, Event-enhanced quantum theory and piecewise deterministic dynamics. Ann. Phys. **4**, 583–599 (1995)
12. J. Fröhlich, *'ETH' in Quantum Mechanics*. Notes of Lectures on the Foundations of Quantum Mechanics, Les Diablerets, 9–14 Jan 2017
13. J. Faupin, J. Fröhlich, B. Schubnel, On the probabilistic nature of quantum mechanics and the notion of 'Closed' systems. Ann. Henri Poincaré **17**, 689–731 (2016)

14. J. Fröhlich, B. Schubnel, The preparation of states in quantum mechanics. J. Math. Phys. **57**, 042101 (2016)
15. D. Buchholz, J.E. Roberts, New light on infrared problems: sectors, statistics, symmetries and spectrum. Commun. Math. Phys. **330**, 935–972 (2014); D. Buchholz, Collision theory for massless bosons. Commun. Math. Phys. **52**, 147–173 (1977)
16. J. Fröhlich, *Relativistic Quantum Theory and Causality*. Talks at the University of Leipzig (2018), TU-Stuttgart (2019), IHES (2019) and at Vietri sul Mare (Italy) (2020) arXiv1912.00726
17. C. Guerlin, J. Bernu, S. Deléglise, C. Sayrin, S. Gleyzes, S. Kuhr, M. Brune, J.M. Raimond, S. Haroche, Progressive field-state collapse and quantum non-demolition photon counting. Nature **448**(7156), 889–893 (2007); S. Haroche, *Controlling Photons in a Box and Exploring the Quantum to Classical Boundary*. Nobel Lecture, 8 Dec 2012. The Nobel Prizes
18. M. Bauer, D. Bernard, Convergence of repeated quantum non-demolition measurements and wave-function collapse. Phys. Rev. A **84**, 044103:1–4 (2011)
19. H. Maassen, B. Kümmerer, *Purification of Quantum Trajectories*. Lecture Notes - Monograph Series, vol. 48 (Springer, Berlin, 2006), pp. 252–261
20. M. Ballesteros, M. Fraas, J. Fröhlich, B. Schubnel, Indirect retrieval of information and the emergence of facts in quantum mechanics. J. Stat. Phys. **162**, 924–958 (2016). arXiv:1506.01213
21. M. Ballesteros, N. Crawford, M. Fraas, J. Fröhlich, B. Schubnel, Non-demolition measurements of observables with general spectra, in: "Mathematical Problems in Quantum Physics," QMATH 13 F. Bonetto et al. (eds.), Contemporary Mathematics **717**, 241–256 (2018)
22. M. Ballesteros, N. Crawford, M. Fraas, J. Fröhlich, B. Schubnel, Perturbation theory for weak measurements in quantum mechanics, I - systems with finite-dimensional states space. Ann. Henri Poincaré **20**, 299–335 (2019). arXiv1700.03149
23. J.S. Bell, *Speakable and Unspeakable in Quantum Mechanics* (Cambridge University Press, Cambridge, 1987). See also: J.A. Wheeler, W.H. Zurek, *Quantum Theory and Measurement* (Princeton University Press, Princeton, 1983); K. Hepp, Quantum theory of measurement and macroscopic observables. Helv. Phys. Acta **45**, 237–248 (1972); H. Primas, Asymptotically disjoint quantum states, in *Decoherence: Theoretical, Experimental and Conceptual Problems*, ed. by Ph. Blanchard, D. Giulini, E. Joos, C. Kiefer, I.-O. Stamatescu (Springer, Berlin, 2000), pp. 161–178
24. D. Frauchiger, R. Renner, Quantum theory cannot consistently describe the use of itself. Nat. Commun. **9**, # 3711 (2018)
25. G. Lüders, Über die Zustandsänderung durch den Messprozess. Ann. Phys. (Leipzig) **443**, 322–328 (1950); J. Schwinger, The algebra of microscopic measurement. Proc. Natl. Acad. Sci. USA **45**, 1542–1553 (1959); E.P. Wigner, *The Collected Works of Eugene Paul Wigner, Part A: The Scientific Papers*, ed. by B.R. Judd, G.W. Mackey (Springer, New York, 1993)
26. M. Takesaki, Conditional expectations in von Neumann algebras. J. Funct. Anal. **9**, 306–321 (1972); F. Combes, Poids et Espérances Conditionnelles dans les Algèbres de von Neumann. Bull. Soc. Math. France **99**, 73–112 (1971)
27. K. Kraus, *States, Effects and Operations* (Springer, Berlin, 1983)
28. B.S. Kay, V. Abyaneh, Expectation values, experimental predictions, events and entropy in quantum gravitationally decohered quantum mechanics. arXiv:0710.0992 (v1), unpublished; B.S. Kay, The matter-gravity entanglement hypothesis. Found. Phys. **48**, 542–557 (2018)

Linear and Nonlinear Harmonic Boundaries of Graphs; An Approach with ℓ^p-Cohomology in Degree One

Antoine Gournay

1 Introduction

Graphs are defined by their vertices (henceforth X) and their edges $E \subset X \times X$. In a sense understanding a graph means to understand how the vertices and edges work together. In a finite graph, it is common to reduce the whole graph to the incidence matrix.

In an oriented graph, the incidence matrix B has $|X|$ lines and $|E|$ columns. Each column contains a -1 and a $+1$ to indicate the source and target of every edge. This matrix not only encodes the whole graph, but also a very familiar operation: the vector space $\mathbb{R}^{|X|}$ is the space of functions on the vertices, $\mathbb{R}^{|E|}$ the space of functions on the edges, and the matrix B is the gradient. More precisely, given a function on the vertices f (that is an element of $\mathbb{R}^{|X|}$), Bf is a function on the edges and its value on the edge (x, y) from x to y is $f(y) - f(x)$.

For infinite graphs, the gradient encompasses also all the information of the graph. Most people would no longer refer to it as a matrix though, but rather as an operator. In short, ℓ^p-cohomology in degree one aims at understanding the image of this operator.

The history of the topic can be split in two "cases". The case $p = 2$ has been largely studied and offers even more connections to other fields of mathematics (see Lück [38] or Eckmann [14] among many references). The case $p \neq 2$ has been introduced through Zucker (see [63] and references therein) to study compactifications of manifolds and Gromov (see [28, §8]) as a large-scale invariant of groups. Since then, applications have been found to harmonic functions, many

A. Gournay (✉)
Department of Mathematics, TU Dresden, Dresden, Germany
e-mail: gournay@tu-dresden.de

N. Anantharaman et al. (eds.), *Frontiers in Analysis and Probability*,
https://doi.org/10.1007/978-3-030-56409-4_3

notions of boundaries, representation theory of groups, quasi-isometry and packing of graphs; see §2 for details.

The main aim of this paper is to present the connection between ℓ^p-cohomology in degree one and harmonic functions, *i.e.* to interpret it as a special subspace of the Poisson boundary. As such the presentation tries to streamline some results of [21, 22] and [24].

Here is a thinned out version of this result (the actual result applies to a larger class of graphs, but the statement becomes technical).

Theorem 1.1 *Let G be the Cayley graph of a group which is not virtually nilpotent. Fix some* $p \in]1,\infty[$ *(and not* $p \in [1\infty[$*). Then* (1) $\implies$ (2) $\implies$ (3) $\implies$ (4) $\implies$ (5) *where*

(1) *The reduced* ℓ^p*-cohomology in degree one vanishes.*
(2) *For any functions* f *with gradient in* ℓ^p *there is a* $c \in \mathbb{R}$ *so that* $\lim f(x_n) = c$ *for any sequence* x_n *going to infinity.*
(3) *There are no non-constant harmonic functions with gradient in* ℓ^p*.*
(4) *There are no non-constant bounded harmonic functions with gradient in* ℓ^p*.*
(5) *For any* $q < p$*, the reduced* ℓ^q*-cohomology in degree one vanishes.*

See §4 and Theorem 4.1 for details. Among others, this has applications to the question whether the Poisson boundary is invariant under quasi-isometries (see Corollary 4.16).

Organisation: §1.2 gives the definition of ℓ^p-cohomology in degree one. §1.3 follows with examples which are not too hard to grasp. §2 presents some applications of ℓ^p-cohomology in degree one to other problems and topics. §3 shows how ℓ^1-cohomology in degree one can be seen as a space of functions on the ends of the group, giving a first sign that ℓ^p-cohomology has to do with ideal boundaries of graphs. §4 tackles the connection between ℓ^p-cohomology in degree one and harmonic functions. Lastly, §5.1 tries to summarises some other results and §5.2 presents some questions. But first, let us start with some preliminaries.

1.1 Conventions and Preliminaries

The conventions are that a graph $\Gamma = (X, E)$ is defined by X, its set of vertices, and E, its set of edges. All graphs will be assumed to be of bounded valency and the set of vertices X will always be assumed to be countable. The set of edges will be thought of as a subset of $X \times X$. The set of edges will be assumed symmetric (*i.e.* $(x, y) \in E \implies (y, x) \in E$). Functions will take value in $\mathbb{R}$ (but we could easily work with $\mathbb{C}$ too). Functions on E will often be anti-symmetric (*i.e.* $f(x, y) = -f(y, x)$). This said $\ell^p(X)$ is the Banach space of functions on the vertices which are p-summable, while $\ell^p(E)$ will be the subspace of functions on the edges which are p-summable.

The gradient $\nabla : \mathbb{R}^X \to \mathbb{R}^E$ is defined by $\nabla g(x, y) = g(y) - g(x)$.

$c_0(X)$ denotes the space of functions f which tend to 0 at infinity. This can be defined as follows: $f \in c_0(X)$, if for any sequence of finite sets $A_n \subset X$ with $\cup A_n = X$ and $A_i \subset A_{i+1}$, $\sup_{x \notin A_n} f(x) \xrightarrow{n\to\infty} 0$. Another possible description is the closure of finitely supported functions in ℓ^∞-norm.

Lastly, p' will denote the Hölder conjugate exponent of p, *i.e.* $p' = p/(p-1)$ (with the usual convention that 1 and ∞ are conjugate).

1.2 ℓ^p-Cohomology in Degree One

So our lofty goal is to understand the gradient map from $\mathbb{R}^X$ to $\mathbb{R}^E$. The first thing is that $\mathbb{R}^E$ is way too big as a space, even if one restricts to anti-symmetric functions. Indeed, any function on the edge which does not sum to 0 along a 2-cycle cannot come from the gradient.

Hence we restrict to the image of the gradient (or the kernel of the second coboundary operator [from edges to cycles] if you are curious about the origins of the name "cohomology").

The next step is to bring some simple functional analysis by restricting to ℓ^p-spaces.

At last we have ℓ^p-cohomology in degree one: given that the gradient of some function is in $\ell^p(E)$, can this gradient be approximated by gradients of functions in $\ell^p(X)$?

More precisely, the ℓ^p-cohomology in degree one of the graph Γ is the quotient

$$\ell^p H^1(\Gamma) := (\ell^p(E) \cap \nabla \mathbb{R}^X)/\nabla \ell^p(X).$$

Unfortunately, the image of ∇ is not always closed. In order to avoid dealing with unseparated space (and space which trivially have lots of things in their ℓ^p cohomology), the focus is usually on the largest separated quotient, the reduced ℓ^p-cohomology:

$$\ell^p \underline{H}^1(\Gamma) := (\ell^p(E) \cap \nabla \mathbb{R}^X)/\overline{\nabla \ell^p(X)}^{\ell^p(E)}.$$

Now, if you are wondering when is the image of ∇ actually closed, then

Theorem 1.2 *Let $p \in [1, \infty[$. The image of $\nabla : \ell^p(X) \to \ell^p(E)$ is closed if and only if the graph is amenable* (i.e. *there is a sequence of finite sets F_n such that $\frac{|\partial F_n|}{|F_n|} \to 0$, where ∂F is the set of edges with only one extremity in F).*

One direction of the proof is straightforward:

Proof of the "easy" Part Assume there is a sequence of sets $F_n \subset X$ so that $\frac{|\partial F_n|}{|F_n|} \to 0$. Take $f_n = \frac{1}{|F_n|^{1/p}} \mathbb{1}_{F_n}$ where $\mathbb{1}_F$ is the characteristic function of the set F (the function which takes value 1 on F and 0 elsewhere). By construction $\|f_n\|_{\ell^p(X)} = 1$.

But ∇f_n takes value $\pm\frac{1}{|F_n|^{1/p}}$ on ∂F_n (the sign depends on whether the edge points towards or away from the set F_n) and 0 elsewhere. Hence (the upcoming factor of 2 comes from the two orientation of the edges)

$$\|\nabla f_n\|_{\ell^p(E)} = \frac{1}{|F_n|^{1/p}}\|\mathbb{1}_{\partial F_n}\|_{\ell^p(E)} = \frac{1}{|F_n|^{1/p}}(2|\partial F_n|)^{1/p} = 2^{1/p}\big(\tfrac{|\partial F_n|}{|F_n|}\big)^{1/p}.$$

By hypothesis, this sequence tends to 0. As a consequence of the closed image theorem (an operator has a closed image if and only if it has a bounded inverse), the image of ∇ is not closed. □

The other direction of the statement is a typical technical slicing argument (given a sequence of functions f_n with norm 1 whose gradient tends to 0, look at "well-chosen" level sets of these functións). As it is quite technical, the proof would bring us off-topic, so the reader is encouraged to look up surveys on amenability for all the details (a very nice book, which is not so easy to find was written by Greenleaf [26]; there are some surveys freely available in Internet).

Most of the times it is much more convenient to think only in term of functions. To this end, introduce the Banach space of p-Dirichlet functions as the space of functions f on X such that $\nabla f \in \ell^p(E)$. It will be denoted $\mathsf{D}^p(\Gamma)$.

In order to introduce the $\mathsf{D}^p(\Gamma)$-norm on $\mathbb{R}^X$, it is necessary to choose a vertex, denoted e_Γ. This said $\|f\|^p_{\mathsf{D}^p(\Gamma)} = \|\nabla f\|^p_{\ell^p(E)} + |f(e_\Gamma)|^p$.

By taking the primitive of these gradients, one may also prefer to think of reduced ℓ^p-cohomology in degree one as:

$$\underline{\ell^p H}^1(\Gamma) := \mathsf{D}^p(\Gamma)/\overline{\ell^p(X) + \mathbb{R}}^{\mathsf{D}^p}.$$

A common abuse of language and notation happens, as one says that the reduced cohomology is equal to the non-reduced one: this means that the "natural" quotient map $\ell^p H^1(\Gamma) \to \underline{\ell^p H}^1(\Gamma)$ is injective. By Theorem 1.2 above, this happens exactly when the graph is non-amenable.

1.3 Some Examples

Before moving on to general statements, the reader might want to look at some simple examples. Since most of our examples come from Cayley graphs, let us also shortly recall their construction.

Given a finitely generated group G and a finite set S, the Cayley graph Cay(G, S) is the graph whose vertices are the element of G and $(\gamma, \gamma') \in E$ if $\exists s \in S$ such that $s^{-1}\gamma = \gamma'$. (This convention might be unusual from the point of view of random walks, but is much more convenient to write convolutions.) In order for the resulting graph to have a symmetric edge set, S is always going to be symmetric (*i.e.* $s \in$

$S \implies s^{-1} \in S$). Also, Cayley graphs are always going to be connected (*i.e.* S is generating).

Example 1.3 The group $\mathbb{Z}$ with its most tempting generating set $\{\pm 1\}$ has the line as its Cayley graph.

Since there are no cycles, the question is: are all elements of $\ell^p(E)$ in the closure of $\nabla \ell^p(X)$? The simplest element of $\ell^p(E)$ is the "Dirac mass" (due to our convention that edges are oriented and function on edges are anti-symmetric, this is $\delta_{(1,0)} - \delta_{(0,1)}$), so that seems a nice place to start.

It is somehow easier to represent it as a function in $\mathsf{D}^p(X)$: namely $f(x) = 0$ for $x \le 0$ and $f(x) = 1$ for $x > 1$.

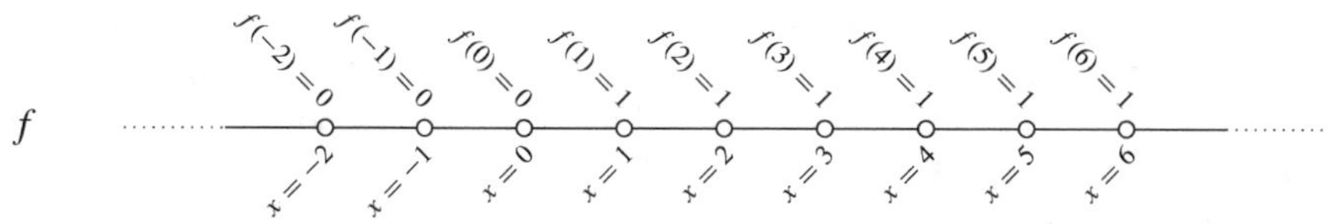

This function looks hard to approximate: it is not even finitely supported. But remember, we are trying to approximate its gradient (not the values the function takes).

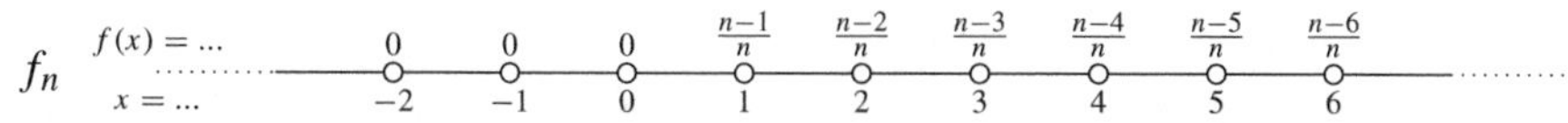

and f_n stays 0 once it reaches 0 (so for $x \ge n$). Now the important point is that we want $\nabla(f - f_n)$ to tend to 0 ($f - f_n$ obviously does not). A quick computation shows that $\nabla(f - f_n)$ takes on $2n$ edges (recall that (0, 1) and (1, 0) are both edges) the value $\pm\frac{1}{n}$. Hence

$$\|\nabla(f - f_n)\|_{\ell^p} = \big(2n\frac{1}{n^p}\big)^{1/p} = \frac{2^{1/p}}{n^{1-1/p}}.$$

This tends to 0 given that $p > 1$.

This shows that the basis of $\ell^p(E)$ is in $\overline{\nabla \ell^p(X)}$. Since this basis is dense in $\ell^p(E)$, we just showed that $\underline{\ell^p H}^1(\mathbb{Z})$ is trivial when $p > 1$.

And what about $p = 1$? Well, there is a trick (see Martin & Valette [40, Example 3 in §4] who mention hearing it from M. Bourdon). Let us quickly outline it here, it will be discussed at length in §3.

Note that any function on the line with gradient in ℓ^1 has a value as $x \to +\infty$ and $x \to -\infty$. For any $g \in \mathsf{D}^1(X)$, define $L(g) = \lim_{x \to +\infty} g(x) - \lim_{x \to -\infty} g(x)$. Then $L : \mathsf{D}^1(X) \to \mathbb{R}$ is a bounded operator. Its kernel contains $\ell^1(X)$ and so it will also contain its closure in the D^1-norm. Since our function f above has $L(f) = 1 \ne 0$, it lies outside of the closure $\ell^1(X)$.

As an upshot, $\ell^1\underline{H}^1(\mathbb{Z})$ is not trivial (in fact, it is a one-dimensional real space). ◊

Example 1.4 Another simple example are the Cayley graphs of free groups on k generators F_k (resp. free products $C_2 * C_2 * \ldots * C_2$ where C_2 is the group with two elements). The Cayley graphs for the "standard" generating sets (*i.e.* the k free generators, resp. the generators of each C_2 factor) are $2k$-regular (resp. k-regular) trees.

Again, since trees have no cycles one gets that $\ell^p(E) = \ell^p(E) \cap \nabla\mathbb{R}^X$. If $k = 1$ (resp. $k = 2$), then we obtain the same graph as in the previous example. So we may assume that $k > 1$ (resp. $k > 2$). Now it comes in very handy to note that these graphs are not amenable. By Theorem 1.2, this means that $\nabla\ell^p(X)$ is closed or, if one thinks in terms of functions, that $\ell^p(X)$ is a closed subspace in the D^p-norm.

But functions in $\ell^p(X)$ also belong to $c_0(X)$ (the space of functions which tend to 0 at ∞, see §1.1). Hence, if we can find a function with gradient in ℓ^p which is not in $c_0(X)$, then we are done.

But this is fairly easy: (a) pick some edge, (b) removing it will disconnect the tree in two components, (c) set f to be identically 0 on one component and identically 1 on the other, (d) the gradient of this function is supported on one edge, so it lies definitively in D^p.

Consequently, $\ell^p\underline{H}^1$ and ℓ^pH^1, are non-trivial for any p. ◊

It is straightforward to generalise this to any tree which is not amenable: one just has to make sure that the edge disconnects the tree in two infinite components. In fact, the argument applies to any non-amenable graph which can be disconnected into two infinite components by removing a finite number of edges.

These two examples are somehow extreme in the sense that $\ell^p\underline{H}^1$ is either trivial for all $p > 1$ or not. However, in the case of hyperbolic space, it turns out the p for which $\ell^p\underline{H}^1$ passes from trivial to non-trivial is a significant number (see §2.2.2 for details).

Also, the last example might make you think that almost all non-amenable graphs have a non-trivial $\ell^p\underline{H}^1$ (for some p). But it turns out it is often hard to construct an element of $\mathsf{D}^p(X) \setminus c_0(X)$. The following proposition can partially explain why (as well as generalising Example 1.3 and introducing some important proof technique).

Proposition 1.5 *Assume Γ is the Cayley graph of a group G whose centre $Z := Z(G)$ is infinite. Then $\ell^p\underline{H}^1(\Gamma) = \{0\}$ for all $p > 1$.*

Proof Needless to say, elements of the centre have the very nice property that, for any $g \in G$, $zg = gz$. This translates in a graph theoretical property. Indeed, the action of an element g of G on the right is a graph automorphism. The action on the left by the same element g means one follows the path labelled by the generators $s_i \in S$ so that $g = s_n s_{n-1} \ldots s_2 s_1$.

So being in the centre means that if you follow (starting at any vertex) a path labelled by $z = s_n s_{n-1} \ldots s_2 s_1$, then this is a graph automorphism.

Here is why elements of the centre are so special for this problem. Let $\rho_z f(g) := f(gz)$. Write $z = s_n s_{n-1} \dots s_2 s_1$ and let $t_i = s_i s_{i-1} \dots s_2 s_1$ (with $t_0 = e$ the identity element in G). Then

$$f(g) - \rho_z f(g) = f(g) - f(gz) = f(g) - f(zg) = \sum_{i=1}^{n} f(t_i g) - f(t_{i-1} g).$$

Note that this last expression is a sum of n values of the gradient of f. Hence, by the triangle inequality, $\|f - \rho_z f\|_{\ell^p(X)} \leq n \|\nabla f\|_{\ell^p(E)}$. This implies that f and $\rho_z f$ belong to the same equivalence class.

This can be used to bring the following plan into action. Given some function f with gradient in ℓ^p, consider $\rho_{z_n} f$ where z_n is some sequence of elements of the centre which goes to infinity. Since $\rho_{z_n} f$ are images under graph automorphisms of f, we are effectively translating the gradient of f to infinity.

Since $\ell^p(E) \subset c_0(E)$, this means that $\nabla \rho_{z_n} f$ tends point-wise to 0. Point-wise convergence is synonymous with weak* convergence. But weak* and weak convergence coincide in the reflexive case. And a classical consequence of the Hahn–Banach theorem is that weak and norm convergence to 0 also coincide.

So we found a way to build a sequence of elements which all belong to the equivalence class of f (in the quotient space $\nabla \mathbb{R}^X \cap \ell^p(E)/\overline{\ell^p(X)}$) and whose gradients tend (in norm) to 0. This shows that 0 is in the (closure) of the class too.

But we made no specific assumption on the function f, hence 0 is in the equivalence class of any function, and $\underline{\ell^p H}^1(\Gamma) = \{0\}$. □

The previous proposition can be found [often with weaker hypothesis] in Kappos [32, Theorem 6.4], Martin & Valette [40, Theorem 4.3], Puls [52, Theorem 5.3], Tessera [59, Proposition 3] or [20, Theorem 3.2].

There are many groups with an infinite centre many of them are not amenable. This hopefully contrasts with Example 1.4.

2 Applications

Before we move to our main focus (which has to do with harmonic functions), here is an overview of the different applications of ℓ^p-cohomology to themes.

2.1 Quasi-Isometries

One of the original motivation of ℓ^p-cohomology was to use it as an invariant of quasi-isometry, see Gromov [28, §8].

Let us briefly recall that a map $f : (X, d_X) \to (Y, d_Y)$ between two metric spaces is a quasi-isometry, if there is a constant $K > 1$ such that:

$$\tfrac{1}{K}d_X(x,x') - K \le d_Y\big(f(x), f(x')\big) \le K d_X(x,x') + K.$$

There are few important "exercises" on this concept, here are two: (1) "being quasi-isometric" is an equivalence relation; (2) a graph (with its combinatorial distance) can be quasi-isometric to a manifold (with its Riemannian metric).

In fact, Kanai has shown [31] that any Riemannian manifold with Ricci curvature and injectivity radius bounded from below is quasi-isometric to a graph (of bounded valency).

Theorem 2.1 (See Élek [15, §3] or Pansu [45]) *If two graphs of bounded valency Γ and Γ' are quasi-isometric, then they have the same ℓ^p-cohomology (in all degrees, reduced or not).*

The result is actually much more powerful, in the sense that it holds in a larger category (measure metric spaces; see above-mentioned references). For shorter proofs in more specific situations see Puls [55, Lemma 6.1] or Bourdon & Pajot [7, Théorème 1.1].

The previous theorem is sometimes very convenient, since it means that results can be transferred between graphs and Riemannian manifolds. This allows for a great flexibility in the methods that can be used to prove the results.

A consequence of 2.1 is that, if G is a finitely generated group, the ℓ^p-cohomology in degree one of any two Cayley graphs (for a finite generating set) is isomorphic. Indeed, the identity map on the vertices is a quasi-isometry between the Cayley graphs (hint: write the generators of one Cayley graph as words in the generators of the other Cayley graph). Thus, one may speak of the ℓ^p-cohomology of a group without making reference to a Cayley graph.

In [47] and [48], Pansu computed the ℓ^p cohomology (in degree 1 and above) of a variety of homogeneous spaces with pinched negative curvature. He then used the triviality or non-triviality of this cohomology to show that many of these spaces are not quasi-isometric, thus answering an old question of Berger.

The study of quasi-isometries also motivated some variants of ℓ^p-cohomology. First, by considering Orlicz spaces (instead of just ℓ^p spaces) Carrasco Piaggio [11] proved a fixed-point result for self-quasi-isometries of (many) Heintze groups.

Second, there is a body of work on the L_{pq}-cohomology (investigations of the quotients of the form $\mathrm{d}^p/(\mathbb{R} + \ell^q)$. The interested reader is encourage to look at Gol'shtein & Troyanov [19], Kopylov [36] and references therein.

2.2 Boundaries

S. Zucker was one of the first person to introduce ℓ^p-cohomology and use it to study manifolds with thin ends (see [63] and references therein). There are however many other applications to other ideal boundaries of spaces.

The ends are another typical "ideal boundary" for a space, and it turns out that the reduced ℓ^1-cohomology in degree one is isomorphic to the space of function on the ends modulo constant functions (see §3 for details).

2.2.1 Poisson Boundary

There is also a strong connection between ℓ^p-cohomology in degree one and harmonic functions. This particular topic will be explained in more details in §4.

The short version is that (if the isoperimetric dimension of the graph is large enough then) a function with a non-trivial cohomology class gives rise to a non-constant bounded harmonic function. This is easier to see in the case $p = 2$, but it extends to other $p \neq 2$ (if the isoperimetric dimension is large enough).

This is interesting since the Poisson boundary (which can be roughly thought of as the space of bounded harmonic function) is not an invariant of quasi-isometry (see, for example, T. Lyons' examples [39]). Namely, there are quasi-isometric graphs one of which has many non-constant bounded harmonic functions, while the other has none.

Theorem 2.1 can be invoked to show that the ℓ^p-cohomology in degree one gives rise to a part of the Poisson boundary which is invariant under quasi-isometries.

2.2.2 Boundary of Hyperbolic Spaces

There are also applications of ℓ^p-cohomology to the boundary of hyperbolic spaces, more precisely to problems which are related to the famous

Conjecture 2.2 (J. Cannon) *Let Γ be a hyperbolic group whose ideal boundary is a 2-sphere. Then Γ is virtually a cocompact lattice in* $\mathrm{PSL}(2, \mathbb{C})$.

Using a result of Keith & Laakso [34, Corollary 1.0.3], Bonk & Kleiner [2] were able to show that if Γ is a hyperbolic group whose ideal boundary is a 2-sphere and the conformal dimension is achieved by some metric, then Γ is virtually a cocompact lattice in $\mathrm{PSL}(2, \mathbb{C})$.

Further results by Bourdon & Pajot [7] show that, for hyperbolic spaces, one can define a L^p-dimension as the infimum over all p for which $\ell^p H^1$ is non-trivial. It turns out that the L^p-dimension coincides with the conformal dimension if there is a metric which achieves the conformal dimension.

Bourdon & Pajot [7] gave examples where these dimensions do not coincide, hence one cannot expect that the strategy from Bonk & Kleiner [2] works out of the box. On the positive side, there has been further work (using ℓ^p-cohomology) by Bourdon & Kleiner [4] which covers the case of Coxeter groups.

For a proof that any hyperbolic space has a non-trivial ℓ^p-cohomology in degree one starting at some p_0 see either Bourdon & Pajot [7], Élek [15] or Puls [54].

M. Bourdon pointed out to the author a very interesting point (see also [3, §2.4.1]). A result of Puls [54, Theorem 1.3] shows that if a group has a non-trivial

Floyd boundary for a Floyd function $\phi(g) = a^{-d(e,g)}$ (where $a > 1$), then its [reduced] ℓ^p-cohomology will be non-trivial for all p such that $\phi \in \ell^p(G)$. A careful reading of the construction of Gerasimov [18] shows that relatively hyperbolic groups will have non-trivial Floyd boundaries satisfying these conditions. Consequently, their reduced ℓ^p cohomology is non-trivial for all p larger than some p_0.

On the other hand, D. Osin pointed out to the author that some acylindrically hyperbolic groups have a trivial ℓ^p-cohomology for all $p \in [1, \infty[$ (these are right-angled Artin groups corresponding to the graph •–•–•–•–•).

Pansu [44] showed that among continuous Lie groups, having non-trivial reduced ℓ^p-cohomology is equivalent to hyperbolicity. This extends to algebraic groups over local fields of characteristic 0 by a result of de Cornulier & Tessera [13].

Lastly, Bourdon & Pajot [7, §3, Proposition 4.1 and the following Remarques] also showed that for p larger than the conformal dimension of the boundary, functions with ℓ^p-gradient, when extended to the boundary of [Gromov] hyperbolic spaces, can separate points of its boundary. In fact, they show that (non-trivial) Lipschitz functions on the boundary give rise to (non-trivial) classes in ℓ^p-cohomology. Bourdon & Kleiner [5, Theorem 3.8(1)] showed that if p is strictly smaller than the conformal boundary, then extensions of ℓ^p classes no longer separate points.

2.2.3 "Nonlinear" Boundaries

Reduced ℓ^p-cohomology (in degree one) is very strongly related to p-harmonic functions. When $p = 2$, this is the same as harmonic functions, but for $p \neq 2$ these are a nonlinear variation of the harmonic equation.

When p is an integer, p-harmonic functions come up naturally when studying a relaxation of conformal maps (called quasi-regular maps). Given two manifolds M and N of dimension p, a map $f : M \to N$ is called quasi-regular if there is a constant C so that $\|\mathrm{d}f\|^p \leq C|\det \mathrm{d}f|$.

When $g : N \to \mathbb{R}$ is a function, the p-Laplacian is $\Delta_p = \mathrm{div}(|\nabla g|^{p-2}\nabla g)$ and p-harmonic functions are functions whose p-Laplacian is 0. A quasi-regular map will allow to pull-back [non-constant] p-harmonic functions, so the existence or absence of [non-constant] p-harmonic functions can be used as obstruction to the existence of quasi-regular maps.

In addition to quasi-regular maps, there are also interesting limiting cases for the p-harmonic equation: when $p \to 1$ this is related to the mean curvature operator and when $p \to \infty$ to Lipschitz extensions.

In the setting of graphs there are two things which might be unclear:

(1) what is the divergence? see §4.
(2) what is a quasi-regular map? see either Benjamini, Schramm & Timàr [10, §1.1] or §2.4.

Furthermore, much like harmonic functions can be used to construct a Royden boundary and a harmonic boundary, p-harmonic functions can be used to construct a p-Royden boundary and a p-harmonic boundary. For the definitions see Puls [55, §2.1]. These boundaries are spaces constructed with the help of the Gelfand transform which can be associated with the classes of the reduced ℓ^p-cohomology in degree one. See paragraph after Lemma 4.7 for details.

The relation between reduced ℓ^p-cohomology in degree one and p-harmonic function is fairly straightforward (see Puls [53] or Martin & Valette [40] for details). Basically, given $f \in \mathsf{D}^p(\Gamma)$, one can try to search for the element which belongs to the same equivalence class as f but whose norm is minimal. For $p \in]1, \infty[$ such an element will exist by convexity of the norm. Furthermore, for all g of finite support on the edges $\frac{\mathrm{d}}{\mathrm{d}t} \|\nabla f + t\nabla g\|^p_{\ell^p(E)}\big|_{t=0} = 0$ (by minimality of the norm of this element). Massaging this last equation (and the fact that g is an arbitrary function of finite support) will show that f is p-harmonic.

Other known consequences of the triviality of the reduced ℓ^p-cohomology in degree one include the triviality of the p-capacity between finite sets and ∞ (see Yamasaki [62] and Puls [55, Corollary 2.3]) and existence of continuous translation invariant linear functionals on $\mathsf{D}^p(\Gamma)/\mathbb{R}$ (see [55, §8]).

2.3 Representation Theory

For infinite groups it is often interesting to look at their representation on infinite dimensional space. For example, Property (T) is defined using the topology on the space of unitary representations in Hilbert spaces. It can also be expressed as a condition on the first cohomology of these representations.

It turns out that ℓ^p-cohomology in degree one (of some Cayley graph of a finitely generated group) is the same thing as the first cohomology of the regular representation (in ℓ^p), see Martin & Valette [40] or Puls [52]. There is also a nice text from Bourdon [3] on the topic (isometric actions on Banach space are equivalent to cohomology linear representations).

Though it might seem a very particular case, it turns out this has a direct and indirect application to Hilbertian representations. The direct application is that triviality of the reduced ℓ^p-cohomology in degree one implies that the reduced first cohomology of any unitary representation with coefficients in ℓ^p is trivial. (The coefficients of a unitary representation π are the functions $\kappa(\gamma) := \langle \pi_\gamma \xi \mid \xi' \rangle$ where ξ, ξ' are elements of the Hilbert space.)

The indirect application is that techniques that are useful to show the vanishing (or non-triviality) of ℓ^p-cohomology may also be applied for unitary representations. See [24] for more details.

2.4 Sphere Packings

A last nice application of reduced ℓ^p-cohomology in degree one is to sphere packings of graphs. Circle packings are a lovely topic which the reader should definitively try to read a survey about (for example, Stephenson [58] and Rohde [56]). The question of realising a graph as the contact graph of some spheres (of varying radius) is a natural generalisation of the circle case.

In fact, one can even relax the hypothesis significantly by requiring that the spheres be some (contractible) domains whose ratio $\frac{\text{outer radius}}{\text{inner radius}}$ is bounded by some constant. With this relaxation, every finite graph can be realised as a contact graph (although the bound on the ratio might get large). But is that true for infinite graphs?

Benjamini & Schramm explore this question in [9] and show that [non-constant] p-harmonic functions can be an obstruction to such packings. Since non-triviality of the reduced ℓ^p-cohomology in degree one is equivalent to the existence of non-constant p-harmonic functions, this gives yet another application of ℓ^p-cohomology.

This topic has been developed further by Pansu in [49].

3 ℓ^1-Cohomology and the Ends

One of the apparent features of Examples 1.3 and 1.4 is that cutting the graph in two infinite components by removing an edge helps a lot to find non-trivial elements of $\underline{\ell^p H^1}$.

This feature will be heavily supported in this section as we show that:

1. a function in $\mathsf{D}^1(\Gamma)$ can be assigned a value on each end of the graph (see below for the definition of the ends of a graph).
2. the function is trivial in reduced ℓ^1-cohomology in degree one if and only if it takes the same value on all the ends.

The ends of a graph are the infinite components of a group which cannot be separated by a finite (*i.e.* compact) set. More precisely, an end ξ is a function from finite sets to infinite connected components of their complement so that $\xi(F) \cap \xi(F') \neq \emptyset$ (for any F and F'). It may also be seen as an equivalence class of (infinite) rays who eventually leave any finite set. Two rays r and r' are equivalent if, for any finite set F, the infinite part of r and r' lie in the same (infinite) connected component.

Thanks to Stallings' theorem, groups with infinitely many ends contain an (non-trivial) amalgamated product or a (non-trivial) HNN extension. Being without ends is equivalent to being finite, and amenable groups may not have infinitely many ends. This may be seen using Stallings' theorem, see also Moon & Valette [41] for a direct proof.

Here is an idea of the proof. Assume there are 3 ends or more, that is upon removing the finite set F, there remains [at least] 3 infinite components, say K_1, K_2 and K_3. By vertex-transitivity, it may be assumed that the identity element belongs to F. Let $c = \max_{f \in F} d(e, f)$ where $d(e, \cdot)$ is the distance to the identity element. Pick elements $h_i \in K_i$ so that $d(e, k_i) > 2c$. Then it is not too hard to check that the set Fh_i (the groups acts on the right by graph automorphism) disconnects K_i in [at least] two infinite components. The technical part comes in when you need to show that Fh_ih_j (for $i \neq j$) further disconnects those components. It then follows that the subgroup generated by $\langle h_1, h_2, h_3 \rangle$ is isomorphic to a free product $H_1 * H_2 * H_3$ where $H_i = \langle h_i \rangle$ is cyclic (finite or infinite). This then implies the group contains a free subgroup and, hence, is not amenable.

Groups with two ends admit $\mathbb{Z}$ as a finite index subgroup. These groups are peculiar, as they have non-trivial reduced ℓ^1-cohomology in degree 1, even if their reduced ℓ^p-cohomology (in all degrees) vanishes for $1 < p < \infty$.

So outside virtually-$\mathbb{Z}$ groups, all infinite amenable groups have one end.

Before moving on, let me mention that the results of this section were first written up in [21, Appendix A]. This result was partially remarked by Pansu (essentially, case where there is one end). As mentioned in Example 1.3, the special case of the group $\mathbb{Z}$ was written down in Martin & Valette [40, Example 3 in §4], who learned it from M. Bourdon (like the author, a former student of Pansu). So the case with two ends was already known to Bourdon. There is only a small step to make to the general case, so that the author is uncertain if he deserves any credit there.

Proposition 3.1 *Let Γ be a connected graph, then $\underline{\ell^1 H}^1(\Gamma) = 0$ if and only if the number of ends of Γ is ≤ 1. More precisely, let $\mathcal{N} = \mathbb{R}^{\text{ends}(\Gamma)}/\mathbb{R}$ be the vector space of functions on ends modulo constants. There is a boundary value map β : $\mathsf{D}^1(\Gamma) \to \mathcal{N}$ such that $\beta(g) = \beta(h) \iff [g] = [h] \in \underline{\ell^1 H}^1(\Gamma)$.*

Note that the isomorphism β between $\underline{\ell^1 H}^1(\Gamma)$ and $\mathbb{R}^{\text{ends}(\Gamma)}/\mathbb{R}$ is in the category of vector spaces, not of normed vector spaces. In a few cases, the norm on $\mathcal{N}$ resembles the norm of the quotient $\ell^\infty(|\text{ends}|)/\mathbb{R}$ (see Question 5.1). The proof is barely different from the argument of M. Bourdon found in Martin & Valette [40, Example 3 in §4].

Proof Note that $\mathsf{D}^1(\Gamma) \subset \ell^\infty(X)$: if $g \in \mathsf{D}^1(\Gamma)$, then, for P a path from x to y,

$$|g(y)| = |g(x) + \sum_{e \in P} g(e)| \leq |g(x)| + \|\nabla g\|_{\ell^1(E)}.$$

In fact, $\|g\|_{\ell^\infty(X)} \leq \|g\|_{\mathsf{D}^1(\Gamma)} + \inf_{x \in X} |g(x)|$. Since functions in ℓ^1 decrease at ∞, if one removes a large enough finite set, the function g on the resulting graph is almost constant. In particular, it is possible to define a value of g on each end: let B_n be the ball of radius n at some fixed vertex (root) o, then

$$\beta g(\xi) := \lim_{n\to\infty} g(x_n) \text{ where } x_n \in \xi(B_n).$$

Alternatively, if $r : \mathbb{Z}_{\geq 0} \to X$ is a ray representing the end ξ, then the value at ξ can also be defined as $\lim_{n\to\infty} g\big(r(n)\big)$. It is fairly straightforward to check these limits do not depend on the choice (of x_n and o or of the ray r).

Fix an end ξ_0. Then, define $\beta : \mathsf{D}^1(\Gamma) \to \mathcal{W}$ by changing with a constant the value of g to be 0 at ξ_0 and then looking at the values at the ends. This map is continuous and trivial on $\ell^1(X) + \mathbb{R}$ (since functions in $\ell^1(X)$ have trivial value at the ends). By continuity, $\overline{\ell^1(X) + \mathbb{R}}^{\mathsf{D}^1(\Gamma)} \subset \ker \beta$.

Assume, $\beta(f) = 0$, this means that, $\forall \epsilon > 0, \exists X_\epsilon \subset X$ a finite set such that $f(X_\epsilon^{\mathsf{c}}) \subset [-\epsilon, \epsilon]$. Set

$$f_\epsilon(\gamma) = \begin{cases} \epsilon f(\gamma)/|f(\gamma)| & \text{if } |f(\gamma)| > \epsilon, \\ f(\gamma) & \text{otherwise.} \end{cases}$$

Then $g_\epsilon := f - f_\epsilon$ is finitely supported, so in $\ell^1(X)$. Furthermore, $\|f - g_\epsilon\|_{\mathsf{D}^1(\Gamma)} = \|f_\epsilon\|_{\mathsf{D}^1(\Gamma)}$.

Let X_ϵ be as before, then

$$\nabla f_\epsilon \text{ is } \begin{cases} \text{equal to } \nabla f & \text{on } E \cap (X_\epsilon^{\mathsf{c}} \times X_\epsilon^{\mathsf{c}}), \\ \text{smaller in } |\cdot| \text{ than } \nabla f & \text{on } \partial X_\epsilon, \\ 0 & \text{on } E \cap (X_\epsilon \times X_\epsilon). \end{cases}$$

But $E \cap (X_\epsilon \times X_\epsilon)$ increases, as $\epsilon \to 0$, to the whole of E. More importantly, the ℓ^1-norm of ∇f outside this set tends to 0. Thus $\|f_\epsilon\|_{\mathsf{D}^1(\Gamma)} \to 0$ as $\epsilon \to 0$, and consequently $g_\epsilon \to f$ as $\epsilon \to 0$. Since g_ϵ are finitely supported, they belongs to $\ell^1(X)$. This shows that $f \in \overline{\ell^1(X)}^{\mathsf{D}^1(\Gamma)}$. □

Groups with two ends step strangely out of the crowd: although their reduced ℓ^p-cohomology is always trivial if $p > 1$, it is non-trivial for $p = 1$ (actually isomorphic to the base field). An amusing corollary is

Corollary 3.2 *Let G be a finitely generated group. G has infinitely many ends if and only if for some (and hence all) Cayley graph Γ, $\forall p \in [1, \infty[$, $\underline{\ell^p H}^1(\Gamma) \neq 0$. G has two ends if and only if for some (and hence all) Cayley graph Γ, $\forall p \in]1, \infty[$, $\underline{\ell^p H}^1(\Gamma) = 0$ but $\underline{\ell^1 H}^1(\Gamma) = \mathbb{R}$.*

Proof Use Proposition 3.1 for reduced ℓ^1-cohomology, use any vanishing theorem on groups of polynomial growth (such groups have an infinite centre, so see Proposition 1.5, Kappos [32] or Tessera [59]) to get the remaining values of p for groups with two ends.

Theorem 4.10 (which we have not discussed yet) will give the conclusion for groups with infinitely many ends (which are in particular non-amenable). □

It is worth noting that Bekka & Valette showed in [1, Lemma 2, p.316] that (for G discrete) the cohomology $H^1(G, \mathbb{C}G)$ is also isomorphic (as a vector space) to $\mathcal{N}$. Furthermore, by [1, Proposition 1], there is an embedding $H^1(G, \mathbb{C}G) \hookrightarrow \ell^1 H^1(G)$. A careful reading would probably reveal this remains injective in reduced cohomology (the only case to check is when G has two ends).

4 ℓ^p-Cohomology and Harmonic Functions

In §3, we dealt with one of the apparent features of Examples 1.3 and 1.4. Another feature which is present in those examples as well as the previous section is that it is very useful to think in terms of values at infinity.

However, for functions with gradient in ℓ^p with $p > 1$ this is somewhat counter intuitive. Indeed, the reader can quickly come up with a function on the graph of the line (a Cayley graph of $\mathbb{Z}$) which grows to ∞ even though its gradient is in ℓ^2. Nevertheless, this obstacle can be overcome.

The main motivation in this section is to show that the reduced ℓ^p-cohomology in degree one can be seen as a space of function on an ideal boundary, namely the Poisson boundary. The oldest result in this direction is a theorem of Lohoué [37] which says that in a non-amenable graph there is exactly one harmonic function in each equivalence class of $\underline{\ell^p H}^1(\Gamma)$. The results presented in this section come from [21], with some simplifications in the presentation coming mostly from [24].

In contrast to the result of Lohoué [37], the amenable case is trickier, so this result can only be generalised to some extent. To say how, some preliminary definitions are required.

Isoperimetric profiles. For $F \subset X$ a subset of the vertices, recall that ∂F is set of edges between F and $F^\complement$. Let $d \in \mathbb{R}_{\geq 1}$. Then, a graph Γ has

$$\begin{aligned}
&\mathrm{IS}_d \text{ if there is a } \kappa > 0 \text{ such that for all finite } F \subset X,\ |F|^{(d-1)/d} \leq \kappa |\partial F|;\\
&\mathrm{IS}_\omega \text{ if there is a } \kappa > 0 \text{ such that for all finite } F \subset X,\ |F| \leq \kappa |\partial F|.
\end{aligned}$$

Quasi-homogeneous graphs with a certain (uniformly bounded below) volume growth in n^d will satisfy these isoperimetric profiles, see Woess' book [61, (4.18) Theorem].

A Cayley graph will satisfy IS_d (for any $d \leq \delta$) if the growth of balls in this Cayley graph is bounded below by Kn^δ (for some $K > 0$). A Cayley graph will not satisfy IS_d (for any $d > \delta$) if the growth of balls in this Cayley graph is bounded above by $K'n^\delta$ (for some $K' > 0$).

Using Gromov's theorem on groups of polynomial growth [27], that the only groups which do not satisfy IS_d for all d are virtually nilpotent groups.

Cayley graphs of a group G does not satisfy IS_ω if and only if G is amenable. (There are many amenable groups which are not virtually nilpotent.) The upcoming result will apply best to groups which are not virtually nilpotent. See [61, §14] for more details.

Values at infinity. It is difficult to speak of a value at infinity, since it is not clear with what we can identify infinity (yet). However it is easy to say if a function is constant at infinity. This means that it belongs to $\mathbb{R} + c_0(X)$, *i.e.* a constant function plus an element of $c_0(X)$.

More precisely, let B_n be a sequence of balls in the graph with the same centre and B_n^{c} the sequence of their complement. On a connected graph, a function $f : X \to \mathbb{R}$ is constant at infinity if $\exists c \in \mathbb{R}$ so that $\forall \epsilon > 0$, $\exists n_\epsilon$ satisfying $f(B_{n_\epsilon}^{\mathsf{c}}) \subset [c - \epsilon, c + \epsilon]$.

Harmonic functions. A function $f : X \to \mathbb{R}$ is harmonic if it satisfies the mean-value property: for any vertex $x \in X$, $\sum_{y \in N(x)} \big(f(y) - f(x)\big) = 0$ (where $N(x)$ denotes the neighbours of x).

Let us define the following spaces of harmonic functions:

- $\mathscr{H}(\Gamma)$ is the space of harmonic functions.
- $\mathscr{H}\mathsf{D}^p(\Gamma) = \mathscr{H}(\Gamma) \cap \mathsf{D}^p(\Gamma)$ is the space of harmonic functions whose gradient is in ℓ^p.
- $\mathsf{B}\mathscr{H}\mathsf{D}^p(\Gamma) = \ell^\infty(X) \cap \mathscr{H}(\Gamma) \cap \mathsf{D}^p(\Gamma)$ is the space of bounded harmonic functions whose gradient is in ℓ^p.

Divergence. There is another way to define harmonic functions by introducing the divergence. For two finitely supported function f and g on a countable set Y, define the pairing $\langle f \mid g \rangle_Y = \sum_{y \in Y} f(y)g(y)$. (The subscript Y will often be dropped.) This allows to define the adjoint of the gradient ∇, denoted ∇^* and called divergence, by $\langle f \mid \nabla g \rangle_E = \langle \nabla^* f \mid g \rangle_X$. More precisely, for $f : E \to \mathbb{R}$, one finds

$$\nabla^* f(x) = \sum_{y \in N(x)} f(y, x) - \sum_{y \in N(x)} f(x, y).$$

In particular

$$\nabla^* \nabla f(x) = 2 \sum_{y \in N(x)} \big(f(y) - f(x)\big).$$

Thus, harmonic functions are exactly the functions for which the divergence of the gradient is trivial.

Four conditions. Define for $p \geq 1$:

(1_p) The reduced ℓ^p-cohomology in degree one vanishes (for short, $\underline{\ell^p H^1} = \{0\}$).
(2_p) All functions in $\mathsf{D}^p(G)$ take only one value at infinity.
(3_p) There are no non-constant functions in $\mathscr{H}\mathsf{D}^p(G)$.
(4_p) There are no non-constant functions in $\mathsf{B}\mathscr{H}\mathsf{D}^p(G)$.

For the record, note that $(1_1) \iff (2_1) \iff$ the number of ends is ≤ 1 (see Proposition 3.1 above).

Here is the best known to date extension of Lohoué's result [37].

Theorem 4.1 *Assume a graph Γ is of bounded valency and has* IS_d. *For* $1 < p < d/2$, $(1_p) \iff (2_p) \implies (3_p) \implies (4_p)$ *and, for* $q \geq \frac{dp}{d-2p}$, $(4_q) \implies (1_p)$.

If Γ has IS_d *for all d, then "*$\forall p \in]1,\infty[$, (i_p) *holds" where* $i \in \{1,2,3,4\}$ *are four equivalent conditions.*

The proof is split as follows: $(1_p) \iff (2_p)$ is the content of §4.2 (see Corollary 4.9). $(2_p) \implies (3_p)$ is a fairly easy consequence of the maximum principle (see Lemma 4.12). $(3_p) \implies (4_p)$ is obvious (since $\mathsf{B}\mathscr{H}\mathsf{D}^p(\Gamma) \subset \mathscr{H}\mathsf{D}^p(\Gamma)$). $(4_p) \implies (1_q)$ is the bulk of §4.3 (see Theorem 4.14).

4.1 Reduction to Bounded Functions

Now the first step in order to associate a value at infinity to any function in $\mathsf{D}^p(\Gamma)$ is to show that one can restrict to bounded functions.

This is basically the content of Lemma 4.4 from Holopainen & Soardi [30]. The Lemma is there stated in terms of p-harmonic functions, but its proof can be adapted without much difficulty.

We will use $[f] \in \underline{\ell^p H}^1(\Gamma)$ to denote the equivalence class of the function f, *i.e.* the closure of $f + \ell^p(X)$ in D^p-norm (or $\overline{f+\ell^p(X)}^{\mathsf{D}^p}$).

Lemma 4.2 (Holopainen & Soardi [30], 1994) *Let* $g \in \mathsf{D}^p(\Gamma)$ *be such that* $g \notin [0] \in \underline{\ell^p H}^1(\Gamma)$. *For* $t \in \mathbb{R}_{>0}$, *let* g_t *be defined as*

$$g_t(x) = \begin{cases} g(x) & \text{if } |g(x)| < t, \\ t\frac{g(x)}{|g(x)|} & \text{if } |g(x)| \geq t. \end{cases}$$

Then there exists t_0 *such that* $g_t \notin [0]$, *for any* $t > t_0$.

In particular, the reduced ℓ^p *cohomology is trivial if and only if all bounded functions in* $\mathsf{D}^p(\Gamma)$ *have trivial classes.*

Proof The proof goes essentially as in Proposition 3.1. Assume without loss of generality that $g(o) = 0$ for some preferred vertex (*i.e.* root) $o \in X$. Since $\|\nabla g\|_{\ell^\infty(E)} \leq \|\nabla g\|_{\ell^p(E)} =: K$, given $x \in X$ and P a path from o to x,

$$|g(x)| = |g(x) - g(o)| = \sum_{e \in P: o \to x} \nabla g(e) \leq d(o,x)\|\nabla g\|_{\ell^p(E)}.$$

In particular, g_t is identical to g on $B_{t/K}$. Hence $\|g - g_t\|_{D^p(\Gamma)} \leq \|\nabla g\|_{\ell^p(B^{\mathsf{c}}_{t/K})}$, where $\ell^p(B^{\mathsf{c}}_{t/K})$ denotes the ℓ^p-norm restricted to edges which are not inside $B_{t/K}$. Because $\nabla g \in \ell^p(E)$, $\|\nabla g\|_{\ell^p(B^{\mathsf{c}}_{t/K})}$ tends to 0, as t tends to ∞.

Now if there is an infinite sequence t_n such that g_{t_n} are in $[0]$ and $t_n \to \infty$, then g_{t_n} is a sequence of functions in $[0]$ which tends (in D^p-norm) to g. This implies $g \in [0]$, a contradiction. Hence, for some t_0, $g_t \notin [0]$ given that $t > t_0$. □

4.2 Values at Infinity

The aim of the current subsection is to show that (if the proper isoperimetric profile is present) functions in $\mathsf{D}^p(\Gamma)$ corresponding to the trivial class are exactly those which are constant at infinity. Some concepts from nonlinear potential theory will also come in handy.

Definition 4.3 Let (X, E) be an infinite connected graph. The **inverse p-capacity** of a vertex $x \in X$ is

$$\operatorname{icp}_p(x) := \big(\inf\{\|\nabla f\|_{\ell^p E} \mid f : X \to \mathbb{C} \text{ is finitely supported and } f(x) = 1\}\big)^{-1}.$$

The graph is called p-**parabolic** if $\operatorname{icp}_p(x) = +\infty$ for some $x \in X$. A graph is called p-**hyperbolic** if it is not p-parabolic.

One might also like to call the inverse p-capacity the "p-resistance to ∞". (When $p = 2$ capacity and resistance are strongly related.)

Recall (see Holopainen [29], Puls [55] or Yamasaki [62]) that if $\operatorname{icp}_p(x_0) = 0$ for some x_0, then $\operatorname{icp}_p(x) = 0$ for all $x \in X$. Recall also that 2-parabolicity is equivalent to recurrence.

Remark 4.4

1. If the graph Γ is vertex-transitive, $\operatorname{icp}_p(x) = \operatorname{icp}_p(y)$ for all $x, y \in X$. Let $\operatorname{icp}_p(\Gamma) := \operatorname{icp}_p(x)$ be this constant. It is also easy to see that if the automorphism group acts co-compactly on the graph, the inverse p-capacity is bounded from below.
2. Note that in the definition of p-capacity, one may also assume that the functions take value only in $\mathbb{R}_{\geq 0}$. Indeed, looking at $|f|$ instead of f reduces the norm of the gradient. Likewise, one can even assume f takes value only in $[0, 1]$ as truncating f at values larger than 1 will again reduce the norm of the gradient. ♢

The following proposition is an adaptation of a result of Keller, Lenz, Schmidt & Wojchiechowski [35, Theorem 2.1].

Proposition 4.5 *Assume Γ is vertex-transitive and has* IS_d. *Let $p < d$. If $f \in \mathsf{D}^p(\Gamma)$ represents a trivial class in $\underline{\ell^p H^1}(\Gamma)$, then f is constant at infinity.*

Furthermore, $c_0(X) \subset \mathsf{D}^p(\Gamma)$ and $\forall f \in c_0(X)$, $\|f\|_{\ell^\infty(X)} \leq \operatorname{icp}_p(\Gamma)\|\nabla f\|_{\ell^p(E)}$.

Proof A consequence of the Sobolev embedding corresponding to IS_d is that the graph is p-hyperbolic. See Troyanov [60, §7] as well as Woess' book [61, §4 and §14] and references therein for details.

As Γ is p-hyperbolic and by Remark 4.4.2, one has $\forall f$ of finite support $|f(x)| \leq \operatorname{icp}_p(x)\|\nabla f\|_p$. However, by Remark 4.4.1, there is no dependence on x on the right-hand side. So $\forall x \in G$, $\forall f$ of finite support $|f(x)| \leq \operatorname{icp}_p\|\nabla f\|_p$ where icp_p is $\operatorname{icp}_p(\Gamma)$. Trivially this implies

$$\forall f : G \to \mathbb{C} \text{ of finite support } \|f\|_\infty \leq \operatorname{icp}_p\|\nabla f\|_p.$$

As a first consequence, assume $f_n \overset{\mathsf{D}^p}{\to} f$ with f_n finitely supported. Then f_n also converge to f in $\ell^\infty(X)$. Since $c_0(X)$ is the closure of finitely supported functions in $\ell^\infty(X)$, this shows that $f \in \overline{\ell^p(X)}^{\mathsf{D}^p}$ implies $f \in c_0(X)$. In other words, if f represents a trivial class in reduced ℓ^p-cohomology, then f is constant at infinity.

As a second consequence, let us show the "Furthermore". Pick some $f \in c_0(X)$. Apply the inequality to $g_\epsilon = f - f_\epsilon$ where f_ϵ is the truncation of f:

$$f_\epsilon(x) = \begin{cases} \epsilon f(x)/|f(x)| & \text{if } |f(x)| > \epsilon \\ f(x) & \text{else.} \end{cases}$$

Indeed, g_ϵ is finitely supported so it satisfies $\|g_\epsilon\|_\infty \leq \mathrm{icp}_p \|\nabla g_\epsilon\|_p$ (recall that $\mathrm{icp}_p = \mathrm{icp}_p(\Gamma)$). Also $\|\nabla g_\epsilon\|_p \leq \|\nabla f\|_p$ and $\|f\|_\infty \leq \epsilon + \|g_\epsilon\|_\infty$. Hence $\|f\|_\infty \leq \epsilon + \mathrm{icp}_p \|\nabla f\|_p$ and the conclusion follows by letting $\epsilon \to 0$. □

The above proposition gives the following very nice characterisation of functions corresponding to the trivial class.

Corollary 4.6 *Assume Γ is vertex-transitive and has* IS_d. *Let $d > p$. $f \in \mathsf{D}^p(\Gamma)$ represents a trivial class in $\underline{\ell^p H}^1(\Gamma)$ if and only if f is constant at infinity.*

Proof Without loss of generality the constant at infinity is 0 (because one may add a constant function to f). Considering again $g_\epsilon = f - f_\epsilon$ (where f_ϵ is the truncation of f as in Lemma 4.2 and Proposition 4.5), one can check that, as $\epsilon \to 0$, $g_\epsilon \overset{\mathsf{D}^p}{\to} f$. Since g_ϵ is finitely supported, it is in $\ell^p(X)$ (and this concludes the proof). □

The above results are very nice, but they do require a fairly strong hypothesis, namely that the graph is vertex-transitive. If the isoperimetric profile is good enough, this can be remedied.

As in Keller, Lenz, Schmidt & Wojchiechowski [35], say that the graph Γ is **uniformly p-hyperbolic** if $\mathrm{icp}_p(\Gamma) := \sup_{x \in X} \mathrm{icp}_p(x)$ is finite. One can show:

Lemma 4.7 *If Γ is a graph of bounded valency with d-dimensional isoperimetry and $d > 2p$, then Γ is uniformly p-hyperbolic.*

Proof First, recall that d-dimensional isoperimetry implies that the Green's kernel ($k_o := \sum_{n \geq 0} P_o^n$ where P_o^n is the random walk distribution at times n starting at the vertex o) has an $\ell^q(X)$-norm (for some $q < p' = \frac{p}{p-1}$) which is bounded independently from o.

Indeed, d-dimensional isoperimetry implies that $\|P_o^n\|_\infty \leq \kappa n^{-d/2}$ (where $\kappa \in \mathbb{R}$ comes from the constant in the isoperimetric profile; see Woess' book [61, (14.5) Corollary] for details). From there, one gets that $\|P_o^n\|_q^q \leq \|P_o^n\|_q^{q-1} \|P_o^n\|_1 \leq \kappa^{q-1} n^{-d(q-1)/2}$. This implies that $\|k_o\|_q \leq \sum_{n \geq 0} \kappa^{1/q'} n^{-d/2q'}$ (a series which converges if $d > 2q'$).

Second, let f be a finitely supported function with $f(o) = 1$, then

$$\langle \nabla f \mid \nabla k_o \rangle = \langle f \mid \nabla^* \nabla k_o \rangle = \langle f \mid \delta_o \rangle = f(o) = 1.$$

Since $\|\nabla f\|_p \geq \|\nabla k_o\|_{p'}^{-1} \langle \nabla f \mid \nabla k_o \rangle$, $\|\nabla k_o\|_{p'} \leq 2\nu \|k_o\|_{p'} \leq 2\nu \|k_o\|_q = 2\nu \kappa_q^{-1}$ (where ν is the maximal valency of a vertex) and there is no dependence in o, this means that $\mathrm{icp}_p(\Gamma) \leq \kappa_q / 2\nu$.

Noting that, for the above, the conditions $q \leq p'$ and $2q' < d$ need to hold, one gets that the bound holds as long as $2p < d$. □

Remark 4.8 Pansu pointed out the following shortcut. The Sobolev (or Nash) inequality corresponding to IS_d and the exponent p is actually: for any finitely supported function f and all $p < d$, $\|f\|_{\frac{pd}{d-p}} \leq K \|\nabla f\|_p$ (where K depends only on the constant in the isoperimetric profile and p). Consequently, $\inf\{\|\nabla f\|_{\ell^p E} \mid f : X \to \mathbb{C}$ is finitely supported and $f(x) = 1\} \geq \frac{1}{K}$. Hence $\mathrm{icp}_p(x) \leq K$ for any x. So the graph is uniformly hyperbolic for any $p < d$. ◊

An amusing corollary is that, most of the time (*i.e.* if the isoperimetric dimension of the graph is large enough), the p-Royden and p-harmonic boundaries are equal. See [24, Corollary 5.10]. However, for our current purpose, only the corollary will be required.

Corollary 4.9 *Assume Γ has IS_d and that $d > p$. Then $f \in \mathsf{D}^p(\Gamma)$ represents a trivial class in $\ell^p \underline{H}^1(\Gamma)$ if and only if f is constant at infinity.*

The proof is essentially the same as Proposition 4.5 and Corollary 4.6 above.

There are two very useful consequences of this result.

Note that for $q < p$, $\mathsf{D}^q(\Gamma) \subset \mathsf{D}^p(\Gamma)$. This means that the identity map $\ell^q \underline{H}^1(\Gamma) \to \ell^p \underline{H}^1(\Gamma)$ is a quotient map (since one quotients out by a larger subspace in $\ell^p \underline{H}^1(\Gamma)$).

Theorem 4.10 *Assume Γ has IS_d and that $d > p$. Then, for $1 \leq q < p$, the natural quotient map $\ell^q \underline{H}^1(\Gamma) \to \ell^p \underline{H}^1(\Gamma)$ is injective.*

Proof According to Corollary 4.9 (or Proposition 3.1 if $q = 1$), if there is a function $f \in \mathsf{D}^q(\Gamma)$ such that $[f] \neq 0 \in \ell^q \underline{H}^1(\Gamma)$, then f is not constant at ∞. But since f is not constant at infinity, Corollary 4.9 implies that f is not in the trivial class in $\ell^p \underline{H}^1(\Gamma)$ too. Consequently, the map is injective. □

This is very effective in the realm of groups since:

- either the group is nilpotent, and in that case Proposition 1.5 shows that $\ell^p \underline{H}^1(\Gamma)$ is trivial for any $p \in]1, \infty[$.
- or the group is not nilpotent, and in that case it has IS_d for any $d \geq 1$. Hence $\ell^q \underline{H}^1(\Gamma) \to \ell^p \underline{H}^1(\Gamma)$ is injective for any $1 \leq q < p$.

This also shows that for any amenable group and for all $p \in]1, 2]$, $\ell^p \underline{H}^1(\Gamma)$ is trivial. Indeed, Cheeger & Gromov [12] showed that $\ell^2 \underline{H}^1(\Gamma)$ is trivial for any amenable group.

Theorem 4.10 is also counter intuitive if one thinks in terms of p-harmonic functions. Indeed, there is *a priori* no reason to believe that the absence of non-constant p-harmonic function implies the absence of non-constant q-harmonic function. These are different nonlinear equations.

Another powerful consequence of this result is

Theorem 4.11 *Assume Γ has a spanning connected subgraph Γ' such that: Γ' has* IS_d *and $\underline{\ell^p H}^1(\Gamma')$ is trivial for some $p < d$. Then $\underline{\ell^p H}^1(\Gamma)$ is trivial for any $q \leq p$.*

Proof It follows from the definition of IS_d that Γ has IS_d too. Take any $f \in \mathsf{D}^q(\Gamma)$. Then $f \in \mathsf{D}^p(\Gamma')$ too (since $q \leq p$ and there are less edges in Γ' so the norm of the gradient can only be smaller). By Corollary 4.9 (applied to Γ'), f is constant at infinity. But then Corollary 4.9 (applied to Γ) tells us that f must have a trivial class in $\underline{\ell^q H}^1(\Gamma)$. □

There are many applications of this simple fact. It can be used to show that many wreath products and Cartesian products of graph have trivial $\underline{\ell^p H}^1$ for all $p \in [1, \infty[$. In fact, the Cartesian product of any two groups $G = G_1 \times G_2$ has trivial $\underline{\ell^p H}^1$ for all $p \in [1, \infty[$. See [22] for details.

4.3 Harmonic Functions

Harmonic functions come naturally into play not only because of the result of Lohoué [37]. Because of the maximum principle, a harmonic function which is constant at infinity is constant. Hence

Lemma 4.12 *Assume Γ has* IS_d. *Assume either that*

- *Γ is vertex-transitive and $d > p$.*
- *or $d > 2p$.*

If $\mathcal{H}\mathsf{D}^p(\Gamma)$ contains a non-constant harmonic function then $\underline{\ell^p H}^1(\Gamma)$ is not trivial.

The reverse implication is essentially a question of Pansu [46, Question 6 in §1.9] (Pansu restricts the question to groups which are not nilpotent). A short answer (and the best to date) is "almost yes, because we lose a bit of regularity":

Lemma 4.13 *Let Γ be a graph with* IS_d *and $1 \leq p < d/2$. For any $g \in \mathsf{D}^p(\Gamma)$, there is a function $\tilde{g}$ such that:*

- *$\tilde{g}$ is harmonic.*
- *$\tilde{g}$ is bounded if g is.*
- *$\tilde{g} - g \in \ell^r(X)$ for any $r > \frac{dp}{d-2p}$.*

Proof It turns out $\tilde{g}$ is in the most obvious function which could fit the bill. Indeed, given g one can make it "more harmonic" by replacing the value at a vertex by the average if its values at neighbouring vertices. Since harmonic functions are exactly those which have the mean-value property, repeating this process infinitely many times, one finds the desired function $\tilde{g}$.

So, let R be the random walk operator, *i.e.* given a function $g : X \to \mathbb{R}$, $Rg(x) = \sum_{y \in N(x)} g(y)$ (where $N(x)$ are the neighbours of x). We want to show that $\tilde{g} = \lim_{n\to\infty} R^n g$ is a well-defined function with all the above properties. Actually the two first properties are essentially automatic (if the limit converges even just in the point-wise sense).

The operator R and its iterations R^n are given by very simple kernels. Recall that $P^n_x(y)$ is the probability that a simple random walk from x lands at y after n steps. Then $R^n g(x) = \sum_{y\in X} P^n_x(y) g(y)$.

Write:

$$\tilde{g} - g = \lim_{n\to\infty} R^n g - g = \sum_{i\geq 0}(R^{i+1}g - R^i g) = \sum_{i \geq 0} R^i (R - \mathrm{Id}) g,$$

where Id is the identity operator. Let $h = (R - \mathrm{Id})g$, then $h(x)$ is a finite average of values of the gradient of g. Since $g \in \mathsf{D}^p(\Gamma)$ then $h \in \ell^p(X)$. In fact $\|h\|_{\ell^p(X)} \leq 2\|\nabla g\|_{\ell^p(E)}$.

Since R^i are operators defined by a kernel one may use Young's inequality (see *e.g.* Sogge's book [57, Theorem 0.3.1]): for $r > p$ and $1 + \frac{1}{r} = \frac{1}{p} + \frac{1}{q}$,

$$\begin{aligned} \|\tilde{g} - g\|_{\ell^r(X)} = \Big\| \big(\sum_{i\geq 0} R^i\big) h \Big\|_{\ell^r(X)} &\leq \sup_{x\in X} \Big\| \sum_{i\geq 0} P^i_x \Big\|_{\ell^q(X)} \|h\|_{\ell^p(X)} \\ &\leq 2 \sup_{x \in X} \Big\| \sum_{i\geq 0} P^i_x \Big\|_{\ell^q(X)} \|\nabla g\|_{\ell^p(E)}. \end{aligned}$$

We are done if one can show that $\sup_{x\in X} \| \sum_{i\geq 0} P^i_x \|_{\ell^q(X)} < +\infty$ for all $q' < d/2$. Indeed this would mean that $\tilde{g} - g \in \ell^r(X)$ (for all $r > \frac{dp}{d-2p}$). This shows the convergence (and existence of $\tilde{g}$) and concludes the proof.

Fortunately, there are very good estimates at hand for $\|P^i_x\|_{\ell^q(X)}$, which rely only on isoperimetric profiles (see proof of Lemma 4.7). Indeed, if Γ has IS_d, then

$$\exists K > 0, \quad \forall x, y \in X, \quad P^n_x(y) \leq K n^{-d/2}.$$

Obviously $\|P^n_x\|_{\ell^1(X)} = 1$ (because it is a probability distribution). By Hölder's inequality,

$$\|P^n_x\|_{\ell^q(X)} \leq \|P^n_x\|^{1/q}_{\ell^1(X)} \|P^n_x\|^{1/q'}_{\ell^\infty(X)}.$$

Hence $\|P^{(n)}_x\|_{\ell^q(X)} \leq K' n^{-d/2q'}$ uniformly in x, for some $K' > 0$. The condition $\frac{d}{2q'} > 1$ translates as $q > \frac{d}{d-2}$. Plunging this in $1 + \frac{1}{r} = \frac{1}{p} + \frac{1}{q}$ yields $r > \frac{pd}{d-2p}$. □

Theorem 4.14 *If Γ has* IS_d, $p < \frac{d}{2}$ *and* $q > \frac{dp}{d-2p}$ *then* $(4_q) \implies (1_p)$.

Proof We will show the contrapositive. So assume $\neg(1_p)$, *i.e.* there is $f \in \mathsf{D}^p(\Gamma)$ which is not trivial in $\underline{\ell^p H}^1(\Gamma)$. Then, by Lemma 4.2 one may assume f is actually bounded (otherwise, consider some truncation of f). By Corollary 4.9, f is not constant at infinity. By Lemma 4.13, there is a harmonic bounded function $\tilde{f}$ which differs from f by an element of $\ell^q(X)$.

Since $\ell^q(X) \subset c_0(X)$, $\tilde{f}$ is not constant at infinity either and hence not constant. Lastly, since $\nabla : \ell^q(X) \to \ell^q(E)$ is bounded, $\tilde{f} \in \mathsf{D}^q(\Gamma)$.

To sum up $\tilde{f}$ is not constant, bounded, harmonic and its gradient is in ℓ^q. So $f \in \mathsf{B}\mathscr{H}\mathsf{D}^q(\Gamma)$. This shows $\neg(4_q)$ as claimed. □

The most effective application of Theorem 4.14 are the two following corollaries:

Corollary 4.15 *Assume a graph has the Liouville property (*i.e. *there are no non-constant bounded harmonic functions) and satisfies* IS_d *for some* $d > 2$. *Then* $\underline{\ell^p H}^1(\Gamma)$ *is trivial for any* $p < \frac{d}{2}$.

This is again very effective in the realm of groups since one may assume IS_d for any d. Also, there are many amenable groups which are known to have the Liouville property (in some Cayley graph). Hence the previous corollary covers a lot of amenable groups (for all p).

Corollary 4.16 *Assume Γ has* IS_d. *If* $\underline{\ell^p H}^1(\Gamma)$ *is not trivial for some* $p < \frac{d}{2}$, *then any graph quasi-isometric to Γ has a non-trivial Poisson boundary.*

As mentioned before, this contrasts with the fact that the triviality of the Poisson boundary is not invariant under quasi-isometries.

5 Epilogue

5.1 Further Results

Let us summarise some of the results in the realm of groups. It is known that the reduced ℓ^p-cohomology in degree one is trivial in degree 1 for the following groups ($1 < p < \infty$):

1. G has an infinite FC-centre (see Kappos [32, Theorem 6.4], Martin & Valette [40, Theorem 4.3], Puls [52, Theorem 5.3], Tessera [59, Proposition 3] or [20, Theorem 3.2])
2. G has a finitely supported measure with the Liouville property, *i.e.* no bounded μ-harmonic functions (see [21, Theorem 1.2 or Corollary 3.14]). This includes all polycyclic groups (for such groups, see also Tessera [59])
3. G is a direct product of two infinite finitely generated groups (see [22, Corollary 3]).

4. G is a wreath product with infinite base group (see [22, Proposition 1] and Martin & Valette [40, Theorem.(iv)]) unless the base group has infinitely many ends and the lamp group is amenable. Arguments from Georgakopoulos [17] show that this also holds for finite lamp groups (even if the base group has infinitely many ends).
5. G is some specific type of semi-direct product $N \rtimes H$ with N not finitely generated (see [23] for the full hypothesis).
6. L^p-cohomology can be defined for groups which are not endowed with the discrete topology. Amenable groups can then be non-unimodular. For such groups results of Tessera show the L^p-cohomology in degree one is trivial, see [59].

It is also trivial in any amenable group for any $1 < p \leq 2$ (see [21] or Theorem 4.10 above).

Lastly:

- (see [24, Corollary 1.3]) if G is finitely generated and there is a finitely generated subgroup K so that (a) either $\underline{\ell^p H}^1(K)$ is trivial or K has an infinite FC-centraliser, (b) K has growth at least polynomial of degree $d > p$, and (c) K is not contained in an almost-malnormal strict subgroup of G, then $\underline{\ell^p H}^1(G)$ is trivial.
- (see [24, Corollary 5.11] or Bourdon, Martin & Valette [6, Theorem 1.1)] for a weaker version) if $K < G$ is an infinite subgroup and $\ell^p H^1(K) = \{0\}$, then either $\ell^p H^1(G) = 0$ or there is an almost-malnormal subgroup $H \lneqq G$ so that $K < H$.

In particular, Baumslag–Solitar groups also have trivial reduced ℓ^p-cohomology for all $p \in [1, \infty[$.

These last two can actually be interpreted as a trichotomy (resp. a dichotomy) which resembles a result of Gaboriau [16, Théorème 6.8] (in the case $p = 2$). Gaboriau presents [16, Théorème 6.8] as a generalisation of a result of Schreier [16, ¶ after Théorème 6.8 in §0]. Gaboriau's result cannot be generalised to $p > 2$: Bourdon in [3, paragraph **4)** in §1.6] gives an example to this effect.

As for groups where the ℓ^p-cohomology is not trivial:

1. any hyperbolic group or relatively hyperbolic group has a p_0 so that $\ell^p H^1(G)$ is not trivial for any $p > p_0$ (see §2.2.2 for details).
2. there are torsion groups (of infinite exponent) for which $\ell^p H^1(G)$ is not trivial for all $p > 2$. These groups have no free subgroups, yet are not amenable. They do not have a finite presentation. See Osin [42].

However, there are acylindrically hyperbolic groups for which $\ell^p H^1$ is trivial for all $p \in [1, \infty[$ (see §2.2.2 for details).

As for graphs it is easy to construct graphs which are amenable and have non-trivial $\underline{\ell^p H}^1$. Indeed, take any graph Γ which has IS_d for some $d > 2$ and more than two ends. By Proposition 3.1, $\underline{\ell^1 H}^1(\Gamma)$ is not trivial. By Theorem 4.10, for any $p \in [1, \frac{d}{2}[$, $\underline{\ell^p H}^1(\Gamma)$ is also non-trivial.

To make the example slightly more specific, take two copies of a Cayley graph of some group which is amenable but not nilpotent. Join these two copies by an edge. Then it fits the description of the previous paragraph and has IS_d for any d.

5.2 *Questions*

It was shown in §3 that the reduced ℓ^1-cohomology in degree one identifies to the space of functions on the ends modulo constant functions. This is an isomorphism of vector space, but the norm on the space of functions is probably related to how "large" the ends are and how they are connected.

Question 5.1 *Describe the norm on the vector space* $\ell^1\underline{H}^1(\Gamma) \simeq \mathcal{N} = \mathbb{R}^{\mathrm{ends}(\Gamma)}/constants.$

A question dating back at least to Gromov [28, §8.A_1.(A_2), p.226]:

Question 5.2 *Let G be an amenable group, is it true that for one (and hence all) Cayley graph* Γ *and all* $1 < p < \infty$, $\ell^p\underline{H}^1(\Gamma) = 0$?

The original question concerns cohomology in all degrees.

Of course, this brings up the question what should be the cohomology of a graph in higher degree. The only results (beyond Cheeger & Gromov [12]) are those of Kappos [32] (in the discrete case) and those of Bourdon & Rémy [8], Pansu & Rumin [50], and Pansu & Tripaldi [51] (in the continuous case).

One of the problems is that there are many possibilities (and that unlike in degree one, they do not coincide). The simplest possibility pops up in the case of groups. One considers the left-regular representation on $\ell^p(G)$. There are standard definitions to speak of the cohomology of this representation in higher degree.

Another simple definition is in the continuous set-up (*i.e.* the cohomology of manifolds). We dealt almost exclusively with the case of graphs, but for manifolds ℓ^p-cohomology in degree k can be defined as "$(k-1)$-forms ω so that $d\omega \in \mathsf{L}^p$"/"$(k-1)$-forms ω in L^p".

In the case of graphs (which are not necessarily Cayley graphs), one possibility (for degree two) is to look at the space of cycles C. There are some technicalities in finding the reasonable (*e.g.* countable) basis of this space so as to make it tractable. For example, in the Cayley graph of a group with a finite presentation, the presentation gives a good basis for the space of cycles. One can then define the "rotational" as follows: if c is a[n oriented] cycle given by following the oriented edges $e_1, \ldots, e_n$ and g is a function on the edges, then $\mathrm{rot}g(c) = \sum_i g(e_i)$. Assuming the rotational gives a bounded operator (in the case groups, this amounts to the fact that the presentation is finite), a possibility for the cohomology in degree two is then given by taking the quotient "functions on the edges with rotational in $\ell^p(C)$"/"functions which are in $\ell^p(E)$".

Question 5.3 *Given a Cayley graph of a finitely presented amenable group, are the triviality of both definitions above equivalent? invariant under quasi-isometry? for which class of groups are they trivial?*

Élek [15] showed that the following three definition of ℓ^p-cohomology coincide for groups which possess a finite $K(\pi, 1)$:

- the coarse ℓ^p-cohomology of finitely generated groups defined by Élek himself in [15];
- the singular ℓ^p-cohomology for any countable group defined as the ℓ^2-cohomology from Cheeger & Gromov [12];
- Pansu's asymptotic L^p-cohomology defined for any measured metric space (see [45]).

Note that Pansu's definition can also be used for graphs (in degree two, it should coincide with the definition given above using the rotational).

Here is a conjecture motivated by Osin [43, Problem 3.3] (do $\underline{\ell^2 H}^1(\Gamma) \neq 0$ and finite presentation imply acylindrically hyperbolic)

Conjecture 5.4 *Assume Γ is a torsion-free finitely presented group. If, for some $p \in]1, \infty[$, $\underline{\ell^p H}^1(\Gamma) \neq \{0\}$ then Γ contains a free subgroup (of rank 2).*

One could also strengthen the hypothesis to "finite $K(\Gamma, 1)$". Osin [42] showed that there are (non-amenable) groups without free subgroups (in fact, infinite torsion groups), whose reduced ℓ^2-cohomology in degree one is not trivial.

Note that these groups also show that groups whose ℓ^p-cohomology is not trivial can have a trivial Floyd boundary (a natural question coming from Puls [54]). Indeed, Karlsson [33] showed that groups with a non-trivial Floyd boundary contain free subgroups.

The next step for a positive answer to question 5.2 would be:

Question 5.5 *If G is a finitely generated solvable group, does $\underline{\ell^p H}^1(G) = \{0\}$ for any $1 < p < \infty$?*

Already the metabelian (derived length 2) case is not clear. In fact the special case "locally nilpotent not finitely generated"-by-Abelian would probably suffice to answer the question.

An interesting strengthening of Question 5.2 is

Question 5.6 *Can an amenable group have a Cayley graph with a non-constant harmonic function with gradient in c_0?*

The case of nilpotent group (more generally, groups with an infinite centre) is already treated in [25, Proposition 1.5 and Lemma 2.7].

Note that this is not the same thing as reduced c_0-cohomology in degree one. In fact, it is not too hard to see that reduced c_0-cohomology in degree one is always trivial, while reduced ℓ^∞-cohomology in degree one is never trivial.

As mentioned in Corollary 4.16, $\underline{\ell^p H}^1(\Gamma)$ can be a great way to see that some harmonic functions may not disappear after a quasi-isometry. However, because there is a small loss in the exponent in Theorem 4.1, the following remains open:

Question 5.7 *Are there two graphs Γ and Γ' which are quasi-isometric but so that $\mathscr{H}\mathsf{D}^p(\Gamma)$ contains only the constant function, while $\mathscr{H}\mathsf{D}^p(\Gamma')$ contains more than just these functions? In other words, is the triviality of $\mathscr{H}\mathsf{D}^p$ invariant under quasi-isometries?*

The same question could be asked with $\mathsf{B}\mathscr{H}\mathsf{D}^p$ (with the chance of a negative answer being higher).

In a similar vein, one could ask, in the spirit of a question of Pansu [46, Question 6 in §1.9], whether there is a harmonic function in each equivalence class of $\underline{\ell^p H}^1(G)$. The uniqueness up to a constant can be easily obtained: if h_1 and h_2 are two such functions, then $h_1 - h_2$ is harmonic and belongs to the trivial class; by Corollary 4.9, $h_1 - h_2$ is harmonic and constant at infinity, hence constant everywhere.

The referee pointed out the following interesting question, related to Corollary 4.16:

Question 5.8 *When* (i.e. *for which groups and which p) can the ℓ^p-cohomology be used to define a boundary of the random walk* (i.e. *a quotient of the Poisson boundary)?*

In the hyperbolic set-up, this question should admit a positive answer. Indeed, Bourdon & Pajot [7] showed that when p is larger than the conformal dimension of the boundary, functions in different cohomology classes can separate points on the boundary. It is to be expected that the Gromov boundary is a quotient of the p-harmonic boundary (see Puls [55])

Simple cases of groups which are not hyperbolic but have non-trivial [reduced] ℓ^p-cohomology are groups of the form $\mathbb{Z}^n * \mathbb{Z}^m$ with $n + m \geq 3$ (and $m, n > 0$). To see that the ℓ^p-cohomology is non-trivial for any $p \in [1, \infty[$, note that there are infinitely many ends and use the embedding of $\underline{\ell^p H}^1$ in $\underline{\ell^q H}^1$ for $p < q$; to see that it is not hyperbolic, use the fact it contains $\mathbb{Z}^2$ as a subgroup.

Let me conclude with a technical question:

Question 5.9 *If Γ is the Cayley graph of a group, can one relax the condition $p < \frac{d}{2}$ in Lemma 4.13 to $p < d$?*

Indeed, Proposition 4.5 shows that the second condition is sufficient. Note that the condition $p < \frac{d}{2}$ of Lemma 4.13 comes from the same estimates as those of Lemma 4.7 (which themselves do not really require $p < \frac{d}{2}$, see Remark 4.8).

On the one hand, the proof of Lemma 4.13 uses very crude estimates, so an improvement seems likely. On the other hand, one really looks for an estimate on the functions $\tau_n \in \mathbb{R}^E$ on the edges so that $\nabla^* \tau_n = \delta_x - P_n^x$. For $p = 2$, the only point where an improvement might occur is to avoid the triangle inequality (as

$\tau_n = \nabla(\sum_{i=0}^{n-1} P_x^i)$ is the function minimising the $\ell^2 E$-norm). For other p, there might be more room for improvement.

Acknowledgments I would like to thank M. Bourdon, P. Pansu as well as the anonymous referee for their comments and corrections.

References

1. M.E.B. Bekka, A. Valette, Group cohomology, harmonic functions and the first L^2-Betti number. Potential Anal. **6**(4), 313–326 (1997)
2. M. Bonk, B. Kleiner, Quasisymmetric parametrizations of two-dimensional metric spheres. Invent. Math. **150**(1), 127–183 (2002)
3. M. Bourdon, Cohomologie et actions isométriques propres sur les espaces L^p, in *Geometry, Topology and Dynamics, Proceedings of the 2010 Bangalore Conference* (to appear). http://math.univ-lille1.fr/~bourdon/papiers/coho.pdf
4. M. Bourdon, B. Kleiner, Combinatorial modulus, the combinatorial Loewner property, and coxeter groups. Groups Geom. Dyn. **7**(1), 39–107 (2013)
5. M. Bourdon, B. Kleiner, Some applications of ℓ^p-cohomology to boundaries of Gromov hyperbolic spaces (2012). arXiv:1203.1233
6. M. Bourdon, F. Martin, A. Valette, Vanishing and non-vanishing of the first L^p-cohomology of groups. Comment. Math. Helv. **80**, 377–389 (2005)
7. M. Bourdon, H. Pajot, Cohomologie ℓ^p et espaces de Besov. J. Reine Angew. Math. **558**, 85–108 (2003)
8. M. Bourdon, B. Rémy, Quasi-isometric invariance of continuous group L_p-cohomology, and first applications to vanishings (2018). arXiv:1803.09284
9. I. Benjamini, O. Schramm, Lack of sphere packing of graphs via nonlinear potential theory. J. Topol. Anal. **5**(1), 1–11 (2013)
10. I. Benjamini, O. Schramm, A. Timár, On the separation profile of infinite graphs (2010). arXiv:1004.0921
11. M. Carrasco Piaggio, Orlicz spaces and the large scale geometry of Heintze groups. Math. Ann. **368**(1), 433–481 (2017)
12. J. Cheeger, M. Gromov, L_2-cohomology and group cohomology. Topology **25**, **25**, 189–215 (1986)
13. Y. de Cornulier, Tessera, Contracting automorphisms and L_p-cohomology in degree one. Ark. Mat. **49**(2), 295–324 (2011)
14. B. Eckmann, Introduction to ℓ_2-methods in topology: reduced ℓ_2-homology, harmonic chains, ℓ_2-Betti numbers. Isr. J. Math. **117**(1), 183–219 (2000)
15. G. Élek, Coarse cohomology and ℓ_p-cohomology. K-Theory **13**, 1–22 (1998)
16. D. Gaboriau, Invariants ℓ^2 de relations d'équivalence et de groupes. Publ. Math. Inst. Hautes Études Sci. **95**, 93–150 (2002)
17. A. Georgakopoulos, Lamplighter graphs do not admit harmonic functions of finite energy. Proc. Am. Math. Soc. **138**(9), 3057–3061 (2010)
18. V. Gerasimov, Floyd maps for relatively hyperbolic groups. Geom. Funct. Anal. **22**(5), 1361–1399 (2012)
19. V.M. Gol'dshtein, M. Troyanov, Distortion of mappings and $L_{q,p}$-cohomology. Math. Z. **264**, 279 (2010). https://doi.org/10.1007/s00209-008-0463-x
20. A. Gournay, Vanishing of ℓ^p-cohomology and transportation cost. Bull. Lond. Math. Soc. **46**(3), 481–490 (2014)
21. A. Gournay, Boundary values of random walks and ℓ^p-cohomology in degree one. Groups Geom. Dyn. **9**(4), 1153–1184 (2015)

22. A. Gournay, Harmonic functions with finite p-energy on lamplighter graphs are constant. C. R. Acad. Sci. Paris **354**, 762–765 (2016)
23. A. Gournay, Absence of harmonic functions with ℓ^p gradient in some semi-direct products. Potential Anal. **45**(1), 619–642 (2016)
24. A. Gournay, Mixing, malnormal subgroups and cohomology in degree one. Groups Geom. Dyn. **12**(4), 1371–1416 (2018)
25. A. Gournay, P.-N. Jolissaint, Functions conditionally of negative type on groups acting on regular trees. J. Lond. Math. Soc. **93**(3), 619–642 (2016)
26. F.P. Greenleaf, *Invariant means on topological groups and their applications*. Van Nostrand Mathematical Studies Series. **16**, 113 (Van Nostrand Reinhold Company, New York 1969)
27. M. Gromov, Groups of Polynomial growth and expanding maps. Publ. Math. Inst. Hautes Études Sci. **53**, 53–73 (1981)
28. M. Gromov, Asymptotic invariants of groups, in *Geometric Group Theory (Vol. 2)*. London Mathematical Society Lecture Note Series, vol. 182 (Cambridge University Press, Cambridge, 1993), viii+295.
29. I. Holopainen, Nonlinear potential theory and quasiregular mappings on Riemannian manifolds. Ann. Acad. Sci. Fenn. Ser. A I Math. Dissertationes **74**, 45pp. (1990)
30. I. Holopainen, P. Soardi, p-harmonic functions on graphs and manifolds. Manuscr. Math. **94**, 95–110 (1997)
31. M. Kanai, Rough isometries and combinatorial approximations of geometries of non-compact Riemannian manifolds. J. Math. Soc. Jpn. **37**, 391–413 (1985)
32. E. Kappos, ℓ^p-cohomology for groups of type FP_n (2005). arXiv:math/0511002
33. A. Karlsson, Free subgroups of groups with nontrivial floyd boundary. Commun. Alg. **31**(11), 5361–5376 (2003)
34. S. Keith , T. Laakso, Conformal Assouad dimension and modulus. Geom. Funct. Anal. **14**, 1278–1321 (2004)
35. M. Keller, D. Lenz, M. Schmidt, R.K. Wojchiechowski, Note on uniformly transient graphs (2014). arXiv:1412.0815
36. Y.A. Kopylov, $L_{p,q}$-cohomology and normal solvability. Arch. Math. **89**(1), 87–96 (2007)
37. N. Lohoué, Remarques sur un théorème de Strichartz. C. R. Acad. Sci. Ser.I **311**, 507–510 (1990)
38. W. Lück, L^2-*Invariants: Theory and Applications to Geometry and K-Theory* (Springer, Berlin, 2002), 595pp.
39. T. Lyons, Instability of the Liouville property for quasi-isometric Riemannian manifolds and reversible Markov chains. J. Diff. Geom. **26**(1), 33–66 (1987)
40. F. Martin, A. Valette, On the first L^p cohomology of discrete groups. Groups Geom. Dyn. **1**, 81–100 (2007)
41. S. Moon, A. Valette, Non-properness of amenable actions on graphs with infinitely many ends, in *Ischia Group Theory 2006* (World Sci. Publ., Hackensack, 2007), pp. 227–233
42. D. Osin, L^2-Betti numbers and non-unitarizable groups without free subgroups. Int. Math. Res. Not. IMRN **22**, 4220–4231 (2009)
43. D. Osin, On acylindrical hyperbolicity of groups with positive first ℓ^2-Betti number. arXiv:1501.03066
44. P. Pansu, Cohomologie L^p des variétés à courbure négative, cas du degré un. *PDE and Geometry 1988, Rend. Sem. Mat. Torino, Fasc. Spez.* (1989), pp. 95–120
45. P. Pansu, Cohomologie ℓ^p: invariance sous quasi-isométrie. Unpublished*, 1995 (updated in 2004)
46. P. Pansu, Cohomologie ℓ^p, espaces homogènes et pincement. Unpublished* (1999)
[*] Both unpublished papers of Pansu are available on P. Pansu's webpage https://www.math.u-psud.fr/~pansu/liste-prepub.html (These papers have been for the most part integrated to other published papers)
47. P. Pansu, Cohomologie L^p en degré 1 des espaces homogènes. Potential Anal. **27**(2), 151–165 (2007)
48. P. Pansu, Cohomologie L^p et pincement. Comment. Math. Helv. **83**(2), 327–357 (2008)

49. P. Pansu, Large-scale conformal maps. arXiv:1604.01195 or hal-01297830 (Version April 2019)
50. P. Pansu, M. Rumin, On the $\ell^{q,p}$ cohomology of Carnot groups (2018). arXiv:1802.07618
51. P. Pansu, F. Tripaldi, Averages and the $\ell^{q,1}$-cohomology of Heisenberg groups (2019). arXiv:1904.06669
52. M. Puls, Group cohomology and L^p-cohomology of finitely generated groups. Can. Math. Bull. **46**(2), 268–276 (2003)
53. M.J. Puls, The first L^p-cohomology of some finitely generated groups and p-harmonic functions. J. Funct. Anal. **237**(2), 391–401 (2006)
54. M.J. Puls, The first L^p-cohomology of some groups with one end. Arch. Math. **88**(6), 500–506 (2007)
55. M.J. Puls, Graphs of bounded degree and the p-harmonic boundary. Pac. J. Math. **248**(2), 429–452 (2010)
56. S. Rohde, Oded Schramm: from circle packing to SLE. Ann. Probab. **39**(5), 1621–1667 (2011)
57. C.D. Sogge, *Fourier Integrals in Classical Analysis*. Cambridge Tracts in Mathematics, vol. 105 (Cambridge University Press, Cambridge, 2008)
58. K. Stephenson, Circle packings: a mathematical tale. Not. Am. Math. Soc. **50**(11):13–1388 (2003)
59. R. Tessera, Vanishing of the first reduced cohomology with values in an L^p-representation. Ann. Inst. Fourier **59**(2), 851–876 (2009)
60. M. Troyanov, Parabolicity of manifolds. Siberian Adv. Math. **9**(4), 125–150 (1999)
61. W. Woess, *Random Walks on Infinite Graphs and Groups*. Cambridge Tracts in Mathematics, vol. 138 (Cambridge University Press, Cambridge, 2000)
62. M. Yamasaki, Parabolic and hyperbolic infinite networks. Hiroshima Math. J. **7**, 135–146 (1977)
63. S. Zucker, L^p-Cohomology and Satake compactifications, in *Prospects in Complex Geometry*, ed. by J. Noguchi, T. Ohsawa. Lecture Notes in Mathematics, vol. 1468 (Springer, Berlin, 1991)

Polyharmonic Functions for Finite Graphs and Markov Chains

Thomas Hirschler and Wolfgang Woess

1 Introduction

In the setting of the classical Laplacian Δ on a Euclidean domain, or the Laplace-Beltrami operator on a Riemannian manifold, a polyharmonic function f is one for which $\Delta^n f = 0$. Their study goes back to work in the 19th century, see, e.g., ALMANSI [1]. A basic reference is the monograph by ARONSZAJN, CREESE AND LIPKIN [3]. A more recent one is the volume by GAZZOLA, GRUNAU AND SWEERS [8], with a nice introduction to classical problems from elasticity where polyharmonic (in fact biharmonic) functions and Δ^2 come up.

While there is a huge body of literature in the smooth case, the literature in the discrete setting is quite restricted: an early reference is VORONKOVA [14], who analysed the discretised version of $\Delta^2 f = 0$ in a half-strip $[0\,,\infty] \times [0\,,\ H]$. Other quite early references are YAMASAKI [16] and KAYANO AND YAMASAKI [10] who investigated the Green kernel for the bi-Laplacian on an infinite network, and a follow-up of this is VENKATARAMAN [13]. Biharmonic Laplacians on trees where also studied by COHEN, COLONNA AND SINGMAN [6, 7], seemingly without link to [16] and [10]. Prior to that, COHEN, COLONNA, GOWRISANKARAN AND SINGMAN [5] were the first to undertake a detailed study of polyharmonic functions on infinite, locally finite trees. In particular, for the standard Laplacian arising from simple random walk on a regular tree, they provided a boundary integral representation which is an analogue of Almansi's expansion of polyharmonic functions on the unit disk. (To get a flavour of the many close analogies between the potential theory of the unit disk and regular trees, the reader is invited to the introductory sections of BOIKO AND WOESS [4].) Recently, PICARDELLO AND

T. Hirschler · W. Woess (✉)
Institut für Diskrete Mathematik, Technische Universität Graz, Steyrergasse 30, A-8010 Graz, Austria
e-mail: thirschler@tugraz.at; woess@tugraz.at

N. Anantharaman et al. (eds.), *Frontiers in Analysis and Probability*,
https://doi.org/10.1007/978-3-030-56409-4_4

Woess [12] extended the study of [5] and proved, among others, a boundary integral representation of λ-polyharmonic functions (see below for more details) for arbitrary nearest neighbour transition operators on countable trees, not necessarily required to be locally finite.

In all this work, *finite* graphs, resp. Markov chains had only marginal appearances: in [16] for the biharmonic Green function of finite subnetworks of an infinite network, and in [5] for finite trees and an associated boundary value problem for biharmonic functions. Anandam [2] also studies polyharmonic functions on finite subtrees of infinite trees.

In the present note, we elaborate a detailed account of the general finite case, in which the mentioned potential theoretic questions turn into issues of linear algebra which can be solved rather easily.

The setting. We start with a finite set X, subdivided into the disjoint union of two non-empty subsets X^o, the *interior*, and ∂X, the *boundary*. On X, we consider a stochastic *transition matrix* $P = \big(p(x,y)\big)_{x,y\in X}$ with the following properties, where $p^{(n)}(x,y)$ denotes the (x,y)-entry of the matrix power P^n.

(i) For all $x \in X^o$, there is $w \in \partial X$ such that $p^{(n)}(x,w) > 0$ for some n.
(ii) For all $w \in \partial X$, we have $p(w,w) = 1$, and thus $p(w,x) = 0$ for all $x \in X \setminus \{w\}$.
(iii) For all $w \in \partial X$, there is $x \in X^o$ such that $p^{(n)}(x,w) > 0$ for some n.

Thus, X can be given the structure of a digraph, where we have an oriented edge $x \to y$ when $p(x,y) > 0$. Then (i) means that the boundary can be reached from any interior point by an oriented path, (ii) means that each boundary point is absorbing, i.e., the only outgoing edge is a loop at that point and (iii) means that every boundary point is active in the sense that it is reached by some oriented path from an interior point. In probabilistic terms, we have a Markov chain (random process) on X, whose evolution is governed by P: if the current position is x, then the next step is from x to y with probability $p(x,y)$.

Example 1.1 The most typical situation is the one where we start with a finite *resistive network*, that is, a connected, non-oriented graph (X,E) where each edge $e = [x,y] = [y,x]$ carries a positive *conductance* $a(e) = a(x,y)$. Then we choose our partition $X = X^o \cup \partial X$, and we set $m(x) = \sum_y a(x,y)$. The transition probabilities become $p(x,y) = a(x,y)/m(x)$, if $x \in X^o$ and $y \in X$, while $p(w,w) = 1$ for $w \in \partial X$. This defines a reversible Markov chain which is absorbed in ∂X, see, e.g., Woess [15, Ch. 4]. In particular, setting all $a(x,y)$ equal to 1, the conductances correspond to the adjacency matrix.

The transition matrix P acts on functions (column vectors) $f : X \to \mathbb{C}$ by

$$Pf(x) = \sum_y p(x,y)f(y),$$

and the (normalised) graph Laplacian is $I - P$, where $I = I_X$ is the identity matrix over X. It is typically defined on X without assigning a boundary ∂X, but the study undertaken here makes sense only in presence of absorbing points. Note that the more direct analogue of the (negative definite) smooth Laplacian would in reality be $P - I$. More generally, we shall work with suitable variant of $\lambda \cdot I - P$ for $\lambda \in \mathbb{C}$.

A *λ-harmonic function* $h : X \to \mathbb{C}$ is one for which

$$Ph(x) = \lambda\, h(x) \quad \text{for every} \quad x \in X^o\,. \tag{1.2}$$

When $\lambda = 1$, we speak of a harmonic function. When speaking of *λ-polyharmonic functions* of order n, we have two possible approaches: one is to look for functions $f : X \to \mathbb{C}$ which satisfy

$$(\lambda \cdot I - P)^n f = 0 \quad \text{on } X. \tag{1.3}$$

These *global* λ-polyharmonic functions can be easily described.

The more interesting version is related with the pre-assignment of boundary values. Let P_{X^o} and Q be the restrictions of P to $X^o \times X^o$ and $X^o \times \partial X$, respectively. Then we define the λ-Laplacian as the matrix given in block-form by

$$\Delta_\lambda = \begin{pmatrix} \lambda \cdot I_{X^o} - P_{X^o} & -Q \\ 0 & 0 \end{pmatrix} = \begin{pmatrix} \lambda \cdot I_{X^o} & 0 \\ 0 & I_{\partial X} \end{pmatrix} - P\,, \tag{1.4}$$

where the 0s stand for the zero matrices in the respective dimensions. Here, the identity matrix over ∂X is *not* multiplied by λ, so that functions annihilated by Δ_λ are λ-harmonic only in X^o.

Our main focus is on polyharmonic functions in the sense that they satisfy

$$\Delta_\lambda^n f = 0 \tag{1.5}$$

on X, or – more reasonably, as we shall see – on the "n-th interior" of X, i.e., all points in X^o from which ∂X cannot be reached in less than n steps. When $\lambda = 1$, the two notions (1.3) and (1.5) coincide.

This note is organised as follows. In Section 2, we first consider ordinary harmonic and polyharmonic functions, that is, the case $\lambda = 1$. After recalling the well-known solution of the *Dirichlet problem* for harmonic functions with preassigned boundary values (Lemma 2.2), we explain why all global harmonic functions in the sense of (1.3) (with $\lambda = 1$) are indeed harmonic (Proposition 2.6). Then we look at all global λ-polyharmonic functions as in (1.3). In this case, λ must belong to the spectrum of P_{X^o}, and the solutions can be described in terms of a Jordan basis (Proposition 2.7).

In Section 3, we turn to studying Δ_λ and its powers, for λ in the resolvent set of P_{X^o} (the spectrum being settled in Section 2). There is a direct analogue to the solution of the Dirichlet problem, and again, any function which satisfies $\Delta_\lambda^n f =$

0 on all of X must be λ-harmonic (Proposition 3.2). Finally, we give the precise formulation of the *Riquier* problem, which consists in assigning boundary functions $g_1, \ldots, g_n$ and – loosely spoken – searching for a function f such that the boundary values of $\Delta_\lambda^{r-1} f$ coincide with g_r for $r = 1, \ldots, n$. That problem for the special case of finite trees is briefly touched in [5]. Here, we provide the general solution (Theorem 3.4).

Finally, in Section 4, we undertake a comparison of those results with the case of infinite trees without leaves, which was studied recently in [12] by use of Martin boundary theory.

All results of this note are achieved by applying basic tools from Linear Algebra in the right way. We believe that this material provides a useful basis, firstly as a link to the classical, smooth case (regarding the Laplacian on bounded domains), and secondly, as a basis for handling and understanding polyharmonic functions not only on infinite trees but also on more general infinite graphs and their boundaries at infinity.

2 The Dirichlet Problem and Global λ-Polyharmonic Functions

We start with some observations on the case $\lambda = n = 1$, that is, ordinary harmonic functions. We start with a simple observation on $\mathsf{spec}(P_{X^o})$, the set of eigenvalues of P_{X^o}.

Lemma 2.1 *The spectral radius* $\rho = \rho(P_{X^o}) = \max\{|\lambda| : \lambda \in \mathsf{spec}(P_{X^o})\}$ *satisfies* $\rho < 1$.

Proof (Outline) Condition (i) on P implies that for each $x \in X^o$, there is n such that $\sum_{v \in X^o} p^{(n)}(x, v) < 1$, that is, $P_{X^o}^n$ is strictly substochastic in the row of x. One easily deduces that there is m such that $P_{X^o}^m$ is strictly substochastic in every row, which yields the claim. □

The following *solution of the Dirichlet problem* is folklore in the Markov chain community; see, e.g., [15, §6.A]. It keeps being "rediscovered" by analysts who deviate into the discrete world, see, for example, KISELMAN [11].

Lemma 2.2 *For every function* $g : \partial X \to \mathbb{C}$ *there is a unique harmonic function* h *on* X *such that* $h|_{\partial X} = g$. *It is given by*

$$h(x) = \sum_{w \in \partial X} F(x, w) g(w) ,$$

where $F(x, w)$ *is the probability that the Markov chain starting at* x *hits* ∂X *in the point* v.

We next want to describe the kernel $F(x, w)$ in matrix terminology. Let $\mathsf{res}(P_{X^o}) = \mathbb{C} \setminus \mathsf{spec}(P_{X^o})$ be the resolvent set of P_{X^o}. For $\lambda \in \mathsf{res}(P_{X^o})$, the *resolvent* is the $X^o \times X^o$-matrix

$$\mathbf{G}(\lambda) = \big(G(x, y|\lambda)\big)_{x,y \in X^o} = (\lambda \cdot I_{X^o} - P_{X^o})^{-1}. \tag{2.3}$$

The kernels $G(x, y|\lambda)$ are called *Green functions*. They are rational functions of λ. Now we define the $X^o \times \partial X$-matrix

$$\mathbf{F}(\lambda) = \big(F(x, w|\lambda)\big)_{x \in X^o, w \in \partial X} = \mathbf{G}(\lambda)\, Q. \tag{2.4}$$

We can extend it to $X \times \partial X$ by setting $F(v, w|\lambda) = \delta_w(v)$ for $v, w \in \partial X$. When $\lambda = 1$, we just write $G(x, y)$ for $G(x, y|1)$ and $F(x, w)$ for $F(x, w|1)$. For $|\lambda| > \rho$, we can expand

$$G(x, y|\lambda) = \sum_{n=0}^{\infty} p^{(n)}(x, y)/\lambda^{n+1} \quad \text{and} \quad F(x, w|\lambda) = \sum_{n=0}^{\infty} f^{(n)}(x, w)/\lambda^{n},$$

where the probabilistic meaning is that for the Markov chain starting at x, the probability to be at y at time n is $p^{(n)}(x, y)$, while $f^{(n)}(x, w)$ is the probability that the first visit in $w \in \partial X$ occurs at time n.

Coming back to the Dirichlet problem, it is a straightforward matrix computation to see that the function h, as defined in Lemma 2.2, is harmonic. Its uniqueness follows from invertibility of $(I_{X^o} - P_{X^o})$. Instead, it may also be instructive to deduce uniqueness from the potential theoretic *maximum principle:* every real-valued harmonic function attains its maximum on ∂X, see [15, §6.A].

This also yields one way to see that the Markov chain must hit the boundary almost surely, that is,

$$\sum_{w \in \partial X} F(x, w) = 1 \quad \text{for every } x \in X^o.$$

Namely, the unique harmonic extension of the constant boundary function $g \equiv 1$ is the constant function $h \equiv 1$ on X. Also, the function $x \mapsto F(x, w)$ provides the unique harmonic extension of the boundary function $g = \mathbf{1}_v$.

Corollary 2.5 *The geometric and the algebraic multiplicity of the eigenvalue $\lambda = 1$ of P coincide and are equal to $|\partial X|$.*

Proof Lemma 2.2 yields that the geometric multiplicity is $|\partial X|$. The characteristic polynomial of the matrix P is

$$\chi_P(\lambda) = \mathsf{det}(\lambda \cdot I - P) = (\lambda - 1)^{|\partial X|} \chi_{P^o}(\lambda).$$

By Lemma 2.1, $\chi_{P^o}(1) \neq 0$. □

Now we can easily describe all *free* polyharmonic functions of order $n \geq 1$, that is, those which satisfy $(I - P)^n f = 0$ on X.

Proposition 2.6 *A function $f : X \to \mathbb{C}$ satisfies $(I - P)^n f = 0$ if and only if f is harmonic.*

Proof Suppose $n \geq 2$, and let $h = (I - P)^{n-1} f$. Then h is harmonic, and $(I - P)f = h$. Since $(I - P)^{n-1} f = 0$ on ∂X, the function h solves the Dirichlet problem with boundary values 0. Therefore $h = 0$, that is, $(I - P)^{n-1} f = 0$. Proceeding by induction, we obtain that f is harmonic. □

Similarly, we can handle the case $(\lambda \cdot I - P)^n f = 0$, when $\lambda \neq 1$. First of all, when $n \geq 2$ then the function $h = (\lambda \cdot I - P)^{n-1} f$ satisfies $Ph = \lambda \cdot h$. Second, we see that $f = 0$ in ∂X, so (by abuse of notation) we consider f as a function on X^o. In other words, $\lambda \in \mathsf{spec}(P_{X^o})$.

Let $\kappa = \kappa(\lambda)$ and $\mu = \mu(\lambda)$ be the algebraic and geometric eigenvalue multiplicities of λ. Let $h_1, \ldots, h_\mu$ be a basis of $\mathsf{ker}(\lambda \cdot I_{X^o} - P_{X^o})$. For each $j \in \{1, \ldots, \mu\}$, let κ_j be the length of the associated Jordan chain (= dimension of the associated Jordan block in the Jordan normal form). That is, $\kappa_1 + \cdots + \kappa_\mu = \kappa$, and we have functions $f_j^{(k)}$, $k = 1, \ldots, \kappa_j$ such that $f_j^{(1)} = h_j$ and $(\lambda \cdot I_{X^o} - P_{X^o}) f_j^{(k)} = f_j^{(k-1)}$ for $k \geq 2$. All those functions are extended to X by assigning value 0 on ∂X. Then it is clear that $\{f_j^{(k)} : k = 1, \ldots, \kappa_j\,,\ j = 1, \ldots, \mu\}$ is a basis of the linear space of all global λ-polyharmonic functions (of arbitrary order). We subsume.

Proposition 2.7 *With the above notation, for $\lambda \in \mathsf{spec}(P_{X^o})$, the space of functions $f : X \to \mathbb{C}$ with $(\lambda \cdot I - P)^n f = 0$ is spanned by*

$$\big\{f_j^{(k)} : k = 1, \ldots, \min\{n, \kappa_j\}\,,\ j = 1, \ldots, \mu\big\}.$$

Corollary 2.8 *For a finite network with boundary as in Example* (1.1), *every global λ-polyharmonic h function satisfies $Ph = \lambda \cdot h$, and $\lambda \in \mathsf{spec}(P) \subset \mathbb{R}$. Furthermore, h vanishes on ∂X when $\lambda \neq 1$.*

Proof If we define the diagonal matrix $M = \mathsf{diag}\big(\sqrt{m(x)}\,\big)_{x \in X^o}$, then $M\, P_{X^o}\, M^{-1}$ is symmetric, so that the spectrum is real and the geometric and algebraic multiplicities of the eigenvalues of P_{X^o} coincide. □

3 Boundary Value Problems for λ-Polyharmonic Functions

In this section, we assume that $\lambda \in \mathsf{res}(P_{X^o})$ and study the operator (resp. matrix) Δ_λ of (1.4) and its powers.

Notation: in accordance with the block form used above, for any function $f : X \to \mathbb{C}$ we write $f = \begin{pmatrix} f^o \\ f^\partial \end{pmatrix}$, where $f^o = f|_{X^o}$ and $f^\partial = f|_{\partial X}$. Also, we write Δ_λ^o for the restriction of the matrix of (1.4) to $X \times X^o$, that is, $\Delta_\lambda^o f = (\Delta_\lambda f)^o$.

First of all, there is an obvious λ-variant of the solution of the Dirichlet problem.

Lemma 3.1 *Let $\lambda \in \mathsf{res}(P_{X_o})$. For every function $g : \partial X \to \mathbb{C}$ there is a unique λ-harmonic function h on X such that $h|_{\partial X} = g$. It is given by*

$$h(x) = \sum_{w \in \partial X} F(x, w|\lambda) g(w)\,, \quad x \in X^o\,,$$

where $F(x, w|\lambda)$ is defined by (2.4).

Proof We write $h = \begin{pmatrix} h^o \\ g \end{pmatrix}$, where $h^o = h|_{X^o}$ and g is the given boundary function. Then the equation $\Delta_\lambda = 0$ transforms into

$$(\lambda \cdot I_{X^o} - P_{X^o})h^o = Qg\,,$$

which has the unique solution $h^o = \mathbf{G}(\lambda)Qg$, as proposed. □

Next, we note that

$$\Delta_\lambda^n = \begin{pmatrix} (\lambda \cdot I_{X^o} - P_{X^o})^n & -(\lambda \cdot I_{X^o} - P_{X^o})^{n-1} Q \\ 0 & 0 \end{pmatrix}.$$

Thus, if we look for a solution of $\Delta_\lambda^n h = 0$ then with $h = \begin{pmatrix} h^o \\ g \end{pmatrix}$ as above, we get the equation

$$(\lambda \cdot I_{X^o} - P_{X^o})^n h^o = (\lambda \cdot I_{X^o} - P_{X^o})^{n-1} Qg\,,$$

which has the same solution as in Lemma 3.1. Thus, we have the following general version of Proposition 2.6.

Proposition 3.2 *A function $f : X \to \mathbb{C}$ satisfies $\Delta_\lambda^n f = 0$ on all of X if and only if f is λ-harmonic.*

For $n \geq 2$, what is more interesting is to assign further boundary conditions. Recall that $\Delta_\lambda f$ always vanishes on ∂X. The analogue of the Dirichlet problem is the *Riquier problem* of order n. We assign n boundary functions $g_1, \ldots, g_n : \partial X \to \mathbb{C}$ and look for a function $f : X \to \mathbb{C}$ such that we have a "tower" of boundary value problems for functions $f_n, f_{n-1}, \ldots, f_1 = f : X \to \mathbb{C}$ as follows:

$$f_r = \begin{pmatrix} f_r^o \\ g_r \end{pmatrix}, \quad \Delta_\lambda^o f_n = 0\,, \quad \text{and} \quad \Delta_\lambda^o f_r = f_{r+1}^o \quad \text{for} \quad r = n-1, n-2, \ldots, 1\,. \tag{3.3}$$

Theorem 3.4 *For $\lambda \in \mathsf{res}(P_{X^o})$, the unique solution $f = f_1$ of* (3.3) *is given by*

$$f(x) = \sum_{r=1}^{n} \big[\mathbf{G}(\lambda)^r \, Q \, g_r\big](x)\,, \quad x \in X^o\,,$$

where $\mathbf{G}(\lambda)^r$ is the r-th matrix power of $\mathbf{G}(\lambda)$.

Proof We use induction on n. For $n = 1$, this is Lemma 3.1. Suppose the statement is true for $n - 1$. The function f_2 is the solution of the Riquier problem of order $n - 1$ for the boundary functions $g_2, \ldots, g_n$. By the induction hypothesis,

$$f_2(x) = \sum_{r=2}^{n} \big[\mathbf{G}(\lambda)^{r-1} \, Q \, g_r\big](x)\,, \quad x \in X^o\,,$$

and this is the unique solution. The last one of the "tower" of Equations (3.3) is

$$\Delta_\lambda^o f = f_2^o\,, \quad \text{where} \quad f = \begin{pmatrix} f^o \\ g_1 \end{pmatrix}.$$

This can be rewritten as

$$(\lambda \cdot I_{X^o} - P_{X^o}) f^o - Q\, g_1 = f_2^o\,.$$

Inserting the solution for f_2^o and multiplying by $\mathbf{G}(\lambda)$, we get the solution for f, and it is unique. □

Note that the solution f does not satisfy (1.5) on all of X^o. This is due to the fact that our discrete Laplacian is not infinitesimal. Let

$$\partial^n X = \{x \in X : p^{(k)}(x, w) > 0 \text{ for some } w \in \partial X \text{ and } k \le n - 1\}\,, \tag{3.5}$$

the set of all points in X from which ∂X can be reached in $n - 1$ or less steps. Then $\Delta_\lambda^n f = 0$ only on the *n-th interior* $X \setminus \partial^n X$, while the values on $\partial^n X$ depend on the boundary functions $g_1, \ldots, g_n$.

The functions $\lambda \mapsto G(x, y|\lambda)$ are rational, and the union of the set of their poles is $\mathsf{spec}(P_{X^o})$. For $\lambda \in \mathsf{res}(P_{X^o})$, we can differentiate the identity $\lambda \cdot \mathbf{G}(\lambda) - P\,\mathbf{G}(\lambda) = I_{X^o}$ k times, and Leibniz' rule yields

$$(\lambda \cdot I_{X^o} - P_{X^o})\, \mathbf{G}^{(r)}(\lambda) = -k \cdot \mathbf{G}^{(r-1)}(\lambda)\,,$$

where $\mathbf{G}^{(r)}(\lambda)$ is the (elementwise) r-th derivative of $\mathbf{G}(\lambda)$ with respect to λ. From this, we get recursively for the matrix powers of $\mathbf{G}(\lambda)$

$$\mathbf{G}(\lambda)^r = \frac{(-1)^{r-1}}{(r-1)!}\, \mathbf{G}^{(r-1)}(\lambda)\,. \tag{3.6}$$

We can insert this in the formula of Theorem 3.4 for an alternative form of the solution of the Riquier problem.

4 Comparison with the Case of Infinite Trees; Examples

We now want to relate the preceding material, and in particular Theorem 3.4, with the potential theory of countable Markov chains, and more specifically, with Martin boundary theory and λ-polyharmonic functions on trees, as studied in [12]. We choose and fix a reference point (origin) $o \in X^o$ and consider the rational functions $\lambda \mapsto F(o, w|\lambda)$ of (2.4) for $\lambda \in \mathsf{res}(P_{X^o})$ and $w \in \partial X$. They have (at most) finitely many zeros. Let

$$\mathsf{res}^*(P_{X^o}) = \mathsf{res}(P_{X^o}) \setminus \{\lambda : F(o, w|\lambda) = 0 \text{ for some } w \in \partial X\}.$$

Every positive real $\lambda > \rho(P)$ belongs to $\mathsf{res}^*(P_{X^o})$, in particular, $\lambda = 1$. For $\lambda \in \mathsf{res}^*(P_{X^o})$, we define the λ-*Martin kernel*

$$K^{(X)}(x, w|\lambda) = \frac{F(x, w|\lambda)}{F(o, w|\lambda)}, \quad x \in X, \ w \in \partial X. \tag{4.1}$$

The function $x \mapsto K^{(X)}(x, w|\lambda)$ is the unique solution of the λ-Dirichlet problem of Lemma 3.1 with value 1 at the root o and the boundary function g_v proportional to δ_w, that is, $g_w(v) = \delta_w(v)/F(o, w|\lambda)$. Thus, for a generic boundary function $g : \partial X \to \mathbb{C}$, we can write the solution of the λ-Dirichlet problem for $x \in X^o$ as

$$\begin{aligned} h(x) &= \sum_{w \in \partial X} K^{(X)}(x, w|\lambda)\nu(w) =: \int_{\partial X} K^{(X)}(x, \cdot\,|\lambda)\, d\nu \quad (x \in X^o), \quad \text{where} \\ \nu(w) &= g(w)\, F(o, w|\lambda). \end{aligned} \tag{4.2}$$

The integral notation indicates that we think of $\nu = \nu_g$ as a complex distribution on ∂X. In the same way, the solution of the Riquier problem in Theorem 3.4 can be written as

$$\begin{aligned} f(x) &= \sum_{r=1}^{n} \int_{\partial X} K_r^{(X)}(x, \cdot\,|\lambda)\, d\nu_r\,, \quad \text{where for } w \in \partial X \\ K_r^{(X)}(\cdot, w|\lambda) &= \mathbf{G}(\lambda)^{r-1} K(\cdot, w|\lambda) \quad \text{and} \quad \nu_r(w) = g_r(w)\, F(o, w|\lambda)\,. \end{aligned} \tag{4.3}$$

Now let us look at the case of a nearest neighbour transition operator $P = P_T$ on a countable tree T without leaves (i.e., vertices distinct from o have more than just one neighbour): there, the geometric boundary is attached to the tree "at infinity", and there is no "interior" of T which appears as a subset of the vertex set: the interior is T itself. The Martin kernel $K^{(T)}(x,\xi|\lambda)$ is defined for $x \in T$ and $\xi \in \partial T$, and it satisfies $(\lambda \cdot I - P)K^{(T)}(\cdot,\xi|\lambda) = 0$, without any restriction to a sub-matrix such as P_{X^o}. In this setting, [12, Thm. 5.4] says that any λ-polyharmonic function f of order n on T has a unique representation of the form

$$f(x) = \sum_{r=1}^{n} \int_{\partial T} K_r^{(T)}(x,\cdot\,|\lambda)\, d\nu_r\,, \qquad \text{where}$$
$$K_r^{(T)}(x,\xi|\lambda) = \frac{(-1)^{r-1}}{(r-1)!}\frac{d^{r-1}}{d\lambda^{r-1}} K(x,\xi|\lambda) \qquad (x \in T,\ \xi \in \partial T), \tag{4.4}$$

and $\nu_1\,,\dots,\,\nu_n$ are distributions on ∂T. The normalisation is slightly different here from the one chosen in [12], and in particular,

$$(\lambda \cdot I_T - P_T)K_r^{(T)}(\cdot,\xi|\lambda) = K_{r-1}^{(T)}(\cdot,\xi|\lambda) \quad \text{for } r \geq 2\,. \tag{4.5}$$

Let us compare the kernels $K_r^{(X)}$ and $K_r^{(T)}$. We have

$$(\lambda \cdot I_{X^o} - P_{X^o})^{r-1} K_r^{(X)}(\cdot,w|\lambda) = K^{(X)}(\cdot,w|\lambda) \quad \text{for} \quad w \in \partial X\,, \quad \text{and}$$
$$(\lambda \cdot I_T - P_T)^{r-1} K_r^{(T)}(\cdot,\xi|\lambda) = K^{(T)}(\cdot,\xi|\lambda) \quad \text{for} \quad \xi \in \partial T\,. \tag{4.6}$$

The only, but crucial difference is that in the first of the two identities, we may multiply from the left by $\mathbf{G}^{(X)}(\lambda)^{r-1} = (\lambda \cdot I_{X^o} - P_{X^o})^{-(r-1)}$. In the second identity, we may *not* multiply by $\mathbf{G}^{(T)}(\lambda)^{r-1}$, where $\mathbf{G}^{(T)}(\lambda) = (\lambda \cdot I_T - P_T)^{-1}$ is the resolvent of P as an operator on the Hilbert space $\ell^2(T,m)$, with the weights $m(x)$ analogous to Example 1.1 above. Indeed, $K^{(T)}(\cdot,\xi|\lambda)$ does in general not belong to $\ell^2(T,m)$.

"Forward only" Laplacians on finite and infinite trees

We now consider a class of examples which constitute the finite analogue of [12, §6]. They were also studied, from the viewpoint of Information Theory, by HIRSCHLER AND WOESS [9].

In order to carry the above comparison with the infinite case a bit further, we need some more details on the geometry of an infinite tree T with root o. We assume that T is locally finite and has no leaves. Each vertex $x \neq o$ has a unique *predecessor* x^-, its neighbour which is closer to o. For each $x \in T$ there is the unique *geodesic path* $\pi(o,x) = [o = x_0\,,x_1\,,\dots,x_n = x]$ from o to x, where $x_k^- = x_{k-1}$ for $k = 1,\dots,n$. In this case, $|x| = n$ is the *length* of x.

The boundary at infinity ∂T of T consists of all geodesic rays $\xi = [o = x_0\,,x_1\,,x_2\,,\dots]$, where $x_k^- = x_{k-1}$ for $k \geq 1$. For a vertex $x \in T$, we define the *boundary arc*

$$\partial_x T = \{\xi \in \partial T : x \in \xi\}\,.$$

The collection of all $\partial_x T$, $x \in T$, is the basis of a topology on ∂T, which thus becomes a compact, totally disconnected space, and each boundary arc is open and compact. We now take a Borel probability measure $\mathbb{P}$ on ∂T which is supported by the entire boundary, that is, $\mathbb{P}(\partial_x T) > 0$ for all $x \in T$. It induces a *forward only* Markov operator on T, as follows:

$$p(x, y) = \begin{cases} \mathbb{P}(\partial_y T)/\mathbb{P}(\partial_x T), & \text{if } y^- = x, \\ 0, & \text{otherwise.} \end{cases} \tag{4.7}$$

Conversely, if we start with transition probabilities $p(x, y)$ such that $p(x, y) > 0$ precisely when $y^- = x$, then we can construct $\mathbb{P}$ on $\partial_x T$ by setting

$$\mathbb{P}(\partial_x T) = p(o, x_1)p(x_1, x_2) \cdots p(x_{n-1}, x), \quad \text{if}$$
$$\pi(o, x) = [o = x_0, x_1, \ldots, x_n = x].$$

This determines $\mathbb{P}$ on the Borel σ-algebra of ∂T.

More generally, a *distribution* on ∂T is a set function

$$\nu : \{\partial_x T : x \in T\} \to \mathbb{C} \quad \text{with} \quad \nu(\partial_x T) = \sum_{y:y^-=x} \nu(\partial_y T) \quad \text{for all } x \in T. \tag{4.8}$$

If ν is non-negative real, then it extends uniquely to a Borel measure on ∂T. A *locally constant function* φ on ∂T is one such that every $\xi \in \partial T$ has a neighbourhood on which φ is constant. Thus, one can write it as a finite linear combination of boundary arcs

$$\varphi = \sum_{j=1}^{m} c_j \, \mathbf{1}_{\partial_{x(j)} T},$$

and we can define

$$\int_{\partial T} \varphi \, d\nu = \sum_{j=1}^{m} c_j \, \nu(\partial_{x(j)} T).$$

Indeed, in this way, the space of all distributions is the dual of the linear space of all locally constant functions on ∂T.

Now take $\lambda \in \mathbb{C} \setminus \{0\}$. Following [12, §6], the λ-Martin kernel on T is

$$K^{(T)}(x, \xi | \lambda) = \begin{cases} \lambda^{|x|}/\mathbb{P}(\partial_x T), & \text{if } \xi \in \partial_x T, \\ 0, & \text{otherwise.} \end{cases}$$

For fixed x, the function $\xi \mapsto K^{(T)}(x, \xi|\lambda)$ and its derivatives with respect to λ are locally constant, whence they can be integrated against distributions on ∂T. According to (4.4), we get

$$K_r^{(T)}(x, \xi|\lambda) = \begin{cases} (-1)^{r-1}\lambda^{|x|-(r-1)}\binom{|x|}{r-1}\dfrac{1}{\mathbb{P}(\partial_x T)}, & \text{if } \xi \in \partial_x T, \\ 0, & \text{otherwise,} \end{cases} \tag{4.9}$$

and every λ-polyharmonic function of order n on T has a unique representation

$$f(x) = \sum_{r=1}^{n}(-1)^{r-1}\lambda^{|x|-(r-1)}\binom{|x|}{r-1}\frac{\nu_r(\partial_x T)}{\mathbb{P}(\partial_x T)}, \tag{4.10}$$

where the $\nu_r = \nu_r^{(T)}$ $(r = 1, \ldots, n)$ are distributions on ∂T.

We now consider the finite situation. The graph X under consideration is a finite subtree of T with the same root o. The boundary consists of the *leaves* of the tree:

$$\partial X = \{w \in X : w \neq o,\ \deg(w) = 1\}.$$

We suppose that ∂X is a *section* of T in the sense of [9]: For every $\xi \in \partial T$, the geodesic ray starting from o that represents ξ intersects ∂X in a unique vertex. (A typical special case is the one where $\partial X = \{x \in T : |x| = L\}$ with $L \in \mathbb{N}$.) For each $x \in X$, we define the finite version of the boundary arc rooted at x as

$$\partial_x X = \{w \in \partial X : x \in \pi(o, w)\}.$$

In particular, $\partial_o X = \partial X$, and $\partial_w X = \{w\}$ for $w \in \partial X$.

We consider the restriction to X of the given forward transition matrix P_T on T. That is,

$$p_X(x, y) = \mathbb{P}(\partial_y T)/\mathbb{P}(\partial_x T), \text{ if } y^- = x \in X^o, \text{ and } p_X(w, w) = 1 \text{ if } w \in \partial X,$$

while $p_X(x, y) = 0$ in all other cases. Exactly as on the whole tree, we have for $x, y \in X$

$$p^{(n)}(x, y) > 0 \iff x \in \pi(o, y) \text{ and } n = |y| - |x|,$$
$$\text{and then} \quad p^{(n)}(x, y) = \mathbb{P}(\partial_y T)/\mathbb{P}(\partial_x T).$$

The matrix P_{X^o} is nilpotent, so that $\mathsf{spec}(P) = \{0, 1\}$, and the algebraic multiplicities of those two eigenvalues are $|X^o|$ and $|\partial X|$, respectively. For $\lambda \in \mathbb{C} \setminus \{0\} = \mathsf{res}(P_{X^o})$ and $x, y \in X^o$, we have

$$G(x, y|\lambda) = \begin{cases} \lambda^{-d(x,y)-1}\,\mathbb{P}(\partial_y T)/\mathbb{P}(\partial_x T)\,, & \text{if } x \in \pi(o, y)\,, \\ 0\,, & \text{otherwise.} \end{cases}$$

Therefore in this example, the right-hand side of (3.6) is obtained by

$$\frac{(-1)^{r-1}}{(r-1)!}\,G^{(r-1)}(x, y|\lambda) = \lambda^{-d(x,y)-r}\binom{d(x, y)+r-1}{r-1}\mathbb{P}(\partial_y T)/\mathbb{P}(\partial_x T)\,,$$

if $x \in \pi(o, y)$. We note that $\mathsf{res}^*(P_{X^o}) = \mathsf{res}(P_{X^o})$ and that $F(o, w|\lambda) = \lambda^{-|w|}\,\mathbb{P}(\partial_w T)$ for $w \in \partial X$. We can now compute the kernels $K_r^{(X)}$ of (4.3) as follows:

$$K_r^{(X)}(x, w|\lambda) = \begin{cases} \lambda^{|x|-r+1}\binom{d(x, w)+r-2}{r-1}\dfrac{1}{\mathbb{P}(\partial_x T)}\,, & \text{if } w \in \partial_x T\,, \\ 0\,, & \text{otherwise.} \end{cases} \tag{4.11}$$

Then, given boundary functions $g_1, \ldots, g_n$, the associated solution of the Riquier problem reads

$$f(x) = \sum_{r=1}^{n} \int_{\partial X} K_r^{(X)}(x, \cdot\,|\lambda)\,d\nu_r^{(X)}\,, \quad \text{with} \quad \nu_r^{(X)}(w) = \lambda^{-|w|}\,g_r(w)\,\mathbb{P}(\partial_w T)\,.$$

Now consider (4.6) and the fact that P_{X^o} is the restriction of P_T to X^o. In spite of this, when $n \geq 2$ we see that for $w \in \partial X$, the function $x \mapsto K_n^{(X)}(x, w|\lambda)$ is *not* the restriction to X^o of $x \mapsto K_n^{(T)}(x, \xi|\lambda)$, where $\xi \in \partial_w X$. (The value is the same for every such ξ, when $x \in X^o$.) For a closer look, fix $\xi \in \partial_w T$ and let $f(x) = K_n^{(T)}(x, \xi|\lambda)$ for $x \in X$. This function solves the Riquier problem on X with boundary functions

$$g_r(v) = K_{n+1-r}^{(T)}(w, \xi|\lambda)\,\delta_w(v)\,, \quad v \in \partial X\,,$$

or, equivalently, with boundary measures on ∂X

$$\nu_r^{(X)} = (-\lambda)^{n-r}\binom{|w|}{n-r}\delta_w\,.$$

Indeed, verification of

$$K_n^{(T)}(x, \xi|\lambda) = \sum_{r=1}^{n} \int_{\partial X} K_r^{(X)}(x, \cdot\,|\lambda)\,d\nu_r^{(X)}$$

leads to known combinatorial identity

$$\binom{|w|}{n-1} = \sum_{r=1}^{n} (-1)^{n-r} \binom{|w| - |x| - r - 2}{r-1} \binom{|w|}{n-r},$$

in which $|w|$ and $|x|$ can be arbitrary integers with $|w| > |x| \geq 0$.

References

1. E. Almansi, Sull'integrazione dell'equazione differenziale $\Delta^{2n} = 0$. Ann. Mat. Serie III **2**, 1–59 (1899)
2. V. Anandam, *Harmonic Functions and Potentials on Finite or Infinite Networks.* Lecture Notes of the Unione Matematica Italiana, vol. 12 (UMI, Bologna, 2011)
3. N. Aronszajn, T.M. Creese, L.J. Lipkin, *Polyharmonic Functions.* Oxford Math. Monographs (Oxford University Press, New York, 1983)
4. T. Boiko, W. Woess, Moments of Riesz measures on Poincaré disk and homogeneous tree – a comparative study. Expo. Math. **33**, 353–374 (2015)
5. J.M. Cohen, F. Colonna, K. Gowrisankaran, D. Singman, Polyharmonic functions on trees. Am. J. Math. **124**, 999–1043 (2002)
6. J.M. Cohen, F. Colonna, D. Singman, Biharmonic Green functions on homogeneous trees. Mediterr. J. Math. **6**, 249–271 (2009)
7. J.M. Cohen, F. Colonna, D. Singman, Biharmonic extensions on trees without positive potentials. J. Math. Anal. Appl. **378**, 710–722 (2011)
8. F. Gazzola, H.-Ch. Grunau, G. Sweers, *Polyharmonic Boundary Value Problems.* Lecture Notes in Mathematics, vol. 1991 (Springer, Berlin, 2010)
9. T. Hirschler, W. Woess, Comparing entropy rates on finite and infinite rooted trees. IEEE Trans. Inf. Theory **64**, 5570–5580 (2018)
10. T. Kayano, M. Yamasaki, Discrete biharmonic Green function β. Mem. Fac. Sci. Shimane Univ. **19**, 1–10 (1985)
11. C.O. Kiselman, Subharmonic functions on discrete structures, in *Harmonic Analysis, Signal Processing, and Complexity*. Progr. Math., vol. 238 (Birkhäuser, Boston, 2005), pp. 67–80
12. M.A. Picardello, W. Woess, Boundary representations of λ-harmonic and polyharmonic functions on trees. Potential Anal. **51**, 541–561 (2019)
13. M. Venkataraman, Laurent decomposition for harmonic and biharmonic functions in an infinite network. Hokkaido Math. J. **42**, 345–356 (2013)
14. A.I. Voronkova, A solution of the first boundary value problem for a biharmonic equation of the halfaxis by means of the network Green's function. Ž. Vyčisl. Mat. i Mat. Fiz. **11**, 667–676 (1971) [English transl. in USSR Comp. Math. Math. Phys. **11**(3), 156–167 (1971)]
15. W. Woess, *Denumerable Markov Chains. Generating functions, Boundary Theory, Random Walks on Trees* (European Math. Soc. Publishing House, Zürich, 2009)
16. M. Yamasaki, Biharmonic Green function of an infinite network. Mem. Fac. Sci. Shimane Univ. **14**, 55–62 (1980)

Interacting Electrons in a Random Medium: A Simple One-Dimensional Model

Frédéric Klopp and Nikolaj A. Veniaminov

2010 Mathematics Subject Classification Primary 81V70, 82B44; Secondary 82D30

1 Introduction: The Model and the Main Results

On $\mathbb{R}$, consider a Poisson point process $d\mu(\omega)$ of intensity μ. Let $(x_k(\omega))_{k\in\mathbb{Z}}$ denote its support (i.e., $d\mu(\omega) = \sum_{k\in\mathbb{Z}} \delta_{x_k(\omega)}$), the points being ordered increasingly.

On $L^2(\mathbb{R})$, define the Luttinger-Sy or pieces model (see e.g. [13, 14]), that is, the random operator

$$H_\omega = \bigoplus_{k\in\mathbb{Z}} -\Delta^D_{|[x_k,x_{k+1}]}$$

where, for an interval I, $-\Delta^D_{|I}$ denotes the Dirichlet Laplacian on I.

Pick $L > 0$ and let $\Lambda = \Lambda_L = [0, L]$. Restrict H_ω to Λ with Dirichlet boundary conditions: on $\mathfrak{H} := L^2(\Lambda)$, define

This work is partially supported by the grant ANR-08-BLAN-0261-01. The authors also acknowledge the support of the IMS (NU Singapore) where part of this work was done. F.K. thanks T. Duquesne for his explaining the Palm formula.

F. Klopp (✉)
Sorbonne Université, Université Paris Diderot, CNRS, Institut de Mathématiques de Jussieu - Paris Rive Gauche, F-75005 Paris, France
e-mail: frederic.klopp@imj-prg.fr

N. A. Veniaminov
CEREMADE, UMR CNRS 7534, Université Paris IX Dauphine, Place du Maréchal De Lattre De Tassigny, F-75775 Paris cedex 16, France
e-mail: veniaminov@ceremade.dauphine.fr

N. Anantharaman et al. (eds.), *Frontiers in Analysis and Probability*,
https://doi.org/10.1007/978-3-030-56409-4_5

$$H_\omega(L) = H_\omega(\Lambda) = \bigoplus_{k_- - 1 \leqslant k \leqslant k_+} -\Delta^D_{|\Delta_k(\omega)} \tag{1.1}$$

where we have defined $\Delta_k(\omega) := [x_k(\omega), x_{k+1}(\omega)]$ to be the *k-th piece* and we have set

$$k_- = \min\{k; x_k > 0\}, \quad x_{k_- - 1} = 0,$$
$$k_+ = \max\{k; x_k < L\}, \quad x_{k_+ + 1} = L.$$

From now on, we let $m(\omega)$ be the number of pieces and renumber them from 1 to $m(\omega)$ (i.e., $k_- = 2$ and $k_+ = m(\omega)$). For L large, with probability $1 - O(L^{-\infty})$, one has $m(\omega) = \mu L + O(L^{2/3})$.

The pieces model admits an integrated density of states that can be computed explicitly (see Section 2.2 or [13, 20]), namely,

$$\begin{aligned} N_\mu(E) &:= \lim_{L \to +\infty} \frac{\#\{\text{eigenvalues of } H_\omega(L) \text{ in } (-\infty, E]\}}{L} \\ &= \frac{\mu \cdot \exp(-\mu \ell_E)}{1 - \exp(-\mu \ell_E)} \mathbf{1}_{E \geqslant 0} \quad \text{where} \quad \ell_E := \frac{\pi}{\sqrt{E}}. \end{aligned} \tag{1.2}$$

1.1 Interacting Electrons

Consider first n free electrons restricted to the box Λ in the background Hamiltonian $H_\omega(\Lambda)$, that is, on the space

$$\mathfrak{H}^n(\Lambda) = \mathfrak{H}^n(\Lambda_L) = \bigwedge_{j=1}^{n} L^2(\Lambda) = L^2_-(\Lambda^n), \tag{1.3}$$

consider the operator

$$H^0_\omega(\Lambda, n) = \sum_{i=1}^{n} \underbrace{\mathbf{1}_{\mathfrak{H}} \otimes \ldots \otimes \mathbf{1}_{\mathfrak{H}}}_{i-1 \text{ times}} \otimes H_\omega(\Lambda) \otimes \underbrace{\mathbf{1}_{\mathfrak{H}} \otimes \ldots \otimes \mathbf{1}_{\mathfrak{H}}}_{n-i \text{ times}}. \tag{1.4}$$

This operator is self-adjoint and lower semi-bounded. Let $E^0_\omega(\Lambda, n)$ be its ground state energy and $\Psi^0_\omega(\Lambda, n)$ be its ground state.

To $H^0_\omega(\Lambda, n)$, we now add a repulsive finite range pair interaction potential. Therefore, pick $U : \mathbb{R} \to \mathbb{R}$ satisfying

(HU): U is a repulsive (i.e., non-negative), even pair interaction potential decaying sufficiently fast at infinity. More precisely, we assume

$$x^3 \int_x^{+\infty} U(t)\mathrm{d}t \xrightarrow[x\to+\infty]{} 0. \tag{1.5}$$

To control the possible local singularities of the interactions, we require that $U \in L^p(\mathbb{R})$ for some $p \in (1, +\infty]$.

On $\mathfrak{H}^n(\Lambda)$, we define

$$H^U_\omega(\Lambda, n) = H^0_\omega(\Lambda, n) + W_n \tag{1.6}$$

where

$$W_n(x^1, \cdots, x^n) := \sum_{i<j} U(x^i - x^j) \tag{1.7}$$

on the domain

$$\mathcal{D}^n(\Lambda) := \mathcal{C}_0^\infty\left(\left(\bigcup_{k=1}^{m(\omega)}]x_k, x_{k+1}[\right)^n\right) \cap \mathfrak{H}^n(\Lambda). \tag{1.8}$$

As U is non-negative, $H^U_\omega(\Lambda, n)$ is non-negative. From now on, we let $H^U_\omega(\Lambda, n)$ be the Friedrichs extension of this operator. As W_n is a sum of pair interactions, the fact that $U \in L^p(\mathbb{R})$ for some $p > 1$ (see assumption **(HU)**) guarantees that W_n is $H^0_\omega(\Lambda, n)$-form bounded with relative form bound 0 (see, e.g., [5, section 1.2]). Thus, the form domain of the operator $H^U_\omega(\Lambda, n)$ is

$$\mathfrak{H}^n_\infty(\Lambda) := \left(H^1_0\left(\bigcup_{k=1}^{m(\omega)}]x_k, x_{k+1}[\right)\right)^{\otimes n} \cap \mathfrak{H}^n(\Lambda). \tag{1.9}$$

Moreover, $H^U_\omega(\Lambda, n)$ admits $\mathcal{D}^n(\Lambda)$ as a form core (see, e.g., [5, section 1.3]) and it has a compact resolvent, thus, only discrete spectrum.

We define $E^U_\omega(\Lambda, n)$ to be its ground state energy, that is,

$$E^U_\omega(\Lambda, n) := \inf_{\substack{\Psi\in\mathcal{D}^n(\Lambda)\\ \|\Psi\|=1}} \langle H^U_\omega(\Lambda, n)\Psi, \Psi\rangle \tag{1.10}$$

and $\Psi^U_\omega(\Lambda, n)$ to be a ground state, i.e., to be an eigenfunction associated to the eigenvalue $E^U_\omega(\Lambda, n)$.

By construction, there is no unique continuation principle for the pieces model (as the union of disjoint non-empty intervals is not connected); so, one should not expect uniqueness for the ground state. Nevertheless due to the properties of the Poisson process, for the non-interacting system, one easily sees that the ground state

$\Psi^0_\omega(\Lambda, n)$ is unique ω almost surely (see Section 2.4). For the interacting system, it is not as clear. Nonetheless, one proves

Theorem 1.1 (Almost Sure Non-degeneracy of the Ground State) *Suppose that U is real analytic. Then, ω-almost surely, for any L and n, the ground state of $H^U_\omega(L, n)$ is non-degenerate.*

For a general U, while we don't know whether the ground state is degenerate or not, our analysis will show where the degeneracy may come from: we shall actually write $\mathfrak{H}^n(\Lambda)$ as an orthogonal sum of subspaces invariant by $H^U_\omega(L, n)$ such that on each such subspace, the ground state of $H^U_\omega(L, n)$ is unique. This will enable us to show that all the ground states of $H^U_\omega(L, n)$ on $\mathfrak{H}^n(\Lambda)$ are very similar to each other, i.e., they differ only by a small number of particles.

The goal of the present paper is to understand the thermodynamic limits of $E^U_\omega(\Lambda, n)$ and $\Psi^U_\omega(\Lambda, n)$. As usual, we define the *thermodynamic limit* to be the limit $L \to \infty$ and $n/L \to \rho$ where ρ is a positive constant. The constant ρ is the *density of particles*.

We will describe the thermodynamic limits of $E^U_\omega(\Lambda, n)$, or rather $n^{-1}E^U_\omega(\Lambda, n)$, and $\Psi^U_\omega(\Lambda, n)$ when ρ is positive and small (but independent of L and n). We will be specially interested in the influence of the interaction U, i.e., we will compare the thermodynamic limits for the non-interacting and the interacting systems.

1.2 The Ground State Energy Per Particle

Our first result describes the thermodynamic limit of $n^{-1}E^U_\omega(\Lambda, n)$ when we assume the density of particles n/L to be ρ. For the sake of comparison, we also included the corresponding result on the ground state energy of the free particles, i.e., on $n^{-1}E^0_\omega(\Lambda, n)$.

We prove

Theorem 1.2 *Under the assumptions made above, the following limits exist ω-almost surely and in L^1_ω*

$$\mathcal{E}^0(\rho, \mu) := \lim_{\substack{L\to+\infty \\ n/L\to\rho}} \frac{E^0_\omega(\Lambda, n)}{n} \quad \text{and} \quad \mathcal{E}^U(\rho, \mu) := \lim_{\substack{L\to+\infty \\ n/L\to\rho}} \frac{E^U_\omega(\Lambda, n)}{n} \tag{1.11}$$

and they are independent of ω.

In [21] (see also [20]), the almost sure existence of the thermodynamic limit of the ground state energy per particle is established for quite general systems of interacting electrons in a random medium if one assumes that the interaction has compact support. For decaying interactions (as in **(HU)**), only the L^2_ω convergence is proved. The improvement needed on the results of [21] to obtain the almost sure convergence is the purpose of Theorem 5.1.

In [3], the authors study the existence of the above limits in the grand canonical ensemble for Coulomb interactions.

The energy $\mathcal{E}^0(\rho,\mu)$ can be computed explicitly for our model (see Section 2.4.1). We shall obtain a two-term asymptotic formula for $\mathcal{E}^U(\rho,\mu)$ in the case when the disorder is not too large and the Fermi length $\ell_{\rho,\mu}$ is sufficiently large.

Define

- the *effective density* is defined as the ratio of the density of particles to the density of impurities, i.e., $\rho_\mu = \dfrac{\rho}{\mu}$,
- the *Fermi energy* $E_{\rho,\mu}$ is the unique solution to $N_\mu(E_{\rho,\mu}) = \rho$,
- the *Fermi length* $\ell_{\rho,\mu} := \ell_{E_{\rho,\mu}}$ where ℓ_E is defined in (1.2); the explicit formula for N_μ yields

$$\ell_{\rho,\mu} = \frac{1}{\mu}\left|\log\frac{\rho_\mu}{1+\rho_\mu}\right| = \frac{1}{\mu}\left|\log\frac{\rho}{\mu+\rho}\right|. \tag{1.12}$$

For the free ground state energy per particle, a direct computation using (1.2) yields

$$\mathcal{E}^0(\rho,\mu) = \frac{1}{\rho}\int_{-\infty}^{E_{\rho,\mu}} E\, dN_\mu(E) = E_{\rho,\mu}\left(1 + O\left(\sqrt{E_{\rho,\mu}}\right)\right) \tag{1.13}$$

We prove

Theorem 1.3 *Under the assumptions made above, for $\mu > 0$ fixed, one computes*

$$\mathcal{E}^U(\rho,\mu) = \mathcal{E}^0(\rho,\mu) + \pi^2\,\gamma_*^\mu\,\mu^{-1}\,\rho_\mu\,\ell_{\rho,\mu}^{-3}\,(1+o(1)) \quad \textit{where} \quad o(1) \underset{\rho_\mu\to 0}{\longrightarrow} 0. \tag{1.14}$$

The positive constant γ_*^μ depends solely on U and μ; it is defined in (1.17) *below.*

At fixed disorder, in the small density regime, the Fermi length is large and the Fermi energy is small. Moreover, the shift of ground state energy (per particle) due to the interaction is exponentially small compared to the free ground state energy: indeed, it is of order $\rho|\log\rho|^{-3}$ while the ground state energy is of order $|\log\rho|^{-2}$.

For fixed μ, a coarse version of (1.14) was established, in the PhD thesis of the second author [20], namely, for ρ sufficiently small, one has

$$\frac{1}{C_\mu}\rho|\log\rho|^{-3} \leqslant \mathcal{E}^U(\rho,\mu) - \mathcal{E}^0(\rho,\mu) \leqslant C_\mu\,\rho|\log\rho|^{-3}.$$

Moreover, from [21, Propositions 3.6 and 3.7], we know that the function $\rho \mapsto \mathcal{E}^U(\rho,\mu)$ is a non-decreasing continuous function and that the function $r \mapsto \mathcal{E}^U(r^{-1},\mu)$ is convex.

Let us now define the constant γ_*^μ. Therefore, we prove

Proposition 1.4 *Consider two electrons in $[0,\ell]$ interacting via an even non-negative pair potential $U \in L^p(\mathbb{R}^+)$ for some $p > 1$ and such that*

$$\int_{\mathbb{R}} x^2 U(x) dx < +\infty.$$

That is, on $\mathfrak{H}^2([0,\ell]) = L^2([0,\ell]) \wedge L^2([0,\ell])$, *consider the Hamiltonian*

$$\left(-\Delta^D_{x_1|[0,\ell]}\right) \otimes \mathbf{1}_{\mathfrak{H}} + \mathbf{1}_{\mathfrak{H}} \otimes \left(-\Delta^D_{x_2|[0,\ell]}\right) + U(x_1 - x_2), \tag{1.15}$$

i.e., the Friedrichs extension of the same differential expression defined on the domain $\mathcal{D}^2([0,\ell])$ *(see* (1.8)*).*

For large ℓ, $E^U([0,\ell],2)$, *the ground state energy of this Hamiltonian admits the following expansion*

$$E^U([0,\ell],2) = \frac{5\pi^2}{\ell^2} + \frac{\gamma}{\ell^3} + o\left(\frac{1}{\ell^3}\right) \tag{1.16}$$

where $\gamma = \gamma(U) > 0$ *when* U *does not vanish a.e.*

Let us first notice that the expansion (1.16) immediately implies that $U \mapsto \gamma(U)$ is a non-decreasing concave function of the (non-negative) interaction potential U such that $\gamma(0) = 0$; for α small positive, one computes

$$\frac{\gamma(\alpha U)}{\alpha} = 10\pi^2 \int_{\mathbb{R}} x^2 U(x) dx \,(1 + O(\alpha)).$$

Concavity and monotony follow immediately from the definition of $E^U([0,\ell],2)$ and the form of (1.16).

In terms of γ, we then define

$$\gamma_*^\mu := 1 - \exp\left(-\frac{\mu\gamma}{8\pi^2}\right). \tag{1.17}$$

1.3 The Ground State: Its One- and Two-Particle Density Matrices

We shall now describe our results on the ground state. We start with a description of the spectral data of the one-particle Luttinger-Sy model. Then, we describe the non-interacting ground state.

1.3.1 The Spectrum of the One-Particle Luttinger-Sy Model

Let $(E^{\Lambda}_{j,\omega})_{j\geqslant 1}$ and $(\varphi^{\Lambda}_{j,\omega})_{j\geqslant 1}$, respectively, denote the eigenvalues (ordered increasingly) and the associated eigenfunctions of $H_\omega(\Lambda)$ (see (1.1)). Clearly, the eigenvalues and the eigenfunctions are explicitly computable from the points $(x_k)_{1\leqslant k\leqslant m(\omega)+1}$. In particular, one sees that the eigenvalues are simple ω almost surely.

As n/L is close to ρ and L is large, the n first eigenvalues are essentially all the eigenvalues below the Fermi energy $E_{\rho,\mu}$. These eigenvalues are the eigenvalues of $-\Delta^D_{|\Delta_k(\omega)}$ below $E_{\rho,\mu}$ for all the pieces $(\Delta_k(\omega))_{k_- -1\leqslant k\leqslant k_+}$ of length at least $\ell_{\rho,\mu}$ (see (1.2) and (1.13)). ω-Almost surely, the number of pieces $(\Delta_k(\omega))_{1\leqslant k\leqslant m(\omega)}$ longer than $\ell_{\rho,\mu}$ is asymptotic to n (see Section 2.3), the number of those longer than $2\ell_{\rho,\mu}$ to $\rho_\mu\, n$, the number of those longer than $3\ell_{\rho,\mu}$ to $\rho_\mu^2\, n$, etc. We refer to Section 2.2 for more details.

1.3.2 The Non-interacting Ground State

The ground state of the non-interacting Hamiltonian $H^0_\omega(\Lambda, n)$ is given by the (normalized) Slater determinant

$$\Psi^0_\omega(\Lambda, n) = \bigwedge_{j=1}^{n} \varphi^{\Lambda}_{j,\omega} = \frac{1}{\sqrt{n!}}\mathrm{Det}\left(\left(\varphi^{\Lambda}_{j,\omega}(x_k)\right)\right)_{1\leqslant j,k\leqslant n}. \tag{1.18}$$

Here and in the sequel, the exterior product is normalized so that the L^2-norm of the product be equal to the product of the L^2-norms of the factors (see (C.2) in Appendix C).

It will be convenient to describe the interacting ground state using its one-particle and two-particle reduced density matrices. Let us define these now (see Section 4 for more details). Let $\Psi \in \mathfrak{H}^n(\Lambda)$ be a normalized n-particle wave function. The corresponding *one-particle density matrix* is an operator on $\mathfrak{H}^1(\Lambda) = L^2(\Lambda)$ with the kernel

$$\gamma_\Psi(x, y) = \gamma^{(1)}_\Psi(x, y) = n \int_{\Lambda^{n-1}} \Psi(x, \tilde{x})\Psi^*(y, \tilde{x})d\tilde{x} \tag{1.19}$$

where $\tilde{x} = (x^2, \dots, x^n)$ and $d\tilde{x} = dx^2 \cdots dx^n$.

The *two-particle density matrix* of Ψ is an operator acting on $\mathfrak{H}^2(\Lambda) = \bigwedge_{j=1}^{2} L^2(\Lambda)$ and its kernel is given by

$$\gamma^{(2)}_\Psi(x^1, x^2, y^1, y^2) = \frac{n(n-1)}{2} \int_{\Lambda^{n-2}} \Psi(x^1, x^2, \tilde{x})\Psi^*(y^1, y^2, \tilde{x})d\tilde{x} \tag{1.20}$$

where $\tilde{x} = (x^3, \dots, x^n)$ and $d\tilde{x} = dx^3 \cdots dx^n$.

Both γ_Ψ and $\gamma_\Psi^{(2)}$ are positive trace class operators satisfying

$$\operatorname{Tr}\gamma_\Psi = n, \quad \text{and} \quad \operatorname{Tr}\gamma_\Psi^{(2)} = \frac{n(n-1)}{2}. \tag{1.21}$$

So, for the non-interacting ground state, using the description of the eigenvalues and eigenvectors of $H_\omega(\Lambda)$ given in Section 1.3.1, as a consequence of Proposition 4.8, we obtain that

$$\gamma_{\Psi_\omega^0(\Lambda,n)} = \sum_{j=1}^{n} \gamma_{\varphi_{j,\omega}^\Lambda} = \sum_{\ell_{\rho,\mu}\leqslant|\Delta_k(\omega)|<3\ell_{\rho,\mu}} \gamma_{\varphi^1_{\Delta_k(\omega)}} + \sum_{2\ell_{\rho,\mu}\leqslant|\Delta_k(\omega)|<3\ell_{\rho,\mu}} \gamma_{\varphi^2_{\Delta_k(\omega)}} + R^{(1)} \tag{1.22}$$

where

- $|\Delta_k(\omega)|$ denotes the length of the piece $\Delta_k(\omega)$;
- $\varphi^j_{\Delta_k(\omega)}$ denotes the j-th normalized eigenvector of $-\Delta^D_{|\Delta_k(\omega)}$;
- the operator $R^{(1)}$ is trace class and $\|R^{(1)}\|_{\mathrm{tr}} \leqslant 2\,n\,\rho_\mu^2$.

Here, $\|\cdot\|_{\mathrm{tr}}$ denotes the trace norm in the ambient space, i.e., in $L^2(\Lambda)$ for the one-particle density matrix, and in $L^2(\Lambda)\wedge L^2(\Lambda)$ for the two-particle density matrix.

For the two-particle density matrix, again as a consequence of Proposition 4.8, we obtain

$$\gamma^{(2)}_{\Psi_\omega^0(\Lambda,n)} = \frac{1}{2}(\mathrm{Id}-\mathrm{Ex})\left[\gamma_{\Psi_\omega^0(\Lambda,n)} \otimes \gamma_{\Psi_\omega^0(\Lambda,n)}\right] + R^{(2)} \tag{1.23}$$

where

- Id is the identity operator, Ex is the exchange operator on a two-particle space:

$$\mathrm{Ex}\,[f\otimes g] = g\otimes f, \quad f,g\in\mathfrak{H},$$

- the operator $R^{(2)}$ is trace class and $\|R^{(2)}\|_{\mathrm{tr}} \leqslant C_{\rho,\mu} n$.

One can represent graphically the ground state of the non- interacting system by representing the distribution of its particles within the pieces: in abscissa, one puts the length of the pieces, in ordinate, the number of particles the ground state puts in a piece of that length. Figure 1 shows the picture thus obtained.

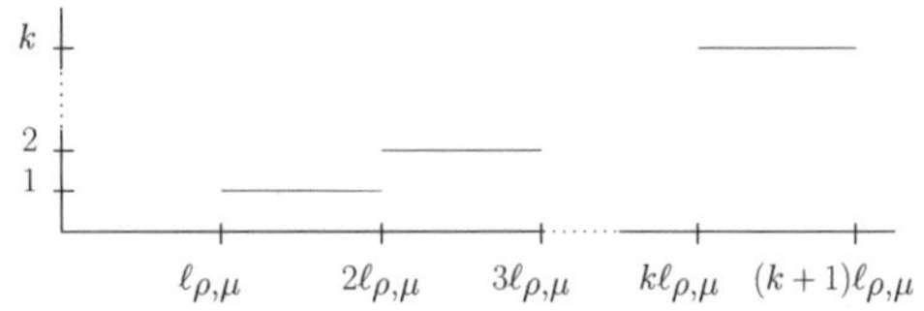

Fig. 1 The distribution of particles in the non-interacting ground state.

1.3.3 The Interacting Ground State

To describe the ground state of the interacting system, we shall describe its one-particle and two-particle reduced density matrices. Therefore, it will be useful to introduce the following approximate one-particle reduced density matrices.

For a piece $\Delta_k(\omega)$, let $\zeta^j_{\Delta_k(\omega)}$ be the j-th normalized eigenvector of $-\Delta^D_{|\Delta_k(\omega)\times\Delta_k(\omega)} + U$ acting on $L^2(\Delta_k(\omega)) \wedge L^2(\Delta_k(\omega))$. We note that, for $U = 0$, the two-particle ground state can be rewritten as $\zeta^{1,U=0}_{\Delta_k(\omega)} = \varphi^1_{\Delta_k(\omega)} \wedge \varphi^2_{\Delta_k(\omega)}$.

Define the following one-particle density matrix

$$\gamma_{\Psi^{\mathrm{opt}}_{\Lambda,n}} = \sum_{\ell_{\rho,\mu}-\rho_\mu\gamma^\mu_* \leqslant |\Delta_k(\omega)| \leqslant 2\ell_{\rho,\mu}-\log(1-\gamma^\mu_*)} \gamma_{\varphi^1_{\Delta_k(\omega)}} + \sum_{2\ell_{\rho,\mu}-\log(1-\gamma^\mu_*) \leqslant |\Delta_k(\omega)|} \gamma_{\zeta^1_{\Delta_k(\omega)}}. \tag{1.24}$$

Because of the possible long range of the interaction U (see the remarks following Theorem 1.5 below), to describe our results precisely, it will be useful to introduce trace norms reduced to certain pieces. For $\ell \geqslant 0$, we define the projection onto the pieces shorter than ℓ

$$\mathbf{1}^1_{<\ell} = \sum_{|\Delta_k(\omega)|<\ell} \mathbf{1}_{\Delta_k(\omega)}. \tag{1.25}$$

We shall use the following function to control remainder terms: define

$$Z(x) = \sup_{x\leqslant v} \left(v^3 \int_v^{+\infty} U(t)dt \right). \tag{1.26}$$

Under assumption (HU), the function Z is continuous and monotonously decreasing on $[0, +\infty)$ and tends to 0 at infinity.

We prove

Theorem 1.5 *Fix $\mu > 0$. Assume (HU) holds. Then, there exist $\rho_0 > 0$ and $C > 0$ such that, for $\rho \in (0, \rho_0)$, ω-a.s., one has*

$$\limsup_{\substack{L\to+\infty \\ n/L\to\rho}} \frac{1}{n} \left\| \left(\gamma_{\Psi^U_\omega(\Lambda,n)} - \gamma_{\Psi^{opt}_{\Lambda,n}}\right) \mathbf{1}^1_{<\ell_{\rho,\mu}+C} \right\|_{tr} \leqslant \frac{1}{\rho_0} \max\left(\frac{\rho_\mu}{\ell_{\rho,\mu}}, \sqrt{\rho_\mu\, Z(\ell_{\rho,\mu})} \right),$$

$$\limsup_{\substack{L\to+\infty \\ n/L\to\rho}} \frac{1}{n} \left\| \left(\gamma_{\Psi^U_\omega(\Lambda,n)} - \gamma_{\Psi^{opt}_{\Lambda,n}}\right) \left(\mathbf{1} - \mathbf{1}^1_{<\ell_{\rho,\mu}+C}\right) \right\|_{tr} \leqslant \frac{1}{\rho_0} \max\left(\frac{\rho_\mu}{\ell_{\rho,\mu}}, \rho_\mu \sqrt{Z(\ell_{\rho,\mu})} \right).$$

Here, $\|\cdot\|_{tr}$ denotes the trace norm in $L^2(\Lambda)$.

This result calls for some comments. Let us first note that, if Z, that is, U, decays sufficiently fast at infinity, typically exponentially fast with a large rate, then the two estimates in Theorem 1.5 can be united into

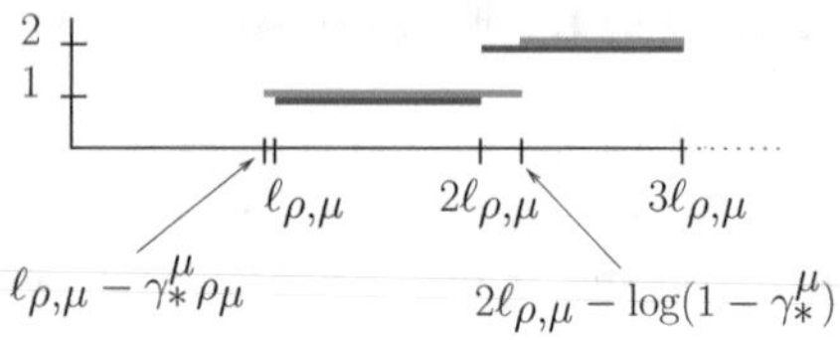

Fig. 2 The distribution of particles in the interacting ground state.

$$\limsup_{\substack{L\to+\infty\\ n/L\to\rho_\mu}}\frac{1}{n}\left\|\gamma_{\Psi^U_\omega(\Lambda,n)}-\gamma_{\Psi^{\text{opt}}_{\Lambda,n}}\right\|_{\text{tr}}\leqslant C\frac{\rho_\mu}{\ell_{\rho,\mu}}.$$

In this case, Theorem 1.5 can be summarized graphically. In Figure 2, using the same representation as in Figure 1, we compare the non-interacting and the interacting ground state. The non-interacting ground state distribution of particles is represented in blue, the interacting one in green. We assume that U has compact support and restrict ourselves to pieces shorter than $3\ell_{\rho,\mu}$.

Indeed, in this case, comparing (1.22) and (1.24), we see

$$\begin{aligned}\gamma_{\Psi^0_\omega(\Lambda,n)}-\gamma_{\Psi^{\text{opt}}_{\Lambda,n}}&=\sum_{2\ell_{\rho,\mu}-\log(1-\gamma_*^\mu)\leqslant|\Delta_k(\omega)|}\left(\gamma_{\varphi^1_{\Delta_k(\omega)}}+\gamma_{\varphi^2_{\Delta_k(\omega)}}-\gamma_{\zeta^1_{\Delta_k(\omega)}}\right)\\&\quad-\sum_{\ell_{\rho,\mu}-\rho_\mu\gamma_*^\mu\leqslant|\Delta_k(\omega)|\leqslant\ell_{\rho,\mu}}\gamma_{\varphi^1_{\Delta_k(\omega)}}\\&\quad+\sum_{2\ell_{\rho,\mu}\leqslant|\Delta_k(\omega)|\leqslant2\ell_{\rho,\mu}-\log(1-\gamma_*^\mu)}\gamma_{\varphi^2_{\Delta_k(\omega)}}+\tilde R^{(1)}\end{aligned}\tag{1.27}$$

where $\tilde R^{(1)}$ satisfies the same properties as $R^{(1)}$ in (1.22).

Thus, to obtain $\gamma_{\Psi^{\text{opt}}_{\Lambda,n}}$ from $\gamma_{\Psi^0_\omega(\Lambda,n)}$, we have displaced (roughly) $\gamma_*^\mu\rho_\mu n$ particles living in pieces of length within $[2\ell_{\rho,\mu},2\ell_{\rho,\mu}-\log(1-\gamma_*^\mu)]$ (i.e., pieces containing exactly two states below energy $E_{\rho,\mu}$ and the energy of the top state stays above $E_{\rho,\mu}\left(1+\frac{\log(1-\gamma_*^\mu)}{\ell_{\rho,\mu}}\right)$ up to smaller order terms in $\ell_{\rho,\mu}^{-1}$) to pieces having lengths within $[\ell_{\rho,\mu}-\rho\gamma_*^\mu,\ell_{\rho,\mu}]$ (i.e., having ground state energy within the interval $\left[E_{\rho,\mu},E_{\rho,\mu}\left(1+\frac{2\rho\gamma_*^\mu}{\ell_{\rho,\mu}}\right)\right]$ up to smaller order terms in $\ell_{\rho,\mu}^{-1}$). In the remaining of (roughly) $(1-\gamma_*^\mu)\rho n$ pieces containing exactly two states below energy $E_{\rho,\mu}$ (that is, pieces of length within $[2\ell_{\rho,\mu}-\log(1-\gamma_*^\mu),3\ell_{E_{\rho,\mu}}]$ or alternatively those with the top state below $E_{\rho,\mu}\left(1+\frac{\log(1-\gamma_*^\mu)}{\ell_{\rho,\mu}}\right)$ (up to smaller order terms in $\ell_{\rho,\mu}^{-1}$), we have substituted the free two-particle ground state (given by the anti-symmetric tensor product of the first two Dirichlet levels in this piece) by the ground state of the interacting system (1.15). In particular, we compute (remark that the first sum in (1.27) contributes only to the error term according to Corollary 6.12)

$$\lim_{\substack{L\to+\infty\\ n/L\to\rho}} \frac{1}{n}\left\|\gamma_{\Psi^0_\omega(\Lambda,n)}-\gamma_{\Psi^{\text{opt}}_{\Lambda,n}}\right\|_{\text{tr}} = 2\gamma_*^\mu\rho_\mu + O\left(\frac{\rho_\mu}{\ell_{\rho,\mu}}\right),$$

and, recalling (1.23), we then compute

$$\lim_{\substack{L\to+\infty\\ n/L\to\rho}} \frac{1}{n^2}\left\|\gamma^{(2)}_{\Psi^0_\omega(\Lambda,n)}-\frac{1}{2}(\text{Id}-\text{Ex})\left[\gamma_{\Psi^{\text{opt}}_{\Lambda,n}}\otimes\gamma_{\Psi^{\text{opt}}_{\Lambda,n}}\right]\right\|_{\text{tr}} = 2\gamma_*^\mu\rho_\mu + O\left(\frac{\rho_\mu}{\ell_{\rho,\mu}}\right). \tag{1.28}$$

So the main effect of the interaction is to shift a macroscopic (though small when ρ_μ is small) fraction of the particles to different pieces.

Let us now discuss what happens when the interaction does not decay so fast, typically, if it decays only polynomially. In this case, Theorem 1.5 tells us that one has to distinguish between short and long pieces. In the long pieces, the description of the ground state is still quite good as the error estimate is still of order $o(\rho_\mu)$. Of course, this result only tells us something for the pieces of length at most $3\ell_{\rho,\mu}$: the larger ones are very few, thus, can only carry so few particles (see Lemma 3.27) that these can be integrated into the remainder term. For short intervals, the situation is quite different. Here, the remainder term becomes much larger, only of order $O\left(\sqrt{\rho_\mu}\ell_{\rho,\mu}^{-k/2}\right)$ if $Z(x)\asymp x^{-k}$ at infinity. This loss is explained in the following way. The short pieces carry the majority of the particles. When U is of longer range, particles in rather distant pieces start to interact in a way that is not negligible with respect to the second term of the expansion (1.14) (which gives an average surplus of energy per particle for the interacting ground state compared to the free one); thus, it may become energetically profitable to relocate some of these particles to new pieces so as to minimize the interaction energy. When the range of the interaction increases, the ground state will relocate more and more particles. Nevertheless, the shift in energy will still be smaller than the correction term obtained by relocating some of the particles living in pairs in not too long intervals; this is going to be the case as long as U satisfies the decay assumption **(HU)**. When U decays slower than that, the main correction to the interacting ground state energy per particle can be expected to be given by the relocation of many particles living alone in their piece to new pieces so as to diminish the interaction energy.

We also obtain an analogue of Theorem 1.5 for the two-particle density matrix of the ground state Ψ^U. We prove

Theorem 1.6 *Fix $\mu>0$. Assume (HU) holds. Then, there exist $\rho_0>0$ such that, for $\rho\in(0,\rho_0)$, ω-a.s., one has*

$$\limsup_{\substack{L\to+\infty\\ n/L\to\rho}} \frac{1}{n^2}\left\|\left(\gamma^{(2)}_{\Psi^U_\omega(\Lambda,n)}-\frac{1}{2}(\text{Id}-\text{Ex})\left[\gamma_{\Psi^{opt}_{\Lambda,n}}\otimes\gamma_{\Psi^{opt}_{\Lambda,n}}\right]\right)\mathbf{1}^2_{<\ell_{\rho,\mu}+C}\right\|_{tr}$$
$$\leqslant \frac{1}{\rho_0}\max\left(\frac{\rho_\mu}{\ell_{\rho,\mu}},\sqrt{\rho_\mu}\,Z(\ell_{\rho,\mu})\right)$$

and

$$\limsup_{\substack{L\to+\infty\\ n/L\to\rho}} \frac{1}{n^2}\left\|\left(\gamma^{(2)}_{\Psi^U_\omega(\Lambda,n)} - \frac{1}{2}(\mathrm{Id}-\mathrm{Ex})\left[\gamma_{\Psi^{opt}_{\Lambda,n}} \otimes \gamma_{\Psi^{opt}_{\Lambda,n}}\right]\right)\left(\mathbf{1} - \mathbf{1}^2_{<\ell_{\rho,\mu}+C}\right)\right\|_{tr}$$
$$\leqslant \frac{1}{\rho_0}\max\left(\frac{\rho_\mu}{\ell_{\rho,\mu}}, \rho_\mu\sqrt{Z(\ell_{\rho,\mu})}\right)$$

where, for $\ell \geqslant 0$, we recall that $\|\cdot\|_{tr}$ denotes the trace norm in $L^2(\Lambda) \wedge L^2(\Lambda)$, recall (1.25) *and define*

$$\mathbf{1}^2_{<\ell} = \mathbf{1}^1_{<\ell} \otimes \mathbf{1}^1_{<\ell}. \tag{1.29}$$

1.4 Discussion and Perspectives

While a very large body of mathematical works has been devoted to one- particle random Schrödinger operators (see, e.g., [10, 16]), there are only few works dealing with many interacting particles in a random medium (for the case of finitely many particles, see, for example, [1] or [4]).

The general Hamiltonian describing n electrons in a random background potential V_ω interacting via a pair potential U can be described as follows. In a d-dimensional domain Λ, consider the operator

$$H_\omega(\Lambda, n) = -\Delta_{nd}\big|_{\Lambda^n} + \sum_{i=1}^{n} V_\omega(x^i) + \sum_{i<j} U(x^i - x^j),$$

where, for $j \in \{1, \dots, n\}$, x^j denotes the coordinates of the j-th particle. The operator $H_\omega(\Lambda, n)$ acts on a space of totally anti-symmetric functions $\bigwedge_{i=1}^{n} L^2(\Lambda)$ which reflects the electronic nature of particles.

The general problem is to understand the behavior of $H_\omega(\Lambda, n)$ in the thermodynamic limit $\Lambda \to \infty$ while $n/|\Lambda| \to \rho > 0$; ρ is the particle density. One of the questions of interest is that of the behavior of the ground state energy, say, $E_\omega(\Lambda, n)$ and of the ground state $\Psi_\omega(\Lambda, n)$.

While the thermodynamic limit is known to exist for various quantities and in various settings (see [21] for the micro-canonical ensemble that we study in the present paper and [3] for the grand canonical ensemble), we don't know of examples, except for the model studied in the present paper, where the limiting quantities have been studied. In particular, it is of interest to study the dependence of these limiting quantities in the different physical parameters like the density of particles, the strength of the disorder or the interaction potential.

As we shall argue now, for these questions to be tractable, one needs a good description of the spectral data of the underlying one-particle random model.

1.4.1 Why the Pieces Model?

In order to tackle the question of the behavior of n-electron ground state, let us first consider the system without interactions. This is not equivalent to a one-particle system as Fermi-Dirac statistics play a crucial role.

Let us assume our one-particle model is ergodic and admits an integrated density of states (see (1.2) and e.g. [11, 16]). As described above for the pieces model, the ground state of the n non-interacting electrons is given by (1.18) and its energy per particle is given by

$$\frac{E^0_\omega(\Lambda, n)}{n} = \frac{1}{n}\sum_{j=1}^{n} E^\Lambda_{j,\omega} = |\Lambda| \int_{-\infty}^{E^\Lambda_{n,\omega}} E\, d\left[\frac{\#\{\text{eigenvalues of } H_\omega(\Lambda) \text{ below } E\}}{|\Lambda|}\right] \tag{1.30}$$

where $E^\Lambda_{n,\omega}$ is the n-th eigenvalue of the one- particle random Hamiltonian $H_\omega(\Lambda)$, i.e., the smallest energy E such that

$$\frac{\#\{\text{eigenvalues of } H_\omega(\Lambda) \text{ below } E\}}{|\Lambda|} = \frac{n}{|\Lambda|}. \tag{1.31}$$

Here, we have kept the notations of the beginning of Section 1.3.

The existence of the density of states, say $N(E)$, (see (1.2)), then, ensures the convergence of $E(\Lambda, n)$ to a solution to the equation $N(E) = \rho$, say E_ρ. Thus, to control the non-interacting ground state, one needs to control all (or at least most of) the energies of the random operator $H_\omega(\Lambda)$ up to some macroscopic energy E_ρ. In particular, one needs to control simultaneously a number of energies of $H_\omega(\Lambda)$ that is of size the volume of Λ.

To the best of our knowledge, up to now, there are no available mathematical results that give the simultaneous control over that many eigenvalues for general random systems. The results dealing with the spectral statistics of (one-particle) random models deal with much smaller intervals: in [15], eigenvalues are controlled in intervals of size $K/|\Lambda|$ for arbitrarily large K if Λ is sufficiently large; in [7, 8], the interval is of size $|\Lambda|^{1-\beta}$ for some not too large positive β.

The second problem is that all these results only give a very rough picture of the eigenfunctions, a picture so rough that it actually is of no use to control the effect of the interaction on such states: the only information is that the eigenstates live in regions of linear size at most $\log|\Lambda|$ and decay exponentially outside such regions (see, e.g., [7] and the references therein).

The pieces model that we deal with in the present paper exhibits the typical behavior of a random system in the localized regime: for $H_\omega(\Lambda)$,

- the eigenfunctions are localized (on a scale $\log|\Lambda|$)
- the localization centers and the eigenvalues satisfy Poisson statistics.

The advantage of the pieces model is that the eigenfunctions and eigenvalues are known explicitly and easily controlled. This is a consequence of the fact that a crucial quantum phenomenon is missing in the pieces model, namely, tunneling. Of course, once the particles do interact with each other, tunneling is again re-enabled.

All of this could lead one to think that the pieces model is very particular. Actually, at low energies, general one-dimensional random models exhibit the same characteristics as the pieces model up to some exponentially small errors which are essentially due to tunneling (see [12]).

It seems reasonable to guess that the behavior will be comparable for general random operators in higher dimensions and, thus, that the results of the present paper on interacting electrons in a random potentials should find their analogues for these models.

1.4.2 Outline of the Paper

In Section 2, after rescaling the parameters of the problem so as to send μ to 1 and ρ to ρ/μ, we first discuss the validity of our results in a more general asymptotic regime in μ and ρ. We, then, gather some basic but crucial statistical properties of the distribution of the pieces. We first describe the free electrons. For the pieces model, a statistical analysis of the distribution of pieces gives exact expressions for the one-particle integrated density of states and the Fermi energy in Proposition 2.6. We also study the non- interacting model and introduce notations for later use.

In Section 3, we first introduce the occupation numbers (i.e., the number of particles a given state puts in each piece); the existence of the occupation numbers is tantamount to the existence of a particular orthogonal sum decomposition of the Hamiltonian $H^U_\omega(\Lambda, n)$. We prove that the ground state of $H^U_\omega(\Lambda, n)$ restricted to a fixed occupation space is non- degenerate and, from this result, derive Theorem 1.1, the almost sure non-degeneracy of the ground state for real analytic interaction.

Next, still in Section 3, we prove the asymptotic formula for the interacting ground state energy per particle. The proof relies essentially on the minimizing properties of the ground state. This minimizing property yields a good description for the occupation numbers associated to a ground state. To get this description, we first study the ground state of the Hamiltonian $H^{U^p}_\omega(\Lambda, n)$ where the interactions have been cut off at infinity (i.e., U^p is compactly supported). We construct an approximate ground state Ψ^{opt} which can essentially be thought of as the ground state for the Hamiltonian $H^{U^p}_\omega(\Lambda, n)$ restricted to the pieces shorter than $3\ell_{\rho,\mu}$. Then, letting $W^r(\Lambda, n) := H^U_\omega(\Lambda, n) - H^{U^p}_\omega(\Lambda, n)$ be the long range behavior of the interactions, one has

$$E^{U^p}_\omega(\Lambda, n) \leqslant E^U_\omega(\Lambda, n) \leqslant \langle H^{U^p}_\omega(\Lambda, n)\Psi^{\text{opt}}, \Psi^{\text{opt}}\rangle + \langle W^r(\Lambda, n)\Psi^{\text{opt}}, \Psi^{\text{opt}}\rangle$$

The minimizing property of Ψ^{opt} yields

$$E_{\omega}^{U^p}(\Lambda, n) \geqslant \langle H_{\omega}^{U^p}(\Lambda, n)\Psi^{\text{opt}}, \Psi^{\text{opt}}\rangle + n\, o(\rho_\mu\, \mu^{-1}\, \ell_{\rho,\mu}^{-3})$$

(see Theorem 3.28).

On the other hand, the decay assumption **(HU)** on U and the explicit construction of Ψ^{opt} yield

$$\langle W^r(\Lambda, n)\Psi^{\text{opt}}, \Psi^{\text{opt}}\rangle = n\, o(\rho_\mu\, \mu^{-1}\, \ell_{\rho,\mu}^{-3})$$

(see Proposition 2.7). This yields the proof of Theorem 1.3.

In the course of these proofs, we also prove a certain number of estimates on the distance between the occupation numbers of the interacting ground state(s) to the state Ψ^{opt}.

Section 4 is devoted to the proofs of Theorems 1.5 and 1.6. Therefore, we transform the bounds of the distance between occupation numbers into bounds on the trace class norms of the difference between the one- (and the two-) particle densities of the interacting ground state(s) and the state Ψ^{opt}.

In Theorems 4.2 (resp. Theorem 4.4), we derive general formulas for the one-particle (resp. two particles) density of a state expressed in a certain well-chosen basis of $\mathfrak{H}^n(\Lambda)$. One of the main steps on the path going from occupation number bounds to the trace class norm bounds is to prove that, in most pieces, once the particle number is known, the state must be in the ground state for the given particle number. This is the purpose of Lemma 4.12; it relies on the minimizing properties of the ground state; actually, it is proved for a larger set of states, states satisfying a certain energy bound.

We then use Theorems 4.2 (resp. Theorem 4.4) to derive Theorems 1.5 (resp. Theorem 1.6).

Section 5 is devoted to the proof of the almost sure convergence of the ground state energy per particle. The proof is essentially identical to that found in [21] except for the sub-additive estimate crucial to the proof. This estimate is provided by Theorem 5.1.

In Section 6, we prove Proposition 1.4 as well as a number of estimates on the ground states and ground state energies for a finite number of electrons living in a fixed number of pieces and interacting.

In three appendices, we gather a number of results used in the main body of paper. In Appendix A, we prove the results on the statistics of the pieces stated in Section 2. Appendix B is devoted to a simple technical lemma used intensively in the derivation of Theorems 1.5 and 1.6 in Section 4. Appendix C is devoted to anti-symmetric tensor products.

2 Preliminary Results

In this section, we state a number of results on the Luttinger-Sy model defined in Section 1 on which our analysis is based. We first recall some results on the thermodynamic limit specialized to the pieces model. Then, we describe the statistics of the eigenvalues and eigenfunctions of the pieces model defined in (1.1); in the case of the pieces model, it suffices therefore to describe the statistics of the pieces (see Section 2.2).

In Section 2.4, we describe the non-interacting system of n electrons.

2.1 Rescaling the Operator

Consider the scaling $\widetilde{x} = \mu x$, that is, define

$$S_\mu : \bigwedge_{j=1}^{n} L^2([0, L]) \to \bigwedge_{j=1}^{n} L^2([0, \widetilde{L}])$$

$$u \mapsto S_\mu u \quad \text{where } (S_\mu u)(x) = \mu^{n/2} u(\mu x) \quad \text{and} \quad \widetilde{L} = \mu L. \tag{2.1}$$

One then computes

$$S_\mu^* H_\omega(L, n) S_\mu = \mu^2 \widetilde{H}_\omega(\widetilde{L}, n)$$

where $\widetilde{H}_\omega(\widetilde{L}, n)$ is the interacting pieces model on the interval $[0, \widetilde{L}]$ defined by a Poisson process of intensity 1 and with pair interaction potential

$$U^\mu(\cdot) = \mu^{-2} U(\mu^{-1} \cdot). \tag{2.2}$$

For $\widetilde{H}_\omega(\widetilde{L}, n)$, the thermodynamic limit becomes

$$\frac{n}{\widetilde{L}} = \frac{n}{\mu L} \to \frac{\rho}{\mu} = \rho_\mu.$$

We shall prove Theorems 1.3, 1.5, and 1.6 under the additional assumption $\mu = 1$. Let us now explain how Theorems 1.3, 1.5, and 1.6 get modified when one goes from $\mu = 1$ to arbitrary μ.

If one denotes by γ^μ the constant defined by Proposition 1.4 applied to the interaction potential U^μ instead of U, a direct computation yields $\gamma^\mu = \mu\gamma$.

In the same way, a direct computation yields that Z^μ, the analogue of Z in assumption **(HU)** for U^μ, is given by $Z^\mu(\cdot) = \mu^2 Z(\mu^{-1}\cdot)$. Thus, for the function f_{Z^μ} (see (1.26), (3.28) and (3.29)) defined for U^μ, see (2.2), one obtains $f_{Z^\mu}(\cdot) =$

$\mu^2 f_Z(\mu^{-1}\cdot)$. This suffices to obtain Theorems 1.5 and 1.6 for μ arbitrary fixed from the case $\mu = 1$.

2.1.1 Other Asymptotic Regimes

In the introduction, for the sake of simplicity we chose to state our results at fixed μ and sufficiently small ρ (depending on μ). Actually, the results that we obtained stay correct under less restrictive conditions on μ and ρ. The conditions that are required are the following. Fix $\mu_0 > 0$; then, Theorems 1.3, 1.5, and 1.6 stay correct as long as $\mu \in (0, \mu_0)$, ρ_μ be sufficiently small and $\ell_{\rho,\mu}$ sufficiently large depending only on μ_0. Let us now explain this.

Therefore, we analyze the remainder terms of (3.80) (thus, of (3.82)). The second term in the last equality in (3.80) multiplied by μ^2 (to rescale energy properly, see above) becomes

$$\pi^2 \mu^2 \gamma_*^\mu \frac{\rho_\mu}{|\log \rho_\mu|^3} = \pi^2 \gamma_*^\mu \mu^{-1} \rho_\mu \ell_{\rho,\mu}^{-3} + o\left(\rho_\mu \ell_{\rho,\mu}^{-3}\right)$$

by (1.12). Note that, by (1.17), $\gamma_*^\mu \mu^{-1}$ stays bound from above and below as $\mu \to 0^+$.

The remainder term in the last equality in (3.80) multiplied by μ^2 (to rescale energy properly, see above) becomes

$$\mu^2 \frac{\rho_\mu}{|\log \rho_\mu|^3} O\left(f_{Z^\mu}(|\log \rho_\mu|)\right) = \frac{\rho_\mu \mu^4}{\ell_{\rho,\mu}^3} O\left(f_Z\left(\ell_{\rho,\mu}(1+o(1))\right)\right) = o\left(\frac{\rho_\mu \mu^{-1}}{\ell_{\rho,\mu}^3}\right)$$

when $\rho_\mu \to 0$ and $\ell_{\rho,\mu} \to +\infty$ while μ stays bounded.

This then yields Theorem 1.3 for (μ, ρ) arbitrary in the regime described above from the case $\mu = 1$ and ρ small.

To obtain Theorems 1.5 and 1.6 for μ arbitrary, we just use $Z^\mu(\cdot) = \mu^2 Z(\mu^{-1}\cdot)$ and the fact that Z is decaying; indeed, this implies that

$$Z^\mu(2|\log \rho_\mu|) = \mu^2 Z(2\ell_{\rho,\mu}(1+o(1))) \leqslant \mu^2 Z(\ell_{\rho,\mu})$$

when $\rho_\mu \to 0$ and $\ell_{\rho,\mu} \to +\infty$ while μ stays bounded.

This suffices to obtain Theorems 1.5 and 1.6 for (μ, ρ) arbitrary in the regime described above from the case $\mu = 1$ and ρ small.

From now on, we fix $\mu = 1$ and assume ρ be small. Thus, we shall drop the sub- or superscript μ and write, e.g., ℓ_ρ for $\ell_{\rho,\mu}$, E_ρ for $E_{\rho,\mu}$, etc. Similarly, the dependence on the random parameter ω will be frequently dropped so as to simplify notations.

2.2 The Analysis of the One-Particle Pieces Model

Most of the proofs of the results stated in the present section can be found in Appendix A.

Recall that we partition $[0, L]$ using a Poisson process of intensity 1 and write

$$[0, L] = \bigcup_{j=1}^{m(\omega)} \Delta_j(\omega). \tag{2.3}$$

Note that, by a standard large deviation principle, for $\beta \in (0, 1/2)$, with probability at least $1 - e^{-L^\beta}$, one has $m = L + O\left(L^{1/2+\beta}\right)$.

Moreover, with probability one,

- $\min\limits_{1 \leqslant j \leqslant m(\omega)} |\Delta_j(\omega)| > 0$,
- if $j \neq j'$ then $\dfrac{|\Delta_j(\omega)|^2}{|\Delta_{j'}(\omega)|^2} \notin \mathbb{Q}$.

Thus, distinct pieces generate distinct Dirichlet Laplacian energy levels. In particular, with probability one, all the eigenfunctions of the one-particle Hamiltonian $H_\omega(L) = H_\omega(L, 1)$ are supported on a single piece $\Delta_j(\omega)$ and the corresponding eigenvalues are simple.

Hence, we will enumerate the eigenvalues and the eigenfunctions of $H_\omega(L)$ using a two-component index (Δ_j, k) where

- Δ_j is the piece of the partition (2.3) on which the eigenfunction is supported,
- k is the index of the eigenvalue within the ordered list of eigenvalues of this piece,

i.e.,

$$\psi_{(\Delta_j,k)}(x) = \sqrt{\frac{2}{|\Delta_j|}} \sin\left(\frac{\pi k(x - \inf \Delta_j)}{|\Delta_j|}\right) \mathbf{1}_{\Delta_j}(x)$$

and the corresponding energy

$$E_{(\Delta_j,k)} = \left(\frac{\pi k}{|\Delta_j|}\right)^2. \tag{2.4}$$

Let $\mathcal{P} = \mathcal{P}(\omega)$ denote the set of all available indices enumerating single-particle states, i.e., $\mathcal{P} = \{\Delta_j\}_{j=1}^{m(\omega)} \times \mathbb{N}$.

In parallel to this two-component enumeration system, we will use a direct indexing procedure: $\{(E_j, \psi_j)\}_{j \in \mathbb{N}}$ are the eigenvalues and associated eigenfunctions of the one-particle Hamiltonian $H_\omega(L)$ counted with multiplicity ordered with increasing energy.

2.3 The Statistics of the Pieces

We first study the statistical distribution of the pieces generated by the Poisson process. We will primarily be interested in the joint distributions of their lengths. These statistics immediately provide the statistics of the eigenvalues and eigenfunctions of the pieces model. These results are presumably well known; as we don't know a convenient reference, we provide their proofs in Appendix A for the sake of completeness.

In the sequel, the probability of the events will typically be $1 - O(L^{-\infty})$: we recall that $A_k = O(k^{-\infty})$ if $\forall N \geqslant 0, \lim_{k \to +\infty} k^N A_k = 0$. Actually, the proofs show that the probabilities lie at an exponentially small distance from 1, i.e., $O(L^{-\infty}) = e^{-L^\beta}$ for some $\beta > 0$.

We prove

Proposition 2.1 *With probability* $1 - O(L^{-\infty})$*, the largest piece has length bounded by* $\log L \cdot \log\log L$*, i.e.,*

$$\max_{1 \leqslant k \leqslant m(\omega)} |\Delta_k(\omega)| \leqslant \log L \cdot \log\log L.$$

On the distribution of the length of the pieces, one proves

Proposition 2.2 *Fix* $\beta \in (2/3, 1)$*. Then, for* L *large, for any* $(a_L, b_L) \in [0, \log L \cdot \log\log L]^2$*, with probability* $1 - O(L^{-\infty})$*, the number of pieces of length contained in* $[a_L, a_L + b_L]$ *is equal to*

$$e^{-a_L}(1 - e^{-b_L}) \cdot L + R_L \cdot L^\beta \quad \textit{where} \quad |R_L| \leqslant \kappa$$

and the positive constant κ *is independent of* a_L, b_L*.*

The proof of Proposition 2.2 is given in Appendix A.

We will also use the joint distributions of pairs and triplets of pieces that are close to each other. We prove

Proposition 2.3 *Fix* $\beta \in (2/3, 1)$*. Then, for any* a, b, c, d, g, f *positive, with probability* $1 - O(L^{-\infty})$*, the number of pairs of pieces such that*

- *the length of the left most piece is contained in* $[a, a + b]$*,*
- *the length of the right most piece is contained in* $[c, c + d]$*,*
- *the distance between the two pieces belongs to* $[g, g + f]$

is equal to

$$f e^{-a-c}(1 - e^{-b})(1 - e^{-d}) \cdot L + R_L \cdot L^\beta \quad \textit{where} \quad |R_L| \leqslant \kappa \tag{2.5}$$

and the positive constant κ *may depend on* (a, b, c, d, f, g)*.*

For pairs of pieces, we shall also use

Proposition 2.4 *For $\ell, \ell', d > 0$, with probability $1 - O(L^{-\infty})$, one has*

$$\#\left\{\begin{array}{r}\textit{pairs of pieces at most at a dis-}\\ \textit{tance } d \textit{ from each other such that}\\ \textit{the left most piece is longer than } \ell,\\ \textit{the right most piece is longer than } \ell'.\end{array}\right\} \leqslant (2+d)e^{-\ell-\ell'}L.$$

Finally, for triplets of pieces, we shall use

Proposition 2.5 *For $\ell, \ell', \ell'', d > 0$, with probability $1 - O(L^{-\infty})$, one has*

$$\#\left\{(\Delta,\ \Delta',\ \Delta'')\ \textit{s.t.}\ \left|\begin{array}{l}\Delta' \textit{ between } \Delta \textit{ and } \Delta''\\ \textit{dist}(\Delta,\Delta') \leqslant d,\ \textit{dist}(\Delta',\Delta'') \leqslant d\\ |\Delta| \geqslant \ell,\ |\Delta'| \geqslant \ell',\ |\Delta''| \geqslant \ell''.\end{array}\right.\right\} \leqslant (2+d^2)e^{-\ell-\ell'-\ell''}L.$$

As a straightforward consequence of Proposition 2.2, exploiting the formula (2.4) for the Dirichlet eigenvalues of the Laplacian on an interval, one obtains the explicit formula (1.2) for the one-particle integrated density of states for the pieces model defined in (1.2) (here, $\mu = 1$) That is, one proves

Proposition 2.6 (The One-Particle IDS) *The one-particle integrated density of states for the pieces model is given by*

$$N(E) = \frac{\exp(-\ell_E)}{1-\exp(-\ell_E)}\mathbf{1}_{E>0} \tag{2.6}$$

where ℓ_E is defined in (1.2).

Formula (2.6) was already obtained in [14]; in Appendix A.1, we give a short proof for the readers convenience.

Recalling the scaling defined in Section 2.1 immediately yields (1.2) for general μ.

2.4 Free Electrons

Understanding the system without interactions will be key to answering the main questions raised in the present work. For free electrons, i.e., when the interactions are absent, $U \equiv 0$, the energy per particle $\mathcal{E}^0(\rho)$ can be expressed in terms of one-particle density of states measure.

2.4.1 The Ground State Energy Per Particle

Recall that (see Theorem 1.3), for a density of particles ρ, the *Fermi energy* E_ρ is a solution of the equation $N(E_\rho) = \rho$. In the present case, as N is continuous and strictly increasing from 0 to $+\infty$, the solution to this equation is unique for any $\rho > 0$. The length of the interval whose Dirichlet Laplacian has the Fermi energy E_ρ as ground state energy is the Fermi length ℓ_ρ given by

$$\ell_\rho := \pi/\sqrt{E_\rho} \tag{2.7}$$

As a direct corollary to (1.2) (recall that $\mu = 1$) or equivalently Proposition 2.6, we see that the Fermi energy is given by

$$E_\rho = \pi^2 \left(\log(\rho^{-1} + 1)\right)^{-2} \sim \pi^2 |\log \rho|^{-2} \quad \text{when} \quad \rho \to 0 \tag{2.8}$$

and the Fermi length by:

$$\ell_\rho = \log\left(\rho^{-1} + 1\right) \sim |\log \rho| \quad \text{when} \quad \rho \to 0. \tag{2.9}$$

We recall

Proposition 2.7 ([21, Theorem 5.13 and Lemma 5.14]) *Let $E^\Lambda_{n,\omega}$ denote the n-th energy level of $H_\omega(L)$ (counting multiplicity). Then, ω-a.s., one has*

$$E^\Lambda_{n,\omega} \underset{\substack{L\to\infty \\ n/L\to\rho}}{\longrightarrow} E_\rho \quad \text{and} \quad \mathcal{E}^0(\rho) = \frac{1}{\rho}\int_{-\infty}^{E_\rho} E\, \mathrm{d}N(E). \tag{2.10}$$

Proposition 2.7 follows easily from Lemma 3.13, (1.30), (1.31), and (A.17).

We see that

- the highest energy level occupied by a system of non-interacting electrons tends to the Fermi energy in the thermodynamic limit;
- the n-electron ground state energy per particle is the energy averaged with respect to the density of states measure of the one-particle system conditioned on energies less than the Fermi energy.

Combining formulas (2.8) and (2.10), one can expand $\mathcal{E}^0(\rho)$ into inverse powers of $\log \rho$ up to an arbitrary order. Taking the scaling defined in Section 2.1 into account, (2.10) immediately implies (1.13).

2.4.2 The Eigenfunctions

Let us now describe the eigenfunctions of $H^0_\omega(L, n)$. Let us recall that $(E_p)_{p\in\mathcal{P}}$ are the eigenvalues of the one-particle operator $H_\omega(L)$ and $(\psi_p)_{p\in\mathcal{P}}$ are the corresponding normalized eigenfunctions; here, p in $\mathcal{P}$ is a (piece–energy level) index. The n-electron eigenstates without interactions are given by the following procedure. Pick a set $\alpha := \{\alpha_1, \dots, \alpha_n\} \subset \mathcal{P}$ of n indices, $\operatorname{card}\alpha = |\alpha| = n$. The normalized eigenstate associated to α is given by the Slater determinant

$$\Psi_\alpha(x^1, x^2, \cdots, x^n) := \psi_{\alpha_1} \wedge \cdots \wedge \psi_{\alpha_n} := \frac{1}{\sqrt{n!}} \det\left(\psi_p(x^j)\right)_{\substack{p\in\alpha \\ 1\leqslant j\leqslant n}}. \tag{2.11}$$

One easily checks that $(H^0_\omega(\Lambda, n) - E_\alpha)\Psi_\alpha = 0$ for the energy E_α defined by

$$E_\alpha = \sum_{p\in\alpha} E_p. \tag{2.12}$$

The subset α indicates which one-particle energy levels are occupied in the multi-particle state Ψ_α. For instance, in the ground state of n electrons, one chooses the states with the lowest possible energy.

Notation 2.8 For a Slater determinant Ψ_α (see (2.11)) and $p \in \alpha$, we will refer to the one-particle functions ψ_p as *particles* that constitute the n-electron state indexed by α. Moreover, with a slight abuse of terminology, we will refer to a multi-index α as an (n-electron) state and to p in α as a particle.

3 The Asymptotics for the Ground State Energy Per Particle

In this section, we prove Theorem 1.3 on the asymptotic expansion of the ground state energy per particle in terms of small particle density. We assume that the pair interaction potential U satisfies condition **(HU)**.

3.1 Decomposition by Occupation Numbers

We give a definition of the number of particles occupying a given piece. Therefore, we shall use the special structure of the Hamiltonian $H^0_\omega(\Lambda, n)$, that is, that of $H_\omega(L)$ (see (1.4) and (1.1)).

Fix ω. Recall that $(\Delta_j(\omega))_{1\leqslant j\leqslant m}$ are the pieces defined in (2.3) ($m = m(\omega)$). The one-particle space is then decomposed into

$$L^2(\Lambda) = L^2([0, L]) = \overset{\perp}{\bigoplus_{1 \leqslant j \leqslant m}} L^2(\Delta_j(\omega)). \tag{3.1}$$

Thus, for the n-particle space $\mathfrak{H}^n$ (see (1.3)), we obtain the decomposition

$$\mathfrak{H}^n = \mathfrak{H}^n(\Lambda) = \bigwedge_{j=1}^{n} L^2(\Lambda) = \bigoplus_{\substack{Q=(Q_1,\cdots,Q_m)\in\mathbb{N}^m \\ Q_1+\cdots+Q_m=n}} \mathfrak{H}_Q \tag{3.2}$$

where we have defined

Definition 3.1 For $Q = (Q_1, \cdots, Q_m) \in \mathbb{N}^m$ s.t. $Q_1 + \cdots Q_m = n$, the space of states of fixed occupation Q denoted by $\mathfrak{H}_Q$ is given by

$$\mathfrak{H}_Q = \bigwedge_{j=1}^{m} \left(\bigwedge_{k=1}^{Q_j} L^2(\Delta_j(\omega)) \right). \tag{3.3}$$

Here, as usual, we set $\bigwedge_{k=1}^{0} L^2(\Delta_j(\omega)) = \mathbb{C}$.

An occupation Q is a multi-index of length m and of "modulus" n. Note that, as $\Delta_j(\omega) \cap \Delta_{j'}(\omega) = \emptyset$ for $j \neq j'$, we can identify

$$\mathfrak{H}_Q = \bigotimes_{j=1}^{m} \left(\bigwedge_{k=1}^{Q_j} L^2(\Delta_j(\omega)) \right).$$

Remark 3.2 The spaces of fixed occupation could also be defined starting from the eigenstates of $H^0_\omega(L, n)$ as in [20]. Indeed, each of the eigenstates of $H^0_\omega(L, n)$, the non-interacting Hamiltonian, belongs to a state of fixed occupation. More precisely, if $\Psi_\alpha \in \mathfrak{H}^n$ is the eigenstate of $H^0_\omega(L, n)$ given by (2.11) where $\alpha \subset \mathcal{P}$, $\operatorname{card}\alpha = n$, then, defining the occupation $Q(\alpha) = (Q_1(\alpha), \cdots, Q_m(\alpha))$ where, for $1 \leqslant j \leqslant m$, $Q_j(\alpha) := \#\left\{p \in \alpha |\ \operatorname{supp}\psi_p = \Delta_j\right\}$, we see that $\Psi_\alpha \in \mathfrak{H}_Q$.

The following lemma is crucial in our analysis as it gives global information on the structure of the ground state of the Hamiltonian $H^U_\omega(L, n) = H^0_\omega(L, n) + W_n$. We prove

Lemma 3.3 *Let ω be fixed and let α and β be two n-electron indices corresponding each to an eigenstate of $H^0_\omega(L, n)$.*

If their occupations are different, then the corresponding n-particle states do not interact:

$$Q(\alpha) \neq Q(\beta) \Rightarrow \langle \Psi_\alpha, W_n \Psi_\beta \rangle = 0.$$

Proof If α and β have different occupation numbers, the supports of Ψ_α and Ψ_β in Λ^n intersect at a set of measure zero: indeed, these supports are obtained by symmetrizing different collections of products of pieces (with repetitions for the pieces that are occupied more than once):

$$Q(\alpha) \neq Q(\beta) \quad \Rightarrow \quad \text{meas}\left(\text{supp}\,\Psi_\alpha \cap \text{supp}\,\Psi_\beta\right) = 0.$$

The latter means that $\Psi_\alpha \cdot \Psi_\beta \equiv 0$ as a function in $L^2(\Lambda^n)$. Then, clearly, by definition, for the matrix elements, one obtains

$$\langle \Psi_\alpha, W_n \Psi_\beta \rangle = \int_{\Lambda^n} W_n(\mathbf{x}) \Psi_\alpha(\mathbf{x}) \Psi_\beta^*(\mathbf{x}) d\mathbf{x} = 0.$$

Lemma 3.3 is proved. □

As an immediate corollary to Lemma 3.3, we obtain

Corollary 3.4 (Decomposition by Occupation) *Fix ω. For any $Q \in \mathbb{N}^m$ ($m = m(\omega)$), the subspace $\mathfrak{H}_Q$ is invariant under the action of the n-particle Hamiltonian $H^U_\omega(L,n) = H^0_\omega(L,n) + W_n$, i.e.,*

$$(H^U_\omega(L,n) + i)^{-1} \mathfrak{H}_Q \subset \mathfrak{H}_Q. \tag{3.4}$$

Thus, the total Hamiltonian $H^U_\omega(L,n)$ is decomposed according to (3.2) *in direct sum of its parts H_Q on subspaces of fixed occupation, i.e.,*

$$H^U_\omega(L,n) = \bigoplus_{\substack{Q \in \mathbb{N}^m \\ Q_1 + \cdots + Q_m = n}} H_Q, \tag{3.5}$$

where $H_Q = H^U_\omega(L,n)\big|_{\mathfrak{H}_Q}$.

Remark 3.5 All terms of this decomposition as well as the number of pieces m depend on the randomness ω, i.e., the configuration of pieces.

***Proof of Corollary* 3.4** Fix ω. The space

$$\mathcal{D}^n_\omega := C_0^\infty\left(\left(\bigcup_{1 \leqslant j \leqslant m} \mathring{\Delta}_j(\omega)\right)^n\right) \bigcap \mathfrak{H}^n$$

is a core for $H^U_\omega(L,n)$. Here, $\mathring{\Delta}_j(\omega)$ denotes the interior of $\Delta_j(\omega)$.

It, thus, suffices to check that, for $H^U_\omega(L,n)\left(\mathfrak{H}_Q \cap \mathcal{D}^n_\omega\right) \subset \mathfrak{H}_Q$; this follows immediately from Lemma 3.3. This ensures the existence of the decomposition (3.5) and completes the proof of Corollary 3.4. □

Corollary 3.4 states that the interaction operator W_n is partially diagonalized in the basis of eigenfunctions of $H^0_\omega(L,n)$, i.e., its matrix representation has a block structure corresponding to the subspaces of constant occupation.

3.2 *Almost Sure Non-degeneracy of the Interacting Ground State*

We first restrict ourselves to spaces with fixed occupation to prove

Lemma 3.6 *Fix an occupation Q. The ground state of $H^U_\omega(L,n)|_{\mathfrak{H}_Q}$ is non-degenerate.*

Proof To simplify notations, let us write $H = H^U_\omega(L,n)$ and $H^0 = H^0_\omega(L,n)$. Let $(\Delta_{j_p})_{1\leqslant p\leqslant n}$ be the pieces such that $Q_{j_p} \geqslant 1$; in the list $(\Delta_{j_p})_{1\leqslant p\leqslant n}$, each piece Δ_{j_p} is repeated Q_{j_p} times. We enumerate the pieces so that their left endpoints are non- decreasing (i.e., from the leftmost piece to the rightmost piece). So, $p \mapsto j_p$ is non-decreasing. Then, the operator H^0_Q is the Dirichlet Laplacian on a space of anti-symmetric functions defined on the symmetrized domain

$$\Delta_Q = \mathrm{Sym}\left(\mathop{\times}_{p=1}^{n} \Delta_{j_p}\right) := \bigcup_{\sigma\in\mathfrak{S}_n} \mathop{\times}_{p=1}^{n} \Delta_{\sigma(j_p)}. \tag{3.6}$$

Anti-symmetric functions on the domain (3.6) that vanish on the boundary $\partial(\Delta_Q)$ are in one-to-one correspondence with functions defined on the domain

$$\delta_Q = \left\{(x^1,\dots,x^n) \text{ s.t. } x^p \in \Delta_{j_p} \text{ and } x^p \leqslant x^q \text{ for } p<q\right\} \tag{3.7}$$

that vanish on $\partial(\delta_Q)$, the boundary of δ_Q. Actually,

$$\Delta_Q = \bigcup_{\sigma\in\mathfrak{S}_n} \sigma(\delta_Q) \quad \text{and, for } (\sigma,\sigma')\in\mathfrak{S}_n^2, \quad \sigma(\delta_Q)\cap\sigma'(\delta_Q) = \emptyset \text{ if } \sigma\neq\sigma'.$$

Here, for $\sigma \in \mathfrak{S}_n$, we have set $\sigma : (x^1,\cdots,x^n) \mapsto (x^{\sigma(1)},\cdots,x^{\sigma(n)})$.

Thus, finding the ground state of $H_Q = H^0 + W$ is equivalent to finding the ground state of the Schrödinger operator $-\Delta + W$ with Dirichlet boundary conditions on the domain δ_Q. As the domain δ_Q is connected and has a piecewise linear boundary, the ground state of $-\Delta + W$ is non-degenerate (see [6, Theorems 1.4.3, 1.8.2 and 3.3.5] and [17, Section XIII.12]). This completes the proof of Lemma 3.6.
□

3.3 The Proof of Theorem 1.1

Considering the decomposition (3.5), Lemma 3.6 implies that the only possible source of degeneracy of the ground state is that different occupations, i.e., distributions of particles in the pieces, provide the same ground state energy. Let us show that, almost surely, this does not happen.

Let Π be the support of $d\mu(\omega)$, the Poisson process of intensity 1 on $\mathbb{R}_+$. Let $\#(\Pi \cap [0, L])$ be the number of points the Poisson process puts into $(0, L)$. Suppose now that the probability that the ground state of $H_\omega^U(L, n)$ is degenerate is positive.

Thus, for some $m \geqslant 0$, conditioned on the fact that the Poisson process puts m points into $(0, L)$ (i.e., $\#(\Pi \cap [0, L]) = m$), the probability that the ground state of $H_\omega^U(L, n)$ be degenerate is positive. Let $(\ell_j)_j$ be the lengths of the pieces $(\Delta_j(\omega))_j$, i.e., the $(\Delta_j)_j$ are connected and $\cup_j \Delta_j(\omega) = (0, L) \setminus (\Pi \cap [0, L])$. Conditioned $\#(\Pi \cap [0, L]) = m$, the joint distribution of the vector $(\ell_j)_j$ is known.

Proposition 3.7 ([9]) *Under the condition* $\#(\Pi \cap [0, L]) = m$*, the vector* $(\ell_1, \ldots, \ell_{m+1})$ *has the same distribution as the random vector*

$$\left(\frac{L \cdot \eta_1}{\eta_1 + \ldots + \eta_{m+1}}, \frac{L \cdot \eta_2}{\eta_1 + \ldots + \eta_{m+1}}, \ldots, \frac{L \cdot \eta_{m+1}}{\eta_1 + \ldots + \eta_{m+1}}\right), \tag{3.8}$$

where $(\eta_i)_{1 \leqslant i \leqslant m}$ *are i.i.d. exponential random variables of parameter* 1.

As the lengths $(\ell_j)_j$ are continuous functions of the parameters $(\eta_j)_j$, we know that there exists an open set in $(\mathbb{R}^+)^{m+1}$, say O, such that, for each $(\ell_j)_{1 \leqslant j \leqslant m+1} \in O$, there are at least two occupations $Q_1((\ell_j)_{1 \leqslant j \leqslant m+1})$ and $Q_2((\ell_j)_{1 \leqslant j \leqslant m+1})$ that have the same ground state energy (which is at the same time the smallest possible among the ground state energies for all the occupations). Let us denote these branches of energy by $(\ell_j)_{1 \leqslant j \leqslant m+1} \mapsto E_1((\ell_j)_{1 \leqslant j \leqslant m+1})$ and $(\ell_j)_{1 \leqslant j \leqslant m+1} \mapsto E_2((\ell_j)_{1 \leqslant j \leqslant m+1})$, respectively.

For a fixed number of pieces, there are finitely many occupations and a change in the number of pieces occurs only when a wall, i.e., an endpoint of a piece, crosses 0 or L. Thus, there exists a non- empty open subset $O_1 \subset O$, such that $Q_1((\ell_j)_{1 \leqslant j \leqslant m+1})$ and $Q_2((\ell_j)_{1 \leqslant j \leqslant m+1})$ are constant on O_1.

Now, let us fix an initial set of lengths $(\ell_j^0)_{1 \leqslant j \leqslant m+1}$ in O_1 and move it continuously inside this exceptional set O_1. This actually corresponds to moving continuously walls inside the interval $(0, L)$. As Q_1 and Q_2 are two different occupations, there exists a piece $[a, b] \subset [0, L]$, such that Q_1 and Q_2 put different number of particles in this piece, i.e., $Q_1([a, b]) \neq Q_2([a, b])$.

Now, we move a continuously towards b; if $a = 0$, we will move b towards a. Let a^0 be the value of a in the configuration $(\ell_j^0)_{1 \leqslant j \leqslant m+1}$. Let $E_1(a)$ and $E_2(a)$ be the ground state energies corresponding to the two different occupations Q_1 and Q_2. In a small neighborhood of a_0, by the definition of O_1, one has

$$E_1(a) = E_2(a).$$

As U is real analytic and as the ground state of H_Q is simple for any occupation Q, the functions $E_1(a)$ and $E_2(a)$ are analytic in the open interval (c, b) where c is the end of the piece $[c, a]$ to the left of the piece $[a, b]$. Indeed, E_1 (and E_2) is analytic around a_0. Assume that $E_1(a)$ stops being analytic somewhere inside (c, b). This would mean that the eigenvalue $E_1(a)$ of H_{Q_1} becomes degenerate, thus, that the ground state of H_{Q_1} becomes degenerate. This was already ruled out.

This immediately implies that $E_1(a) = E_2(a)$ for all $a \in (c, b)$.

But this cannot be. Indeed, if Q_1 puts k_1 particles in the piece $[a, b]$, and Q_2 puts k_2 particles in the piece $[a, b]$ with $k_1 \neq k_2$, the functions E_1 and E_2 have different asymptotics as a approaches b, indeed,

$$E_i(a) \sim k_i^3/(b-a)^2 \quad \text{as} \quad a \to b.$$

This contradicts the fact that the two functions agree on the whole interval. This completes the proof of Theorem 1.1. □

Finally, we use the results from Section 3.1 together with Theorem 1.1 to obtain the following

Corollary 3.8 *Assume U is real analytic. Then, ω-almost surely, for any L and n, the ground state of $H_\omega^U(L, n)$ belongs to a unique occupation subspace $\mathfrak{H}_Q$.*

Proof Consider the orthogonal decomposition (3.5). As any projection of $\Psi_\omega(L, n)$ on $\mathfrak{H}_Q$ is either a ground state or zero and as the ground state is ω-a.s. simple, only one of the projections of the ground state on a space of fixed occupation is different from zero. Thus, $\Psi_\omega(L, n)$ belongs to one of the subspaces $\mathfrak{H}_Q$. This completes the proof of Corollary 3.8. □

3.4 The Approximate Ground State Ψ^{opt}

The basic idea of the construction of Ψ^{opt} is to find the optimal configuration with respect to different occupations. All the n-electron states are considered as deformations of the unperturbed ground state Ψ^0 which, we recall (2.11), is given by the Slater determinant:

$$\Psi^0 = \psi_1 \wedge \psi_2 \wedge \ldots \wedge \psi_n.$$

When the interactions are turned on, the particles in the state Ψ^0 start to interact. For some particles, these interactions may be quite large. In particular, it may become energetically favorable to "decouple" some particles by moving them apart from each other to unoccupied pieces; obviously, it is better to move the more excited particles. One, thus, reduces the interaction energy but this will necessarily result in an increase of the "non-interaction" energy of the state, i.e., of $\langle H_\omega^0(L, n)\Psi, \Psi\rangle$: indeed, in the non-interacting ground state, the n particles occupy the n lowest levels of the system. Nevertheless the decrease of the interaction energy, i.e.,

$\langle W_n \Psi, \Psi \rangle$ may compensate the increase in "non-interacting" energy. The "optimal" configuration then arises through the optimization on the occupation governed by the interplay between the loss of interaction energy and the gain of "non-interacting" energy: it is achieved when loss and gain balance.

Let us note that a ground state Ψ is obviously the ground state of the Hamiltonian restricted to the appropriate fixed occupation subspace, i.e., Ψ is the ground state of $H_{Q(\Psi)}$ (see (3.5)). This corresponds to writing the minimization problem in the form

$$\inf_{\substack{\Phi \in \mathfrak{H}^n \\ \|\Phi\|=1}} \langle H_\omega(L,n)\Phi, \Phi \rangle = \inf_{\substack{Q \in \mathbb{N}^m \\ |Q|=n}} \inf_{\substack{\Phi \in \mathfrak{H}_Q \\ \|\Phi\|=1}} \langle H_Q \Phi, \Phi \rangle. \tag{3.9}$$

This reduces the problem to finding the optimal occupations rather than the optimal n-electron state itself.

Recalling that the constant γ is defined in Proposition 1.4, we set

$$A_* := \frac{\gamma}{8\pi^2}, \qquad x_* := 1 - e^{-\frac{\gamma}{8\pi^2}}. \tag{3.10}$$

Note that

$$A_* = -\log(1 - x_*).$$

Let us now define Ψ^{opt}. Therefore, recall that the pieces in the model are denoted by $(\Delta_k(\omega))_{1 \leqslant k \leqslant m(\omega)}$ (see Section 1) and that for $\Delta_k(\omega)$, a piece, we define (see Sections 1.3.2 and 1.3.3)

- $\varphi^j_{\Delta_k(\omega)}$ to be the j-th normalized eigenvector of $-\Delta^D_{|\Delta_k(\omega)}$,
- $\zeta^j_{\Delta_k(\omega)}$ to be the j-th normalized eigenvector of $-\Delta^D_{|\Delta_k(\omega)^2} + U$ acting on $\bigwedge_{j=1}^{2} L^2(\Delta_k(\omega))$.

We will define the state Ψ^{opt} in two steps. We first define Ψ^{opt}_m: it will contain less than n particles and will be the main part of Ψ^{opt}. We, then, add the missing particles to get the n-particle state Ψ^{opt}.

Definition 3.9 Consider all the pieces in $[0, L]$. For each piece, depending on its length, do one of the following:

(a) keep the pieces of length in $[0, \ell_\rho - \rho x_*) \cup [3\ell_\rho, \infty)$ empty;
(b) put one particle in its ground state in each piece of length in $[\ell_\rho - \rho x_*, 2\ell_\rho + A_*)$;
(c) in pieces of length in $[2\ell_\rho + A_*, 3\ell_\rho)$, put the ground state of a two-particle system with interactions (see Proposition 1.4 and Section 6.1);

We define the state $\Psi^{\mathrm{opt}}_m = \Psi^{\mathrm{opt}}_m(L,n)$ to be the anti-symmetric tensor product of the thus constructed one- and two-particle sub-states, that is,

$$\Psi_m^{\mathrm{opt}}(L,n) = \bigwedge_{|\Delta_j(\omega)|\in[\ell_\rho-\rho x_*, 2\ell_\rho+A_*)} \varphi^1_{\Delta_j(\omega)} \wedge \bigwedge_{|\Delta_j(\omega)|\in[2\ell_\rho+A_*, 3\ell_\rho)} \zeta^1_{\Delta_j(\omega)}. \tag{3.11}$$

Note that, as the $(\zeta^1_{\Delta_j(\omega)})_j$ carry two particles, $\Psi_m^{\mathrm{opt}}(L,n)$ is not given by a Slater determinant; an explicit formula for such an anti-symmetric tensor product is given in (C.2) in Appendix C.

Remark 3.10 Note that, in step (c) of Definition 3.9, we put two interacting particles within these pieces. Because of the interactions, this is different from putting separately two particles on the two lowest one-particle energy levels (see Section 6).

Let us now compute the total number of particles contained in Ψ_m^{opt}. We prove

Lemma 3.11 *With probability* $1-O(L^{-\infty})$*, for L sufficiently large, in the thermodynamic limit, the total number of particles in* Ψ_m^{opt} *constructed in Definition 3.9 is given by*

$$\mathcal{N}(\Psi_m^{\mathrm{opt}}) = n\left[1-\rho^2\left(3-x_*-\frac{x_*^2}{2}\right)+O(\rho^3)\right].$$

Proof It suffices to count the number of pieces of each type and multiply by the corresponding number of particles. We recall that, by (3.10), one has $\exp(-\ell_\rho) = \frac{\rho}{1+\rho}$ and $\exp(-A_*) = 1-x_*$. Thus, for $\beta\in(0,1/2)$, using Proposition 2.2 and the second equation in (3.10), with probability $1-O(L^{-\infty})$, one computes

$$\begin{aligned}
\mathcal{N}(\Psi_m^{\mathrm{opt}}) &= \sharp\{l\in[\ell_\rho-\rho x_*, 2\ell_\rho+A_*)\} + 2\cdot\sharp\{l\in[2\ell_\rho+A_*, 3\ell_\rho)\} \\
&= L\left[e^{-(\ell_\rho-\rho x_*)} - e^{-(2\ell_\rho+A_*)} + 2e^{-(2\ell_\rho+A_*)} - 2e^{-3\ell_\rho}\right] + O\left(L^{1/2+\beta}\right) \\
&= \frac{L\rho}{1+\rho}\left[e^{\rho x_*} + \rho e^{-A_*} - \rho^2 e^{-A_*} - 2\rho^2 + O(\rho^3)\right] \\
&= \frac{L\rho}{1+\rho}\left[1+\rho-\rho^2\left(e^{-A_*}+2-\frac{x_*^2}{2}\right)+O(\rho^3)\right] \\
&= n\left[1-\rho^2\left(3-x_*-\frac{x_*^2}{2}\right)+O(\rho^3)\right].
\end{aligned}$$

This completes the proof of Lemma 3.11. □

Lemma 3.11 shows that, for ρ small, Ψ_m^{opt} contains less than n particles. Let us now add particles to Ψ_m^{opt} to complete it into Ψ^{opt}. Therefore, we prove

Lemma 3.12 *Let* $(\widetilde{\varphi}_k)_{1\leqslant k\leqslant k_\rho(\omega)}$ *be the particles that* Ψ^0*, the non-interacting ground state, puts in the pieces longer than* $3\ell_\rho$ *ordered by increasing energy.*

With probability $1-O(L^{-\infty})$*, for L sufficiently large, one has* $k_\rho(\omega)\geqslant n\rho^2(3-18\rho)$.

Proof By Proposition 2.2, with probability $1 - O(L^{-\infty})$, the number of pieces of length in $\ell_\rho[3+\rho, 4)$ is equal to

$$n\frac{\rho^2}{(1+\rho)^3}\left(e^{-\rho} - \frac{\rho}{1+\rho}\right) + o(L) \geqslant n\rho^2\,(1-6\rho)$$

for L large.

To complete the proof of Lemma 3.12, let us now establish some auxiliary results. By (2.10) in Proposition 2.7, we know that $E^{\Lambda}_{n,\omega}$ converges to E_ρ in the thermodynamic limit. We will first investigate the rate of convergence in (2.10).

Lemma 3.13 *Denote by $\ell_{n,L}$ the length of an interval having a ground state energy equal to $E^{\Lambda}_{n,\omega}$, i.e.,*

$$\ell_{n,L} = \frac{\pi}{\sqrt{E^{\Lambda}_{n,\omega}}}.$$

Let $\rho > 0$ be fixed. For any $\delta > 0$, in the thermodynamic limit $L \to \infty$, $n/L \to \rho$, with probability $1 - O(L^{-\infty})$, one has

$$\ell_{n,L} = \ell_\rho + O(L^{-(1/2-\delta)}) + O\left(\left|\frac{n}{L} - \rho\right|\right),$$

$$E^{\Lambda}_{n,\omega} = E_\rho + O(L^{-(1/2-\delta)}) + O\left(\left|\frac{n}{L} - \rho\right|\right).$$

In view of Lemma 3.13 and by the definition of Ψ^0, for L sufficiently large, each piece of length in $\ell_\rho[3+\rho, 4)$ contains at least 3 particles of Ψ^0. This completes the proof of Lemma 3.12. □

Proof of Lemma 3.13 By (A.17), with probability $1 - O(L^{-\infty})$, the normalized counting function for the Dirichlet eigenvalues of $H_\omega(L, 1)$ (see (2.4)) satisfies

$$\frac{n}{L} = N^D_L(E^{\Lambda}_{n,\omega}) = \frac{\exp(-\ell_{n,L})}{1-\exp(-\ell_{n,L})} + O(L^{-(1/2-\delta)}).$$

Taking into account the fact that

$$\rho = N(E_\rho) = \frac{\exp(-\ell_\rho)}{1-\exp(-\ell_\rho)},$$

we deduce that

$$\frac{\exp(-\ell_{n,L})}{1-\exp(-\ell_{n,L})} = \frac{\exp(-\ell_\rho)}{1-\exp(-\ell_\rho)} + O(L^{-(1/2-\delta)}) + O\left(\left|\frac{n}{L} - \rho\right|\right).$$

This immediately yields

$$\exp(-\ell_{n,L}) = \exp(-\ell_\rho) + O(L^{-(1/2-\delta)}) + O\left(\left|\frac{n}{L} - \rho\right|\right).$$

The proof of Lemma 3.13 is complete. □

For ρ small, by Lemmas 3.11 and 3.12, one has $n - \mathcal{N}(\Psi_m^{\text{opt}}) < k_\rho(\omega)$. Thus, to construct Ψ^{opt}, we just add $n - \mathcal{N}(\Psi_m^{\text{opt}})$ particles of Ψ^0 living in pieces of length in $\ell_\rho[3+\rho, 4)$ to Ψ_m^{opt}.

Definition 3.14 We define

$$\Psi^{\text{opt}} = \Psi^{\text{opt}}(L, n) := \Psi_m^{\text{opt}}(L, n) \wedge \bigwedge_{k=1}^{n-\mathcal{N}(\Psi_m^{\text{opt}})} \widetilde{\varphi}_k. \tag{3.12}$$

Remark 3.15 Let us give an alternative approach to defining Ψ^{opt} which does not result in exactly the same Ψ^{opt} but which can serve exactly the same purpose in the subsequent arguments.

We start with the non-interacting ground state Ψ^0 and describe how it is modified:

- for pairs of particles living in the same piece, the modification depends on the length of this piece:
  - for the pieces of length between $2\ell_\rho$ and $2\ell_\rho + A_*$, remove the more excited particle and put it into an unoccupied piece of length between $\ell_\rho - \rho x_*$ and ℓ_ρ;
  - for the remaining pieces, i.e., the pieces of length between $2\ell_\rho + A_*$ and $3\ell_\rho$, the factorized two-particle state corresponding to Ψ^0 should be replaced by a true ground state of a two-particle system with interaction in this piece (see Section 6.1 for a description of such a two-particle state);
- do not modify any of the particles in Ψ^0 that are either alone or live in groups of three or more pieces.

One can easily verify that, in the above procedure, up to a small relative error, the number of pieces to which the excited particles are displaced is equal to the number of pieces where we decouple the particles. Indeed, according to Proposition 2.2, with probability at least $1 - O(L^{-\infty})$, for the former, one has

$$\begin{aligned} \sharp\{l \in (2\ell_\rho, 2\ell_\rho - \log(1-x_*))\} &= L\exp(-2\ell_\rho)x_* + O(L^{1/2+\beta}) \\ &= n\rho x_*(1+O(\rho)), \end{aligned} \tag{3.13}$$

and, for the latter, one has

$$\begin{aligned} \sharp\{l \in (\ell_\rho - \rho x_*, \ell_\rho)\} &= L\exp(-\ell_\rho)(\exp(\rho x_*) - 1) + O(L^{1/2+\beta}) \\ &= n\rho x_*(1+O(\rho)). \end{aligned} \tag{3.14}$$

Thus, both sets contain the same number of pieces (up to an error of order $n\rho^2$). This completes the construction of Ψ^{opt}.

3.5 Comparing Ψ^{opt} with the Ground State of the Interacting System

Our goal in the sections to come is to estimate how much Ψ^{opt} differs from a true ground state $\Psi^U = \Psi^U_\omega(L, n)$ (and to show that it doesn't differ much). This will be done through the comparison of their occupation numbers. We shall see that the ground states of the interacting Hamiltonian must live in subspaces with special occupation numbers (see Corollary 3.32).

To compare occupation numbers, we introduce the distance dist_1.

Definition 3.16 Let $m = m(\omega)$ be the number of pieces in $[0, L]$. For $j \in \{1, 2\}$, pick an occupation

$$Q^j = (Q^j_1, Q^j_2, \ldots, Q^j_m) \in \mathbb{N}^m, \quad |Q^j| = n.$$

Define

$$\text{dist}_1(Q^1, Q^2) = \sum_{i=1}^{m} |Q^1_i - Q^2_i|.$$

Remark 3.17 Recall that the non-interacting ground state Ψ^0 has a single occupation $Q(\Psi^0)$: all the states with energy below $E^\Lambda_{n,\omega}$ (where we recall that $E^\Lambda_{n,\omega}$ denote the n-th (counting multiplicity) energy level of the one-particle Hamiltonian $H_\omega(L)$); moreover, only those states are occupied. In [20], for U compactly supported, for Ψ^U an interacting ground state, it was proved that

$$C^{-1} n\rho \leqslant \text{dist}_0(Q(\Psi^U), Q(\Psi^0)) \leqslant Cn\rho. \tag{3.15}$$

where dist_0 is defined by $\text{dist}_0(Q^1, Q^2) = \sum_{i=1}^{m} \mathbf{1}_{Q^1_i \neq Q^2_i}$. Clearly, one has $\text{dist}_0 \leqslant \text{dist}_1$.

In the sequel, we shall prove that Ψ^{opt} is a better approximation of a ground state of the interacting system than is the non-interacting ground state Ψ^0 (compare (3.83) with (3.15)).

For interaction potentials that decrease at infinity sufficiently fast (see **(HU)**), we will prove that the main modification to the ground state energy comes from U restricted to some (sufficiently large) compact set.

Fix a constant $B > 2$. We decompose the interaction potential in the sum of the "principal" and "residual" parts that is, write $U = U^p + U^r$ where

$$U^p := \mathbf{1}_{[-B\ell_\rho, B\ell_\rho]}U \quad \text{and} \quad U^r := \mathbf{1}_{(-\infty,-B\ell_\rho)\cup(B\ell_\rho,+\infty)}U. \tag{3.16}$$

As the sum of pair interactions W_n is linear in U, this yields the following decomposition for the full Hamiltonian:

$$H^U = H^0 + W_n = H^0 + W_n^{U^p} + W_n^{U^r} = H^{U^p} + W_n^r. \tag{3.17}$$

Our analysis is done in the following steps:

(a) first, we prove that Ψ^{opt} approximates well the ground state for the system with compactified interactions Ψ^{U^p};
(b) second, we show that the quadratic form of the residual interactions W_n^r on Ψ^{opt} contributes only to the error term; this will imply (1.16);
(c) finally, we will conclude that the same Ψ^{opt} gives also a good approximation for the full Hamiltonian H^U ground state Ψ^U in terms of the distance dist_1 for the respective occupations.

Remark 3.18 Let us clarify a point of terminology: we will minimize the quadratic form $\langle H_Q\Psi, \Psi\rangle = \langle H_Q^0\Psi, \Psi\rangle + \langle W_n\Psi, \Psi\rangle$; the term $\langle H_Q^0\Psi, \Psi\rangle$ is referred to as the "non- interacting energy" term and $\langle W_n\Psi, \Psi\rangle$ the "interaction energy" term; we use the same decomposition and terminology for smaller groups of particles or at the single particle level.

3.6 The Analysis of H^{U^p}

We start with the analysis of H^{U^p}, in particular, of its ground state energy and ground state(s). Later, we show that the addition of W_n^r will not change much in the ground state energy and ground state(s). First, we compute the energy of Ψ^{opt}. We prove

Theorem 3.19 *There exists $\rho_0 > 0$ such that, for $\rho \in (0, \rho_0)$, in the thermodynamic limit, with probability* 1, *one has*

$$\lim_{\substack{L\to\infty \\ n/L\to\rho}} \frac{1}{n}\big\langle H_\omega^{U^p}(L,n)\Psi^{\text{opt}}(L,n), \Psi^{\text{opt}}(L,n)\big\rangle$$
$$= \mathcal{E}^0(\rho) + \pi^2\gamma_*\rho|\log\rho|^{-3}\left(1 + O\left(f_Z(|\log\rho|)\right)\right) \tag{3.18}$$

where γ_* is defined in (1.17) *and f_Z is a continuous function satisfying $f_Z(x) \to 0$ as $x \to +\infty$ no faster than $1/x$ (for more details, see* (3.29)).

Proof To shorten the notations, we will frequently drop the arguments L, n, and the subscript ω in this proof. We will show that, up to error terms, the only terms that contribute to $\langle H^{U^p}\Psi^{\text{opt}}, \Psi^{\text{opt}}\rangle - \langle H^0\Psi^0, \Psi^0\rangle$ are those due to

(a) the interactions between two particles in the same piece,
(b) the decoupling of a fraction of these particles following the construction of Ψ^{opt}.

In (3.18), the interactions between neighboring distinct pieces will be shown to contribute only to the error term where we have defined

Definition 3.20 A pair of *neighboring* or *interacting* pieces is a pair of distinct pieces at distance at most $B\ell_\rho$ from one another; in particular, particles in two such pieces can still interact via the potential U^p.

Let us now outline the main idea of the proof of Theorem 3.19. The pieces longer than $2\ell_\rho + A_*$ contain two particles both in Ψ^0 and Ψ^{opt}. Hence, for each piece of this type, the energy difference is given by the second term in the asymptotics (1.16) in Proposition 1.4. On the contrary, in pieces of length between $2\ell_\rho$ and $2\ell_\rho + A_*$, in Ψ^0, the two particles were decoupled in order to construct Ψ^{opt}, keeping one intact and displacing another to a piece of length between $\ell_\rho - \rho x_*$ and ℓ_ρ. In this case, the energy difference is given by the increase of non- interacting energy of the second (displaced) particle. The single particles in Ψ^0 remain untouched in Ψ^{opt} and groups of three and more particles contribute only to the error term (as they carry only a small number of particles).

To put the above arguments into a rigorous form, we will use the following partition of the set of available pieces according to their length. Choose K large but independent of L. For $k \in \{1, \dots, K\}$, consider the sets of pieces

$$\mathcal{L}_k^1 = \left\{\text{pieces of length in } \left[\ell_\rho - \tfrac{k}{K}\rho, \ell_\rho - \tfrac{k-1}{K}\rho\right)\right\},$$
$$\mathcal{L}_k^2 = \left\{\text{pieces of length in } \left[2\ell_\rho - \log\left(1 - \tfrac{k-1}{K}\right), 2\ell_\rho - \log\left(1 - \tfrac{k}{K}\right)\right)\right\}.$$

As K is independent of L, with probability $1 - O(L^{-\infty})$, the number of pieces in the classes $((\mathcal{L}_k^j))_{\substack{j\in\{1,2\}\\ k\in\{1,\dots,K\}}}$ is given by Proposition 2.2. We will, henceforth, use these estimates without reference to probabilities.

As in (3.13) and (3.14), one shows that these two sets map one-to-one onto one another up to an error estimated as follows:

$$\operatorname{card} \mathcal{L}_k^1 = \operatorname{card} \mathcal{L}_k^2 + O(n\rho^2K^{-1}) = n\rho K^{-1}(1 + O(\rho)).$$

Recall that x_* is defined in (3.10). For $k \leqslant Kx_*$, according to our scheme, the pairs of particles in pieces belonging to $\mathcal{L}_k^2$ get decoupled, one of the particles being sent to occupy a piece belonging to $\mathcal{L}_k^1$. For $k > Kx_*$, the pairs of particles in the pieces of $\mathcal{L}_k^2$ are kept untouched. The latter pieces are those of size at least $2\ell_\rho + A_*$. It is easily seen that the number of such pieces is given by

$$\sharp\{j : |\Delta_j(\omega)| \geqslant 2\ell_\rho + A_*\} = n\rho e^{-A_*}(1 + O(\rho)) = n\rho(1 - x_*) + O(n\rho^2).$$

The majority of these pieces is smaller than $2\ell_\rho + A_* + \log \ell_\rho$; indeed,

$$\sharp\{j : |\Delta_j(\omega)| \in 2\ell_\rho + A_* + [0, \log \ell_\rho]\} = n\rho(1 - x_*) + O(n\rho|\log \rho|^{-1}).$$

By Proposition 1.4, for a piece of length ℓ in $2\ell_\rho + A_* + [0, \log \ell_\rho]$, the interaction energy of the two-particle system is given by

$$\frac{\gamma}{\ell^3} + o(\ell^{-3}) = \frac{\gamma}{8\ell_\rho^3} + o(\ell_\rho^{-3}).$$

For the difference of energies, this yields

$$\begin{aligned}
&\langle H^{U^p}\Psi^{\mathrm{opt}}, \Psi^{\mathrm{opt}}\rangle - \langle H^0\Psi^0, \Psi^0\rangle \\
&= \frac{n\rho}{K}\sum_{k=1}^{Kx_*}\left[\frac{\pi^2}{\left(\ell_\rho - \frac{k}{K}\rho\right)^2} - \frac{4\pi^2}{\left(2\ell_\rho - \log\left(1 - \frac{k}{K}\right)\right)^2}\right] \\
&\qquad\qquad + \frac{\gamma}{8\ell_\rho^3}n\rho(1 - x_*) + o\left(n\rho|\log \rho|^{-3}\right).
\end{aligned} \tag{3.19}$$

Taking K large, we approximate the Riemann sum in the last expression by an integral

$$\begin{aligned}
&\frac{1}{K}\sum_{k=1}^{Kx_*}\left[\frac{\pi^2}{\left(\ell_\rho - \frac{k}{K}\rho\right)^2} - \frac{4\pi^2}{\left(2\ell_\rho - \log\left(1 - \frac{k}{K}\right)\right)^2}\right] \\
&= x_* \int_0^1 \left[\frac{\pi^2}{\left(\ell_\rho - tx_*\rho\right)^2} - \frac{\pi^2}{\left(\ell_\rho - \frac{1}{2}\log(1 - tx_*)\right)^2}\right] \mathrm{d}t + O\left(\frac{1}{K}\right) \\
&= x_* \left(-\int_0^1 \frac{\pi^2}{\ell_\rho^3}\log(1 - tx_*)\mathrm{d}t + o(\ell_\rho^{-3})\right) + O\left(\frac{1}{K}\right) \\
&= \pi^2\ell_\rho^{-3}(x_* - (1 - x_*)A_*)(1 + o(1)) + O\left(\frac{1}{K}\right).
\end{aligned}$$

Picking $\delta \in (0, 1)$, letting $K = \rho^{-\delta}$, and recalling (3.10) for A_* and x_*, for δ small, we get

$$\begin{aligned}\langle H^{U^p}\Psi^{\text{opt}}, \Psi^{\text{opt}}\rangle - \langle H^0\Psi^0, \Psi^0\rangle &= n\rho\ell_\rho^{-3}\left(\pi^2(x_* - (1-x_*)A_*) + \frac{\gamma}{8}(1-x_*)\right) \\ &\quad + o\left(n\rho\ell_\rho^{-3}\right) \\ &= n\rho\ell_\rho^{-3}\pi^2\left(1 - e^{-\frac{\gamma}{8\pi^2}}\right) + o\left(n\rho\ell_\rho^{-3}\right).\end{aligned} \tag{3.20}$$

In order to finish the proof of (3.18) and, thus, of Theorem 3.19, it suffices to upper bound the interactions between distinct pieces. Recall that Ψ^{opt} is an anti-symmetric exterior product of one- and two-particle eigenstates (see (3.11) and (3.12)):

$$\Psi^{\text{opt}} = \bigwedge_{i=1}^{\hat{k}_1} \varphi_i \wedge \bigwedge_{j=1}^{k_2} \zeta_j \wedge \bigwedge_{i=1}^{\widetilde{k}_1} \widetilde{\phi}_i, \tag{3.21}$$

where the numbers of sub-states in each group are, respectively,

$$\begin{aligned}\hat{k}_1 &= n\left(1 - 2\rho(1-x_*) + \rho^2\left(3(1-x_*) + \frac{x_*^2}{2}\right) + O(\rho^3)\right), \\ k_2 &= n\rho(1 - x_* - \rho(3 - 2x_*) + O(\rho^2)), \\ \widetilde{k}_1 &= n - \mathcal{N}(\Psi_m^{\text{opt}}) = n\rho^2\left(3 - x_* - \frac{x_*^2}{2}\right)(1 + O(\rho)).\end{aligned}$$

The functions φ_i and $\widetilde{\phi}_i$ are one-particle ground states in certain and the functions ζ_j are two-particle ground states in certain pieces. Of course, $\hat{k}_1 + k_2 + \widetilde{k}_1 = n$. As in what follows we will only need to distinguish between one- and two-particle states, let us put the two groups of one-particle sub-states from (3.21) together, i.e. write

$$\Psi^{\text{opt}} = \bigwedge_{i=1}^{k_1} \phi_i \wedge \bigwedge_{j=1}^{k_2} \zeta_j, \tag{3.22}$$

where $k_1 = \hat{k}_1 + \widetilde{k}_1$ and $\{\phi_i\}_{i=1}^{k_1} = \{\varphi_i\}_{i=1}^{\hat{k}_1} \cup \{\widetilde{\phi}_i\}_{i=1}^{\widetilde{k}_1}$. As W^p is a totally symmetric sum of pair interaction potentials, one computes

$$\begin{aligned}\langle W^p\Psi^{\text{opt}}, \Psi^{\text{opt}}\rangle &= \sum_{1\leqslant i<j\leqslant n} \int_{[0,L]^n} U(x_i - x_j)\left|\Psi^{\text{opt}}(x)\right|^2 dx \\ &= \frac{n(n-1)}{2}\int_{[0,L]^n} U(x_1 - x_2)\left|\Psi^{\text{opt}}(x)\right|^2 dx = \operatorname{Tr}\left(U^p\gamma^{(2)}_{\Psi^{\text{opt}}}\right).\end{aligned} \tag{3.23}$$

According to Proposition 4.8, for Ψ^{opt} having the structure (3.22), its two-particle density matrix is given by

$$\gamma^{(2)}_{\Psi^{\mathrm{opt}}} = \sum_{j=1}^{k_2} \gamma^{(2)}_{\zeta_j} + (\mathrm{Id} - \mathrm{Ex}) \sum_{\substack{i,j=1,\dots,k_1 \\ i<j}} \gamma_{\phi_i} \otimes^s \gamma_{\phi_j} + (\mathrm{Id} - \mathrm{Ex}) \sum_{i=1}^{k_1} \sum_{j=1}^{k_2} \gamma_{\phi_i} \otimes^s \gamma_{\zeta_j}$$
$$+ (\mathrm{Id} - \mathrm{Ex}) \sum_{\substack{i,j=1,\dots,k_2 \\ i<j}} \gamma_{\zeta_i} \otimes^s \gamma_{\zeta_j}. \tag{3.24}$$

As ζ_j is a two-particle state and ϕ_j is a one-particle state, one has

$$\gamma^{(2)}_{\zeta_j} = \langle \cdot, \zeta_j \rangle \zeta_j \quad \text{and} \quad \gamma_{\phi_j} = \langle \cdot, \phi_j \rangle \phi_j .$$

The decomposition (3.24) being plugged in the r.h.s. of (3.23) reads as follows:

(a) the first term corresponds to the interaction of two particles living in the same piece; this term is the leading one in the difference (3.19) and has been already taken into account in the first part of the proof;
(b) the second term is the interaction between two one-particle sub-states living in distinct pieces;
(c) the third term is due to the interaction between a one-particle sub-state in one piece and a two-particle sub-state (represented by its one-particle reduced density matrix) in another piece;
(d) finally, the last term describes the interaction between two distinct two-particle sub-states.

Thus, we are interested in upper bounds on $\mathrm{Tr}(U^p \beta)$ where β is any of the last three terms in (3.24). Let γ_1 and γ_2 be two arbitrary one-particle density matrices encountered in the above expressions. Then, the kernel of $(\mathrm{Id} - \mathrm{Ex})\gamma_1 \otimes^s \gamma_2$ is given by

$$(\mathrm{Id} - \mathrm{Ex})(\gamma_1 \otimes^s \gamma_2)(x, y, x', y') = \frac{1}{2}\big(\gamma_1(x, x')\gamma_2(y, y') + \gamma_2(x, x')\gamma_1(y, y')$$
$$- \gamma_1(y, x')\gamma_2(x, y') - \gamma_2(y, x')\gamma_1(x, y')\big). \tag{3.25}$$

Taking into account the fact that in our case γ_1 and γ_2 live on distinct pieces Δ_1 and Δ_2, respectively, (3.25) implies

$$\mathrm{Tr}\left(U^p (\mathrm{Id} - \mathrm{Ex})\gamma_1 \otimes^s \gamma_2\right) = \int_{\mathbb{R}^2} U^p(x - y)(\mathrm{Id} - \mathrm{Ex})(\gamma_1 \otimes^s \gamma_2)(x, y, x, y)\mathrm{d}x\mathrm{d}y$$
$$= \int_{\Delta_1} \int_{\Delta_2} U^p(x - y)\gamma_1(x, x)\gamma_2(y, y)\mathrm{d}x\mathrm{d}y. \tag{3.26}$$

To upper bound the last expression, we use the estimates proved in Section 6.2. We now study the different sums in (3.24).

For pairs of one-particle states, we estimate the number of pairs of pieces at a certain distance by Proposition 2.3 and we bound individual terms by Lemma 6.18. We compute that, for any $\eta > 0$ and $\varepsilon > 0$, for L sufficiently large, with probability $1 - O(L^{-\infty})$, one has

$$\begin{aligned}
&\operatorname{Tr}\Big(U^p(\mathrm{Id}-\mathrm{Ex}) \sum_{\substack{i,j=1,\dots,k_1\\ i<j}} \gamma_{\phi_i} \otimes^s \gamma_{\phi_j}\Big)\\
&\leqslant \sum_{\substack{|\Delta_i|\geqslant \ell_\rho-\rho x_*\\ |\Delta_j|\geqslant \ell_\rho-\rho x_*\\ \mathrm{dist}(\Delta_i,\Delta_j)\leqslant B\ell_\rho}} \int_{\Delta_i\times\Delta_j} U(x-y)|\varphi^1_{\Delta_i}(x)|^2|\varphi^1_{\Delta_j}(y)|^2 \mathrm{d}x\mathrm{d}y\\
&\leqslant \sum_{k=0}^{B\ell_\rho/\eta} \sum_{\substack{|\Delta_i|\geqslant \ell_\rho-\rho x_*\\ |\Delta_j|\geqslant \ell_\rho-\rho x_*\\ k\eta\leqslant \mathrm{dist}(\Delta_i,\Delta_j)<(k+1)\eta}} \int_{\Delta_i\times\Delta_j} U(x-y)|\varphi^1_{\Delta_i}(x)|^2|\varphi^1_{\Delta_j}(y)|^2 \mathrm{d}x\mathrm{d}y\\
&\leqslant C\sum_{k=0}^{B\ell_\rho/\eta} \#\left\{\begin{array}{c} |\Delta_i|\geqslant \ell_\rho-\rho x_*,\\ |\Delta_j|\geqslant \ell_\rho-\rho x_*\\ k\eta\leqslant \mathrm{dist}(\Delta_i,\Delta_j)<(k+1)\eta \end{array}\right\} \ell_\rho^{-4+\varepsilon}\,((k+1)\eta)^{-\varepsilon} Z((k+1)\eta)\\
&\leqslant CLe^{-2\ell_\rho}\,\ell_\rho^{-4+\varepsilon} \sum_{k=0}^{B\ell_\rho/\eta} ((k+1)\eta)^{-\varepsilon}\, Z((k+1)\eta)\eta.
\end{aligned}$$

Here, to get line three from line two, we have used Lemma 6.18, and to get line four from line three, we have used Proposition 2.3 to bound the counting function with a probability $1 - O(L^{-\infty})$.

Thus, by the continuity and local integrability of $x \mapsto x^{-\varepsilon}Z(x)$, choosing η small and $\varepsilon \in [0, 1)$, we obtain that, for L sufficiently large, with probability $1 - O(L^{-\infty})$, one has

$$\operatorname{Tr}\Big(U^p(\mathrm{Id}-\mathrm{Ex}) \sum_{\substack{i,j=1,\dots,k_1\\ i<j}} \gamma_{\phi_i} \otimes^s \gamma_{\phi_j}\Big) \leqslant C\,n\,\rho\,\ell_\rho^{-4+\varepsilon} \int_0^{B\ell_\rho} a^{-\varepsilon} Z(a)\mathrm{d}a. \tag{3.27}$$

Let us now estimate the last integral. For $\varepsilon \in [0, 1)$ and $0 \leqslant Y < X$, one computes

$$\begin{aligned}\int_0^X a^{-\varepsilon} Z(a)\mathrm{d}a &\leqslant \left(\int_0^Y + \int_Y^X\right) a^{-\varepsilon} Z(a)\mathrm{d}a \\ &\leqslant (1-\varepsilon)^{-1}\left[Z(0)Y^{1-\varepsilon} + Z(Y)X^{1-\varepsilon} - Z(Y)Y^{1-\varepsilon}\right] \\ &= (1-\varepsilon)^{-1}X^{1-\varepsilon}\left[(Y/X)^{1-\varepsilon}(Z(0)-Z(Y)) + Z(Y)\right].\end{aligned}$$

Let us now optimize the last expression with respect to $\alpha = Y/X \in [0, 1]$. Consider

$$f(X,\alpha) := \alpha^{1-\varepsilon}(Z(0) - Z(\alpha X)) + Z(\alpha X). \tag{3.28}$$

In general, the more rapidly Z goes to zero at infinity, the smaller the optimal α and, thus, the smaller is the minimal value. Let us define the following functional of Z (depending also on X):

$$f_Z(X) = \inf_{\alpha\in[0,1]} f(X,\alpha). \tag{3.29}$$

Obviously, as soon as $Z(X) = o(1)$ for $X \to +\infty$, one finds that $f_Z(X) = o(1)$ for $X \to +\infty$. Then, plugging this into the estimate (3.27), we obtain

$$\operatorname{Tr}\Big(U^p(\mathrm{Id}-\mathrm{Ex}) \sum_{\substack{i,j=1,\dots,k_1\\ i<j}} \gamma_{\phi_i} \otimes^s \gamma_{\phi_j}\Big) \leqslant C_1\, n\, \rho\, \ell_\rho^{-3} \cdot f_Z(B\ell_\rho). \tag{3.30}$$

In particular, the last expression is $o(n\rho\ell_\rho^{-3})$. Note also that it can never be made better than $O(n\rho\ell_\rho^{-4})$ as there is no control of the size of Z near the origin.

To estimate the interactions between a one-particle state and a one-particle density matrix of a two-particle state, we use the bound derived in Lemma 6.20. We estimate the number of pairs of pieces of this type at a certain distance by Proposition 2.4 (in this case, there is no need for a more precise Proposition 2.3 as in the derivation of (3.30) above). This yields

$$\begin{aligned}&\operatorname{Tr}\Big(U^p(\mathrm{Id}-\mathrm{Ex})\sum_{i=1}^{k_1}\sum_{j=1}^{k_2}\gamma_{\phi_i}\otimes^s\gamma_{\zeta_j}\Big) \\ &\leqslant \sum_{\substack{|\Delta_i|\geqslant \ell_\rho-\rho x_*\\ |\Delta_j|\in[2\ell_\rho+A_*,3\ell_\rho)\\ \operatorname{dist}(\Delta_i,\Delta_j)\leqslant B\ell_\rho}} \int_{\Delta_i\times\Delta_j} U(x-y)|\varphi^1_{\Delta_i}(x)|^2\gamma_{\zeta^1_{\Delta_j}}(y,y)\mathrm{d}x\mathrm{d}y \\ &\leqslant C\,n\,\rho^2\,\ell_\rho\,\ell_\rho^{-7/2+\varepsilon}\int_0^{B\ell_\rho} a^{-\varepsilon}Z(a)da.\end{aligned} \tag{3.31}$$

Finally, for interactions between two reduced density matrices of two-particle sub-states, we proceed as before; using Lemma 6.21 for each term, we compute

$$\begin{aligned}
&\operatorname{Tr}\Big(U^p(\mathrm{Id}-\mathrm{Ex})\sum_{\substack{i,j=1,\dots,k_2\\ i<j}}\gamma_{\zeta_i}\otimes^s\gamma_{\zeta_j}\Big)\\
&=\sum_{\substack{|\Delta_i|,|\Delta_j|\in[2\ell_\rho+A_*,3\ell_\rho)\\ i<j\\ \operatorname{dist}(\Delta_i,\Delta_j)\leqslant B\ell_\rho}}\int_{\Delta_i\times\Delta_j}U(x-y)\gamma_{\zeta^1_{\Delta_i}}(x,x)\gamma_{\zeta^1_{\Delta_j}}(y,y)\mathrm{d}x\mathrm{d}y \qquad (3.32)\\
&\leqslant C\,n\,\rho^3\int_0^{B\ell_\rho}\min(1,a^{-2}Z(a))da.
\end{aligned}$$

Summing (3.30), (3.31), (3.32), we obtain

$$\langle W^p\Psi^{\mathrm{opt}},\Psi^{\mathrm{opt}}\rangle\leqslant C\,n\,\rho\,\ell_\rho^{-3}\cdot f_Z(B\ell_\rho). \tag{3.33}$$

Taking (3.20) into account, this completes the proof of Theorem 3.19. □

To formulate our next result, we will first need to define the notion of occupation restricted to a subset of the total set of pieces.

Definition 3.21 Let $\mathcal{P}_\omega=\{\Delta_k(\omega)\}_{k=1}^{m(\omega)}$ be the total set of pieces and let $Q\in\mathbb{N}^m$ be an occupation. For $P\subset\mathcal{P}_\omega$ a subset of pieces, define the corresponding sub-occupation (or a restriction of occupation) as an occupation vector containing only those components that are singled out by P:

$$Q|_P=(Q_k)_{k:\,\Delta_k\in P}.$$

When the subset P is defined by a condition on the length of the pieces, we will use a shorthand notation involving only this condition, e.g., $Q|_{>\ell_\rho}$ stands for the occupation Q restricted to the pieces of length greater than the Fermi length ℓ_ρ.

Recall that Ψ^{opt} is constructed in Definition 3.14.

Theorem 3.22 *For any non-negative function $r:[0,\rho_0]\to\mathbb{R}^+$ such that $r(\rho)=o(1)$ when $\rho\to0^+$, there exist $C>0$ and $\rho_r>0$ such that, for $\rho\in(0,\rho_r)$, in the thermodynamic limit, with probability $1-O(L^{-\infty})$, if Ψ is a normalized n-particles state in $\mathfrak{H}_{Q(\Psi)}\cap\mathfrak{H}^n_\infty([0,L])$ (see (3.3)) satisfying*

$$\frac{1}{n}\big\langle H_\omega^{U^p}(L,n)\Psi,\Psi\big\rangle\leqslant\frac{1}{n}\big\langle H_\omega^{U^p}(L,n)\Psi^{\mathrm{opt}},\Psi^{\mathrm{opt}}\big\rangle+\rho|\log\rho|^{-3}(r(\rho))^2, \tag{3.34}$$

then

$$\mathrm{dist}_1\left(Q|_{\geqslant \ell_\rho + C}(\Psi), Q|_{\geqslant \ell_\rho + C}(\Psi^{\mathrm{opt}})\right) \leqslant Cn\rho \cdot \max(r(\rho), |\log\rho|^{-1}),$$
$$\mathrm{dist}_1\left(Q|_{< \ell_\rho + C}(\Psi), Q|_{< \ell_\rho + C}(\Psi^{\mathrm{opt}})\right) \leqslant Cn \max(\sqrt{\rho}\cdot r(\rho), \rho|\log\rho|^{-1}). \tag{3.35}$$

***Proof of Theorem* 3.22** First of all, taking into account the form of the first inequality in (3.35), while dealing with its proof we may suppose without loss of generality that $|\log\rho|^{-1}$ is asymptotically bounded by $r(\rho)$, i.e., for ρ small,

$$|\log\rho|^{-1} \lesssim r(\rho). \tag{3.36}$$

For the proof of the second inequality in (3.35), we will no longer assume (3.36).

Consider now the pieces $(\Delta_k(\omega))_{1\leqslant k\leqslant m(\omega)}$ (see Section 1). Fix $\varepsilon > 0$. We say that a piece $\Delta_k(\omega)$ is of ε-type

(a) if $|\Delta_k(\omega)| \geqslant 3\ell_\rho(1-\varepsilon)$, that is, it has length at least $3\ell_\rho(1-\varepsilon)$;
(b) if $|\Delta_k(\omega)| \geqslant 2\ell_\rho(1-\varepsilon)$ and $\Delta_k(\omega)$ has at least one neighbor (in the sense of interactions U^p from (3.16)) of length at least $\ell_\rho(1-\varepsilon)$;
(c) if $|\Delta_k(\omega)| \geqslant \ell_\rho(1-\varepsilon)$ and $\Delta_k(\omega)$ has at least two neighbors, each of length at least $\ell_\rho(1-\varepsilon)$.

Note that, by (3.16), as U^p is of compact support of radius at most $B\ell_\rho$, there exists $\rho_0 > 0$ such that for $\rho \in (0, \rho_0)$ and $\varepsilon \in (0, 1/2)$, a given piece can have at most $2B$ neighbors of length at least $\ell_\rho(1-\varepsilon)$.

We first prove that "exceptional" pieces contribute only to the error term.

Lemma 3.23 *Fix $\eta \in (0, 1/3)$. There exists $\varepsilon \in (0, 1/2)$ and $\rho_0 > 0$ such that, for $\rho \in (0, \rho_0)$, in the thermodynamic limit, with probability $1 - O(L^{-\infty})$, if $\Psi \in \mathfrak{H}_{Q(\Psi)} \cap \mathfrak{H}^n_\infty([0, L])$ satisfies*

$$\langle H_\omega^{U^p}(L, n)\Psi, \Psi\rangle \leqslant 2\mathcal{E}^0(\rho)n\|\Psi\|^2, \tag{3.37}$$

then

$$\sum_{\bullet\in\{a,b,c\}} \sum_{\Delta_k(\omega) \text{ of } \varepsilon\text{-type } (\bullet)} Q_k(\Psi) \leqslant n\rho^{1+\eta}/2. \tag{3.38}$$

and

$$\sum_{\Delta_k(\omega) \text{ of } \varepsilon\text{-type } (a)} [Q_k(\Psi)]^2 \lesssim \mathcal{E}^0(\rho)n\cdot\log n\cdot\log\log n. \tag{3.39}$$

Let us postpone the proof of this result for a while and continue with the proof of Theorem 3.22. The following lemma estimates the total contribution of "normal" pieces (i.e., that are not of ε-type) that carry too many particles.

Lemma 3.24 *Recall that $\{\Delta_k\}_{k=1}^{m(\omega)}$ denote the pieces.*

There exists $C > 0$ such that, for L sufficiently large, with probability $1 - O(L^{-\infty})$, for a normalized n-state Ψ in $\mathfrak{H}_{Q(\Psi)} \cap \mathfrak{H}^n_\infty([0, L])$ satisfying (3.34) and $Q(\Psi) = (Q_k)_{1\leqslant k\leqslant m(\omega)}$, the occupation number of the state Ψ, one has

$$\sum_{\substack{|\Delta_k|\leqslant \ell_\rho(1-\rho^2)\\ Q_k\geqslant 2}} Q_k + \sum_{\substack{|\Delta_k|\in[\ell_\rho(1-\rho^2),2\ell_\rho(1-\rho^2))\\ Q_k\geqslant 3}} Q_k + \sum_{\substack{|\Delta_k|\in[2\ell_\rho(1-\rho^2),3\ell_\rho(1-\rho^2))\\ Q_k\geqslant 4}} Q_k \leqslant Cn\rho\ell_\rho^{-1} \tag{3.40}$$

and

$$\sum_{|\Delta_k|\leqslant 3\ell_\rho(1-\rho^2)} Q_k^2 \leqslant Cn\rho\ell_\rho^{-1} \tag{3.41}$$

and, for $\varepsilon \in (\rho^2, 1/4)$,

$$\sum_{\substack{|\Delta_k|\leqslant \ell_\rho(1-\varepsilon)\\ Q_k\geqslant 1}} Q_k + \sum_{\substack{|\Delta_k|\leqslant 2\ell_\rho(1-\varepsilon)\\ Q_k\geqslant 2}} Q_k + \sum_{\substack{|\Delta_k|\leqslant 3\ell_\rho(1-\varepsilon)\\ Q_k\geqslant 3}} Q_k \leqslant Cn\frac{\rho}{\varepsilon-\rho^2}\ell_\rho^{-1}. \tag{3.42}$$

Proof First, note that by Theorem 3.19 and (3.34), there exists a constant $\widetilde{C}$ such that

$$\langle H^{U^p}_\omega \Psi, \Psi\rangle \leqslant \langle H^{U^p}_\omega \Psi^{\text{opt}}, \Psi^{\text{opt}}\rangle + n\rho|\log\rho|^{-3}(r(\rho))^2 \leqslant \langle H^0_\omega\Psi^0, \Psi^0\rangle + \widetilde{C}n\rho\ell_\rho^{-3}. \tag{3.43}$$

Moreover, if $-\Delta^{Q_k}_{\Delta_k}$ denotes the Laplacian with Dirichlet boundary conditions on $\bigwedge^{Q_k} L^2(\Delta_k)$, one has

$$(H^{U^p}_\omega)_{\mathfrak{H}_{Q(\Psi)}} \geqslant (H^0_\omega)_{\mathfrak{H}_{Q(\Psi)}} \geqslant \sum_{k=1}^{m(\omega)} \inf(\sigma(-\Delta^{Q_k}_{\Delta_k})) = \sum_{k=1}^{m(\omega)}\sum_{j=1}^{Q_k} \frac{\pi^2 j^2}{|\Delta_k|^2} = \sum_{k=1}^{m(\omega)} \frac{\pi^2 P(Q_k)}{|\Delta_k|^2} \tag{3.44}$$

where $P(X) := \dfrac{(2X+1)(X+1)X}{6}$.

On the other hand, by the description of Ψ^0, for some $C > 0$, one has

$$\langle H^0_\omega\Psi^0, \Psi^0\rangle \leqslant \sum_{|\Delta_k|\in[\ell_\rho(1-\rho^2),2\ell_\rho(1-\rho^2))} \frac{P(1)\,\pi^2}{|\Delta_k|^2} + \sum_{|\Delta_k|\in[2\ell_\rho(1-\rho^2),3\ell_\rho(1-\rho^2))} \frac{P(2)\,\pi^2}{|\Delta_k|^2} + Cn\rho^2$$

Plugging this and (3.44) into (3.43), we obtain

$$\sum_{|\Delta_k|\leqslant\ell_\rho(1-\rho^2)}\frac{\pi^2}{|\Delta_k|^2}P(Q_k)+\sum_{|\Delta_k|\in[\ell_\rho(1-\rho^2),2\ell_\rho(1-\rho^2))}\frac{\pi^2}{|\Delta_k|^2}(P(Q_k)-P(1))$$
$$+\sum_{|\Delta_k|\in[2\ell_\rho(1-\rho^2),3\ell_\rho(1-\rho^2))}\frac{\pi^2}{|\Delta_k|^2}(P(Q_k)-P(2))\leqslant Cn\rho\ell_\rho^{-3}. \tag{3.45}$$

By Lemma 3.23 and the explicit description of the non- interacting ground state Ψ^0 (see the beginning of Section 3.5), for some $C>0$ and ρ sufficiently small, for L sufficiently large, with probability $1-O(L^{-\infty})$, one has

$$\begin{aligned}\sum_{|\Delta_k|\leqslant\ell_\rho(1-\rho^2)}Q_k+&\sum_{|\Delta_k|\in[\ell_\rho(1-\rho^2),2\ell_\rho(1-\rho^2))}Q_k+\sum_{|\Delta_k|\in[2\ell_\rho(1-\rho^2),3\ell_\rho(1-\rho^2))}Q_k\\ &\geqslant n(1-C\rho^2)\\ &\geqslant\left[\sum_{|\Delta_k|\in[\ell_\rho(1+\rho^2),2\ell_\rho(1-\rho^2))}1+\sum_{|\Delta_k|\in[2\ell_\rho(1+\rho^2),3\ell_\rho(1-\rho^2))}2\right]-2Cn\rho^2\\ &\geqslant\left[\sum_{|\Delta_k|\in[\ell_\rho(1-\rho^2),2\ell_\rho(1-\rho^2))}1+\sum_{|\Delta_k|\in[2\ell_\rho(1-\rho^2),3\ell_\rho(1-\rho^2))}2\right]-3Cn\rho^2\end{aligned} \tag{3.46}$$

as

$$\#\{k;\ |\Delta_k|\in[\ell_\rho(1-\rho^2),\ell_\rho(1+\rho^2))\cup[2\ell_\rho(1-\rho^2),2\ell_\rho(1+\rho^2))\}\leqslant Cn\rho^2.$$

Thus, (3.46) yields

$$\begin{aligned}&\sum_{\substack{|\Delta_k|\leqslant\ell_\rho(1-\rho^2)\\ Q_k\geqslant1}}Q_k+\sum_{\substack{|\Delta_k|\in[\ell_\rho(1-\rho^2),2\ell_\rho(1-\rho^2))\\ Q_k\geqslant2}}(Q_k-1)+\sum_{\substack{|\Delta_k|\in[2\ell_\rho(1-\rho^2),3\ell_\rho(1-\rho^2))\\ Q_k\geqslant3}}(Q_k-2)\\ &\geqslant\left[\sum_{\substack{|\Delta_k|\in[\ell_\rho(1-\rho^2),2\ell_\rho(1-\rho^2))\\ Q_k=0}}1+\sum_{\substack{|\Delta_k|\in[2\ell_\rho(1-\rho^2),3\ell_\rho(1-\rho^2))\\ Q_k\leqslant1}}2\right]-3n\rho^{1+\eta}\end{aligned} \tag{3.47}$$

Rewrite (3.45) as

$$
\begin{aligned}
Cn\rho\ell_\rho^{-1} \geqslant & \sum_{\substack{|\Delta_k|\leqslant \ell_\rho(1-\rho^2)\\ Q_k\geqslant 1}} \frac{\pi^2}{|\Delta_k|^2} P(Q_k) + \sum_{\substack{|\Delta_k|\in[\ell_\rho(1-\rho^2),2\ell_\rho(1-\rho^2))\\ Q_k\geqslant 2}} \frac{\pi^2}{|\Delta_k|^2}(P(Q_k)-P(1)) \\
& + \sum_{\substack{|\Delta_k|\in[2\ell_\rho(1-\rho^2),3\ell_\rho(1-\rho^2))\\ Q_k\geqslant 3}} \frac{\pi^2}{|\Delta_k|^2}(P(Q_k)-P(2)) \\
& - \sum_{\substack{|\Delta_k|\in[\ell_\rho(1-\rho^2),2\ell_\rho(1-\rho^2))\\ Q_k=0}} \frac{P(1)\pi^2}{|\Delta_k|^2} \\
& - \sum_{\substack{|\Delta_k|\in[2\ell_\rho(1-\rho^2),3\ell_\rho(1-\rho^2))\\ Q_k\leqslant 1}} \frac{(P(2)-P(Q_k))\pi^2}{|\Delta_k|^2} \\
\geqslant & \sum_{\substack{|\Delta_k|\leqslant \ell_\rho(1-\rho^2)\\ Q_k\geqslant 1}} \frac{\pi^2}{|\Delta_k|^2} P(Q_k) + \sum_{\substack{|\Delta_k|\in[\ell_\rho(1-\rho^2),2\ell_\rho(1-\rho^2))\\ Q_k\geqslant 2}} \frac{\pi^2}{|\Delta_k|^2}(P(Q_k)-P(1)) \\
& + \sum_{\substack{|\Delta_k|\in[2\ell_\rho(1-\rho^2),3\ell_\rho(1-\rho^2))\\ Q_k\geqslant 3}} \frac{\pi^2}{|\Delta_k|^2}(P(Q_k)-P(2)) \\
& - P(1)\left(\sum_{\substack{|\Delta_k|\in[\ell_\rho(1-\rho^2),2\ell_\rho(1-\rho^2))\\ Q_k=0}} \frac{\pi^2}{|\Delta_k|^2} \right) - \\
& P(2)\left(\sum_{\substack{|\Delta_k|\in[2\ell_\rho(1-\rho^2),3\ell_\rho(1-\rho^2))\\ Q_k\leqslant 1}} \frac{\pi^2}{|\Delta_k|^2} \right).
\end{aligned}
$$

Hence,

$$
\begin{aligned}
Cn\rho\ell_\rho^{-1} \geqslant & \sum_{\substack{|\Delta_k|\leqslant \ell_\rho(1-\rho^2)\\ Q_k\geqslant 1}} \frac{\pi^2}{|\Delta_k|^2}P(Q_k) + \sum_{\substack{|\Delta_k|\in[\ell_\rho(1-\rho^2),2\ell_\rho(1-\rho^2))\\ Q_k\geqslant 2}} \frac{\pi^2}{|\Delta_k|^2}(P(Q_k)-P(1)) \\
& + \sum_{\substack{|\Delta_k|\in[2\ell_\rho(1-\rho^2),3\ell_\rho(1-\rho^2))\\ Q_k\geqslant 3}} \frac{\pi^2}{|\Delta_k|^2}(P(Q_k)-P(2)) \\
& - \frac{\pi^2}{|\ell_\rho(1-\rho^2)|^2}\left[\sum_{|\Delta_k|\in[\ell_\rho(1-\rho^2),2\ell_\rho(1-\rho^2))} 1 + \sum_{|\Delta_k|\in[2\ell_\rho(1-\rho^2),3\ell_\rho(1-\rho^2))} 2\right]
\end{aligned}
$$

as $P(1) = 1$ and $P(2) = 5 \leqslant 8 = 2^3 P(1)$.

Using (3.47), we then obtain

$$
\begin{aligned}
Cn\rho\ell_\rho^{-1} \geqslant & \sum_{\substack{|\Delta_k|\leqslant \ell_\rho(1-\rho^2)\\ Q_k\geqslant 1}} \left(\frac{\pi^2}{|\Delta_k|^2}P(Q_k) - \frac{\pi^2}{|\ell_\rho(1-\rho^2)|^2}Q_k\right) \\
& + \sum_{\substack{|\Delta_k|\in[\ell_\rho(1-\rho^2),2\ell_\rho(1-\rho^2))\\ Q_k\geqslant 2}} \left(\frac{\pi^2}{|\Delta_k|^2}(P(Q_k)-P(1)) - \frac{\pi^2}{|\ell_\rho(1-\rho^2)|^2}(Q_k-1)\right) \\
& + \sum_{\substack{|\Delta_k|\in[2\ell_\rho(1-\rho^2),3\ell_\rho(1-\rho^2))\\ Q_k\geqslant 3}} \left(\frac{\pi^2}{|\Delta_k|^2}(P(Q_k)-P(2)) - \frac{\pi^2}{|\ell_\rho(1-\rho^2)|^2}(Q_k-2)\right).
\end{aligned}
\tag{3.48}
$$

Now, we note that, for $X \geqslant n+1$, X integer, one has

$$
P(X) - P(n) = \sum_{k=n+1}^{X} k^2 \geqslant (n+1)^2(X-n). \tag{3.49}
$$

This yields

- for $Q_k \geqslant 1$ and $|\Delta_k| \leqslant \ell_\rho(1-\rho^2)$, one has

$$
\frac{\pi^2}{|\Delta_k|^2}P(Q_k) - \frac{\pi^2}{|\ell_\rho(1-\rho^2)|^2}Q_k > \frac{\pi^2 Q_k(Q_k-1)(2Q_k+3)}{6|\ell_\rho(1-\rho^2)|^2} \geqslant 0; \tag{3.50}
$$

  if, moreover, $|\Delta_k| \leqslant \ell_\rho(1-\varepsilon)$ $(\rho^2 < \varepsilon < 1/2)$, by (3.49), one has

$$
\frac{\pi^2}{|\Delta_k|^2}P(Q_k) - \frac{\pi^2}{|\ell_\rho(1-\rho^2)|^2}Q_k \geqslant \left(\frac{\pi^2}{|\Delta_k|^2} - \frac{\pi^2}{|\ell_\rho(1-\rho^2)|^2}\right)Q_k \geqslant \frac{(8\pi)^2(\varepsilon-\rho^2)}{|\ell_\rho|^2}Q_k; \tag{3.51}
$$

- for $Q_k \geqslant 2$ and $|\Delta_k| \leqslant 2\ell_\rho(1-\rho^2)$, one has

$$\frac{\pi^2}{|\Delta_k|^2}(P(Q_k)-P(1))-\frac{\pi^2}{|\ell_\rho(1-\rho^2)|^2}(Q_k-1) > \frac{\pi^2(2Q_k+9)(Q_k-2)(Q_k-1)}{24|\ell_\rho(1-\rho^2)|^2} \geqslant 0; \tag{3.52}$$

if, moreover, $|\Delta_k| \leqslant 2\ell_\rho(1-\varepsilon)$ ($\rho^2 < \varepsilon < 1/2$), by (3.49), one has

$$\frac{\pi^2}{|\Delta_k|^2}(P(Q_k)-P(1))-\frac{\pi^2}{|\ell_\rho(1-\rho^2)|^2}(Q_k-1) \geqslant \left(\frac{4\pi^2}{|\Delta_k|^2}-\frac{\pi^2}{|\ell_\rho(1-\rho^2)|^2}\right)(Q_k-1) \geqslant \frac{(8\pi)^2(\varepsilon-\rho^2)}{|\ell_\rho|^2}(Q_k-1); \tag{3.53}$$

- for $Q_k \geqslant 3$ and $|\Delta_k| \leqslant 3\ell_\rho(1-\rho^2)$, one has

$$\frac{\pi^2}{|\Delta_k|^2}(P(Q_k)-P(2))-\frac{\pi^2}{|\ell_\rho(1-\rho^2)|^2}(Q_k-2) > \frac{\pi^2(2Q_k+13)(Q_k-3)(Q_k-2)}{|\ell_\rho(1-\rho^2)|^2} \geqslant 0; \tag{3.54}$$

if, moreover, $|\Delta_k| \leqslant 3\ell_\rho(1-\varepsilon)$ ($\rho^2 < \varepsilon < 1/2$), by (3.49), one has

$$\frac{\pi^2}{|\Delta_k|^2}(P(Q_k)-P(2))-\frac{\pi^2}{|\ell_\rho(1-\rho^2)|^2}(Q_k-2) \geqslant \left(\frac{9\pi^2}{|\Delta_k|^2}-\frac{\pi^2}{|\ell_\rho(1-\rho^2)|^2}\right)(Q_k-9) \geqslant \frac{(8\pi)^2(\varepsilon-\rho^2)}{|\ell_\rho|^2}(Q_k-2). \tag{3.55}$$

Plugging (3.50)–(3.55) into (3.48) immediately yields (3.40) and (3.42), thus, completes the proof of (3.40) and (3.42) in Lemma 3.24.

To derive (3.41), we proceed as follows. Clearly, for $Q_k \geqslant 4$, the right-hand sides of (3.50), (3.52), and (3.54) are larger than $\delta \cdot Q_k^2$ (for some $\delta \in (0,1)$). Thus, (3.48) implies

$$\sum_{\substack{|\Delta_k|\leqslant 3\ell_\rho(1-\rho^2)\\ Q_k\geqslant 4}} Q_k^2 \leqslant Cn\rho\ell_\rho^{-1}.$$

On the other hand, by (3.40), one clearly has

$$\sum_{\substack{|\Delta_k|\leqslant 3\ell_\rho(1-\rho^2)\\ Q_k\leqslant 3}} Q_k^2 \leqslant 3 \sum_{\substack{|\Delta_k|\leqslant 3\ell_\rho(1-\rho^2)\\ Q_k\leqslant 3}} Q_k \leqslant Cn\rho\ell_\rho^{-1}.$$

Thus, the proof of (3.41) is complete. This completes the proof of Lemma 3.24. □

We also remark the following

Lemma 3.25 *Consider $\Psi_\omega^{U^p}$, the ground state of $H_\omega^{U^p}(L, n)$.*
There exists $C > 0$ such that for L sufficiently large, with probability at least $1 - O(L^{-\infty})$, no piece of length smaller than

$$\ell_{min} = \ell_\rho - C\rho\ell_\rho \tag{3.56}$$

is occupied by particles of Ψ^{U^p}.

Remark 3.26 The proof of Lemma 3.25 shows that it suffices to take $C > 4B + 4$ for ρ sufficiently small; here, B is the constant defining U^p (see (3.16)).

Proof Suppose that the claim of the lemma is false. Then, a piece shorter than ℓ_{min} is occupied.

Let us show now that, as there are too many such pieces, pieces longer than ℓ_{min} cannot be all in interaction with n particles, no matter where these n particles are.

First of all, according to Proposition 2.2, the total number of pieces longer than ℓ_{min} is

$$\begin{aligned}\sharp\{j\,:\,|\Delta_j(\omega)| \geqslant \ell_{min}\} &= Le^{-\ell_{min}} + O(L^{1/2+0}) = L\frac{\rho}{1+\rho}(1 + C\rho\ell_\rho + O(\rho^2\ell_\rho^2)) \\ &= n(1 + C\rho\ell_\rho + O(\rho)).\end{aligned}$$

The number of pieces of length larger than $2\ell_\rho$ is $n\rho(1 + O(\rho))$. If a particle lies in one of these pieces, it can interact with at most $2B$ other pieces of length greater than ℓ_{min}.

For pieces smaller than $2\ell_\rho$ (but as always larger than ℓ_{min}), we remark that if two such pieces are at a distance greater than $(2B + 2)\ell_\rho$ from one another then they cannot interact with the same particle, except for the cases already taken into account above.

Moreover, according to Proposition 2.3, the number of pairs of such pieces at distance at most $(2B + 2)\ell_\rho$ is given by

$$\begin{aligned}&\sharp\{(\Delta_i, \Delta_j), |\Delta_i| > \ell_{min}, |\Delta_j| > \ell_{min}, \mathrm{dist}(\Delta_i, \Delta_j) \leqslant (2B + 2)\ell_\rho\} \\ &= 2(2B + 2)\ell_\rho L\left(e^{-\ell_{min}}\right)^2 + O(L^{3/4}) \\ &= (4B + 4)n\rho\ell_\rho(1 + O(\rho\ell_\rho)).\end{aligned}$$

Consequently, the rest of these pieces are at larger distances from each other. This leaves at least

$$\begin{aligned}n(1 + C\rho\ell_\rho + O(\rho)) - (2B + 1)n\rho(1 + O(\rho)) - (4B + 4)n\rho\ell_\rho(1 + O(\rho\ell_\rho)) \\ = n(1 + (C - 4B - 4)\rho\ell_\rho + O(\rho))\end{aligned}$$

pieces such that no two of them can interact with the same particle. Remark that it suffices to take $C > 4B + 4$ to ensure that this number is larger than n for ρ small. This proves that there exists at least one piece longer than ℓ_{min} which is neither occupied nor interacting with any particle in a ground state $\Psi_\omega^{UP}(L,n)$.

This leads to a contradiction with the fact that the ground state $\Psi_\omega^{UP}(L,n)$ puts at least one particle in a piece smaller than ℓ_{min}: indeed, moving this particle to the piece longer than ℓ_{min} which was singled out just above would result in a decrease of energy as no interaction energy would be added and non-interacting energy would obviously decrease with the increase of the piece's length. This completes the proof of Lemma 3.25. □

Let us now resume the proof of Theorem 3.22. In what follows, Ψ is a function satisfying condition (3.34). By Theorem 3.19, using $\Psi^{\mathrm{opt}}(L,n)$ as a trial function, we see that both Ψ and $\Psi^{\mathrm{opt}}(L,n)$ satisfy the assumptions of Lemma 3.23. Thus, picking $\eta \in (0, 1/3)$ and ε sufficiently small, by Lemma 3.23, for ρ sufficiently small and L sufficiently large, with probability $1 - O(L^{-\infty})$, we have

$$\sum_{\bullet\in\{a,b,c\}} \sum_{\Delta_k(\omega) \text{ of } \varepsilon\text{-type } (\bullet)} \left(Q_k(\Psi^{\mathrm{opt}}(L,n)) + Q_k(\Psi)\right) \leqslant n\rho^{1+\eta}. \tag{3.57}$$

We will now reason on the particles in $\Psi_\omega^{UP}(L,n)$ that live in pieces that are not of ε-type (a), (b) or (c).

Recall that, by definition (see Definitions 3.9 and 3.14), $\Psi^{\mathrm{opt}}(L,n)$ puts

- no particle in each piece of length in $(0, \ell_\rho - x_*\rho)$;
- one particle in each piece of length in $[\ell_\rho - x_*\rho, 2\ell_\rho + A_*)$;
- two particles (as a true two-particle state) in each piece of length in $[2\ell_\rho + A_*, 3\ell_\rho)$;

Let C be the constant from the claim of Theorem 3.22 that we will fix later on. Define

- n_0^+ to be the total number of pieces of length in $(0, \ell_\rho - x_*\rho)$ where Ψ puts exactly 1 particle;
- n_1^- to be the total number of pieces of length in $[\ell_\rho - x_*\rho, \ell_\rho + C)$ where Ψ puts no particle;
- n_1^+ to be the total number of pieces of length in $[\ell_\rho - x_*\rho, \ell_\rho + C)$ where Ψ puts exactly 2 particles;
- $\widetilde{n}_1^-$ to be the total number of pieces of length in $[\ell_\rho + C, 2\ell_\rho + A_*)$ where Ψ puts no particle;
- $\widetilde{n}_1^+$ to be the total number of pieces of length in $[\ell_\rho + C, 2\ell_\rho + A_*)$ where Ψ puts exactly 2 particles;
- n_2^- to be the total number of pieces of length in $[2\ell_\rho + A_*, 3\ell_\rho(1-\varepsilon))$ where Ψ puts exactly 1 particle;
- n_2^+ to be the total number of pieces of length in $[2\ell_\rho + A_*, 3\ell_\rho(1-\varepsilon))$ where Ψ puts exactly 3 particles.

The general idea of the forthcoming proof is the following. On the one hand, Lemma 3.23 tells that pieces with too many neighbors are a sort of exception in a sense that they occur relatively rarely and carry relatively few particles. On the other hand, according to Lemma 3.24, pieces with too many particles are also relatively exceptional.

Finally, let us complement these two observations by noting that no particle in a piece of length in $[2\ell_\rho + A_*, 3\ell_\rho(1-\varepsilon))$ can also occur for a small fraction of them. Therefore, we first note that it is sufficient to argue for pieces that are not of ε-type (as those of ε-type are already handled by Lemma 3.23). Let us now take a look at the distribution of particles in the state Ψ^{opt} in the pieces of length in $[2\ell_\rho + A_*, 3\ell_\rho(1-\varepsilon))$ that have no particles and no neighbors (as they are not of ε-type) in Ψ. Obviously, moving a particle from a piece of length greater than $2\ell_\rho + A_*$ to a smaller piece induces an increase of the non-interacting energy of order ℓ_ρ^{-2} just because the pieces longer than $\ell_\rho - \rho x_*$ are already occupied by at least one particle (thus, the non-interacting energy of a second particle is at best $4\pi^2/(2\ell_\rho + A_*)^2$ and $\pi^2/(\ell_\rho - \rho x_*)^2$ if a particle is placed in a non- occupied piece). Thus, the total number of pieces of length greater than $2\ell_\rho + A_*$ with no particles is bounded by $O(n\rho\ell_\rho^{-1})$.

The last three arguments together prove essentially that the distances dist_0 and dist_1 coincide for the matter of the current proof up to an admissible error, i.e. of size $O(n\rho\ell_\rho^{-1})$. Namely, by the definition of the distance dist_1, one has

$$\begin{aligned} \mathrm{dist}_1(Q|_{<\ell_\rho+C}(\Psi), Q|_{<\ell_\rho+C}(\Psi^{\mathrm{opt}})) &= n_0^+ + n_1^+ + n_1^- + r, \\ \mathrm{dist}_1(Q|_{\geqslant\ell_\rho+C}(\Psi), Q|_{\geqslant\ell_\rho+C}(\Psi^{\mathrm{opt}})) &= \widetilde{n}_1^+ + \widetilde{n}_1^- + n_2^+ + n_2^- + r', \end{aligned} \tag{3.58}$$

and, by the fact that the total number of particles in both states is the same, one gets

$$n_0^+ + n_1^+ + \widetilde{n}_1^+ + n_2^+ + r'' = n_1^- + \widetilde{n}_1^- + n_2^- + r''' \tag{3.59}$$

where

$$\max(r, r', r'', r''') \leqslant Cn\rho\ell_\rho^{-1}. \tag{3.60}$$

Recall that $r(\rho)$ is of order at most $|\log\rho|^{-1}$. Hence, if (3.35) does not hold, for any constant C_1, if L is large enough, either one has

$$\widetilde{n}_1^+ + \widetilde{n}_1^- + n_2^+ + n_2^- \geqslant C_1 n\rho \cdot r(\rho) \tag{3.61}$$

or one has

$$n_0^+ + n_1^+ + n_1^- \geqslant C_1 n\sqrt{\rho} \cdot r(\rho). \tag{3.62}$$

First, we simplify (3.61). Suppose that, for some C_1 large, one has

$$n_2^+ \geqslant \frac{C_1}{4} n\rho \cdot r(\rho). \tag{3.63}$$

The number of pieces of length in $\left[\frac{5}{2}\ell_\rho, 3\ell_\rho(1-\varepsilon)\right)$ is given by

$$\sharp\left\{j : |\Delta_j(\omega)| \in \left[\frac{5}{2}\ell_\rho, 3\ell_\rho(1-\varepsilon)\right)\right\} = O(n\rho^{3/2}).$$

Thus, at least $\frac{C_1}{5} n\rho \cdot r(\rho)$ of the pieces with three particles (as given by (3.63)) have their length in $\left[2\ell_\rho + A_*, \frac{5}{2}\ell_\rho\right)$. Hence, the non-interacting energy excess (compared to the non-interacting energy in the ground state) for each of these pieces is lower bounded by $O(\ell_\rho^{-2})$ which, in turn, being multiplied by their total number, contradicts (3.34). This simplifies (3.61) into

$$\widetilde{n}_1^+ + \widetilde{n}_1^- + n_2^- \geqslant C_1 n\rho \cdot r(\rho). \tag{3.64}$$

The conditions (3.59), (3.60) and either (3.62) or (3.64) lead us to a number of possibilities that we will now study one by one. More precisely, there are nine possible variants as at least one among n_1^-, $\tilde{n}_1^-$ and n_2^- should be "large" and the same is true for either n_0^+, n_1^+, n_2^+ and $\tilde{n}_1^+$. We now discuss these cases.

(a) Consider first the case when

$$\min(\widetilde{n}_1^+, n_2^-) \geqslant C_2 n\rho \cdot r(\rho) \tag{3.65}$$

with $C_2 < C_1/3$.

This corresponds to taking the same configuration of particles as in Ψ^{opt} and move some of them from pieces of length in $[2\ell_\rho + A_*, 3\ell_\rho(1-\varepsilon))$ to pieces of length in $[\ell_\rho + C, 2\ell_\rho + A_*)$ that already contain one particle each. As we are now dealing only with pieces that are not of ε-type, this implies in particular that the pieces of length in $[2\ell_\rho + A_*, 3\ell_\rho(1-\varepsilon))$ from which we withdraw particles and that originally contain 2 particles do not have any neighbors.

Taking the smallest available pieces for particle donors and the largest available for particle acceptors gives a lower bound on the total energy increase induced by this operation. Suppose that $C_2 n\rho r(\rho)$ smallest pieces have their length between $2\ell_\rho + A_*$ and $2\ell_\rho + A_* + \delta$. Then, choosing C_1 (thus, C_2) much larger than the constant in Lemma 3.24 for the case when $r(\rho) \asymp |\log\rho|^{-1}$, we obtain

$$Le^{-2\ell_\rho - A_*}(1 - e^{-\delta}) \geqslant \frac{C_2}{2} n\rho \cdot r(\rho),$$

which yields

$$\delta \geqslant \frac{C_2 e^{A_*}}{2} r(\rho). \tag{3.66}$$

Moreover, analogous calculations show that at least $\frac{C_2}{3} n\rho r(\rho)$ of these pieces have length in $(2\ell_\rho + A_* + \delta/2, 2\ell_\rho + A_* + \delta)$. For the particles in these pieces, the increase of energy is lower bounded by

$$\frac{4\pi^2}{(2\ell_\rho + A_* + \delta/2)^2} + \frac{\gamma}{(2\ell_\rho + A_* + \delta/2)^3} - \frac{4\pi^2}{(2\ell_\rho + A_*)^2} - \frac{\gamma}{(2\ell_\rho + A_*)^3} + O(\ell_\rho^{-4}) \geqslant C_3 r(\rho) \ell_\rho^{-3}, \tag{3.67}$$

where $C_3 > 0$. Multiplying the number of pieces by the lower bound (3.67) gives a total energy excess that contradicts (3.34) if we choose C_2 (hence, C_1) sufficiently large.

(b) The case

$$\min(n_1^+, n_2^-) \geqslant C_2 n\rho \cdot r(\rho)$$

is even simpler than the previous one. Indeed, in Ψ^{opt}, the occupations of the pieces of length in $[\ell_\rho - \rho x_*, \ell_\rho + C)$ and in $[\ell_\rho + C, 2\ell_\rho + A_*)$ are the same but the lengths considered in the previous case are smaller. Hence, the arguments developed in point (a) above enable one to conclude with the only difference that the increase of energy is even larger. Moreover, there is no need to remove the small interval of size δ.

(c) Next, the situation when

$$\min(n_0^+, n_2^-) \geqslant C_2 n\rho \cdot r(\rho) \tag{3.68}$$

corresponds to moving excited particles, i.e., particles occupying the second energy level, from pieces of length in $[2\ell_\rho + A_*, 3\ell_\rho(1-\varepsilon))$ to empty pieces of length smaller than $\ell_\rho - \rho x_*$. Recall that actually the approximate equilibrium between the gain in interaction energy due to decoupling and the increase of non-interaction energy was part of the definition of values of x_* and A_*, i.e.,

$$\frac{4\pi^2}{(2\ell_\rho + A_*)^2} + \gamma \ell_\rho^{-3} = \frac{\pi^2}{(\ell_\rho - \rho x_*)^2} + O(\ell_\rho^{-4}). \tag{3.69}$$

Obviously, the smaller the piece we choose to remove the second particle from, the more energy one gains. On the other hand, the larger the piece where one puts the particle, the smaller the non-interacting energy increase, thus, the better.

According to these two observations, we choose to move particles from the $C_2 n\rho \cdot r(\rho)$ smallest pieces longer than $2\ell_\rho + A_*$. Suppose that the largest of these pieces has length $2\ell_\rho + A_* + B_2$. Then, by Proposition 2.2, B_2 satisfies

$$Le^{-2\ell_\rho - A_*}(1 - e^{-B_2}) + O(L^{1/2+0}) = C_2 n\rho \cdot r(\rho).$$

Hence, $B_2 = C_2 e^{A_*} r(\rho)(1 + O(r(\rho)))$. Moreover, the number of such pieces with length in $[2\ell_\rho + A_* + B_2/2, 2\ell_\rho + A_* + B_2)$ is

$$\begin{aligned} &\sharp\{k;\ |\Delta_k(\omega)| - 2\ell_\rho - A_* \in [B_2/2, B_2)\} \\ &= Le^{-2\ell_\rho - A_*}(e^{-B_2/2} - e^{-B_2}) + O(L^{1/2+0}) \geqslant \frac{C_2}{3} n\rho \cdot r(\rho). \end{aligned} \tag{3.70}$$

Clearly, for all these $\frac{C_2}{3} n\rho\ell_\rho^{-1}$ pieces, the non-interacting energy excess is proportional to $C_2\ell_\rho^{-3} r(\rho)$; thus, multiplied by their total number (3.70), for large C_2, this energy excess does not fit within the margin allowed by (3.34).

(d) Yet another possibility for (3.64) is that

$$\min(\max(n_1^+, \widetilde{n}_1^+), \max(n_1^-, \widetilde{n}_1^-)) \geqslant C_2 n\rho \cdot r(\rho).$$

Obviously, the variant

$$\min(\widetilde{n}_1^+, n_1^-) \geqslant C_2 n\rho \cdot r(\rho).$$

is more advantageous from the energetic point of view. The question here is whether it is worth moving a particle from a piece of length close to the lower bound of the corresponding group, i.e., $\ell_\rho - \rho x_*$, to another piece (but as the second particle because there is already another particle in that piece) of length close to the upper bound, i.e., $2\ell_\rho + A_*$. In a certain sense, this is the opposite to the case (c) as the latter tells that the threshold value A_* is not too small, while the current case will explain why A_* is not too big.

As above, one shows that, in order to choose the $C_2 n\rho \cdot r(\rho)$ largest pieces of length in $[\ell_\rho - \rho x_*, 2\ell_\rho + A_*)$, it is sufficient to solve

$$Le^{-2\ell_\rho - A_*}(e^{B_1} - 1) + O(L^{1/2+0}) = C_2 n\rho \cdot r(\rho),$$

which also implies $B_1 = C_2 e^{A_*} r(\rho)(1 + O(r(\rho)))$. Then, as above, the energy excess is proportional to $C_2\ell_\rho^{-3} r(\rho)$ (where the constant C_2 can be chosen arbitrarily large) whereas the interaction terms are uniformly bounded by $O(\ell_\rho^{-4+0})$. Thus, the total energy gained by such an operation exceeds the limits imposed by (3.34).

(e) The next possible option is that

$$\min(n_0^+, \widetilde{n}_1^-) \geqslant C_2 n\rho \cdot r(\rho). \tag{3.71}$$

This corresponds to moving particles in Ψ^{opt} from pieces longer than $\ell_\rho + C$ to pieces shorter than $\ell_\rho - \rho x_*$. Remark first that the increase of non- interacting energy is at least

$$\frac{\pi^2}{(\ell_\rho - \rho x_*)^2} - \frac{\pi^2}{(\ell_\rho + C)^2} \geqslant \frac{2\pi^2 C}{\ell_\rho^3}, \tag{3.72}$$

which always dominates the possible interaction with a particle in a neighboring piece: this interaction is $O(\ell_\rho^{-4+0})$ by Lemma 6.18. Multiplying the left- hand sides of (3.71) and (3.72) gives a lower estimate on the energy excess that contradicts (3.34) because $r(\rho) = o(1)$.

(f) Finally, the only case left is when

$$\min(n_0^+, n_1^-) \geqslant C_2 n\sqrt{\rho} \cdot r(\rho). \tag{3.73}$$

Informally speaking, this is about the question if the threshold $\ell_\rho - \rho x_*$ between occupation zero and occupation one is placed correctly.

It is also remarkable that the allowed number of particle displacements for this case is much larger than in the other cases: one has to compare $o(n\sqrt{\rho})$ to $o(n\rho)$. This is due to the following mechanism. First, note that moving a particle that interacts with another particle in a neighboring piece may result in a decrease of the total energy. Obviously, the contribution of the displacement of such particles is upper bounded by $O(n\rho\ell_\rho^{-4+0})$ because there are at most $O(n\rho)$ neighboring particles and the size of interaction is $O(\ell_\rho^{-4+0})$ by Lemma 6.18. Thus, these particles may be neglected for the precision of the current proof.

Then, reasoning as we did many times above, we observe that at least $\frac{C_2}{2} n\sqrt{\rho} r(\rho)$ of particles that are removed from pieces of length in $[\ell_\rho - \rho x_*, \ell_\rho + C)$ have their length greater than $\ell_\rho + C_3\sqrt{\rho} r(\rho)$, where the constant C_3 grows together with C_2. But, for each of these particles the non-interacting energy increase is of order $C_3 \ell_\rho^{-3}\sqrt{\rho} \cdot r(\rho)$. As above, multiplying the number of involved particles by the lower bound on the energy change, we get a contradiction with (3.34).

This completes the proof of Theorem 3.22. □

We are now left with proving Lemma 3.23.

***The Proof of Lemma* 3.23** We first prove the estimate (3.38). It will be a consequence of the fact that the number of pieces in any of the three type is small and of the following

Lemma 3.27 *Pick k pieces of respective lengths* $l_1 \leqslant l_2 \leqslant \cdots \leqslant l_k$. *Assume that, for* $1 \leqslant i \leqslant k$, *the state* $\Psi \in \mathfrak{H}^n_{Q(\Psi)} \cap \mathfrak{H}^n_\infty([0, L])$ *puts exactly* ν_i *particles in the piece i so that* $\nu_1 + \cdots + \nu_k = \nu$. *Then, one has*

$$\frac{\pi^2\nu^3}{3l_k^2k^2} \leqslant \langle H^0(L,n)\Psi,\Psi\rangle \leqslant \langle H_\omega^{U^p}(L,n)\Psi,\Psi\rangle \leqslant \langle H_\omega^{U}(L,n)\Psi,\Psi\rangle. \tag{3.74}$$

Let us postpone the proof of this result for a while and complete the proof of Lemma 3.23. We shall write out the proof for pieces of type (a). Those for pieces of type (b) and (c) is similar.

Pick $\eta \in (0,1)$ and $\varepsilon > 0$ such that $\eta + 2\varepsilon < 1/6$. The proofs of Propositions 2.2 and 2.1 show that there exists $\rho_\varepsilon > 0$ such that, for $\rho \in (0,\rho_\varepsilon)$, for L sufficiently large, with probability $1 - O(L^{-\infty})$, one has

$$\#\left\{k;\ |\Delta_k(\omega)| \in [3\ell_\rho(1-\varepsilon), 4\ell_\rho)\right\} \leqslant n\rho^{2-3\varepsilon} \tag{3.75}$$

and, for $4 \leqslant k \leqslant \log L \cdot \log\log L$,

$$\#\left\{k;\ |\Delta_k(\omega)| \in [k\ell_\rho, (k+1)\ell_\rho)\right\} \leqslant n\rho^{k-1-\varepsilon}. \tag{3.76}$$

Now, if Ψ places more than $n\rho^{1+\eta}$ particles in pieces of type a then

- either it places at least $2^{-1}n\rho^{1+\eta}$ particles in pieces of length in $[3\ell_\rho(1-\varepsilon), 4\ell_\rho)$; in this case, by Lemma 3.27, as $3(\eta + 2\varepsilon) < 1$, we know that

$$\langle H^0(L,n)\Psi,\Psi\rangle \geqslant \frac{\pi^2(n\rho^{1+\eta})^3}{8(4\ell_\rho)^2(n\rho^{2-3\varepsilon})^2} \gtrsim n\ell_\rho^{-2}\rho^{-1+3(\eta+2\varepsilon)} \gg n\ell_\rho^{-2} \tag{3.77}$$

  for ρ small;
- or, for some $4 \leqslant k \leqslant \log L$, it places at least $n\rho^{1+\eta}2^{-k+2}$ particles in pieces of length in $[k\ell_\rho, (k+1)\ell_\rho)$; in this case, by Lemma 3.27, we know that

$$\langle H^0(L,n)\Psi,\Psi\rangle \gtrsim \frac{n\rho^{3+3\eta-2k-2\varepsilon}}{((k+1)\ell_\rho)^2 2^{3k}} \gtrsim n\ell_\rho^{-2}\rho^{-1}\frac{(8\rho)^{-k}}{(k+1)^2} \geqslant n\ell_\rho^{-2}\rho^{-1} \tag{3.78}$$

  for ρ sufficiently small.

Hence, for ρ sufficiently small, recalling (1.13) and (2.7) (and that here $\mu = 1$), one has $\langle H^0(L,n)\Psi,\Psi\rangle > 2\mathcal{E}^0(\rho)n$.

This completes the proof of (3.38) in Lemma 3.23 for particles of type (a).

To deal with the particles of type (b) (resp. (c)), we replace the upper bounds (3.75) and (3.76) obtained using Proposition 2.2 by analogous upper bounds on the numbers of pieces of type (b) (resp. (c)) obtained through Proposition 2.4 (resp. Proposition 2.5).

This completes the proof of (3.38) in Lemma 3.23.

Let us now prove (3.39). By (3.44), one has

$$\sum_{k=1}^{m(\omega)} \frac{\pi^2 P(Q_k(\Psi))}{|\Delta_k|^2} \leqslant \langle H_\omega^{U^p}(L,n)\Psi, \Psi\rangle \leqslant 2\mathcal{E}^0(\rho)n$$

where P is defined in (3.44).

Taking Proposition 2.1 into account immediately yields (3.39) and completes the proof of Lemma 3.23. □

The Proof of Lemma 3.27 The form of the Hamiltonians (1.4), (3.16) (the definition of U^p), (1.6) and the non- negativity of the interactions guarantee that

$$\langle H_\omega^{U^p}(L,n)\Psi, \Psi\rangle \geqslant \langle H^0(L,n)\Psi, \Psi\rangle \geqslant \sum_{i=1}^{k}\sum_{m=1}^{\nu_i}\left(\frac{\pi\alpha_m^i}{l_i}\right)^2$$

where $(\alpha_m^i)_{1\leqslant m\leqslant \nu_i} \in (\mathbb{N}^*)^{\nu_i}$ and $\alpha_1^i < \alpha_2^i < \cdots < \alpha_{\nu_i}^i$.

Thus

$$\langle H^0(L,n)\Psi, \Psi\rangle \geqslant \sum_{i=1}^{k}\sum_{m=1}^{\nu_i}\left(\frac{\pi m}{l_i}\right)^2 \geqslant \frac{\pi^2}{3l_k^2}\sum_{i=1}^{k}\nu_i^3 \geqslant \frac{\pi^2\nu^3}{3l_k^2k^2}$$

as $\nu_1 + \cdots + \nu_k = \nu$.

This completes the proof of Lemma 3.27. □

Theorem 3.28 *For ρ sufficiently small, in the thermodynamic limit, with probability $1 - O(L^{-\infty})$, for any function $\Psi \in \mathfrak{H}^n \cap \mathfrak{H}_\infty^n([0, L])$,*

$$\frac{1}{n}\langle H_\omega^{U^p}(L,n)\Psi, \Psi\rangle \geqslant \frac{1}{n}\langle H_\omega^{U^p}(L,n)\Psi^{\text{opt}}, \Psi^{\text{opt}}\rangle - o(\rho|\log\rho|^{-3}). \tag{3.79}$$

Proof This result can easily be traced throughout the proof of Theorem 3.22 by considering each of the cases. Before doing so, let us give some preliminary remarks that correspond exactly to the three remarks found in the beginning of the proof of Theorem 3.22.

First, the energy gain due to moving a single particle is always bounded by $O(\ell_\rho^{-2})$ just because each individual particle in Ψ^{opt} brings to the system at most this amount of energy.

Next, the number of pieces of ε-type is $O(n\rho^{1+\eta})$ (see Lemma 3.23); thus, the energy gain due to them is at most $O(n\rho^{1+\eta}\ell_\rho^{-2})$.

The pieces with too many particles are also rare by Lemma 3.24. Moreover, the many particles in these pieces always bring an excess of energy and never an energy gain.

Finally, the analysis of n_2^+ large (see (3.63)) shows that moving an extra particle to the majority of these pieces results in an energy increase of order of $O(\ell_\rho^{-2})$, whereas for only $O(n\rho^{3/2})$ of them adding a particle may be energetically favorable.

We treat now the cases from (a) to (f) of the last part of the proof of Theorem 3.22. For the matter of the current proof we shall put $r(\rho) = 0$ (because we are interested only in those states that have the energy smaller than Ψ^{opt}), thus, reducing the claim of Theorem 3.22 to

$$\mathrm{dist}_1(Q(\Psi), Q(\Psi^{\mathrm{opt}})) = O(n\rho\ell_\rho^{-1}).$$

- For those displacements when the possible energy gain is due to removing interaction with neighbors (this includes the cases (d), (e), and (f)), it suffices to remark that, by Lemma 6.18, the size of the interacting energy is bounded by $O(\ell_\rho^{-4+0})$. Combined with the fact that, in total, there are $O(n\rho)$ pairs of neighboring particles, this yields a total energy gain of size $O(n\rho\ell_\rho^{-4+0})$.
- For those displacements when the possible energy gain is due to decoupling particles living in the same piece (cases (a), (b) and (c)), the individual interacting energy is of size $O(\ell_\rho^{-3})$ while their total number is $O(n\rho\ell_\rho^{-1})$. This yields a total energy gain of size $O(n\rho\ell_\rho^{-4})$.
- Finally, when the energy gain results from a non-interacting energy decrease (like in the case (d)), it is at most $O(\ell_\rho^{-3})$ and the total number of displacements that result in energy decrease is $O(n\rho\ell_\rho^{-1})$. This again yields a total energy gain of size $O(n\rho\ell_\rho^{-4})$.

This concludes the proof of (3.79). □

Corollary 3.29 *There exists $\rho_0 > 0$ such that for $\rho \in (0, \rho_0)$, in the thermodynamic limit, with probability $1 - O(L^{-\infty})$,*

$$\begin{aligned}\frac{1}{n}\langle H_\omega^{U^p}(L,n)\Psi^{U^p}, \Psi^{U^p}\rangle &= \frac{1}{n}\langle H_\omega^{U^p}(L,n)\Psi^{\mathrm{opt}}, \Psi^{\mathrm{opt}}\rangle + O(\rho|\log\rho|^{-4})\\ &= \mathcal{E}^0(\rho) + \pi^2\gamma_*\frac{\rho}{|\log\rho|^3} + \frac{\rho}{|\log\rho|^3}O\left(f_Z(|\log\rho|)\right),\end{aligned} \tag{3.80}$$

where the constant γ_* is given in (1.17)*, Z describes the behavior of U at infinity and f_Z is defined in Theorem* 3.19.

Proof The upper bound is given by the fact that Ψ^{U^p} is the ground state of $H_\omega^{U^p}$. The lower bound is a direct consequence of (3.79) and (3.18). This proves (3.80). □

Remark 3.30 The ground state Ψ^{U^p} satisfies the conditions of Theorem 3.22. Hence, the inequalities (3.35) hold for the distance between the occupations of Ψ^{U^p} and Ψ^{opt}.

3.7 The Proof of Theorem 1.3

Theorem 3.22 and Theorem 3.28 give a rather complete description of the ground state for the operator with compactified interactions $H_\omega^{U^p}(L,n)$. The description is given in terms of comparison with Ψ^{opt} (see Definitions 3.9 and 3.14). In this section, we complement it with estimates on the residual part of interactions W^r (see (3.16)).

Proposition 3.31 *There exists ρ_0 such that, for $\rho \in (0, \rho_0)$, in the thermodynamic limit, for L sufficiently large, with probability $1 - O(L^{-\infty})$, one has*

$$\frac{1}{n}\langle W^r \Psi^{\mathrm{opt}}, \Psi^{\mathrm{opt}}\rangle = O(\rho|\log\rho|^{-3} Z(2|\log\rho|)). \tag{3.81}$$

Proof We will mostly follow the lines of the second part of the proof of Theorem 3.19 (see formula (3.21) and what follows). First, as in (3.23), one computes

$$\langle W^r \Psi^{\mathrm{opt}}, \Psi^{\mathrm{opt}}\rangle = \mathrm{Tr}\left(U^r \gamma^{(2)}_{\Psi^{\mathrm{opt}}}\right)$$

where $\gamma^{(2)}_{\Psi^{\mathrm{opt}}}$ is given by (3.24). Let us treat here only the contribution of the second sum (3.24). It corresponds to interactions between single particles in pieces of length in $[\ell_\rho - \rho x_*, 2\ell_\rho + A_*)$. The other three sums only contribute error terms as the number of two-particle sub-states in Ψ^{opt} is by a factor ρ smaller than that of single-particle sub-states. For the second sum in (3.24)., using Lemma 6.17, one obtains

$$\begin{aligned}
&\mathrm{Tr}\Bigg(U^r(\mathrm{Id}-\mathrm{Ex}) \sum_{\substack{i,j=1,\dots,k_1\\ i<j}} \gamma_{\phi_i}\otimes^s \gamma_{\phi_j}\Bigg)\\
&\leqslant \sum_{\substack{|\Delta_i|,|\Delta_j|\in[\ell_\rho-\rho x_*,2\ell_\rho+A_*)\cup[3\ell_\rho,+\infty)\\ i<j\\ \mathrm{dist}(\Delta_i,\Delta_j)>B\ell_\rho}} \int_{\Delta_i\times\Delta_j} U(x-y)|\varphi^1_{\Delta_i}(x)|^2|\varphi^1_{\Delta_j}(y)|^2 \mathrm{d}x\mathrm{d}y\\
&\leqslant C_1 n\rho \int_{B\ell_\rho}^{+\infty} \ell_\rho^{-1} a^{-3} Z(a)\mathrm{d}a.
\end{aligned}$$

Recall that Z is defined in (1.26).

We compute next

$$\int_{B\ell_\rho}^{+\infty} a^{-3}Z(a)\mathrm{d}a = \int_{B\ell_\rho}^{+\infty}\int_a^{+\infty} U(x)\mathrm{d}x\mathrm{d}a \leqslant \int_{B\ell_\rho}^{+\infty} xU(x)\mathrm{d}x \leqslant C\ell_\rho^{-2}Z(B\ell_\rho),$$

where the last inequality is just (6.61) for $\varepsilon = 2$. This completes the proof of (3.81).
□

***Proof of Theorem* 1.3** Proposition 3.31 immediately entails the asymptotics of the interacting ground state energy $\mathcal{E}^U(\rho)$. Indeed, as $H^{U^p} \leqslant H^U$, one has $\mathcal{E}^{U^p}(\rho) \leqslant \mathcal{E}^U(\rho)$; thus, the announced lower bound is given by (3.80). On the other hand, by Theorem 3.19 and Proposition 3.31, one has

$$\begin{aligned}\langle H^U \Psi^U, \Psi^U\rangle \leqslant \langle H^U \Psi^{\text{opt}}, \Psi^{\text{opt}}\rangle &= \langle H^{U^p} \Psi^{\text{opt}}, \Psi^{\text{opt}}\rangle + \langle W^r \Psi^{\text{opt}}, \Psi^{\text{opt}}\rangle \\ &= \mathcal{E}^0(\rho) + \pi^2 \gamma_* \rho |\log \rho|^{-3} \left(1 + O\left(f_Z(|\log \rho|)\right)\right),\end{aligned} \tag{3.82}$$

which gives the announced upper bound.

This, the facts that $B > 2$ and that Z is decreasing complete the proof of Theorem 1.3. □

Our analysis yields the following description for the possible occupations of the ground state of the full Hamiltonian.

Corollary 3.32 *There exists $C > 0$ such that, ω almost surely, in the thermodynamic limit, with probability $1 - O(L^{-\infty})$, for any Ψ^U, ground state of the full Hamiltonian of fixed occupation $Q(\Psi^U)$, one has*

$$Q(\Psi^U) \in \mathcal{Q}_\rho := \left\{ Q \text{ occ.}; \begin{aligned} &\operatorname{dist}_1\left(Q|_{\geqslant \ell_\rho + C}, Q|_{\geqslant \ell_\rho + C}(\Psi^{\text{opt}})\right) \\ &\qquad \leqslant C n\rho \max\left(\sqrt{Z(2|\log \rho|)}, |\log \rho|^{-1}\right), \\ &\operatorname{dist}_1\left(Q|_{< \ell_\rho + C}, Q|_{< \ell_\rho + C}(\Psi^{\text{opt}})\right) \\ &\qquad \leqslant C n \max\left(\sqrt{\rho\, Z(2|\log \rho|)}, \rho|\log \rho|^{-1}\right). \end{aligned} \right\} \tag{3.83}$$

Proof Note that

$$\langle H^{U^p} \Psi^U, \Psi^U\rangle \leqslant \langle H^U \Psi^U, \Psi^U\rangle \leqslant \langle H^{U^p} \Psi^{\text{opt}}, \Psi^{\text{opt}}\rangle + \langle W^r \Psi^{\text{opt}}, \Psi^{\text{opt}}\rangle.$$

Thus, according to Proposition 3.31, Ψ^U satisfies the condition (3.34) with

$$r(\rho) = C\sqrt{Z(2|\log \rho|)}$$

for some $C > 0$ sufficiently large.

Then, Theorem 3.22 is applicable and yields (3.83). This completes the proof of Corollary 3.32. □

4 From the Occupation and Energy Bounds to the Control of the Density Matrices

In this section, we will derive Theorem 1.5 from Theorem 1.3, Corollary 3.32 and a computation of the reduced one-particle and two-particle density matrix of a (non-factorized) state. More precisely, from Theorem 1.3 and Corollary 3.32, we will infer a description of the ground state Ψ^U in most of the pieces: roughly, in most of the pieces, the only occupied state is the ground state (up to a controllable error). We then use this knowledge to compute the reduced one-particle and two-particle density matrix of Ψ^U (up to a controllable error).

4.1 From the Occupation Decomposition to the Reduced Density Matrices

Fix a configuration of the Poisson points, say, ω, and a state $\Psi \in \mathfrak{H}^n(\Lambda)$. Recall that, in the configuration ω, the pieces are denoted by $(\Delta_j(\omega))_{1\leqslant j\leqslant m} = (\Delta_j)_{1\leqslant j\leqslant m}$ (where $m = m(\omega)$, see Section 2.2). For $1 \leqslant j \leqslant m$ and $q \geqslant 1$, let $(E^j_{q,n})_{1\leqslant n}$ be the eigenvalues (ordered increasingly) and $(\varphi^j_{q,n})_{1\leqslant n}$ be the associated eigenvectors of q interacting electronic particles in the piece $\Delta_j(\omega)$, i.e. the eigenvalues and eigenvectors of the Hamiltonian

$$H^q_{\Delta_j(\omega)} = -\sum_{l=1}^{q} \frac{d^2}{dx_l^2} + \sum_{1\leqslant l<l'\leqslant q} U^p(x_l - x_{l'}) \tag{4.1}$$

acting on $\bigwedge_{l=1}^{q} L^2(\Delta_j(\omega))$ with Dirichlet boundary conditions. Recall that U^p is defined in Section 3.5 (see (3.16)).

The occupation number decomposition (see Section 3.1) implies that one can write

$$\Psi = \sum_{Q} \Psi_Q \quad \text{and} \quad \Psi_Q = \sum_{\overline{n}\in\mathbb{N}^m} a^Q_{\overline{n}} \Phi^Q_{\overline{n}} = \sum_{\substack{(n_j)_{1\leqslant j\leqslant m}\\ \forall j,\, n_j\geqslant 1}} a^Q_{n_1,\cdots,n_m}(\Psi) \bigwedge_{j=1}^{m} \varphi^j_{Q_j,n_j} \tag{4.2}$$

where

- the first sum is taken over the occupation number $Q = (Q_j)_{1\leqslant j\leqslant m}$; recall $\sum_{j=1}^{m} Q_j = n$;

- we have defined $\Phi_{\overline{n}}^{Q} := \bigwedge_{j=1}^{m} \varphi^{j}_{Q_j,n_j}$; we refer to (C.2) in Appendix C for an explicit description of the anti-symmetric tensor product.

Remark 4.1 In (4.2), the convention in the exterior product is that, if $Q_j = 0$, then the corresponding basis vector drops out of the exterior product. Thus, the product is only at most n fold. Moreover, in this case, $a^{Q}_{n_1,\cdots,n_m} = 0$ if $n_j \geqslant 2$.

For $\overline{n} = (n_1, \cdots, n_m) \in \mathbb{N}^m$, we write $a^{Q}_{\overline{n}} = a^{Q}_{n_1,\cdots,n_m} = a^{Q}_{n_1,\cdots,n_m}(\Psi)$. These coefficients are uniquely determined by Ψ.

4.1.1 The One-Particle Density Matrix

We shall first compute the one-particle reduced density matrix in terms of the coefficients $(a^{Q}_{\overline{n}})_{Q,\overline{n}}$ coming up in the occupation number decomposition (4.2). We prove

Theorem 4.2 *The one-particle density $\gamma^{(1)}_{\Psi}$ (see (1.19)) is written as $\gamma^{(1)}_{\Psi} = \gamma^{(1),d}_{\Psi} + \gamma^{(1),o}_{\Psi}$ where*

$$\gamma^{(1),d}_{\Psi} = \sum_{j=1}^{m} \sum_{\substack{Q \text{ occ.} \\ Q_j \geqslant 1}} \sum_{\substack{n_j \geqslant 1 \\ n'_j \geqslant 1}} \sum_{\tilde{n} \in \mathbb{N}^{m-1}} a^{Q}_{\tilde{n}_j} \overline{a^{Q}_{\tilde{n}'_j}} \gamma^{(1)}_{\substack{Q_j \\ n_j, n'_j}} \tag{4.3}$$

$$\gamma^{(1),o}_{\Psi} = \sum_{\substack{i,j=1 \\ i \neq j}}^{m} \sum_{\substack{Q,\ \text{occ.}\ Q_j \geqslant 1 \\ Q':\ Q'_k = Q_k \text{ if } k \notin \{i,j\} \\ Q'_i = Q_i + 1 \\ Q'_j = Q_j - 1}} C_1(Q,i,j) \sum_{\tilde{n} \in \mathbb{N}^{m-2}} \sum_{\substack{n_i, n_j \geqslant 1 \\ n'_i, n'_j \geqslant 1}} a^{Q}_{\tilde{n}_{i,j}} \overline{a^{Q'}_{\tilde{n}'_{i,j}}} \gamma^{(1)}_{\substack{Q_i, Q_j \\ n_i, n_j \\ n'_i, n'_j}} \tag{4.4}$$

and

- *we have used the shorthands*
  - $\tilde{n}_j$ *for the vector* $(\tilde{n}_1 \cdots, \tilde{n}_{j-1}, n_j, \tilde{n}_j, \cdots, \tilde{n}_{m-1})$ *when* $\tilde{n} = (\tilde{n}_1, \cdots, \tilde{n}_{m-1})$,
  - $\tilde{n}_{i,j}$ *for* $(\tilde{n}_1, \cdots, \tilde{n}_{i-1}, n_i, \tilde{n}_i, \cdots, \tilde{n}_{j-2}, n_j, \tilde{n}_{j-1}, \cdots, \tilde{n}_{m-2})$ *when* $i < j$ *and* $\tilde{n} = (\tilde{n}_1, \cdots, \tilde{n}_{m-2})$,
- *the trace class operator* $\gamma^{(1)}_{\substack{Q_j \\ n_j, n'_j}} : L^2(\Delta_j) \to L^2(\Delta_j)$ *has the kernel*

$$\gamma^{(1)}_{\substack{Q_j \\ n_j, n'_j}}(x,y) = Q_j \int_{\Delta_j^{Q_j - 1}} \varphi^{j}_{Q_j,n_j}(x,z) \overline{\varphi^{j}_{Q_j,n'_j}(y,z)} dz,$$

- $C_1(Q,i,j) = \dfrac{(n-Q_j-Q_i-1)!Q_i!Q_j!}{(n-1)!}$;
- *the rank 1 operator* $\gamma^{(1)}_{\substack{Q_i,Q_j\\ n_i,n_j\\ n'_i,n'_j}} : L^2(\Delta_i) \to L^2(\Delta_j)$ *has the kernel*

$$\gamma^{(1)}_{\substack{Q_i,Q_j\\ n_i,n_j\\ n'_i,n'_j}}(x,y) = \int_{\Delta_j^{Q_j-1}} \varphi^j_{Q_j,n_j}(x,z)\overline{\varphi^j_{Q_j-1,n'_j}(z)}dz \int_{\Delta_i^{Q_i}} \varphi^i_{Q_i,n_i}(z)\overline{\varphi^i_{Q_i+1,n'_i}(y,z)}dz. \tag{4.5}$$

Theorem 4.2 follows from a direct computation that we perform in Appendix D.1.

Remark 4.3 In (4.5), in accordance with Remark 4.1, we use the following convention

- if $Q_j = 1$ and $Q_i = 0$, then $n'_j = 1$ and $n_i = 1$ (i.e. for different indices, the coefficient $a^Q_{\tilde{n}_{i,j}}\overline{a^{Q'}_{\tilde{n}'_{i,j}}}$ vanishes) and

$$\gamma^{(1)}_{\substack{Q_i,Q_j\\ 1,n_j\\ n'_i,1}}(x,y) = \varphi^j_{1,n_j}(x)\cdot\overline{\varphi^i_{1,n'_i}(y)}, \tag{4.6}$$

- if $Q_j \geqslant 2$ and $Q_i = 0$, then $n_i = 1$ and

$$\gamma^{(1)}_{\substack{Q_i,Q_j\\ 1,n_j\\ n'_i,n'_j}}(x,y) = \overline{\varphi^i_{1,n'_i}(y)} \int_{\Delta_j^{Q_j-1}} \overline{\varphi^j_{Q_j-1,n'_j}(z)}\varphi^j_{Q_j,n_j}(x,z)dz, \tag{4.7}$$

- if $Q_j = 1$ and $Q_i \geqslant 1$, then $n'_j = 1$ and

$$\gamma^{(1)}_{\substack{Q_i,Q_j\\ n_i,n_j\\ n'_i,1}}(x,y) = \varphi^j_{1,n_j}(x)\int_{\Delta_i^{Q_i}} \varphi^i_{Q_i,n_i}(z)\overline{\varphi^i_{Q_i+1,n'_i}(y,z)}dz. \tag{4.8}$$

4.1.2 The Two-Particle Density Matrix

We shall now compute the two-particle reduced density matrix in terms of the coefficients $(a^Q_{\overline{n}})_{Q,\overline{n}}$ coming up in the occupation number decomposition (4.2). We prove

Theorem 4.4 *The two-particle density* $\gamma^{(2)}_\Psi$ *(see* (1.19)*) is written as*

$$\gamma^{(2)}_\Psi = \gamma^{(2),d,d}_\Psi + \gamma^{(2),d,o}_\Psi + \gamma^{(2),2}_\Psi + \gamma^{(2),4,2}_\Psi + \gamma^{(2),4,3}_\Psi + \gamma^{(2),4,3'}_\Psi + \gamma^{(2),4,4}_\Psi \tag{4.9}$$

where

$$\gamma_\Psi^{(2),d,d} = \sum_{j=1}^{m} \sum_{\substack{Q \text{ occ.} \\ Q_j \geqslant 2}} \sum_{\substack{n_j \geqslant 1 \\ n'_j \geqslant 1}} \sum_{\tilde{n} \in \mathbb{N}^{m-1}} a^Q_{\tilde{n}_j} \overline{a^Q_{\tilde{n}'_j}} \gamma^{(2),d,d}_{\substack{Q_j \\ n_j, n'_j}} \tag{4.10}$$

$$\gamma_\Psi^{(2),d,o} = \sum_{1 \leqslant i < j \leqslant m} \sum_{\substack{Q \text{ occ.} \\ Q_i \geqslant 1 \\ Q_j \geqslant 1}} \sum_{\tilde{n} \in \mathbb{N}^{m-2}} \sum_{\substack{n_j, n'_j \geqslant 1 \\ n_i, n'_i \geqslant 1}} a^Q_{\tilde{n}_{i,j}} \overline{a^Q_{\tilde{n}'_{i,j}}} \gamma^{(2),d,o}_{\substack{Q_i, Q_j \\ n_i, n_j \\ n'_i, n'_j}} \tag{4.11}$$

$$\gamma_\Psi^{(2),2} = \sum_{\substack{i,j=1 \\ i \neq j}}^{m} \sum_{\substack{Q,\ \text{occ. } Q_j \geqslant 1 \\ Q':\ Q'_k = Q_k \text{ if } k \notin \{i,j\} \\ Q'_i = Q_i + 1 \\ Q'_j = Q_j - 1}} \sum_{\tilde{n} \in \mathbb{N}^{m-1}} C_2(Q, i, j) \sum_{\substack{n_j, n'_j \geqslant 1 \\ n_i, n'_i \geqslant 1}} a^Q_{\tilde{n}_{i,j}} \overline{a^{Q'}_{\tilde{n}'_{i,j}}} \gamma^{(2),2}_{\substack{Q_i, Q_j \\ n_i, n_j \\ n'_i, n'_j}} \tag{4.12}$$

$$\gamma_\Psi^{(2),4,2} = \sum_{i \neq j} \sum_{\tilde{n} \in \mathbb{N}^{m-2}} \sum_{\substack{Q \text{ occ.} \\ Q_j \geqslant 2 \\ Q':\ Q'_k = Q_k \text{ if } k \notin \{i,j\} \\ Q'_i = Q_i + 2 \\ Q'_j = Q_j - 2}} C_2(Q, i, j) \sum_{\substack{n_j, n'_j \geqslant 1 \\ n_i, n'_i \geqslant 1}} a^Q_{\tilde{n}_{i,j}} \overline{a^{Q'}_{\tilde{n}'_{i,j}}} \gamma^{(2),4,2}_{\substack{Q_i, Q_j \\ n_i, n_j \\ n'_i, n'_j}} \tag{4.13}$$

$$\gamma_\Psi^{(2),4,3} = \sum_{\substack{i,j,k \\ \text{distinct}}} \sum_{\tilde{n} \in \mathbb{N}^{m-3}} \sum_{\substack{Q \text{ occ.} \\ Q_j \geqslant 2 \\ Q':\ Q'_l = Q_l \text{ if } l \notin \{i,j,k\} \\ Q'_i = Q_i + 1 \\ Q'_j = Q_j - 2 \\ Q'_k = Q_k + 1}} C_3(Q, i, j, k) \sum_{\substack{n_i, n_j, n_k \geqslant 1 \\ n'_i, n'_j, n'_k \geqslant 1}} a^Q_{\tilde{n}_{i,j,k}} \overline{a^{Q'}_{\tilde{n}'_{i,j,k}}} \gamma^{(2),4,3}_{\substack{Q_i, Q_j, Q_k \\ n_i, n_j, n_k \\ n'_i, n'_j, n'_k}} \tag{4.14}$$

$$\gamma_\Psi^{(2),4,3'} = \sum_{\substack{i,j,k \\ \text{distinct}}} \sum_{\tilde{n} \in \mathbb{N}^{m-3}} \sum_{\substack{Q \text{ occ.} \\ Q_i \geqslant 1,\ Q_k \geqslant 1 \\ Q':\ Q'_l = Q_l \text{ if } l \notin \{i,j,k\} \\ Q'_i = Q_i - 1 \\ Q'_j = Q_j + 2 \\ Q'_k = Q_k - 1}} C_3(Q, i, j, k) \sum_{\substack{n_i, n_j, n_k \geqslant 1 \\ n'_i, n'_j, n'_k \geqslant 1}} a^Q_{\tilde{n}_{i,j,k}} \overline{a^{Q'}_{\tilde{n}'_{i,j,k}}} \gamma^{(2),4,3'}_{\substack{Q_i, Q_j, Q_k \\ n_i, n_j, n_k \\ n'_i, n'_j, n'_k}}, \tag{4.15}$$

and

$$\gamma_{\Psi}^{(2),4,4} = \sum_{\substack{i,j,k,l,\tilde{n}\in\mathbb{N}^{m-4}\\ \text{distinct}}} \sum_{\substack{Q \text{ occ.}\\ Q_i\geqslant 1,\ Q_j\geqslant 1\\ Q':\ Q'_p=Q_p \text{ if } p\notin\{i,j,k,l\}\\ Q'_i=Q_i-1,\ Q'_j=Q_j-1\\ Q'_k=Q_k+1,\ Q'_l=Q_l+1}} C_4(Q,i,j,k,l) \sum_{\substack{n_i,n_j,n_k,n_l\geqslant 1\\ n'_i,n'_j,n'_k,n'_l\geqslant 1}} a^{Q}_{\tilde{n}_{i,j,k,l}} \overline{a^{Q'}_{\tilde{n}'_{i,j,k,l}}} \gamma^{(2),4,4}_{\substack{Q_i,Q_j,Q_k,Q_l\\ n_i,n_j,n_k,n_l\\ n'_i,n'_j,n'_k,n'_l}}, \tag{4.16}$$

where

- *we have used the shorthands defined in Theorem 4.2 and defined*
  - $\tilde{n}_{i,j,k}$ *for* $(\tilde{n}_1,\cdots,\tilde{n}_{i-1},n_i,\tilde{n}_i,\cdots,\tilde{n}_{j-2},n_j,\tilde{n}_{j-1},\cdots,\tilde{n}_{k-3},n_k,\tilde{n}_{k-2},\cdots,\tilde{n}_{m-3})$ *when* $i<j<k$ *and* $\tilde{n}=(\tilde{n}_1,\cdots,\tilde{n}_{m-3})$,
  - $\tilde{n}_{i,j,k,l}$ *for* $(\tilde{n}_1,\cdots,\tilde{n}_{i-1},n_i,\tilde{n}_i,\cdots,\tilde{n}_{j-2},n_j,\tilde{n}_{j-1},\cdots,\tilde{n}_{k-3},n_k,\tilde{n}_{k-2},\cdots,\tilde{n}_{l-4},n_l,\tilde{n}_{l-3},\cdots,\tilde{n}_{m-4})$ *when* $i<j<k<l$ *and* $\tilde{n}=(\tilde{n}_1,\cdots,\tilde{n}_{m-4})$,
- *the trace class operator* $\gamma^{(2),d,d}_{\substack{Q_j\\ n_j,n'_j}} : L^2(\Delta_j)\bigwedge L^2(\Delta_j)\to L^2(\Delta_j)\bigwedge L^2(\Delta_j)$ *has the kernel*

$$\gamma^{(2),d,d}_{\substack{Q_j\\ n_j,n'_j}}(x,x',y,y') = \frac{Q_j(Q_j-1)}{2}\int_{\Delta_j^{Q_j-2}} \varphi^j_{Q_j,n_j}(x,x',z)\overline{\varphi^j_{Q_j,n'_j}(y,y',z)}dz \tag{4.17}$$

- *the trace class operator* $\gamma^{(2),d,o}_{\substack{Q_i,Q_j\\ n_i,n_j\\ n'_i,n'_j}} : L^2(\Delta_i)\bigwedge L^2(\Delta_j)\to L^2(\Delta_i)\bigwedge L^2(\Delta_j)$ *has the kernel*

$$\gamma^{(2),d,o}_{\substack{Q_i,Q_j\\ n_i,n_j\\ n'_i,n'_j}}(x,x',y,y') = Q_iQ_j\int_{\Delta_i^{Q_i-1}\times\Delta_j^{Q_j-1}} dzdz'$$

$$\begin{vmatrix} \varphi^i_{Q_i,n_i}(x,z) & \varphi^i_{Q_i,n_i}(x',z)\\ \varphi^j_{Q_j,n_j}(x,z') & \varphi^j_{Q_j,n_j}(x',z')\end{vmatrix} \cdot \overline{\begin{vmatrix} \varphi^i_{Q_i,n'_i}(y,z) & \varphi^i_{Q_i,n'_i}(y',z)\\ \varphi^j_{Q_j,n'_j}(y,z') & \varphi^j_{Q_j,n'_j}(y',z')\end{vmatrix}} \tag{4.18}$$

- $C_2(Q,i,j) = \dfrac{(n-Q_j-Q_i-2)!Q_i!Q_j!}{2\,(n-2)!}$;
- *the trace class operator* $\gamma^{(2),2}_{\substack{Q_i,Q_j\\ n_i,n_j\\ n'_i,n'_j}} : L^2(\Delta_j)\bigwedge L^2(\Delta_j)\to L^2(\Delta_i)\bigwedge L^2(\Delta_i)$ *has the kernel*

$$\gamma^{(2),2}_{\substack{Q_i,Q_j\\ n_i,n_j\\ n'_i,n'_j}}(x,y) = \mathbf{1}_{Q_j\geqslant 2}\int_{\Delta_j^{Q_j-2}\times\Delta_i^{Q_i}} \varphi^j_{Q_j,n_j}(x,x',z)\varphi^i_{Q_i,n_i}(z')$$
$$\times\overline{\begin{vmatrix}\varphi^j_{Q_j-1,n'_j}(y',z) & \varphi^j_{Q_j-1,n'_j}(y,z)\\ \varphi^i_{Q_i+1,n'_i}(y',z') & \varphi^i_{Q_i+1,n'_i}(y,z')\end{vmatrix}}\,dzdz'$$
$$+\mathbf{1}_{Q_i\geqslant 1}\int_{\Delta_j^{Q_j-1}\times\Delta_i^{Q_i-1}}\begin{vmatrix}\varphi^j_{Q_j,n_j}(x',z) & \varphi^j_{Q_j,n_j}(x,z)\\ \varphi^i_{Q_i,n_i}(x',z') & \varphi^i_{Q_i,n_i}(x,z')\end{vmatrix}$$
$$\times\overline{\varphi^j_{Q_j-1,n'_j}(z)\varphi^i_{Q_i+1,n'_i}(y,y',z')}\,dzdz', \tag{4.19}$$

- *the rank 1 operator* $\gamma^{(2),4,2}_{\substack{Q_i,Q_j\\ n_i,n_j\\ n'_i,n'_j}} : L^2(\Delta_j)\bigwedge L^2(\Delta_j)\to L^2(\Delta_i)\bigwedge L^2(\Delta_i)$ *has the kernel*

$$\gamma^{(2),4,2}_{\substack{Q_i,Q_j\\ n_i,n_j\\ n'_i,n'_j}}(x,x',y,y') = \int_{\Delta_j^{Q_j-2}}\varphi^j_{Q_j,n_j}(x,x',z)\overline{\varphi^j_{Q_j-2,n'_j}(z)}dz\int_{\Delta_i^{Q_i}}\varphi^i_{Q_i,n_i}(z)$$
$$\times\overline{\varphi^i_{Q_i+2,n'_i}(y,y',z)}dz. \tag{4.20}$$

- *the rank 2 operator* $\gamma^{(2),4,3}_{\substack{Q_i,Q_j,Q_k\\ n_i,n_j,n_k\\ n'_i,n'_j,n'_k}} : L^2(\Delta_i\cup\Delta_k)\bigwedge L^2(\Delta_i\cup\Delta_k)\to L^2(\Delta_j)\bigwedge L^2(\Delta_j)$ *has the kernel*

$$\gamma^{(2),4,3}_{\substack{Q_i,Q_j,Q_k\\ n_i,n_j,n_k\\ n'_i,n'_j,n'_k}}(x,x',y,y') = \int_{\Delta_j^{Q_j-2}}\varphi^j_{Q_j,n_j}(x,x',z)\overline{\varphi^j_{Q_j-2,n'_j}(z)}dz$$
$$\times\begin{vmatrix}\int_{\Delta_i^{Q_i}}\varphi^i_{Q_i,n_i}(z)\overline{\varphi^i_{Q_i+1,n'_i}(y,z)}dz & \int_{\Delta_i^{Q_i}}\varphi^i_{Q_i,n_i}(z)\overline{\varphi^i_{Q_i+1,n'_i}(y',z)}dz\\ \int_{\Delta_k^{Q_k}}\varphi^k_{Q_k,n_k}(z)\overline{\varphi^k_{Q_k+1,n'_k}(y,z)}dz & \int_{\Delta_k^{Q_k}}\varphi^k_{Q_k,n_k}(z)\overline{\varphi^k_{Q_k+1,n'_k}(y',z)}dz\end{vmatrix}, \tag{4.21}$$

- $C_3(Q,i,j,k) = \dfrac{(n-Q_i-Q_j-Q_k-2)!Q_i!Q_j!Q_k!}{2(n-2)!}$;
- *the rank 2 operator* $\gamma^{(2),4,3'}_{\substack{Q_i,Q_j,Q_k\\ n_i,n_j,n_k\\ n'_i,n'_j,n'_k}} : L^2(\Delta_j)\bigwedge L^2(\Delta_j)\to L^2(\Delta_i\cup\Delta_k)\bigwedge L^2(\Delta_i\cup\Delta_k)$ *has the kernel*

$$\gamma^{(2),4,3'}_{\substack{Q_i,Q_j,Q_k\\ n_i,n_j,n_k\\ n'_i,n'_j,n'_k}}(x,x',y,y') =$$

$$\begin{vmatrix} \int_{\Delta_i^{Q_i-1}} \varphi^i_{Q_i,n_i}(x,z)\overline{\varphi^i_{Q_i-1,n'_i}(z)}dz & \int_{\Delta_i^{Q_i-1}} \varphi^i_{Q_i,n_i}(x',z)\overline{\varphi^i_{Q_i-1,n'_i}(z)}dz \\ \int_{\Delta_k^{Q_k-1}} \varphi^k_{Q_k,n_k}(x,z)\overline{\varphi^k_{Q_k-1,n'_k}(z)}dz & \int_{\Delta_k^{Q_k-1}} \varphi^k_{Q_k,n_k}(x',z)\overline{\varphi^k_{Q_k-1,n'_k}(z)}dz \end{vmatrix}$$

$$\times \int_{\Delta_j^{Q_j}} \varphi^j_{Q_j,n_j}(z)\overline{\varphi^j_{Q_j+2,n'_j}(y,y',z)}dz, \tag{4.22}$$

- *the operator* $\gamma^{(2),4,4}_{\substack{Q_i,Q_j,Q_k,Q_l\\ n_i,n_j,n_k,n_l\\ n'_i,n'_j,n'_k,n'_l}} : L^2(\Delta_l \cup \Delta_k) \bigwedge L^2(\Delta_l \cup \Delta_k) \to L^2(\Delta_i \cup \Delta_j) \bigwedge L^2(\Delta_i \cup \Delta_j)$ *is rank 4 and has the kernel*

$$\gamma^{(2),4,4}_{\substack{Q_i,Q_j,Q_k,Q_l\\ n_i,n_j,n_k,n_l\\ n'_i,n'_j,n'_k,n'_l}}(x,x',y,y') =$$

$$\begin{vmatrix} \int_{\Delta_i^{Q_i-1}} \varphi^i_{Q_i,n_i}(x,z)\overline{\varphi^i_{Q_i-1,n'_i}(z)}dz & \int_{\Delta_i^{Q_i-1}} \varphi^i_{Q_i,n_i}(x',z)\overline{\varphi^i_{Q_i-1,n'_i}(z)}dz \\ \int_{\Delta_j^{Q_j-1}} \varphi^j_{Q_j,n_j}(x,z)\overline{\varphi^j_{Q_j-1,n'_j}(z)}dz & \int_{\Delta_j^{Q_j-1}} \varphi^j_{Q_j,n_j}(x',z)\overline{\varphi^j_{Q_j-1,n'_j}(z)}dz \end{vmatrix}$$

$$\times \begin{vmatrix} \int_{\Delta_k^{Q_k}} \varphi^k_{Q_k,n_k}(z)\overline{\varphi^k_{Q_k+1,n'_k}(y,z)}dz & \int_{\Delta_k^{Q_k}} \varphi^k_{Q_k,n_k}(z)\overline{\varphi^k_{Q_k+1,n'_k}(y',z)}dz \\ \int_{\Delta_l^{Q_l}} \varphi^l_{Q_l,n_l}(z)\overline{\varphi^l_{Q_l+1,n'_l}(y,z)}dz & \int_{\Delta_l^{Q_l}} \varphi^l_{Q_l,n_l}(z)\overline{\varphi^l_{Q_l+1,n'_l}(y',z)}dz \end{vmatrix} \tag{4.23}$$

- $C_4(Q,i,j,k,l) = \dfrac{(n-Q_i-Q_j-Q_k-Q_l-2)!Q_i!Q_j!Q_k!Q_l!}{2\,(n-2)!}$;

Theorem 4.4 follows from a direct computation that we perform in Appendix D.1.

Remark 4.5 In (4.17)–(4.23), in accordance with Remark 4.1, in the degenerate cases, we use the conventions derived from those in Remark 4.3 in an obvious way.

For example, in (4.18), if $Q_i = Q_j = 1$, one has

$$\gamma^{(2),d,o}_{\substack{Q_i,Q_j\\ n_i,n_j\\ n'_i,n'_j}}(x,x',y,y') = Q_i Q_j \begin{vmatrix} \varphi^i_{Q_i,n_i}(x) & \varphi^i_{Q_i,n_i}(x') \\ \varphi^j_{Q_j,n_j}(x) & \varphi^j_{Q_j,n_j}(x') \end{vmatrix} \cdot \overline{\begin{vmatrix} \varphi^i_{Q_i,n'_i}(y) & \varphi^i_{Q_i,n'_i}(y') \\ \varphi^j_{Q_j,n'_j}(y) & \varphi^j_{Q_j,n'_j}(y') \end{vmatrix}}. \tag{4.24}$$

4.1.3 A Particular Case

Let us now explain how the structure of the one-particle and two-particle density matrices may be simplified in the particular case when the ground state is factorized. This in particular immediately yields the expansions (1.22) and (1.23) for the one-particle and two-particle density matrices of the non- interacting ground state.

Definition 4.6 Let $\alpha \in \mathfrak{H}^i(L)$ and $\beta \in \mathfrak{H}^j(L)$ be two states describing i and j electrons, respectively. We say α and β *do not interact* if for all $(x^2, \dots, x^i, y^2, \dots, y^j) \in [0, L]^{i+j-2}$,

$$\int_0^L \alpha(x^1, \dots, x^i)\beta^*(y^1, \dots, y^j)\big|_{x^1=y^1} \mathrm{d}x^1 = 0. \tag{4.25}$$

To denote this complete orthogonality, we will write $\alpha \perp\!\!\!\perp \beta$.

Remark 4.7 Because of the anti-symmetric nature of the states α and β in the above definition, it is sufficient to impose the orthogonality only on the first variables. Thus, an integral of the type (4.25) vanishes for any pair of coordinates $x^{i_1} = y^{j_1}$ for $i_1 \in \{1, \dots, i\}$, and $j_1 \in \{1, \dots, j\}$.

We prove

Proposition 4.8 *Suppose that an n-particle state* $\Psi \in \mathfrak{H}^n(L)$ *is decomposed in its non-interacting parts:*

$$\Psi = \bigwedge_{j=1}^{k} \zeta_j,$$

where each $\zeta_j \in \mathfrak{H}^{k_j}(L)$ *is a* k_j*-particle state describing a packet of particles that do not interact with other packets, i.e., for* $i \neq j$, $\zeta_i \perp\!\!\!\perp \zeta_j$ *in the sense of Definition* 4.6. *Then*

$$\gamma_\Psi = \sum_{j=1}^{k} \gamma_{\zeta_j} \tag{4.26}$$

and

$$\gamma_\Psi^{(2)} = \sum_{j=1}^{k} \left[\gamma_{\zeta_j}^{(2)} - \frac{1}{2}(\mathrm{Id} - \mathrm{Ex})\gamma_{\zeta_j} \otimes \gamma_{\zeta_j} \right] + \frac{1}{2}(\mathrm{Id} - \mathrm{Ex})\gamma_\Psi \otimes \gamma_\Psi, \tag{4.27}$$

where Id *is the identity,* Ex *is the exchange operator on the two-particle space defined as*

$$\mathrm{Ex}\, f \otimes g = g \otimes f, \quad f, g \in \mathfrak{H},$$

and with the obvious convention that $\gamma_{\zeta_j}^{(2)} = 0$ *if* ζ_j *is a one-particle state.*

While Proposition 4.8 could be obtained as a consequence of Theorems 4.2 and 4.4, we will derive it from the following auxiliary lemma.

Lemma 4.9 *Let $\alpha \in \mathfrak{H}^n(L)$ and $\beta \in \mathfrak{H}^m(L)$ be two vectors describing n and m electrons, respectively. Suppose that α and β do not interact:*

$$\alpha \perp\!\!\!\perp \beta.$$

Then,

$$\gamma_{\alpha\wedge\beta} = \gamma_\alpha + \gamma_\beta \tag{4.28}$$

and

$$\gamma^{(2)}_{\alpha\wedge\beta} = \gamma^{(2)}_\alpha + \gamma^{(2)}_\beta + (\mathrm{Id} - \mathrm{Ex})\gamma_\alpha \otimes^s \gamma_\beta \tag{4.29}$$

where $\otimes^s$ denotes the symmetrized tensor product:

$$A \otimes^s B = \frac{1}{2}(A \otimes B + B \otimes A).$$

Proof Define $\mathbb{N}_n := \{1, \dots, n\}$. Consider the two-particle density matrix. By (C.2), the anti-symmetrized product of two eigenfunctions in, respectively, n and m variables is given by

$$(\alpha \wedge \beta)(x^1, \dots, x^{n+m}) = \frac{1}{\sqrt{\binom{n+m}{n}}} \sum_{\substack{J\cup J'=\mathbb{N}_{n+m} \\ J\cap J'=\emptyset,\ |J|=n}} (-1)^{\mathrm{sign}\, J} \alpha(x^J)\beta(x^{J'}).$$

where sign J is the signature of the unique permutation σ of $\{1, \cdots, n+m\}$ such that, if we write $J = \{a_i;\ 1 \leqslant i \leqslant n\}$ and $J' = \{a'_i;\ 1 \leqslant i \leqslant m\}$, both ordered increasingly, then $\sigma(a_i) = i$ and $\sigma(a'_i) = n + i$ (see Appendix C).

Thus, the corresponding two-particle density matrix can be written as

$$\begin{aligned}
&\gamma^{(2)}_{\alpha\wedge\beta}(x^1, x^2, y^1, y^2) \\
&= \frac{(n+m)(n+m-1)}{2} \int_{[0,L]^{n+m-2}} (\alpha\wedge\beta)(x^1, x^2, \overline{x})\,(\alpha\wedge\beta)^*(y^1, y^2, \overline{x})d\overline{x} \\
&= \frac{(n+m)(n+m-1)}{2\binom{n+m}{n}} \\
&\quad \sum_{\substack{I\cup I'=\mathbb{N}_{n+m} \\ I\cap I'=\emptyset,\ |I|=n \\ J\cup J'=\mathbb{N}_{n+m} \\ J\cap J'=\emptyset,\ |J|=n}} \int_{[0,L]^{n+m-2}} (-1)^{\mathrm{sign}\, I + \mathrm{sign}\, J} \alpha(x^I)\beta(x^{I'})\alpha^*(y^J)\beta^*(y^{J'})\Big|_{\substack{y^j = x^j \\ j\in\{3,\dots,n+m\}}} d\overline{x}.
\end{aligned} \tag{4.30}$$

As α and β do not interact, the integrals in the sum in the last part of (4.30) vanish if I differs from J by more than two elements, i.e., $|I\backslash J| \geqslant 2$. Moreover, if $|I\backslash J| \leqslant 1$, such an integral does not vanish if and only if

(a) if $\{1, 2\} \subset I$, then $I = J$; indeed, otherwise J would contain an index in I' and the integration of $\beta(x^{I'})\alpha^*(y^J)\Big|_{\substack{y^j=x^j\\ j\in\{3,\dots,n+m\}}}$ over the corresponding variable would produce zero because $\alpha \perp\!\!\!\perp \beta$.
(b) if $\{1, 2\} \subset J$, then $I = J$.
(c) if $(1, 2) \in (I \times I') \cup (I' \times I)$, then $(1, 2) \in (J \times J') \cup (J' \times J)$ by the same argument as above.

As the functions α and β are completely anti-symmetric under permutations of variables, the terms of the sums over I and J corresponding to different cases described above are all the same. If we denote $\hat{x}^k = x^3, \dots, x^k$ and $d\hat{x}^k = dx^3 \dots dx^k$ for $k \in \{n, m, n+n\}$, this finally yields

$$\gamma^{(2)}_{\alpha\wedge\beta}(x^1, x^2, y^1, y^2) = A + B + C$$

where

$$\begin{aligned} A &:= \frac{(n+m)(n+m-1)}{2}\frac{1}{\binom{n+m}{n}}\binom{n+m-2}{n-2}\int_{[0,L]^{n-2}}\alpha(x^1, x^2, \hat{x}^n)\alpha^*(y^1, y^2, \hat{x}^n)d\hat{x}^n \\ &= \gamma^{(2)}_\alpha(x^1, x^2, y^1, y^2), \end{aligned}$$

$$\begin{aligned} B &:= \frac{(n+m)(n+m-1)}{2}\frac{1}{\binom{n+m}{n}}\binom{n+m-2}{m-2}\int_{[0,L]^{m-2}}\beta(x^1, x^2, \hat{x}^m)\beta^*(y^1, y^2, \hat{x}^m)d\hat{x}^m \\ &= \gamma^{(2)}_\beta(x^1, x^2, y^1, y^2) \end{aligned}$$

and

$$\begin{aligned} C &:= \frac{(n+m)(n+m-1)}{2}\frac{1}{\binom{n+m}{n}}\binom{n+m-2}{m-1}\int_{[0,L]^{n+m-2}} d\hat{x}^{n+m} \\ &\quad \Big(\alpha(x^1, \dots)\beta(x^2, \dots)\alpha^*(y^1, \dots)\beta^*(y^2, \dots) \\ &\qquad - \alpha(x^1, \dots)\beta(x^2, \dots)\alpha^*(y^2, \dots)\beta^*(y^1, \dots) \\ &\qquad - \alpha(x^2, \dots)\beta(x^1, \dots)\alpha^*(y^1, \dots)\beta^*(y^2, \dots) \\ &\qquad + \alpha(x^2, \dots)\beta(x^1, \dots)\alpha^*(y^2, \dots)\beta^*(y^1, \dots)\Big) \\ &= \frac{1}{2}\Big(\gamma_\alpha(x^1, y^1)\gamma_\beta(x^2, y^2) - \gamma_\alpha(x^1, y^2)\gamma_\beta(x^2, y^1) \\ &\qquad - \gamma_\alpha(x^2, y^1)\gamma_\beta(x^1, y^2) + \gamma_\alpha(x^2, y^2)\gamma_\beta(x^1, y^1)\Big). \end{aligned}$$

This completes the proof of (4.29). The proof for the one-particle density matrix (4.28) is done similarly and is even simpler. This completes the proof of Lemma 4.9. □

***Proof of Proposition* 4.8** The identity (4.26) for one-particle density matrix is a direct consequence of (4.28). We prove (4.27) by induction on k.

For $k = 2$, (4.27) is equivalent to (4.29) after noting that

$$A \otimes^s B = \frac{1}{2}\left((A+B)\otimes(A+B) - A\otimes A - B\otimes B\right).$$

This remark also proves that

$$\gamma_\Psi^{(2)} = \sum_{j=1}^{k} \gamma_{\zeta_j}^{(2)} + (\mathrm{Id}-\mathrm{Ex}) \sum_{i<j} \gamma_{\zeta_i} \otimes^s \gamma_{\zeta_j} \tag{4.31}$$

which is equality (4.27).

Let us prove (4.31) inductively. Suppose now that (4.31) holds true and consider

$$\Psi_{k+1} = \bigwedge_{j=1}^{k+1} \zeta_j = \left(\bigwedge_{j=1}^{k} \zeta_j\right) \wedge \zeta_{k+1} = \Psi_k \wedge \zeta_{k+1}.$$

By (4.29), we get

$$\begin{aligned}
\gamma_{\Psi_{k+1}}^{(2)} &= \gamma_{\Psi_k}^{(2)} + \gamma_{\zeta_{k+1}}^{(2)} + (\mathrm{Id}-\mathrm{Ex})\gamma_{\Psi_k} \otimes^s \gamma_{\zeta_{k+1}} \\
&= \sum_{j=1}^{k} \gamma_{\zeta_j}^{(2)} + (\mathrm{Id}-\mathrm{Ex}) \left(\sum_{\substack{i<j \\ i,j=1,\ldots,k}} \gamma_{\zeta_i} \otimes^s \gamma_{\zeta_j} \right) + \gamma_{\zeta_{k+1}}^{(2)} \\
&\qquad + (\mathrm{Id}-\mathrm{Ex}) \left(\sum_{j=1}^{k} \gamma_{\zeta_j} \right) \otimes^s \gamma_{\zeta_{k+1}} \\
&= \sum_{j=1}^{k+1} \gamma_{\zeta_j}^{(2)} + (\mathrm{Id}-\mathrm{Ex}) \sum_{\substack{i<j \\ i,j=1,\ldots,k+1}} \gamma_{\zeta_i} \otimes^s \gamma_{\zeta_j}.
\end{aligned}$$

This completes the proof of Proposition 4.8. □

4.2 The Proof of Theorem 1.5

The proof of Theorem 1.5 will rely on Theorem 4.2 and the analysis of $\Psi^U_\omega(L,n)$ performed in Section 3. The two sums in (4.3) will be analyzed separately and will be split into various components according to the lengths of the pieces coming into play in each component.

As in the beginning of Section 4.1 (see (4.2)), write $\Psi^U_\omega(L,n) = \sum_{\substack{Q \text{ occ.}\\ \overline{n}\in\mathbb{N}^m}} a^Q_{\overline{n}} \Phi_{Q,\overline{n}}$.

We will first transform the results on the ground state obtained in Section 3 into a statement on the coefficients $((a^Q_{\overline{n}}))_{Q,\overline{n}}$, namely,

Proposition 4.10 *There exists $\rho_0 > 0$ such that, for $\rho \in (0,\rho_0)$ and $\varepsilon \in (0,1/10)$, ω almost surely, in the thermodynamic limit, with probability $1 - O(L^{-\infty})$, one has*

(a) *for an occupation $Q \notin \mathcal{Q}_\rho$ (see (3.83)) and any $\overline{n} \in \mathbb{N}^m$, one has $a^Q_{\overline{n}} = 0$;*
(b) *let $\mathcal{P}_-$ be the (indices j of the) pieces $(\Delta_j(\omega))_j$ of lengths less than $3\ell_\rho(1-\varepsilon)$, and, for Q an occupation, let $\mathcal{P}^Q_-$ be the (indices j of the) pieces in $\mathcal{P}_-$ such that $Q_j \leqslant 3$.*

Then, for Q, an occupation number of a ground state $\Psi^U_\omega(L,n)$, letting $(a^Q_{\overline{n}})_{Q,\overline{n}}$ be its coefficients in the decomposition (4.2)*, one has*

$$\sum_{\substack{Q \text{ occ.}\\ \overline{n}\in\mathbb{N}^m}} \#\{j \in \mathcal{P}^Q_-; n_j \geqslant 2\} \left|a^Q_{\overline{n}}\right|^2 \leqslant o\left(\frac{n\cdot\rho}{|\log\rho|}\right). \tag{4.32}$$

The second part of Proposition 4.10 controls the excited particles in the ground state $\Psi^U_\omega(L,n)$. Actually, as the proof shows, we shall prove (4.32) not only for a ground state of $H^U_\omega(L,n)$, but, also for any state Ψ satisfying

$$\frac{1}{n}\langle H^{U^p}_\omega(L,n)\Psi,\Psi\rangle \leqslant \mathcal{E}^0(\rho) + \pi^2\gamma_*\frac{\rho}{|\log\rho|^{-3}} + o\left(\frac{\rho}{|\log\rho|^{-3}}\right). \tag{4.33}$$

Proof of Proposition 4.10 Point (a) is a rephrasing of Corollary 3.32.

Let us prove point (b). Pick an n-state Ψ and decompose it as $\Psi^U_\omega(L,n) = \sum_{Q\in\mathcal{Q}_\rho}\Psi_Q$. Then, if $E^{j,U^p}_{Q_j,n_j}$ denotes the n_j-th eigenvalue of $-\sum_{l=1}^{Q_j}\frac{d^2}{dx_l^2} + \sum_{1\leqslant k<l\leqslant Q_j} U^p(x_k - x_l)$ acting on $\bigwedge_{l=1}^{Q_j} L^2(\Delta_j(\omega))$ with Dirichlet boundary conditions (if $Q_j = 0$, we set $E^{j,U^p}_{Q_j,n_j} = 0$ for all n_j), as $H^U \geqslant H^{U^p}$ (see (3.17)), by (3.82), one has

$$n\left(\mathcal{E}^0(\rho)+\pi^2\gamma_\star\rho|\log\rho|^{-3}\left(1+O\left(f_Z(|\log\rho|)\right)\right)\right)\geqslant\langle H^{U^p}\Psi^{U^p},\Psi^{U^p}\rangle$$
$$\geqslant\sum_{\substack{Q\text{ occ.}\\ \overline{n}\in\mathbb{N}^m}}\left(\sum_{\substack{j\in\mathcal{P}_-^Q\\ Q_j\geqslant 1}}E^{j,U^p}_{Q_j,n_j}\right)\left|a^Q_{\overline{n}}\right|^2. \tag{4.34}$$

One proves

Lemma 4.11 *There exists $C>0$ such that, for $j\in\mathcal{P}_-^Q$, $Q_j\geqslant 1$ and $n_j\geqslant 2$, one has*

$$E^{j,U^p}_{Q_j,n_j}\geqslant E^{j,U^p}_{Q_j,1}+\frac{1}{C\ell_\rho^2}. \tag{4.35}$$

Plugging (4.35) into (4.34) yields

$$\sum_{\substack{Q\text{ occ.}\\ \overline{n}\in\mathbb{N}^m}}\left(\sum_{\substack{j\in\mathcal{P}_-^Q\\ Q_j\geqslant 1}}E^{j,U^p}_{Q_j,1}\right)\left|a^Q_{\overline{n}}\right|^2+\sum_{\substack{Q\text{ occ.}\\ \overline{n}\in\mathbb{N}^m}}\frac{\#\{j\in\mathcal{P}_-^Q;\,n_j\geqslant 2\}}{C\ell_\rho^2}\left|a^Q_{\overline{n}}\right|^2$$
$$\leqslant n\left(\mathcal{E}^0(\rho)+\pi^2\gamma_\star\rho|\log\rho|^{-3}\left(1+O\left(f_Z(|\log\rho|)\right)\right)\right) \tag{4.36}$$

We prove

Lemma 4.12 *There exists $\rho_0>0$ such that, for $\rho\in(0,\rho_0)$, $\varepsilon\in(0,1)$ and ω almost surely, for L sufficiently large and $|n/L-\rho|$ sufficiently small, if Q is an occupation such that*

$$\sum_{j\in\mathcal{P}_-}E^{j,U^p}_{Q_j,1}\leqslant n\left(\mathcal{E}^0(\rho)+\rho|\log\rho|^{-3}\left(\pi^2\gamma_\star+\varepsilon\right)\right) \tag{4.37}$$

then

$$\sum_{\substack{j\in\mathcal{P}_-^Q\\ Q_j\geqslant 1}}E^{j,U^p}_{Q_j,1}\geqslant n\left(\mathcal{E}^0(\rho)+\rho|\log\rho|^{-3}\left(\pi^2\gamma_\star-\frac{1}{\rho_0}\left(\varepsilon+f_Z(|\log\rho|)\right)\right)\right). \tag{4.38}$$

Lemma 4.12 shows that, for low energy states, most of the energy is carried by pieces carrying three particles and less (compare the set $\mathcal{P}_-$ and $\mathcal{P}_-^Q$).

Let us postpone the proof of this result for a while and complete the proof of Proposition 4.10. From (4.38) and (4.36), as $\sum_{\substack{Q \text{ occ.} \\ \overline{n}\in\mathbb{N}^m}} \left|a_{\overline{n}}^{Q}\right|^2 = 1$ and $f_Z(|\log\rho|) = o(1)$, we get that

$$\sum_{\substack{Q \text{ occ.} \\ \overline{n}\in\mathbb{N}^m}} \frac{\#\{j \in \mathcal{P}_-^Q; n_j \geqslant 2\}}{C\ell_\rho^2} \left|a_{\overline{n}}^{Q}\right|^2 \leqslant o\left(n\rho|\log\rho|^{-3}\right).$$

As $\ell_\rho \asymp |\log\rho|$, this immediately yields (4.32) and completes the proof of Proposition 4.10. □

Proof of Lemma 4.12 By Theorem 3.19, for L large and n/L close to ρ, we have

$$\left\langle H_\omega^{U^p}\Psi^{\text{opt}}, \Psi^{\text{opt}}\right\rangle \geqslant n\left(\mathcal{E}^0(\rho) + \pi^2\gamma_\star\rho|\log\rho|^{-3}\left(1 + O\left(f_Z(|\log\rho|)\right)\right)\right).$$

Recall that the occupation Q^{opt} of Ψ^{opt} satisfies

$$Q_j^{\text{opt}} = \begin{cases} 0 \text{ if } |\Delta_j(\omega)| \in [0, \ell_\rho - \rho x_*), \\ 1 \text{ if } |\Delta_j(\omega)| \in [\ell_\rho - \rho x_*, 2\ell_\rho + A_*), \\ 2 \text{ if } |\Delta_j(\omega)| \in [2\ell_\rho + A_*, 3\ell_\rho(1-\varepsilon)). \end{cases} \tag{4.39}$$

Theorem 3.19 shows that

$$\left|\left\langle H_\omega^{U^p}\Psi^{\text{opt}}, \Psi^{\text{opt}}\right\rangle - \sum_{\substack{j\in\mathcal{P}_- \\ Q_j^{\text{opt}}=1}} E_{1,1}^{j,U^p} - \sum_{\substack{j\in\mathcal{P}_- \\ Q_j^{\text{opt}}=2}} E_{2,1}^{j,U^p}\right| \lesssim n\frac{\rho}{|\log\rho|^3} f_Z(|\log\rho|). \tag{4.40}$$

Let

$$\Delta E := \sum_{j\in\mathcal{P}_-} E_{Q_j,1}^{j,U^p} - \sum_{\substack{j\in\mathcal{P}_- \\ Q_j^{\text{opt}}=1}} E_{1,1}^{j,U^p} - \sum_{\substack{j\in\mathcal{P}_- \\ Q_j^{\text{opt}}=2}} E_{2,1}^{j,U^p}. \tag{4.41}$$

Then, (4.40) and assumption (4.37) imply that

$$|\Delta E| \leqslant \frac{C n \rho}{|\log\rho|^3}\left(f_Z(|\log\rho|) + \varepsilon\right). \tag{4.42}$$

Moreover, one has

$$
\begin{aligned}
\Delta E &\geqslant \sum_{\substack{j\in\mathcal{P}_-\\ Q_j^{\mathrm{opt}}=0}} E^{j,U^p}_{Q_j,1} + \sum_{\substack{j\in\mathcal{P}_-\\ Q_j^{\mathrm{opt}}=1}} (E^{j,U^p}_{Q_j,1} - E^{j,U^p}_{1,1}) + \sum_{\substack{j\in\mathcal{P}_-\\ Q_j^{\mathrm{opt}}=2}} (E^{j,U^p}_{Q_j,1} - E^{j,U^p}_{2,1}) \\
&= \sum_{\substack{j\in\mathcal{P}_-\\ Q_j^{\mathrm{opt}}=0\\ Q_j\geqslant 1}} E^{j,U^p}_{Q_j,1} + \sum_{\substack{j\in\mathcal{P}_-\\ Q_j^{\mathrm{opt}}=1\\ Q_j\geqslant 2}} (E^{j,U^p}_{Q_j,1} - E^{j,U^p}_{1,1}) + \sum_{\substack{j\in\mathcal{P}_-\\ Q_j^{\mathrm{opt}}=2\\ Q_j\geqslant 3}} (E^{j,U^p}_{Q_j,1} - E^{j,U^p}_{2,1}) \\
&\qquad - \sum_{\substack{j\in\mathcal{P}_-\\ Q_j^{\mathrm{opt}}=1\\ Q_j=0}} E^{j,U^p}_{1,1} - \sum_{\substack{j\in\mathcal{P}_-\\ Q_j^{\mathrm{opt}}=2\\ Q_j\leqslant 1}} (E^{j,U^p}_{2,1} - E^{j,U^p}_{Q_j,1}).
\end{aligned}
\tag{4.43}
$$

On the other hand, as $|Q| = n = |Q^{\mathrm{opt}}|$, using Lemma 3.23 as $\Psi^U_\omega(L,n)$ satisfies (4.33), we know that

$$
\sum_{\substack{j\in\mathcal{P}_-\\ Q_j^{\mathrm{opt}}=1\\ Q_j=0}} 1 + \sum_{\substack{j\in\mathcal{P}_-\\ Q_j^{\mathrm{opt}}=2\\ Q_j\leqslant 1}} (2-Q_j) = \sum_{\substack{j\in\mathcal{P}_-\\ Q_j^{\mathrm{opt}}=0\\ Q_j\geqslant 1}} Q_j + \sum_{\substack{j\in\mathcal{P}_-\\ Q_j^{\mathrm{opt}}=1\\ Q_j\geqslant 2}} (Q_j-1) + \sum_{\substack{j\in\mathcal{P}_-\\ Q_j^{\mathrm{opt}}=2\\ Q_j\geqslant 3}} (Q_j-2) + O(n\rho^{1+\eta}).
\tag{4.44}
$$

Define

$$
B := \max\left(\max_{\substack{j;\ Q_j=0\\ Q_j^{\mathrm{opt}}=1}} E^{j,U^p}_{1,1},\ \max_{\substack{j;\ Q_j^{\mathrm{opt}}=2\\ 0\leqslant Q_j\leqslant 1}} \frac{E^{j,U^p}_{2,1} - E^{j,U^p}_{Q_j,1}}{2-Q_j} \right).
$$

Then, (4.43) implies that

$$
\begin{aligned}
\Delta E &\geqslant \sum_{\substack{j\in\mathcal{P}_-\\ Q_j^{\mathrm{opt}}=0\\ Q_j\geqslant 1}} E^{j,U^p}_{Q_j,1} + \sum_{\substack{j\in\mathcal{P}_-\\ Q_j^{\mathrm{opt}}=1\\ Q_j\geqslant 2}} (E^{j,U^p}_{Q_j,1} - E^{j,U^p}_{1,1}) + \sum_{\substack{j\in\mathcal{P}_-\\ Q_j^{\mathrm{opt}}=2\\ Q_j\geqslant 3}} (E^{j,U^p}_{Q_j,1} - E^{j,U^p}_{2,1}) \\
&\qquad - B \sum_{\substack{j\in\mathcal{P}_-\\ Q_j^{\mathrm{opt}}=1\\ Q_j=0}} 1 - B \sum_{\substack{j\in\mathcal{P}_-\\ Q_j^{\mathrm{opt}}=2\\ Q_j\leqslant 1}} (2-Q_j).
\end{aligned}
$$

Hence, (4.44) implies that, for some $C > 0$, for ρ sufficiently small, one has

$$\Delta E + Cn\rho^{1+\eta} \geqslant \sum_{\substack{j\in\mathcal{P}_- \\ Q_j^{\text{opt}}=0 \\ Q_j\geqslant 1}} (E_{Q_j,1}^{j,U^p} - B) + \sum_{\substack{j\in\mathcal{P}_- \\ Q_j^{\text{opt}}=1 \\ Q_j\geqslant 2}} (E_{Q_j,1}^{j,U^p} - E_{1,1}^{j,U^p} - B(Q_j-1))$$
$$+ \sum_{\substack{j\in\mathcal{P}_- \\ Q_j^{\text{opt}}=2 \\ Q_j\geqslant 3}} (E_{Q_j,1}^{j,U^p} - E_{2,1}^{j,U^p} - B(Q_j-2)). \tag{4.45}$$

Let us upper bound B. Recalling that for a single particle in a piece there is no interaction, a direct computation and (4.39) show that

$$\max_{\substack{j;\ Q_j=0 \\ Q_j^{\text{opt}}=1}} E_{1,1}^{j,U^p} \leqslant \frac{\pi^2}{(\ell_\rho - \rho x_*)^2}. \tag{4.46}$$

Proposition 1.4 and (4.39) show that, for ρ sufficiently small, one has

$$\max_{\substack{j;\ Q_j=0 \\ Q_j^{\text{opt}}=2}} \frac{E_{2,1}^{j,U^p} - E_{Q_j,1}^{j,U^p}}{2-Q_j} \leqslant \frac{5\pi^2}{2(2\ell_\rho + A^*)^2} + \frac{2\gamma}{(2\ell_\rho + A^*)^3} \leqslant \frac{\pi^2}{(\ell_\rho - \rho x_*)^2}$$

$$\max_{\substack{j;\ Q_j=1 \\ Q_j^{\text{opt}}=2}} \frac{E_{2,1}^{j,U^p} - E_{Q_j,1}^{j,U^p}}{2-Q_j} \leqslant \frac{4\pi^2}{(2\ell_\rho + A^*)^2} + \frac{2\gamma}{(2\ell_\rho + A^*)^3} \leqslant \frac{\pi^2}{(\ell_\rho - \rho x_*)^2}.$$

Thus,

$$B \leqslant \frac{\pi^2}{(\ell_\rho - \rho x_*)^2}. \tag{4.47}$$

Now, notice that

- for j s.t. $Q_j^{\text{opt}} = 0$ (see (4.39)):
  - if $Q_j = 1$, one has

$$E_{Q_j,1}^{j,U^p} - \frac{\pi^2}{(\ell_\rho - \rho x_*)^2} \geqslant \frac{\pi^2}{|\Delta_j(\omega)|^2} - \frac{\pi^2}{(\ell_\rho - \rho x_*)^2} \geqslant 0;$$

  - if $Q_j \geqslant 2$, one has

$$E_{Q_j,1}^{j,U^p} - \frac{\pi^2}{(\ell_\rho - \rho x_*)^2} \geqslant \frac{1}{2}E_{Q_j,1}^{j,U^p} + \frac{5\pi^2}{2|\Delta_j(\omega)|^2} - \frac{\pi^2}{(\ell_\rho - \rho x_*)^2} \geqslant \frac{1}{2}E_{Q_j,1}^{j,U^p};$$

- for j s.t. $Q_j^{\text{opt}} = 1$ (see (4.39)):
  - if $Q_j = 2$, one has

$$\begin{aligned}E^{j,U^p}_{Q_j,1} - E^{j,U^p}_{1,1} - \frac{\pi^2}{(\ell_\rho - \rho x_*)^2} &\geqslant \frac{4\pi^2}{|\Delta_j(\omega)|^2} + \frac{\gamma}{|\Delta_j(\omega)|^3} + o(\ell_\rho^{-3}) - \frac{\pi^2}{(\ell_\rho - \rho x_*)^2}\\ &\geqslant \frac{4\pi^2}{|2\ell_\rho+A_*+\varepsilon_\rho|^2}+\frac{\gamma}{|2\ell_\rho+A_*+\varepsilon_\rho|^3}+o(\ell_\rho^{-3})-\frac{\pi^2}{(\ell_\rho-\rho x_*)^2}\\ &\geqslant \frac{\pi^2}{\ell_\rho^2}-\frac{A_*\pi^2}{2\ell_\rho^3}+\frac{\gamma}{4\ell_\rho^3}+\frac{\pi^2\varepsilon_\rho}{2\ell_\rho^3}+o(\ell_\rho^{-3})-\frac{\pi^2}{(\ell_\rho-\rho x_*)^2}\\ &\geqslant \frac{\pi^2\varepsilon_\rho}{2\ell_\rho^3} + o(\ell_\rho^{-3}) \geqslant 0\end{aligned}$$

  if ρ sufficiently small (see (3.10)) and $|\Delta_j(\omega)| \leqslant 2\ell_\rho + A_* - \varepsilon_\rho$; here, $\varepsilon_\rho \to 0^+$ (but not too fast) as $\rho \to 0^+$; on the other hand, the number of pieces of length in $2\ell_\rho + A_* + [-\varepsilon_\rho, 0]$ is bounded by $C\rho n\varepsilon_\rho$ (see Proposition 2.2) and for such pieces, one has

$$\left|E^{j,U^p}_{2,1} - E^{j,U^p}_{1,1} - \frac{\pi^2}{(\ell_\rho - \rho x_*)^2}\right| = o(\ell_\rho^{-3}); \tag{4.48}$$

  - if $Q_j \geqslant 3$, one has

$$\begin{aligned}E^{j,U^p}_{Q_j,1}-E^{j,U^p}_{1,1}-\frac{\pi^2}{(\ell_\rho-\rho x_*)^2}(Q_j-1) &\geqslant \frac{1}{2}E^{j,U^p}_{Q_j,1}+\frac{1}{2}E^{j,0}_{Q_j,1}-\frac{\pi^2}{(\ell_\rho-\rho x_*)^2}(Q_j-1)\\ &\geqslant \frac{1}{2}E^{j,U^p}_{Q_j,1}+\frac{\pi^2}{4\ell_\rho^2}\frac{5}{12}(Q_j-1) \geqslant \frac{1}{2}E^{j,U^p}_{Q_j,1}\end{aligned}$$

- for j s.t. $Q_j^{\text{opt}} = 2$ (see (4.39)):
  - if $Q_j \geqslant 3$, one has

$$\begin{aligned}E^{j,U^p}_{Q_j,1}-E^{j,U^p}_{2,1}-\frac{\pi^2}{(\ell_\rho-\rho x_*)^2}(Q_j-2) &\geqslant \frac{1}{3}E^{j,U^p}_{Q_j,1}+\frac{2}{3}E^{j,0}_{Q_j,1}-\frac{\pi^2}{(\ell_\rho-\rho x_*)^2}(Q_j-2)\\ &\geqslant \frac{1}{3}E^{j,U^p}_{Q_j,1}+\frac{\pi^2}{9(1-\varepsilon)^2\ell_\rho^2}\left(\frac{102}{9}-9\right)(Q_j-2)\\ &\geqslant \frac{1}{3}E^{j,U^p}_{Q_j,1}.\end{aligned}$$

Plugging these estimates and (4.47) into (4.45), we get that, for ρ sufficiently small,

$$\Delta E + \sum_{|\Delta_j(\omega)|\in 2\ell_\rho + A_* + [-\varepsilon_\rho, 0]} \left| E^{j,U^p}_{2,1} - E^{j,U^p}_{1,1} - \frac{\pi^2}{(\ell_\rho - \rho x_*)^2} \right| + C\, n\, \rho^{1+\eta}$$
$$\geqslant \frac{1}{2} \sum_{\substack{j\in\mathcal{P}_- \\ Q_j^{\mathrm{opt}}=0 \\ Q_j\geqslant 2}} E^{j,U^p}_{Q_j,1} + \frac{1}{2} \sum_{\substack{j\in\mathcal{P}_- \\ Q_j^{\mathrm{opt}}=1 \\ Q_j\geqslant 3}} E^{j,U^p}_{Q_j,1} + \frac{1}{3} \sum_{\substack{j\in\mathcal{P}_- \\ Q_j^{\mathrm{opt}}=2 \\ Q_j\geqslant 3}} E^{j,U^p}_{Q_j,1}.$$

Hence, in view of (4.48) and the estimate on the number of terms in the sum in the left-hand side, one gets

$$3\left(\Delta E + o\left(n\rho\ell_\rho^{-3}\right)\right) \geqslant \sum_{\substack{j\in\mathcal{P}_- \\ Q_j^{\mathrm{opt}}=0 \\ Q_j\geqslant 2}} E^{j,U^p}_{Q_j,1} + \sum_{\substack{j\in\mathcal{P}_- \\ Q_j^{\mathrm{opt}}=1 \\ Q_j\geqslant 3}} E^{j,U^p}_{Q_j,1} + \sum_{\substack{j\in\mathcal{P}_- \\ Q_j^{\mathrm{opt}}=2 \\ Q_j\geqslant 3}} E^{j,U^p}_{Q_j,1} \geqslant 0. \tag{4.49}$$

This implies that

$$o\left(n\rho\ell_\rho^{-3}\right) \leqslant \Delta E = \sum_{j\in\mathcal{P}_-} E^{j,U^p}_{Q_j,1} - \sum_{\substack{j\in\mathcal{P}_- \\ Q_j^{\mathrm{opt}}=1}} E^{j,U^p}_{1,1} - \sum_{\substack{j\in\mathcal{P}_- \\ Q_j^{\mathrm{opt}}=2}} E^{j,U^p}_{2,1}$$

hence, by (4.40), that, for some $C > 0$ and ρ sufficiently small, one has

$$\sum_{j\in\mathcal{P}_-} E^{j,U^p}_{Q_j,1} \geqslant n\left(\mathcal{E}^0(\rho) + \pi^2\gamma_\star\rho|\log\rho|^{-3}\left(1 - C\, f_Z(|\log\rho|)\right)\right) \tag{4.50}$$

We complete the proof of Lemma 4.12 by noting that, by the definition of $\mathcal{P}_-^Q$, one has

$$\sum_{\substack{j\in\mathcal{P}_-^Q \\ Q_j\geqslant 1}} E^{j,U^p}_{Q_j,1} = \sum_{j\in\mathcal{P}_-} E^{j,U^p}_{Q_j,1} - \left(\sum_{\substack{j\in\mathcal{P}_- \\ Q_j^{\mathrm{opt}}=0 \\ Q_j\geqslant 3}} E^{j,U^p}_{Q_j,1} + \sum_{\substack{j\in\mathcal{P}_- \\ Q_j^{\mathrm{opt}}=1 \\ Q_j\geqslant 3}} E^{j,U^p}_{Q_j,1} + \sum_{\substack{j\in\mathcal{P}_- \\ Q_j^{\mathrm{opt}}=2 \\ Q_j\geqslant 3}} E^{j,U^p}_{Q_j,1} \right)$$
$$\geqslant n\left(\mathcal{E}^0(\rho) + \pi^2\gamma_\star\rho|\log\rho|^{-3}\left(1 - C(\varepsilon + f_Z(|\log\rho|))\right)\right)$$

where the last lower bound follows from (4.42) and (4.49).

This completes the proof of Lemma 4.12. □

Let us resume the proof of Theorem 1.5. Recall Theorem 4.2; we analyze the two components $\gamma^{(1),d}_{\Psi^U_\omega(L,n)}$ and $\gamma^{(1),o}_{\Psi^U_\omega(L,n)}$ separately.

Let us start with the analysis of $\gamma^{(1),o}_{\Psi^U_\omega(L,n)}$. We prove

Lemma 4.13 *Under the assumptions of Theorem 4.2, in the thermodynamic limit, with probability* $1 - O(L^{-\infty})$, *one has*

$$\left\| \gamma^{(1),o}_{\Psi^U_\omega(L,n)} \right\|_{tr} \leqslant 3. \tag{4.51}$$

Proof We recall (4.4) from Theorem 4.2 and write

$$\gamma^{(1),o}_{\Psi^U_\omega(L,n)} = \sum_{Q \text{ occ.}} \sum_{\substack{i,j=1 \\ i\neq j \\ Q_j \geqslant 1}}^{m} C_1(Q,i,j) \sum_{\tilde{n}\in\mathbb{N}^{m-1}} \sum_{\substack{n_i,n_j\geqslant 1 \\ n'_i,n'_j\geqslant 1}} a^{Q}_{\tilde{n}_{i,j}} \overline{a^{Q'}_{\tilde{n}'_{i,j}}} \gamma^{(1)}_{\substack{Q_i,Q_j \\ n_i,n_j \\ n'_i,n'_j}}$$

where, by definition, in the above sums, Q' satisfies $Q'_k = Q_k$ if $k \notin \{i, j\}$, $Q'_i = Q_i + 1$ and $Q'_j = Q_j - 1$.

Note that, by point (a) of Proposition 4.10, here and in the sequel when summing over the occupations Q, we can always restrict ourselves to the occupations in $\mathcal{Q}_\rho$.

Decompose

$$\gamma^{(1),o}_{\Psi^U_\omega(L,n)} = \gamma^{(1),o,+,+}_{\Psi^U_\omega(L,n)} + \gamma^{(1),o,+,-}_{\Psi^U_\omega(L,n)} + \gamma^{(1),o,-,+}_{\Psi^U_\omega(L,n)} + \gamma^{(1),o,-,-}_{\Psi^U_\omega(L,n)} \tag{4.52}$$

where (see (4.5), (4.6), (4.7), and (4.8))

$$\gamma^{(1),o,+,+}_{\Psi^U_\omega(L,n)} := \sum_{\substack{Q \text{ occ.} \\ \tilde{n}\in\mathbb{N}^{m-1} \\ i\neq j \\ Q_j\geqslant 2 \\ Q_i\geqslant 1}} C_1(Q,i,j) a^{Q}_{\tilde{n}_{i,j}} \overline{a^{Q'}_{\tilde{n}'_{i,j}}} \gamma^{(1),1,+,+}_{Q,Q',i,j,\tilde{n}}, \quad \gamma^{(1),o,+,-}_{\Psi^U_\omega(L,n)}$$

$$:= \sum_{\substack{Q \text{ occ.} \\ \tilde{n}\in\mathbb{N}^{m-1} \\ i\neq j \\ Q_j\geqslant 2 \\ Q_i=0}} C_1(Q,i,j) a^{Q}_{\tilde{n}_{i,j}} \overline{a^{Q'}_{\tilde{n}'_{i,j}}} \gamma^{(1),1,+,-}_{Q,Q',i,j,\tilde{n}},$$

$$\gamma^{(1),o,-,+}_{\Psi^U_\omega(L,n)} := \sum_{\substack{Q \text{ occ.} \\ \tilde{n}\in\mathbb{N}^{m-1} \\ i\neq j \\ Q_j=1 \\ Q_i\geqslant 1}} C_1(Q,i,j) a^{Q}_{\tilde{n}_{i,j}} \overline{a^{Q'}_{\tilde{n}'_{i,j}}} \gamma^{(1),1,-,+}_{Q,Q',i,j,\tilde{n}}, \quad \gamma^{(1),o,-,-}_{\Psi^U_\omega(L,n)}$$

$$:= \sum_{\substack{Q \text{ occ.} \\ \tilde{n}\in\mathbb{N}^{m-1} \\ i\neq j \\ Q_j=1 \\ Q_i=0}} C_1(Q,i,j) a^{Q}_{\tilde{n}_{i,j}} \overline{a^{Q'}_{\tilde{n}'_{i,j}}} \gamma^{(1),1,-,-}_{Q,Q',i,j,\tilde{n}}$$

and

$$\gamma^{(1),1,+,+}_{Q,Q',i,j,\tilde{n}}(x,y) := \int_{\Delta_i^{Q_i}\times\Delta_j^{Q_j-1}} \left(\sum_{\substack{n_i\geqslant 1 \\ n_j\geqslant 1}} a^{Q}_{\tilde{n}_{i,j}} \varphi^i_{Q_i,n_i}(z) \varphi^j_{Q_j,n_j}(x,z') \right) \left(\overline{\sum_{\substack{n'_j\geqslant 1 \\ n'_j\geqslant 1}} a^{Q'}_{\tilde{n}'_{i,j}} \varphi^i_{Q'_i,n'_i}(y,z) \varphi^j_{Q'_j,n'_j}(z')} \right) dz dz',$$

$$\gamma^{(1),1,+,-}_{Q,Q',i,j,\tilde{n}}(x,y) := \int_{\Delta_j^{Q_j-1}} \left(\sum_{\substack{n_i=1 \\ n_j\geqslant 1}} a^{Q}_{\tilde{n}_{i,j}} \varphi^j_{Q_j,n_j}(x,z') \right) \left(\overline{\sum_{\substack{n'_i\geqslant 1 \\ n'_j\geqslant 1}} a^{Q}_{\tilde{n}'_{i,j}} \varphi^i_{1,n'_i}(y) \varphi^j_{Q_j-1,n'_j}(z')} \right) dz',$$

$$\gamma^{(1),1,-,+}_{Q,Q',i,j,\tilde{n}}(x,y) := \int_{\Delta_i^{Q_i}} \left(\sum_{\substack{n_i\geqslant 1 \\ n_j\geqslant 1}} a^{Q}_{\tilde{n}_{i,j}} \varphi^j_{1,n_j}(x) \varphi^i_{Q_i,n_i}(z) \right) \left(\overline{\sum_{\substack{n'_j=1 \\ n'_i\geqslant 1}} a^{Q}_{\tilde{n}'_{i,j}} \varphi^i_{Q_i+1,n'_i}(y,z)} \right) dz,$$

$$\text{and} \quad \gamma^{(1),1,-,-}_{Q,Q',i,j,\tilde{n}}(x,y) := \left(\sum_{\substack{n_i=1 \\ n_j\geqslant 1}} a^{Q}_{\tilde{n}_{i,j}} \varphi^i_{1,n_j}(x) \right) \left(\overline{\sum_{\substack{n'_j=1 \\ n'_i\geqslant 1}} a^{Q'}_{\tilde{n}'_{i,j}} \varphi^i_{1,n'_i}(y)} \right).$$

Let us first analyze $\gamma^{(1),o,+,+}_{\Psi^U_\omega(L,n)}$. By Lemma B.1, using the orthonormality of the families $(\varphi^j_{Q_j,n_j})_{n_j\in\mathbb{N}}$ (see the beginning of Section 4.1), we know that

$$\left\|\gamma^{(1),1,+,+}_{Q,Q',i,j,\tilde{n}}\right\|_{\mathrm{tr}} \leqslant \left\|\sum_{n_i,n_j} a^{Q}_{\tilde{n}_{i,j}}\varphi^{i}_{Q_i,n_i}\otimes\varphi^{j}_{Q_j,n_j}\right\|\cdot\left\|\sum_{n'_i,n'_j} a^{Q'}_{\tilde{n}'_{i,j}}\varphi^{i}_{Q'_i,n'_i}\otimes\varphi^{j}_{Q'_j,n'_j}\right\|$$
$$\leqslant \frac{1}{2}\left(\sum_{n_i,n_j}\left|a^{Q}_{\tilde{n}_{i,j}}\right|^2+\sum_{n_i,n_j}\left|a^{Q'}_{\tilde{n}_{i,j}}\right|^2\right).$$

Hence, by definition (see the formula following (4.52)) and the symmetry of $C_1(Q,i,j)$ in i and j, we have

$$\left\|\gamma^{(1),o,+,+}_{\Psi^U_\omega(L,n)}\right\|_{\mathrm{tr}} \leqslant \sum_{\substack{i,j=1\\ i\neq j}}^{m}\sum_{\substack{Q \text{ occ.}\\ Q_j\geqslant 2\\ Q_i\geqslant 1}} C_1(Q,i,j)\sum_{\overline{n}\in\mathbb{N}^m}\left|a^{Q}_{\overline{n}}\right|^2.$$

Now, by definition (see Theorem 4.2), for $Q_j\geqslant 2$ and $Q_i\geqslant 1$, one has

$$C_1(Q,i,j)\leqslant\frac{Q_i\,Q_j}{(n-1)(n-2)(n-3)}.$$

Thus,

$$\left\|\gamma^{(1),o,+,+}_{\Psi^U_\omega(L,n)}\right\|_{\mathrm{tr}} \leqslant \frac{1}{(n-1)(n-2)(n-3)}\sum_{\substack{Q \text{ occ.}\\ Q_j\geqslant 2\\ Q_i\geqslant 1}}\left(\sum_j Q_j\right)^2\sum_{\overline{n}\in\mathbb{N}^m}\left|a^{Q}_{\overline{n}}\right|^2$$
$$\leqslant \frac{n^2}{(n-1)(n-2)(n-3)}\sum_{Q,\,\overline{n}\in\mathbb{N}^m}\left|a^{Q}_{\overline{n}}\right|^2=\frac{n^2}{(n-1)(n-2)(n-3)}. \tag{4.53}$$

Let us now analyze $\gamma^{(1),o,-,-}_{\Psi^U_\omega(L,n)}$. By the definition of $C_1(Q,i,j)$, we write

$$\gamma^{(1),o,-,-}_{\Psi^U_\omega(L,n)}(x,y)=\frac{1}{n-1}\sum_{\substack{\tilde{n}\in\mathbb{N}^{m-1}\\ Q\text{ occ.}}}\sum_{\substack{i,j=1\\ i\neq j\\ Q_j=1\\ Q_i=0}}^{m}\left(\sum_{n_i=1,n_j} a^{Q}_{\tilde{n}_{i,j}}\varphi^{j}_{1,n_j}(x)\right)\left(\overline{\sum_{n'_j=1,n'_i} a^{Q'}_{\tilde{n}'_{i,j}}\varphi^{i}_{1,n'_i}(y)}\right).$$

Thus, by Lemma B.1, one has

$$
\begin{aligned}
\left\|\gamma^{(1),o,-,-}_{\Psi^U_\omega(L,n)}\right\|_{\mathrm{tr}} &\leqslant \frac{1}{n-1}\sum_{\substack{\tilde n\in\mathbb{N}^{m-1}\\ Q\text{ occ.}}}\left\|\sum_{j,\ Q_j=1}\sum_{n_i=1,n_j} a^Q_{\tilde n_{i,j}}\varphi^j_{1,n_j}\right\|\cdot\left\|\sum_{i,\ Q_i=0}^{m}\sum_{n'_j=1,n'_i} a^{Q'}_{\tilde n'_{i,j}}\varphi^i_{1,n'_i}\right\|\\
&\leqslant \frac{1}{2n-2}\sum_{\substack{\tilde n\in\mathbb{N}^{m-1}\\ Q\text{ occ.}}}\left\|\sum_{j,\ Q_j=1}\sum_{n_i=1,n_j} a^Q_{\tilde n_{i,j}}\varphi^j_{1,n_j}\right\|^2+\left\|\sum_{i,\ Q_i=0}\sum_{n'_j=1,n'_i} a^{Q'}_{\tilde n'_{i,j}}\varphi^i_{1,n'_i}\right\|^2\\
&\leqslant \frac{1}{2n-2}\sum_{\substack{n\in\mathbb{N}^m\\ Q\text{ occ.}}}\left|a^Q_{\tilde n}\right|^2=\frac{1}{2n-2}.
\end{aligned}
\tag{4.54}
$$

Let us now analyze $\gamma^{(1),o,+,-}_{\Psi^U_\omega(L,n)}$. One has

$$
\begin{aligned}
\gamma^{(1),o,+,-}_{\Psi^U_\omega(L,n)} &= \sum_{\substack{\tilde n\in\mathbb{N}^{m-1}\\ Q\text{ occ.}}}\sum_{\substack{i\neq j\\ Q_j\geqslant 2\\ Q_i=0}}\frac{(n-Q_j-1)!Q_j!}{(n-1)!}\int_{\Delta_j^{Q_j-1}}\left(\sum_{n_i=1,n_j} a^Q_{\tilde n_{i,j}}\varphi^j_{Q_j,n_j}(x,z')\right)\times\\
&\qquad\left(\overline{\sum_{n'_i,n'_j} a^Q_{\tilde n'_{i,j}}\varphi^i_{1,n'_i}(y)\varphi^j_{Q_j-1,n'_j}(z')}\right)dz'\\
&= \sum_{\substack{\tilde n\in\mathbb{N}^{m-1}\\ Q\text{ occ.}}}\sum_{j;\ Q_j\geqslant 2}\frac{(n-Q_j-1)!Q_j!}{(n-1)!}\int_{\Delta_j^{Q_j-1}}\left(\sum_{n_i=1,n_j} a^Q_{\tilde n_{i,j}}\varphi^j_{Q_j,n_j}(x,z')\right)\times\\
&\qquad\left(\overline{\sum_{i;\ Q_i=0}\sum_{n'_i,n'_j} a^Q_{\tilde n'_{i,j}}\varphi^i_{1,n'_i}(y)\varphi^j_{Q_j-1,n'_j}(z')}\right)dz'.
\end{aligned}
$$

Thus, using Lemma B.1 and the orthonormality properties of the families $(\varphi^j_{Q_j,n_j})_{n_j\in\mathbb{N}}$, as $(n-Q_j)!\,Q_j!\leqslant n!$ and $\sum_j Q_j=n$, we get

$$
\left\|\gamma^{(1),o,+,-}_{\Psi^U_\omega(L,n)}\right\|_{\mathrm{tr}}\leqslant\frac{1}{n-1}\sum_{\substack{\tilde n\in\mathbb{N}^{m-1}\\ Q\text{ occ.}}}\sum_{j=1}^{m}Q_j\sum_{n_i=1,n_j}\left|a^Q_{\tilde n_{i,j}}\right|^2\leqslant\frac{n}{n-1}\sum_{\substack{\overline{n}\in\mathbb{N}^m\\ Q\text{ occ.}}}\left|a^Q_{\overline{n}}\right|^2.
\tag{4.55}
$$

The term $\gamma^{(1),o,-,+}_{\Psi^U_\omega(L,n)}$ is analyzed in the same way. Gathering (4.53), (4.54), (4.55) and using (4.52), we obtain (4.51) and, thus, complete the proof of Lemma 4.13. □

Let us now turn to the analysis of $\gamma^{(1),d}_{\Psi^U_\omega(L,n)}$. Therefore, we write

$$\gamma^{(1),d}_{\Psi^U_\omega(L,n)} = \gamma^{(1),d,-}_{\Psi^U_\omega(L,n)} + \gamma^{(1),d,+}_{\Psi^U_\omega(L,n)} \text{ where } \gamma^{(1),d,-}_{\Psi^U_\omega(L,n)} := \sum_{\substack{Q \text{ occ.} \\ \tilde{n}\in\mathbb{N}^{m-1}}} \sum_{j\in\mathcal{P}^Q_-} \sum_{\substack{n_j\geqslant 1 \\ n'_j\geqslant 1}} a^Q_{\tilde{n}_j} \overline{a^Q_{\tilde{n}'_j}} \gamma^{(1)}_{\substack{Q_j \\ n_j,n'_j}}. \tag{4.56}$$

We prove

Lemma 4.14 *Under the assumptions of Theorem 4.2, for* $\eta \in (0, 1)$*, there exists* $\varepsilon_0 > 0$ *and* $C > 1$ *such that, for* $\varepsilon \in (0, \varepsilon_0)$*, in the thermodynamic limit, with probability* $1 - O(L^{-\infty})$*, one has*

$$\left\| \gamma^{(1),d,+}_{\Psi^U_\omega(L,n)} \right\|_{tr} \leqslant Cn\frac{\rho}{\ell_\rho}. \tag{4.57}$$

Proof Define

$$\gamma^{(1),d,+}_{\Psi^U_\omega(L,n)} = \gamma^{(1),d,+,+}_{\Psi^U_\omega(L,n)} + \gamma^{(1),d,+,0}_{\Psi^U_\omega(L,n)} \tag{4.58}$$

where

$$\gamma^{(1),d,+,+}_{\Psi^U_\omega(L,n)} = \sum_{\substack{Q \text{ occ.} \\ \tilde{n}\in\mathbb{N}^{m-1}}} \sum_{j\notin\mathcal{P}_-} \sum_{\substack{n_j\geqslant 1 \\ n'_j\geqslant 1}} a^Q_{\tilde{n}_j} \overline{a^Q_{\tilde{n}'_j}} \gamma^{(1)}_{\substack{Q_j \\ n_j,n'_j}} \quad \text{and}$$

$$\gamma^{(1),d,+,0}_{\Psi^U_\omega(L,n)} = \sum_{\substack{Q \text{ occ.} \\ \tilde{n}\in\mathbb{N}^{m-1}}} \sum_{\substack{j\in\mathcal{P}^Q_- \\ Q_j\geqslant 4}} \sum_{\substack{n_j\geqslant 1 \\ n'_j\geqslant 1}} a^Q_{\tilde{n}_j} \overline{a^Q_{\tilde{n}'_j}} \gamma^{(1)}_{\substack{Q_j \\ n_j,n'_j}}.$$

One computes

$$\begin{aligned}
\gamma^{(1),d,+,+}_{\Psi^U_\omega(L,n)}(x,y) &= \sum_{Q \text{ occ.}} \sum_{j\notin\mathcal{P}_-} \sum_{\substack{n_j\geqslant 1 \\ n'_j\geqslant 1}} \sum_{\tilde{n}\in\mathbb{N}^{m-1}} a^Q_{\tilde{n}_j} \overline{a^Q_{\tilde{n}'_j}} \gamma^{(1)}_{\substack{Q_j \\ n_j,n'_j}}(x,y) \\
&= \sum_{\substack{Q \text{ occ.} \\ \tilde{n}\in\mathbb{N}^{m-1}}} \sum_{j\notin\mathcal{P}_-} Q_j \int_{\Delta_j^{Q_j-1}} \left(\sum_{n_j=1}^{+\infty} a^Q_{\tilde{n}_j} \varphi^j_{Q_j,n_j}(x,z) \right) \\
&\quad \times \overline{\left(\sum_{n_j=1}^{+\infty} a^Q_{\tilde{n}_j} \varphi^j_{Q_j,n_j}(y,z) \right)} dz.
\end{aligned}$$

Thus, by Lemma B.1, we get

$$\begin{aligned}\left\|\gamma^{(1),d,+,+}_{\Psi^U_\omega(L,n)}\right\|_{\mathrm{tr}} &\leqslant \sum_{\substack{Q\text{ occ.}\\ Q\in\mathcal{Q}_\rho\\ \tilde n\in\mathbb{N}^{m-1}}}\sum_{j\notin\mathcal{P}_-} Q_j\sum_{n_j=1}^{+\infty}\left|a^Q_{\tilde n_j}\right|^2 \leqslant \sum_{Q\text{ occ. in }\mathcal{Q}_\rho}\left(\sum_{j\notin\mathcal{P}_-}Q_j\right)\sum_{\overline{n}\in\mathbb{N}^m}\left|a^Q_{\overline{n}}\right|^2\\ &\leqslant \max_{Q\text{ occ. in }\mathcal{Q}_\rho}\left(\sum_{j\notin\mathcal{P}_-}Q_j\right)\leqslant Cn\rho^{1+\eta}\end{aligned} \tag{4.59}$$

by Lemma 3.23.

Finally, one has

$$\gamma^{(1),d,+,0}_{\Psi^U_\omega(L,n)} = \sum_{Q\text{ occ.}}\sum_{\substack{j\in\mathcal{P}_-\\ Q_j\geqslant 4}}\sum_{\substack{n_j\geqslant 1\\ n'_j\geqslant 1}}\sum_{\tilde n\in\mathbb{N}^{m-1}} a^Q_{\tilde n_j}\overline{a^Q_{\tilde n'_j}}\gamma^{(1)}_{\substack{Q_j\\ n_j,n'_j}}.$$

Thus, the same computation as above yields

$$\left\|\gamma^{(1),d,+,0}_{\Psi^U_\omega(L,n)}\right\|_{\mathrm{tr}} \leqslant \sum_{\substack{Q\text{ occ.}\\ Q\in\mathcal{Q}_\rho\\ \tilde n\in\mathbb{N}^{m-1}}}\left(\sum_{\substack{j,\ |\Delta_j(\omega)|\leqslant 3\ell_\rho(1-\varepsilon)\\ Q_j\geqslant 4}} Q_j\right)\sum_{n_j=1}^{+\infty}\left|a^Q_{\tilde n_j}\right|^2\leqslant Cn\frac{\rho}{\ell_\rho}$$

by Lemma 3.24.

This completes the proof of Lemma 4.14. □

Let us now analyze $\gamma^{(1),d,-}_{\Psi^U_\omega(L,n)}$. We recall and compute

$$\gamma^{(1),d,-}_{\Psi^U_\omega(L,n)} := \sum_{\substack{Q\text{ occ.}\\ \tilde n\in\mathbb{N}^{m-1}}}\sum_{j\in\mathcal{P}^Q_-}\sum_{\substack{n_j\geqslant 1\\ n'_j\geqslant 1}} a^Q_{\tilde n_j}\overline{a^Q_{\tilde n'_j}}\gamma^{(1)}_{\substack{Q_j\\ n_j,n'_j}} = \sum_{\substack{Q\text{ occ.}\\ \tilde n\in\mathbb{N}^{m-1}}}\sum_{j\in\mathcal{P}^Q_-} Q_j\left|\varphi^{\tilde n}_j\right\rangle\left\langle\varphi^{\tilde n}_j\right|$$

where $\varphi^{\tilde n}_j = \sum_{n_j\geqslant 1} a^Q_{\tilde n_j}\varphi^j_{Q_j,n_j}$.

For $\tilde n$ and Q given, define the two sets

$$\mathcal{P}^{Q,\tilde n}_{-,+} := \left\{j\in\mathcal{P}^Q_-;\ a^Q_{\tilde n_j}=0 \text{ if } n_j\geqslant 2\right\} \text{ and } \mathcal{P}^{Q,\tilde n}_{-,-} := \left\{j\in\mathcal{P}^Q_-;\ \exists n_j\geqslant 2 \text{ s.t. } a^Q_{\tilde n_j}\not=0\right\}. \tag{4.60}$$

Define also

$$\tilde\varphi^{\tilde n}_j = \begin{cases}\varphi^{\tilde n}_j & \text{if } n_j = 1,\\ \|\varphi^{\tilde n}_j\|\varphi^j_{Q_j,1} & \text{if } n_j\geqslant 2.\end{cases} \tag{4.61}$$

Then, we compute

$$\begin{aligned}\gamma^{(1),d,-}_{\Psi^U_\omega(L,n)} &= \sum_{\substack{Q \text{ occ.}\\ \tilde{n}\in\mathbb{N}^{m-1}}} \sum_{j\in\mathcal{P}^{Q,\tilde{n}}_{-,-}} Q_j \left|\varphi^{\tilde{n}}_j\right\rangle\left\langle\varphi^{\tilde{n}}_j\right| + \sum_{\substack{Q \text{ occ.}\\ \tilde{n}\in\mathbb{N}^{m-1}}} \sum_{j\in\mathcal{P}^{Q,\tilde{n}}_{-,-}} Q_j \left|\varphi^{\tilde{n}}_j\right\rangle\left\langle\varphi^{\tilde{n}}_j\right| \\ &= \sum_{\substack{Q \text{ occ.}\\ \tilde{n}\in\mathbb{N}^{m-1}}} \sum_{j\in\mathcal{P}^{Q}_{-}} Q_j \left|\tilde{\varphi}^{\tilde{n}}_j\right\rangle\left\langle\tilde{\varphi}^{\tilde{n}}_j\right| + \sum_{\substack{Q \text{ occ.}\\ \tilde{n}\in\mathbb{N}^{m-1}}} \sum_{j\in\mathcal{P}^{Q,\tilde{n}}_{-,-}} Q_j \left(\left|\varphi^{\tilde{n}}_j\right\rangle\left\langle\varphi^{\tilde{n}}_j\right| - \left|\tilde{\varphi}^{\tilde{n}}_j\right\rangle\left\langle\tilde{\varphi}^{\tilde{n}}_j\right|\right).\end{aligned} \tag{4.62}$$

The second term in the sum above we estimate by

$$\begin{aligned}\left\| \sum_{\substack{Q \text{ occ.}\\ \tilde{n}\in\mathbb{N}^{m-1}}} \sum_{j\in\mathcal{P}^{Q,\tilde{n}}_{-,-}} Q_j \left(\left|\varphi^{\tilde{n}}_j\right\rangle\left\langle\varphi^{\tilde{n}}_j\right| - \left|\tilde{\varphi}^{\tilde{n}}_j\right\rangle\left\langle\tilde{\varphi}^{\tilde{n}}_j\right|\right)\right\|_{\text{tr}} &\leqslant \sum_{\substack{Q \text{ occ.}\\ \tilde{n}\in\mathbb{N}^{m-1}}} \sum_{j\in\mathcal{P}^{Q,\tilde{n}}_{-,-}} Q_j \left(\left\|\varphi^{\tilde{n}}_j\right\|^2 + \left\|\tilde{\varphi}^{\tilde{n}}_j\right\|^2\right) \\ &\leqslant \sum_{\substack{Q \text{ occ.}\\ \overline{n}\in\mathbb{N}^{m}}} \sum_{j\in\mathcal{P}^{Q}_{-}} \#\{j;\ n_j \geqslant 2\} \left|a^Q_{\overline{n}}\right|^2 \\ &\leqslant n \frac{\rho}{\rho_0 |\log\rho|} f_Z(|\log\rho|).\end{aligned} \tag{4.63}$$

by Lemma 4.11.

As for the first term in the second equality in (4.62), letting $\mathcal{P}_{\text{opt}}$ be the pieces of length less than $3\ell_\rho(1-\varepsilon)$ where Ψ^{opt} puts at least one particle, we write

$$\sum_{\substack{Q \text{ occ.}\\ \tilde{n}\in\mathbb{N}^{m-1}}} \sum_{j\in\mathcal{P}^{Q}_{-}} Q_j \left|\tilde{\varphi}^{\tilde{n}}_j\right\rangle\left\langle\tilde{\varphi}^{\tilde{n}}_j\right| = \sum_{\substack{Q \text{ occ.}\\ \tilde{n}\in\mathbb{N}^{m-1}}} \left(\sum_{j\in\mathcal{P}_{\text{opt}}} + \sum_{j\in\mathcal{P}^{Q}_{-}\setminus\mathcal{P}_{\text{opt}}} - \sum_{j\in\mathcal{P}_{\text{opt}}\setminus\mathcal{P}^{Q}_{-}}\right) Q_j \left|\tilde{\varphi}^{\tilde{n}}_j\right\rangle\left\langle\tilde{\varphi}^{\tilde{n}}_j\right| \tag{4.64}$$

One computes

$$\begin{aligned}\sum_{\substack{Q \text{ occ.}\\ \tilde{n}\in\mathbb{N}^{m-1}}} \sum_{j\in\mathcal{P}_{\text{opt}}} Q_j \left|\tilde{\varphi}^{\tilde{n}}_j\right\rangle\left\langle\tilde{\varphi}^{\tilde{n}}_j\right| &= \sum_{j\in\mathcal{P}_{\text{opt}}} Q_j \left(\sum_{\substack{Q \text{ occ.}\\ \overline{n}\in\mathbb{N}^{m}}} \left|a^Q_{\overline{n}}\right|^2\right) \left|\varphi^j_{Q_j,1}\right\rangle\left\langle\varphi^j_{Q_j,1}\right| \\ &= \sum_{j\in\mathcal{P}_{\text{opt}}} Q_j \left|\varphi^j_{Q_j,1}\right\rangle\left\langle\varphi^j_{Q_j,1}\right| = \gamma_{\Psi^{\text{opt}}} + R\end{aligned} \tag{4.65}$$

where $\|R\|_{\text{tr}} \leqslant Cn\rho^{1+\eta}$.

By Corollary 3.32, we know that

$$\begin{aligned}
&\left\| \sum_{\substack{Q \text{ occ.} \\ \tilde{n}\in\mathbb{N}^{m-1}}} \left(\sum_{\substack{j\in\mathcal{P}_-^Q\setminus\mathcal{P}_{\text{opt}} \\ |\Delta_j(\omega)|\geqslant \ell_\rho + C}} - \sum_{\substack{j\in\mathcal{P}_{\text{opt}}\setminus\mathcal{P}_-^Q \\ |\Delta_j(\omega)|\geqslant \ell_\rho + C}} \right) Q_j \left|\varphi_j^{\tilde{n}}\right\rangle\left\langle\varphi_j^{\tilde{n}}\right| \right\|_{\text{tr}} \\
&\leqslant \sum_{Q \text{ occ.}} \left(\sum_{\substack{j\in\mathcal{P}_-^Q\setminus\mathcal{P}_{\text{opt}} \\ |\Delta_j(\omega)|\geqslant \ell_\rho + C}} + \sum_{\substack{j\in\mathcal{P}_{\text{opt}}\setminus\mathcal{P}_-^Q \\ |\Delta_j(\omega)|\geqslant \ell_\rho + C}} \right) Q_j \sum_{\overline{n}\in\mathbb{N}^m} \left| a_{\overline{n}}^Q \right|^2 \\
&\leqslant Cn\rho \max\left(\sqrt{Z(2|\log\rho|)}, \ell_\rho^{-1}\right) \sum_{\substack{Q \text{ occ.} \\ \overline{n}\in\mathbb{N}^m}} \left| a_{\overline{n}}^Q \right|^2 = Cn\rho \max\left(\sqrt{Z(2|\log\rho|)}, \ell_\rho^{-1}\right)
\end{aligned}$$

and, in the same way,

$$\begin{aligned}
&\left\| \sum_{\substack{Q \text{ occ.} \\ \tilde{n}\in\mathbb{N}^{m-1}}} \left(\sum_{\substack{j\in\mathcal{P}_{\text{opt}}\setminus\mathcal{P}_-^Q \\ |\Delta_j(\omega)|< \ell_\rho + C}} - \sum_{\substack{j\in\mathcal{P}_-^Q\setminus\mathcal{P}_{\text{opt}} \\ |\Delta_j(\omega)|< \ell_\rho + C}} \right) Q_j \left|\varphi_j^{\tilde{n}}\right\rangle\left\langle\varphi_j^{\tilde{n}}\right| \right\|_{\text{tr}} \\
&\qquad\leqslant Cn \max\left(\sqrt{\rho Z(2|\log\rho|)}, \rho|\log\rho|^{-1}\right).
\end{aligned}$$

Plugging this and (4.65) into (4.64) and then into (4.62), using (4.63), we obtain

$$\begin{aligned}
\left\| \gamma^{(1),d,-}_{\Psi_\omega^U(L,n)} - \gamma^{(1)}_{\Psi^{\text{opt}}} \right\|_{\text{tr},<\ell_\rho+C} &\leqslant Cn \max\left(\sqrt{\rho Z(2|\log\rho|)}, \rho|\log\rho|^{-1}\right) \\
\left\| \gamma^{(1),d,-}_{\Psi_\omega^U(L,n)} - \gamma^{(1)}_{\Psi^{\text{opt}}} \right\|_{\text{tr},\geqslant\ell_\rho+C} &\leqslant Cn\rho \max\left(\sqrt{Z(2|\log\rho|)}, \ell_\rho^{-1}\right).
\end{aligned}$$

Taking into account the decomposition (4.56), Theorem 4.2 and Lemmas 4.13 and 4.14 then completes the proof of Theorem 1.5. □

4.3 The Proof of Theorem 1.6

We proceed as in the proof of Theorem 1.5: for $\Psi_\omega^U(L,n)$ a ground state of the Hamiltonian $H_\omega^U(L,n)$, we analyze each of the components of the decomposition (4.9) separately.

We prove

Lemma 4.15 *Under the assumptions of Theorem 4.2, in the thermodynamic limit, with probability* $1 - O(L^{-\infty})$*, one has*

$$\left\|\gamma^{(2),d,d}_{\Psi^U_\omega(L,n)}\right\|_{tr} \lesssim n\log n \cdot \log\log n.$$

Proof Using Lemma B.1 and the orthonormality properties of the families $(\varphi^j_{Q_j,n_j})_{n_j\in\mathbb{N}}$, we compute

$$\begin{aligned}\left\|\gamma^{(2),d,d}_{\Psi^U_\omega(L,n)}\right\|_{\mathrm{tr}} &\leqslant \sum_{Q \text{ occ. for } \Psi^U_\omega(L,n)} \sum_{\substack{j=1\\ Q_j\geqslant 2}}^{m} \frac{Q_j(Q_j-1)}{2} \sum_{\tilde{n}\in\mathbb{N}^{m-1}} \sum_{n_j\geqslant 1} \left|a^Q_{\tilde{n}_j}\right|^2 \\ &\leqslant \sum_{Q \text{ occ. for } \Psi^U_\omega(L,n)} \sum_{\substack{j=1\\ Q_j\geqslant 2}}^{m} \frac{Q_j(Q_j-1)}{2} \sum_{\tilde{n}\in\mathbb{N}^{m}} \left|a^Q_{\tilde{n}}\right|^2.\end{aligned}$$

Applying Lemmas 3.23 and 3.24 yields that, in the thermodynamic limit, with probability $1 - O(L^{-\infty})$, one has

$$\max_{Q \text{ occ. for } \Psi^U_\omega(L,n)} \sum_{\substack{j=1\\ Q_j\geqslant 2}}^{m} \frac{Q_j(Q_j-1)}{2} \lesssim n\log n \cdot \log\log n.$$

This completes the proof of Lemma 4.15 as $\displaystyle\sum_{Q,\ \tilde{n}\in\mathbb{N}^m} \left|a^Q_{\tilde{n}}\right|^2 = 1.$ □

Lemma 4.16 *Under the assumptions of Theorem 4.2, in the thermodynamic limit, with probability* $1 - O(L^{-\infty})$*, one has*

$$\left\|\gamma^{(2),2}_{\Psi^U_\omega(L,n)}\right\|_{tr} \leqslant 2.$$

Proof Using Lemma B.1 and the orthonormality properties of the families $(\varphi^j_{Q_j,n_j})_{n_j\in\mathbb{N}}$, we compute

$$\left\|\gamma^{(2),2}_{\Psi^U_\omega(L,n)}\right\|_{\mathrm{tr}} \leqslant \sum_{i\neq j}\left(\sum_{\substack{Q \text{ occ.}\\ Q_j\geqslant 2}} + \sum_{\substack{Q \text{ occ.}\\ Q_i\geqslant 1\\ Q_j\geqslant 1}}\right) \sum_{\tilde{n}\in\mathbb{N}^{m-2}} C_2(Q,i,j) \sum_{n_i,n_j\geqslant 1} \left|a^Q_{\tilde{n}_{i,j}}\right|^2.$$

For $Q_j \geqslant 1$ and $Q_i \geqslant 1$, one has

$$\begin{aligned} C_2(Q,i,j) &= \frac{(n-Q_j-Q_i-2)!Q_i!Q_j!}{2\,(n-2)!} \\ &= \frac{(Q_i+Q_j-2)!(n-(Q_j+Q_i-2)-4)!}{(n-4)!} \frac{(Q_i-1)!(Q_j-1)!}{(Q_i+Q_j-2)!} \frac{Q_i Q_j}{2(n-2)(n-3)} \\ &\leqslant \frac{Q_i Q_j}{2(n-2)(n-3)}. \end{aligned}$$

For $Q_j \geqslant 2$, one has

$$\begin{aligned} C_2(Q,i,j) &= \frac{Q_i!(Q_j-2)!(n-4-(Q_j+Q_i-2))!}{(n-4)!} \frac{Q_j(Q_j-1)}{2(n-2)(n-3)} \\ &\leqslant \frac{Q_j(Q_j-1)}{2(n-2)(n-3)}. \end{aligned} \tag{4.66}$$

Thus, as $\sum_j Q_j = n$, one estimates

$$\left\| \gamma^{(2),2}_{\Psi^U_\omega(L,n)} \right\|_{\mathrm{tr}} \leqslant \frac{2}{2(n-2)(n-3)} \sum_{\substack{Q \text{ occ.} \\ \tilde{n}\in\mathbb{N}^m}} \left(\sum_j Q_j \right)^2 \left| a^Q_{\tilde{n}} \right|^2 \leqslant \frac{n^2}{(n-2)(n-3)}.$$

This proves Lemma 4.16. □

Lemma 4.17 *Under the assumptions of Theorem 4.2, in the thermodynamic limit, with probability* $1 - O(L^{-\infty})$*, one has*

$$\left\| \gamma^{(2),4,2}_{\Psi^U_\omega(L,n)} \right\|_{tr} \leqslant 1.$$

Proof Using Lemma B.1 and the orthonormality properties of the families $(\varphi^j_{Q_j,n_j})_{n_j\in\mathbb{N}}$, we compute

$$\left\| \gamma^{(2),4,2}_{\Psi^U_\omega(L,n)} \right\|_{\mathrm{tr}} \leqslant \sum_{i\neq j} \sum_{\tilde{n}\in\mathbb{N}^{m-2}} \sum_{\substack{Q \text{ occ.} \\ Q_j\geqslant 2 \\ Q':\ Q'_k=Q_k \text{ if } k\notin\{i,j\} \\ Q'_i=Q_i+2 \\ Q'_j=Q_j-2}} C_2(Q,i,j) \sum_{n_i,n_j\geqslant 1} \left| a^Q_{\tilde{n}_{i,j}} \right|^2.$$

The bound (4.66) then yields

$$\left\|\gamma^{(2),4,2}_{\Psi^U_\omega(L,n)}\right\|_{\mathrm{tr}} \leqslant \frac{2}{2(n-2)(n-3)} \sum_{\substack{Q \text{ occ.} \\ \overline{n}\in\mathbb{N}^m}} \left(\sum_j Q_j\right)^2 \left|a^Q_{\overline{n}}\right|^2 \leqslant \frac{n^2}{2(n-2)(n-3)}.$$

This proves Lemma 4.17. □

Lemma 4.18 *Under the assumptions of Theorem 4.2, in the thermodynamic limit, with probability* $1 - O(L^{-\infty})$*, one has*

$$\left\|\gamma^{(2),4,3}_{\Psi^U_\omega(L,n)}\right\|_{tr} + \left\|\gamma^{(2),4,3'}_{\Psi^U_\omega(L,n)}\right\|_{tr} \leqslant \frac{2n}{\rho}.$$

Proof Using Lemma B.1 and the orthonormality properties of the families $(\varphi^j_{Q_j,n_j})_{n_j\in\mathbb{N}}$, we compute

$$\left\|\gamma^{(2),4,3}_{\Psi^U_\omega(L,n)}\right\|_{\mathrm{tr}} \leqslant \sum_{\substack{i,j,k \\ \text{distinct}}} \sum_{\tilde{n}\in\mathbb{N}^{m-3}} \sum_{\substack{Q \text{ occ.} \\ Q_j\geqslant 2 \\ Q':\ Q'_l=Q_l \text{ if } l\notin\{i,j,k\} \\ Q'_i=Q_i+1 \\ Q'_j=Q_j-2 \\ Q'_k=Q_k+1}} C_3(Q,i,j,k) \sum_{n_i,n_j,n_k\geqslant 1} \left|a^Q_{\tilde{n}_{i,j,k}}\right|^2.$$

For $Q_j \geqslant 2$, one has

$$\begin{aligned} C_3(Q,i,j,k) &= \frac{Q_k!Q_i!(Q_j-2)!(n-(Q_k+Q_i+Q_j-2)-4)!}{(n-4)!} \frac{Q_j(Q_j-1)}{2(n-2)(n-3)} \\ &\leqslant \frac{Q_j(Q_j-1)}{2(n-2)(n-3)}. \end{aligned} \tag{4.67}$$

Hence, by Proposition 2.2, one has

$$\begin{aligned} \left\|\gamma^{(2),4,3}_{\Psi^U_\omega(L,n)}\right\|_{\mathrm{tr}} &\leqslant \frac{1}{2(n-2)(n-3)} \sum_{\substack{\overline{n}\in\mathbb{N}^m \\ Q \text{ occ.}}} \left(\sum_j 1\right)\left(\sum_j Q_j\right)^2 \left|a^Q_{\overline{n}}\right|^2 \\ &\leqslant \frac{Ln^2}{2(n-2)(n-3)} \leqslant \frac{n}{\rho}. \end{aligned}$$

The computation for $\gamma^{(2),4,3'}_{\Psi^U_\omega(L,n)}$ is the same except that, instead of (4.67), one uses, for $Q_k \geqslant 1$ and $Q_i \geqslant 1$,

$$C_3(Q,i,j,k) = \frac{(Q_k-1)!(Q_i-1)!(Q_j)!(n-(Q_j+Q_i+Q_k-2)-4)!}{(n-4)!}\frac{Q_k Q_i}{2(n-2)(n-3)}$$
$$\leqslant \frac{Q_k Q_i}{2(n-2)(n-3)}.$$

This proves Lemma 4.17. □

Lemma 4.19 *Under the assumptions of Theorem 4.2, in the thermodynamic limit, with probability* $1 - O(L^{-\infty})$*, one has*

$$\left\| \gamma^{(2),4,4}_{\Psi^U_\omega(L,n)} \right\|_{tr} \leqslant n^{-1}.$$

Proof As in the proof of Lemma 4.13, we will have to deal with the degenerate cases separately (see Remarks 4.3 and 4.5).

Recall (4.16) and write

$$\gamma^{(2),4,4}_{\Psi^U_\omega(L,n)} = \sum_{\sigma\in\{\pm\}^4} \gamma^{(2),4,\sigma}_{\Psi^U_\omega(L,n)} \tag{4.68}$$

where $\sigma = (\sigma_i, \sigma_j, \sigma_k, \sigma_l) \in \{\pm 1\}^4$,

$$\gamma^{(2),4,\sigma}_{\Psi^U_\omega(L,n)} = \sum_{\substack{i,j,k,l_{\tilde{n}}\in\mathbb{N}^{m-4}\\ \text{distinct}}} \sum \sum_{\substack{Q \text{ occ.}\\ (Q_i,Q_j,Q_k,Q_l)\in\mathcal{Q}_\sigma\\ Q':\, Q'_o=Q_o \text{ if } o\notin\{i,j,k,l\}\\ Q'_i=Q_i-1,\ Q'_j=Q_j-1\\ Q'_k=Q_k+1,\ Q'_l=Q_l+1}} C_4(Q,i,j,k,l) \sum_{\substack{n_i,n_j,n_k,n_l\geqslant 1\\ n'_i,n'_j,n'_k,n'_l\geqslant 1}} a^Q_{\tilde{n}_{i,j,k,l}} \overline{a^{Q'}_{\tilde{n}'_{i,j,k,l}}} \gamma^{(2),4,4}_{\substack{Q_i,Q_j,Q_k,Q_l,\\ n_i,n_j,n_k,n_l\\ n'_i,n'_j,n'_k,n'l}} \tag{4.69}$$

and

$$\mathcal{Q}_\sigma = \left\{ Q_i \geqslant 1 \text{ and } \sigma_i(Q_i-1) \geqslant \frac{\sigma_i+1}{2} \right\} \cap \left\{ Q_j \geqslant 1 \text{ and } \sigma_j(Q_j-1) \geqslant \frac{\sigma_j+1}{2} \right\}$$
$$\cap \left\{ Q_k \geqslant 0 \text{ and } \sigma_k Q_k \geqslant \frac{\sigma_k+1}{2} \right\} \cap \left\{ Q_l \geqslant 0 \text{ and } \sigma_l Q_l \geqslant \frac{\sigma_l+1}{2} \right\}.$$

A term in the right-hand side of (4.68) degenerates if some $\sigma_\bullet$ takes the value -1.

Assume now $\sigma = (1, 1, 1, 1)$. Then,

$$\gamma^{(2),4,(1,1,1,1)}_{\Psi^U_\omega(L,n)} = \sum_{\substack{i,j,k,l_{\tilde{n}}\in\mathbb{N}^{m-4}\\ \text{distinct}}} \sum \sum_{\substack{Q \text{ occ.}\\ Q_i,Q_j\geqslant 2,\ Q_k,Q_l\geqslant 1\\ Q':\, Q'_o=Q_o \text{ if } o\notin\{i,j,k,l\}\\ Q'_i=Q_i-1,\ Q'_j=Q_j-1\\ Q'_k=Q_k+1,\ Q'_l=Q_l+1}} C_4(Q,i,j,k,l) \sum_{\substack{n_i,n_j,n_k,n_l\geqslant 1\\ n'_i,n'_j,n'_k,n'_l\geqslant 1}} a^Q_{\tilde{n}_{i,j,k,l}} \overline{a^{Q'}_{\tilde{n}'_{i,j,k,l}}} \gamma^{(2),4,4}_{\substack{Q_i,Q_j,Q_k,Q_l\\ n_i,n_j,n_k,n_l\\ n'_i,n'_j,n'_k,n'_l}}.$$

Using Lemma B.1 and the orthonormality properties of the families $(\varphi^j_{Q_j,n_j})_{n_j\in\mathbb{N}}$, we compute

$$\left\|\gamma^{(2),4,(+,+,+,+)}_{\Psi^U_\omega(L,n)}\right\|_{\mathrm{tr}} \leqslant 4 \sum_{\substack{i,j,k,l\\ \text{distinct}}} \sum_{\tilde{n}\in\mathbb{N}^{m-4}} \sum_{\substack{Q \text{ occ.}\\ Q_i,Q_j\geqslant 2,\ Q_k,Q_l\geqslant 1\\ Q':\ Q'_o=Q_o \text{ if } o\notin\{i,j,k,l\}\\ Q'_i=Q_i-1,\ Q'_j=Q_j-1\\ Q'_k=Q_k+1,\ Q'_l=Q_l+1}} C_4(Q,i,j,k,l) \sum_{n_i,n_j,n_k,n_l\geqslant 1} \left|a^Q_{\tilde{n}_{i,j,k,l}}\right|^2.$$

When $Q_i \geqslant 2$, $Q_j \geqslant 2$, $Q_k \geqslant 1$ and $Q_l \geqslant 1$ one has

$$C_4(Q,i,j,k,l) \leqslant \frac{Q_i(Q_i-1)Q_j(Q_j-1)Q_kQ_l}{2n(n-2)(n-3)(n-4)(n-5)(n-6)(n-7)}.$$

Thus, by Lemma 3.23, we obtain

$$\begin{aligned}\left\|\gamma^{(2),4,(1,1,1,1)}_{\Psi^U_\omega(L,n)}\right\|_{\mathrm{tr}} &\leqslant \frac{2}{(n-5)^6} \sum_{\overline{n}\in\mathbb{N}^m} \sum_{Q \text{ occ.}} \left(\sum_j Q_j\right)^2 \left(\sum_j Q_j^2\right)^2 \left|a^Q_{\overline{n}}\right|^2 \\ &\leqslant \frac{n^4(\log n)^4}{2(n-7)^6} \leqslant n^{-1}\end{aligned} \tag{4.70}$$

for n large.

Assume now $\sigma = (-1,-1,-1,-1)$. Then,

$$\gamma^{(2),4,(-1,-1,-1,-1)}_{\Psi^U_\omega(L,n)} = \sum_{\substack{i,j,k,l\\ \text{distinct}}} \sum_{\tilde{n}\in\mathbb{N}^{m-4}} \sum_{\substack{Q \text{ occ.}\\ Q_i=Q_j=1,\ Q_k=Q_l=0\\ Q':\ Q'_o=Q_o \text{ if } o\notin\{i,j,k,l\}\\ Q'_i=Q_i-1,\ Q'_j=Q_j-1\\ Q'_k=Q_k+1,\ Q'_l=Q_l+1}} C_4(Q,i,j,k,l) \sum_{\substack{n_i,n_j\geqslant 1\\ n_k=n_l=1\\ n'_i=n'_j=1\\ n'_k,n'_l\geqslant 1}} a^Q_{\tilde{n}_{i,j,k,l}} \overline{a^{Q'}_{\tilde{n}'_{i,j,k,l}}}\, \gamma^{(2),4,4}_{\substack{1,1,0,0\\ n_i,n_j,1,1\\ 1,1,n'_k,n'_l}}$$

where

$$\begin{aligned}\gamma^{(2),4,4}_{\substack{1,1,0,0\\ n_i,n_j,1,1\\ 1,1,n'_k,n'_l}}(x,x',y,y') &= \varphi^i_{1,n_i}(x)\varphi^j_{1,n_j}(x')\overline{\varphi^k_{1,n'_k}(y)\varphi^l_{1,n'_l}(y')} + \varphi^i_{1,n_i}(x')\varphi^j_{1,n_j}(x) \\ &\quad\times \overline{\varphi^k_{1,n'_k}(y)\varphi^l_{1,n'_l}(y')} \\ &\quad + \varphi^i_{1,n_i}(x)\varphi^j_{1,n_j}(x')\overline{\varphi^k_{1,n'_k}(y')\varphi^l_{1,n'_l}(y)} + \varphi^i_{1,n_i}(x')\varphi^j_{1,n_j}(x) \\ &\quad\times \overline{\varphi^k_{1,n'_k}(y')\varphi^l_{1,n'_l}(y)}.\end{aligned}$$

As in the derivation of (4.54), using Lemma B.1 and the orthonormality properties of the families $(\varphi^j_{Q_j,n_j})_{n_j\in\mathbb{N}}$, we compute

$$\left\|\gamma^{(2),4,(-1,-1,-1,-1)}_{\Psi^U_\omega(L,n)}\right\|_{\mathrm{tr}} \leqslant \frac{2}{(n-2)(n-3)} \sum_{\substack{\tilde{n}\in\mathbb{N}^{m-4}\\ Q\text{ occ.}}} \left\| \sum_{\substack{(i,j)\\ Q_i=Q_j=1}} \sum_{\substack{n_i=1\\ n_j=1\\ n_k,n_l}} a^Q_{\tilde{n}_{i,j,k,l}} \varphi^k_{1,n_k}\otimes\varphi^l_{1,n_l} \right\|^2$$

$$+ \left\| \sum_{\substack{(k,l)\\ Q_k=Q_l=0}} \sum_{\substack{n_k=1\\ n_l=1\\ n_i,n_j}} a^Q_{\tilde{n}_{i,j,k,l}} \varphi^i_{1,n_i}\otimes\varphi^j_{1,n_j} \right\|^2$$

$$\leqslant \frac{4}{(n-3)^2} \sum_{\substack{\overline{n}\in\mathbb{N}^m\\ Q\text{ occ.}}} \left|a^Q_{\overline{n}}\right|^2 = \frac{4}{(n-3)^2}.$$

Assume now $\sigma=(-1,1,1,1)$. Then,

$$\gamma^{(2),4,(-1,1,1,1)}_{\Psi^U_\omega(L,n)} = \sum_{\substack{i,j,k,l\\ \text{distinct}}} \sum_{\tilde{n}\in\mathbb{N}^{m-4}} \sum_{\substack{Q\text{ occ.}\\ Q_i=1,\ Q_j\geqslant 2\\ Q_k,Q_l\geqslant 1\\ Q':\ Q'_o=Q_o \text{ if } o\notin\{i,j,k,l\}\\ Q'_i=Q_i-1,\ Q'_j=Q_j-1\\ Q'_k=Q_k+1,\ Q'_l=Q_l+1}} C_4(Q,i,j,k,l) \sum_{\substack{n_i,n_j,n_k,n_l\geqslant 1\\ n'_j,n'_k,n'_l\geqslant 1\\ n'_i=1}} a^Q_{\tilde{n}_{i,j,k,l}} \overline{a^{Q'}_{\tilde{n}'_{i,j,k,l}}}\, \gamma^{(2),4,4}_{\substack{1,Q_j,Q_k,Q_l\\ n_i,n_j,n_k,n_l\\ 1,n'_j,n'_k,n'_l}}$$

where

$$\begin{aligned} C_4(Q,i,j,k,l) &= \frac{(n-Q_j-Q_k-Q_l-3)!Q_j!Q_k!Q_l!}{2\,(n-2)!} \\ &\leqslant \frac{Q_j(Q_j-1)Q_kQ_l}{2(n-2)(n-3)(n-4)(n-5)(n-6)}. \end{aligned} \tag{4.71}$$

The operator $\gamma^{(2),4,4}_{\substack{1,Q_j,Q_k,Q_l\\ n_i,n_j,n_k,n_l\\ 1,n'_j,n'_k,n'_l}}$ is given by (4.23) and

$$\sigma(x,x',y,y') = \varphi^i_{1,n_i}(x) \int_{\Delta_j^{Q_j-1}} \varphi^j_{Q_j,n_j}(x',z)\overline{\varphi^j_{Q_j-1,n'_j}(z)}dz$$

$$\times \int_{\Delta_k^{Q_k}} \varphi^k_{Q_k,n_k}(z)\overline{\varphi^k_{Q_k+1,n'_k}(y,z)}dz \int_{\Delta_l^{Q_l}} \varphi^l_{Q_l,n_l}(z)\overline{\varphi^l_{Q_l+1,n'_l}(y',z)}dz.$$

Hence, as in the derivation of (4.55), using Lemma B.1, (4.71) and the orthonormality properties of the families $(\varphi^j_{Q_j,n_j})_{n_j\in\mathbb{N}}$, we compute

$$\begin{aligned}\left\|\gamma^{(2),4,(-1,1,1,1)}_{\Psi^U_\omega(L,n)}\right\|_{\text{tr}} &\leqslant \frac{2}{(n-2)(n-3)(n-4)(n-5)(n-6)}\\ &\quad\times\sum_{\substack{\tilde{n}\in\mathbb{N}^m\\ Q\text{ occ.}}}\left(\sum_{j=1}^m Q_j^2\right)\left(\sum_{j=1}^m Q_j\right)^2\left|a^Q_{\tilde{n}_{i,j}}\right|^2\\ &\leqslant \frac{n^{10/3}(\log n)^{2/3}}{(n-6)^5}\leqslant n^{-3/2}.\end{aligned}$$

In the same way, we obtain that, if σ contains a least one -1 then $\left\|\gamma^{(2),4,\sigma}_{\Psi^U_\omega(L,n)}\right\|_{\text{tr}}\leqslant n^{-1}$.

This completes the proof of Lemma 4.19. □

Let us now turn to the analysis of $\gamma^{(2),d,o}_{\Psi^U_\omega(L,n)}$, the main term of $\gamma^{(2)}_{\Psi^U_\omega(L,n)}$. The analysis will be similar to that of $\gamma^{(1),d}_{\Psi}$ in the proof of Theorem 4.2.

Recall that $\mathcal{P}^Q_-$ is defined in Proposition 4.10 and write

$$\gamma^{(2),d,o}_{\Psi^U_\omega(L,n)}=\gamma^{(2),d,o,-}_{\Psi^U_\omega(L,n)}+\gamma^{(2),d,o,+}_{\Psi^U_\omega(L,n)} \tag{4.72}$$

where

$$\gamma^{(2),d,o,-}_{\Psi^U_\omega(L,n)}=\sum_{\substack{Q\text{ occ.}\\ Q_i\geqslant 1\\ Q_j\geqslant 1}}\sum_{\tilde{n}\in\mathbb{N}^{m-2}}\sum_{\substack{1\leqslant i<j\leqslant m\\ (i,j)\in(\mathcal{P}^Q_-)^2}}\sum_{\substack{n_j,n'_j\geqslant 1\\ n_i,n'_i\geqslant 1}} a^Q_{\tilde{n}_{i,j}}\overline{a^Q_{\tilde{n}'_{i,j}}}\,\gamma^{(2),d,o}_{\substack{Q_i,Q_j\\ n_i,n_j\\ n'_i,n'_j}}. \tag{4.73}$$

We prove

Lemma 4.20 *Under the assumptions of Theorem 4.4, for $\eta\in(0,1)$, there exists $\varepsilon_0>0$ such that, for $\varepsilon\in(0,\varepsilon_0)$, in the thermodynamic limit, with probability $1-O(L^{-\infty})$, one has*

$$\left\|\gamma^{(2),d,o,+}_{\Psi^U_\omega(L,n)}\right\|_{tr}\leqslant n^2\frac{\rho}{\ell_\rho}.$$

Proof The proof follows that of Lemma 4.14. One estimates

$$\left\|\gamma^{(2),d,o,+}_{\Psi^U_\omega(L,n)}\right\|_{\text{tr}} = \left\| \sum_{\substack{Q \text{ occ.} \\ Q_i \geqslant 1 \\ Q_j \geqslant 1}} \sum_{\substack{1\leqslant i<j\leqslant m \\ (i,j)\notin(\mathcal{P}^Q_-)^2}} \sum_{\tilde{n}\in\mathbb{N}^{m-2}} \sum_{\substack{n_j,n'_j\geqslant 1 \\ n_i,n'_i\geqslant 1}} a^Q_{\tilde{n}_{i,j}} \overline{a^Q_{\tilde{n}'_{i,j}}} \gamma^{(2),d,o}_{\substack{Q_i,Q_j \\ n_i,n_j \\ n'_i,n'_j}} \right\|_{\text{tr}}$$

$$\leqslant \sum_{\substack{Q \text{ occ.} \\ Q_i \geqslant 1 \\ Q_j \geqslant 1}} \sum_{\tilde{n}\in\mathbb{N}^{m-2}} \left(\sum_{\substack{1\leqslant i<j\leqslant m \\ i\notin\mathcal{P}^Q_-}} + \sum_{\substack{1\leqslant i<j\leqslant m \\ j\notin\mathcal{P}^Q_-}} \right) \left\| \sum_{\substack{n_j,n'_j\geqslant 1 \\ n_i,n'_i\geqslant 1}} a^Q_{\tilde{n}_{i,j}} \overline{a^Q_{\tilde{n}'_{i,j}}} \gamma^{(2),d,o}_{\substack{Q_i,Q_j \\ n_i,n_j \\ n'_i,n'_j}} \right\|_{\text{tr}}. \tag{4.74}$$

Let us analyze the first sum in the right-hand side above. Using (4.18), Lemma B.1, and the orthonormality properties of the families $(\varphi^j_{Q_j,n_j})_{n_j\in\mathbb{N}}$, we compute

$$\sum_{\substack{Q \text{ occ.} \\ Q_i \geqslant 1 \\ Q_j \geqslant 1}} \sum_{\substack{1\leqslant i<j\leqslant m \\ i\notin\mathcal{P}^Q_-}} \sum_{\tilde{n}\in\mathbb{N}^{m-2}} \left\| \sum_{\substack{n_j,n'_j\geqslant 1 \\ n_i,n'_i\geqslant 1}} a^Q_{\tilde{n}_{i,j}} \overline{a^Q_{\tilde{n}'_{i,j}}} \gamma^{(2),d,o}_{\substack{Q_i,Q_j \\ n_i,n_j \\ n'_i,n'_j}} \right\|_{\text{tr}}$$

$$\leqslant \sum_{\substack{Q \text{ occ.} \\ Q_i \geqslant 1 \\ Q_j \geqslant 1}} \sum_{\substack{1\leqslant i<j\leqslant m \\ i\notin\mathcal{P}^Q_-}} \sum_{\tilde{n}\in\mathbb{N}^{m-2}} \frac{Q_i Q_j}{2} \sum_{n_i,n_j\geqslant 1} \left| a^Q_{\tilde{n}_{i,j}} \right|^2$$

$$\leqslant \frac{1}{2} \sum_{\substack{\tilde{n}\in\mathbb{N}^m \\ Q \text{ occ.}}} \left(\sum_{i\notin\mathcal{P}^Q_-} Q_i \right) \left(\sum_j Q_j \right) \left| a^Q_{\tilde{n}} \right|^2$$

$$\leqslant C n^2 \frac{\rho}{\ell_\rho}$$

as in the proof of Lemma 4.14 by Lemmas 3.23 and 3.24.

The other sum in the right-hand side of (4.74) is analyzed in the same way. This completes the proof of Lemma 4.20. □

Let us now analyze $\gamma^{(2),d,o,-}_{\Psi^U_\omega(L,n)}$. We proceed as in the analysis of $\gamma^{(1),d}_{\Psi^U_\omega(L,n)}$ (see (4.56) and Lemma 4.14). We recall and compute

$$\gamma^{(2),d,o,-}_{\Psi^U_\omega(L,n)} = \sum_{\substack{Q \text{ occ.} \\ Q_i \geqslant 1 \\ Q_j \geqslant 1 \\ \tilde{n}\in\mathbb{N}^{m-2}}} \sum_{\substack{1\leqslant i<j\leqslant m \\ (i,j)\in(\mathcal{P}^Q_-)^2}} \sum_{\substack{n_j,n'_j\geqslant 1 \\ n_i,n'_i\geqslant 1}} a^Q_{\tilde{n}_{i,j}} \overline{a^Q_{\tilde{n}'_{i,j}}} \gamma^{(2),d,o}_{\substack{Q_i,Q_j \\ n_i,n_j \\ n'_i,n'_j}}$$

$$= \sum_{\substack{Q \text{ occ.} \\ \tilde{n}\in\mathbb{N}^{m-2}}} \sum_{\substack{1\leqslant i<j\leqslant m \\ (i,j)\in(\mathcal{P}_-^Q)^2}} \frac{Q_i Q_j}{2}(\mathrm{Id}-\mathrm{Ex})\varphi_{i,j}^{\tilde{n}} \otimes^s \varphi_{i,j}^{\tilde{n}}.$$

where $\varphi_{i,j}^{\tilde{n}} := \sum_{\substack{n_i\geqslant 1 \\ n_j\geqslant 1}} a_{\tilde{n}_{i,j}}^Q \varphi_{Q_i,n_i}^i \wedge \varphi_{Q_j,n_j}^j$ and the operators Ex and $\otimes^s$ are defined in Proposition 4.8.

Define also

$$\tilde{\varphi}_{i,j}^{\tilde{n}} = \begin{cases} \varphi_{i,j}^{\tilde{n}} & \text{if } n_i + n_j = 2 \\ \|\varphi_{i,j}^{\tilde{n}}\| \varphi_{Q_i,1}^i \wedge \varphi_{Q_j,1}^j & \text{if } n_i + n_j \geqslant 3. \end{cases} \tag{4.75}$$

Then, recalling (4.60), we compute

$$\begin{aligned}
\gamma_{\Psi_\omega^U(L,n)}^{(2),d,o,-} &= \sum_{\substack{Q \text{ occ.} \\ \tilde{n}\in\mathbb{N}^{m-2}}} \sum_{\substack{1\leqslant i<j\leqslant m \\ (i,j)\in(\mathcal{P}_{-,-}^Q)^2}} \frac{Q_i Q_j}{2}(\mathrm{Id}-\mathrm{Ex})\varphi_{i,j}^{\tilde{n}} \otimes^s \varphi_{i,j}^{\tilde{n}} \\
&\quad + \sum_{\substack{Q \text{ occ.} \\ \tilde{n}\in\mathbb{N}^{m-2}}} \sum_{\substack{1\leqslant i<j\leqslant m \\ i\in\mathcal{P}_{-,+}^Q \\ \text{or } j\in\mathcal{P}_{-,+}^Q}} \frac{Q_i Q_j}{2}(\mathrm{Id}-\mathrm{Ex})\varphi_{i,j}^{\tilde{n}} \otimes^s \varphi_{i,j}^{\tilde{n}} \\
&= \sum_{\substack{Q \text{ occ.} \\ \tilde{n}\in\mathbb{N}^{m-2}}} \sum_{\substack{1\leqslant i<j\leqslant m \\ (i,j)\in(\mathcal{P}_-^Q)^2}} \frac{Q_i Q_j}{2}(\mathrm{Id}-\mathrm{Ex})\tilde{\varphi}_{i,j}^{\tilde{n}} \otimes^s \tilde{\varphi}_{i,j}^{\tilde{n}} \\
&\quad + \sum_{\substack{Q \text{ occ.} \\ \tilde{n}\in\mathbb{N}^{m-2}}} \sum_{\substack{1\leqslant i<j\leqslant m \\ i\in\mathcal{P}_{-,+}^Q \\ \text{or } j\in\mathcal{P}_{-,+}^Q}} \frac{Q_i Q_j}{2}(\mathrm{Id}-\mathrm{Ex})\left(\varphi_{i,j}^{\tilde{n}} \otimes^s \varphi_{i,j}^{\tilde{n}} - \tilde{\varphi}_{i,j}^{\tilde{n}} \otimes^s \tilde{\varphi}_{i,j}^{\tilde{n}}\right).
\end{aligned} \tag{4.76}$$

The second term in the sum above we estimate by

$$\begin{aligned}
&\left\| \sum_{\substack{Q \text{ occ.}\\ \tilde n\in\mathbb{N}^{m-2}}} \sum_{\substack{1\leqslant i<j\leqslant m\\ i\in\mathcal{P}^Q_{-,+}\\ \text{or } j\in\mathcal{P}^Q_{-,+}}} \frac{Q_iQ_j}{2}(\mathrm{Id}-\mathrm{Ex})\left(\varphi^{\tilde n}_{i,j}\otimes^s\varphi^{\tilde n}_{i,j}-\tilde\varphi^{\tilde n}_{i,j}\otimes^s\tilde\varphi^{\tilde n}_{i,j}\right)\right\|_{\mathrm{tr}}\\
&\quad\lesssim \sum_{\substack{Q \text{ occ.}\\ \tilde n\in\mathbb{N}^{m-2}}} \sum_{\substack{1\leqslant i<j\leqslant m\\ i\in\mathcal{P}^Q_{-,+}\\ \text{or } j\in\mathcal{P}^Q_{-,+}}} Q_iQ_j\left(\left\|\varphi^{\tilde n}_{i,j}\right\|^2+\left\|\tilde\varphi^{\tilde n}_{i,j}\right\|^2\right) \qquad (4.77)\\
&\quad\lesssim \sum_{\substack{Q \text{ occ.}\\ \overline{n}\in\mathbb{N}^{m}}}\left(\sum_{j\in\mathcal{P}^Q_-}\#\{j;\ n_j\geqslant 2\}\right)\left(\sum_{j\in\mathcal{P}^Q_-}Q_j\right)\left|a^Q_{\overline{n}}\right|^2\\
&\quad\lesssim n^2\frac{\rho}{\rho_0|\log\rho|}f_Z(2|\log\rho|).
\end{aligned}$$

by Lemma 4.11.

As for the first term in the second equality in (4.76), letting $\mathcal{P}_{\mathrm{opt}}$ be the pieces of length less than $3\ell_\rho(1-\varepsilon)$ where Ψ^{opt} puts at least one particle, we write

$$\begin{aligned}
&\sum_{\substack{Q \text{ occ.}\\ \tilde n\in\mathbb{N}^{m-2}}} \sum_{\substack{1\leqslant i<j\leqslant m\\ (i,j)\in(\mathcal{P}^Q_-)^2}} \frac{Q_iQ_j}{2}(\mathrm{Id}-\mathrm{Ex})\tilde\varphi^{\tilde n}_{i,j}\otimes^s\tilde\varphi^{\tilde n}_{i,j}\\
&=\sum_{\substack{Q \text{ occ.}\\ \tilde n\in\mathbb{N}^{m-2}}}\left(\sum_{\substack{1\leqslant i<j\leqslant m\\ (i,j)\in(\mathcal{P}_{\mathrm{opt}})^2}}+\sum_{\substack{1\leqslant i<j\leqslant m\\ i \text{ or } j \text{ in } \mathcal{P}^Q_-\setminus\mathcal{P}_{\mathrm{opt}}}}-\sum_{\substack{1\leqslant i<j\leqslant m\\ i \text{ or } j \text{ in } \mathcal{P}_{\mathrm{opt}}\setminus\mathcal{P}^Q_-}}\right)\frac{Q_iQ_j}{2}(\mathrm{Id}-\mathrm{Ex})\tilde\varphi^{\tilde n}_{i,j}\otimes^s\tilde\varphi^{\tilde n}_{i,j}
\end{aligned} \qquad (4.78)$$

For the first of the three sums above, one computes

$$\sum_{\substack{Q \text{ occ.} \\ \tilde{n}\in\mathbb{N}^{m-2}}} \sum_{\substack{1\leqslant i<j\leqslant m \\ (i,j)\in(\mathcal{P}_{\text{opt}})^2}} \frac{Q_i Q_j}{2}(\text{Id}-\text{Ex})\tilde{\varphi}^{\tilde{n}}_{i,j} \otimes^s \tilde{\varphi}^{\tilde{n}}_{i,j}$$

$$= \sum_{\substack{1\leqslant i<j\leqslant m \\ (i,j)\in(\mathcal{P}_{\text{opt}})^2}} \left(\sum_{\substack{Q \text{ occ.} \\ \overline{n}\in\mathbb{N}^m}} \left|a^Q_{\overline{n}}\right|^2 \right) \frac{Q_i Q_j}{2}(\text{Id}-\text{Ex})\tilde{\varphi}^{\tilde{n}}_{i,j} \otimes^s \tilde{\varphi}^{\tilde{n}}_{i,j}$$

$$= \sum_{\substack{1\leqslant i<j\leqslant m \\ (i,j)\in(\mathcal{P}_{\text{opt}})^2}} \frac{Q_i Q_j}{2}(\text{Id}-\text{Ex})\gamma^{(1)}_{\varphi^i_{Q_i,1}} \otimes^s \gamma^{(1)}_{\varphi^j_{Q_j,1}}$$

$$= \gamma^{(2)}_{\Psi^{\text{opt}}} + R \tag{4.79}$$

where $\|R\|_{\text{tr}} \leqslant Cn^2\rho^{1+\eta}$.

In the last line of (4.79), we have used Proposition 4.8, the definition of Ψ^{opt} (3.12) and Lemma 3.23 to obtain the bound on R.

To estimate the remaining two sums in (4.77), we split them into sums where the summation over pieces is restricted to pieces either longer than $\ell_\rho + C$ or shorter than $\ell_\rho + C$ (C is given by Corollary 3.32).

By Corollary 3.32, we know that

$$\left\| \sum_{\substack{Q \text{ occ.} \\ \tilde{n}\in\mathbb{N}^{m-2}}} \left(\sum_{\substack{1\leqslant i<j\leqslant m \\ i\in\mathcal{P}^Q_-\setminus\mathcal{P}_{\text{opt}} \\ \text{and } |\Delta_i(\omega)|<\ell_\rho+C}} - \sum_{\substack{1\leqslant i<j\leqslant m \\ i\in\mathcal{P}^Q_-\setminus\mathcal{P}_{\text{opt}} \\ \text{and } |\Delta_i(\omega)|<\ell_\rho+C}} \right) \frac{Q_i Q_j}{2}(\text{Id}-\text{Ex})\tilde{\varphi}^{\tilde{n}}_{i,j} \otimes^s \tilde{\varphi}^{\tilde{n}}_{i,j} \right\|_{\text{tr}}$$

$$\leqslant \sum_{\substack{Q \text{ occ.} \\ \tilde{n}\in\mathbb{N}^{m-2}}} \left(\sum_{\substack{1\leqslant i<j\leqslant m \\ i\in\mathcal{P}^Q_-\setminus\mathcal{P}_{\text{opt}} \\ \text{and } |\Delta_i(\omega)|<\ell_\rho+C}} + \sum_{\substack{1\leqslant i<j\leqslant m \\ i\in\mathcal{P}^Q_-\setminus\mathcal{P}_{\text{opt}} \\ \text{and } |\Delta_i(\omega)|<\ell_\rho+C}} \right) \frac{Q_i Q_j}{2} \left\|(\text{Id}-\text{Ex})\tilde{\varphi}^{\tilde{n}}_{i,j} \otimes^s \tilde{\varphi}^{\tilde{n}}_{i,j}\right\|_{\text{tr}}$$

$$\leqslant Cn^2\rho \max\left(\sqrt{Z(2|\log\rho|)}, \ell_\rho^{-1}\right) \sum_{\substack{Q \text{ occ.} \\ \overline{n}\in\mathbb{N}^m}} \left|a^Q_{\overline{n}}\right|^2 = Cn^2 \max\left(\sqrt{\rho Z(2|\log\rho|)}, \rho|\log\rho|^{-1}\right).$$

In the same way, we estimate

$$\left\| \sum_{\substack{Q \text{ occ.} \\ \tilde{n}\in\mathbb{N}^{m-2}}} \left(\sum_{\substack{1\leqslant i<j\leqslant m \\ i\in\mathcal{P}_{\text{opt}}\setminus\mathcal{P}^Q_- \\ \text{and } |\Delta_i(\omega)|\geqslant \ell_\rho+C}} - \sum_{\substack{1\leqslant i<j\leqslant m \\ i\in\mathcal{P}_{\text{opt}}\setminus\mathcal{P}^Q_- \\ \text{and } |\Delta_i(\omega)|\geqslant \ell_\rho+C}} \right) \frac{Q_i Q_j}{2} (\text{Id} - \text{Ex}) \tilde{\varphi}^{\tilde{n}}_{i,j} \otimes^s \tilde{\varphi}^{\tilde{n}}_{i,j} \right\|_{\text{tr}}$$

$$\leqslant C n^2 \rho \max\left(\sqrt{Z(2|\log\rho|)}, \ell_\rho^{-1}\right)$$

and one has the same estimates when i is replaced by j.

Plugging these estimates, (4.77) and (4.78) into (4.72), recalling (1.29), we obtain

$$\left\| \left(\gamma^{(2),d,o,-}_{\Psi^U_\omega(L,n)} - \gamma^{(2)}_{\Psi^{\text{opt}}} \right) \mathbf{1}^2_{<\ell_\rho+C} \right\|_{\text{tr}} \leqslant C n^2 \max\left(\sqrt{\rho Z(2|\log\rho|)}, \rho|\log\rho|^{-1}\right)$$

$$\left\| \left(\gamma^{(2),d,o,-}_{\Psi^U_\omega(L,n)} - \gamma^{(2)}_{\Psi^{\text{opt}}} \right) \left(\mathbf{1} - \mathbf{1}^2_{<\ell_\rho+C}\right) \right\|_{\text{tr},\geqslant \ell_\rho+C} \leqslant C n^2 \rho \max\left(\sqrt{Z(2|\log\rho|)}, \ell_\rho^{-1}\right).$$

Taking into account the decomposition (4.9) and Lemmas 4.15, 4.16, 4.17, 4.18, 4.19 then completes the proof of Theorem 1.5. □

5 Almost Sure Convergence for the Ground State Energy Per Particle

In this section, we prove that, if interactions decay sufficiently fast at infinity, then the convergence in the thermodynamic limit of the ground state energy per particle $E^U_\omega(L, n)/n$ to $\mathcal{E}^U(\rho)$ holds not only in L^2_ω (see [21, Theorem 3.5]) but also ω-almost surely.

From the proof of [21, Theorem 3.5], one clearly sees that it suffices to improve upon the sub-additive estimate given in [21, Lemma 4.1]. We prove

Theorem 5.1 *Assume that the pair potential U be even and such that $U \in L^r(\mathbb{R})$ for some $r > 1$ and that for some $\alpha > 2$, one has $\int_0^{+\infty} x^\alpha U(x)dx < +\infty$.*

In the thermodynamic limit, for disjoint intervals Λ_1 and Λ_2 with n_1 and n_2 electrons, respectively, for $\min(|\Lambda_1|, |\Lambda_2|)$ sufficiently large, with probability $1 - O(\min(|\Lambda_1|, |\Lambda_2|)^{-\infty})$, one has

$$E^U_\omega(\Lambda_1 \cup \Lambda_2, n_1 + n_2) \leqslant E^U_\omega(\Lambda_1, n_1) + E^U_\omega(\Lambda_2, n_2) + o(n_1 + n_2). \quad (5.1)$$

Here, $E^U_\omega(\Lambda, n)$ denotes the ground state energy of $H^U_\omega(\Lambda, n)$ (see Section 1.1).

To apply this result to U satisfying **(HU)**, it suffices to check

Lemma 5.2 *If U satisfies* **(HU)** *then for any $0 < \alpha < 3$, one has* $\int_0^{+\infty} x^\alpha U(x)dx < +\infty$.

Proof Clearly, for $n \geqslant 0$, one has

$$\int_{2^n}^{2^{n+1}} x^\alpha U(x)dx \leqslant 2^{\alpha(n+1)} \int_{2^n}^{2^{n+1}} U(x)dx \leqslant 2^{(\alpha-3)n+\alpha} Z(2^n).$$

As Z is bounded, summing over n yields

$$\int_1^{+\infty} x^\alpha U(x)dx \lesssim \sum_{n\geqslant 1} 2^{(\alpha-3)n+\alpha} < +\infty.$$

This completes the proof of Lemma 5.2. □

Thus, the sub-additive estimate (5.1) holds for our model and, following the analysis provided in [21], we obtain Theorem 1.2.

Proof of Theorem 5.1 Without loss of generality, let us assume that $\Lambda_1 = [-L_1, 0]$ and $\Lambda_2 = [0, L_2]$. For $i \in \{1, 2\}$, we denote by Ψ_i^U ground states of $H_\omega^U(\Lambda_i, n_i)$. In case of degeneracy, we may additionally choose particular ground states Ψ_i^U, $i \in \{1, 2\}$ such that each of them belongs to a fixed occupation subspace. Thus, occupation is well defined for Ψ_i^U. As usual, we will implicitly suppose that Ψ_i^U is extended by zero outside $\Lambda_i^{n_i}$. Consider now

$$\Psi = \Psi_1^U \wedge \Psi_2^U.$$

Then,

$$\begin{aligned} E_\omega^U(\Lambda_1 \cup \Lambda_2, n_1 + n_2) &\leqslant \left\langle H_\omega^U(\Lambda_1 \cup \Lambda_2, n_1 + n_2)\Psi, \Psi \right\rangle \\ &= E_\omega^U(\Lambda_1, n_1) + E_\omega^U(\Lambda_2, n_2) + \mathrm{Tr}(U\gamma_{\Psi_1^U}^{(1)} \otimes^s \gamma_{\Psi_2^U}^{(2)}) \\ &= E_\omega^U(\Lambda_1, n_1) + E_\omega^U(\Lambda_2, n_2) \\ &\quad + \int_{\Lambda_1 \times \Lambda_2} U(x-y)\rho_{\Psi_1^U}(x)\rho_{\Psi_2^U}(y)\mathrm{d}x\mathrm{d}y \end{aligned}$$

The proof will be accomplished by the following

Lemma 5.3 *Under the assumptions of Theorem 5.1, one has*

$$\int_{\Lambda_1 \times \Lambda_2} U(x-y)\rho_{\Psi_1^U}(x)\rho_{\Psi_2^U}(y)\mathrm{d}x\mathrm{d}y = o(n_1 + n_2). \tag{5.2}$$

Proof By Proposition 2.1, with probability $1 - O(\min(|\Lambda_1|, |\Lambda_2|)^{-\infty})$, for $i \in \{1, 2\}$, the largest piece in Λ_i is of length bounded by $\log|\Lambda_i| \cdot \log\log|\Lambda_i|$. This implies that one can partition Λ_i into sub-intervals each containing an integer number of original pieces (i.e., the extremities of these sub-intervals coincide with the extremities of pieces given by the Poisson random process) of length between $\log^2|\Lambda_i|$ and $2\log^2|\Lambda_i|$. Let these new sub-intervals be denoted by Λ_i^j, $j \in \{1, \dots, m_i\}$; we order the intervals in such a way that their distance to 0 increases with j. Thus,

$$\Lambda_i = \bigcup_{j=1}^{m_i} \Lambda_i^j$$

and

$$\log^2|\Lambda_i| \leqslant |\Lambda_i^j| \leqslant 2\log^2|\Lambda_i|. \tag{5.3}$$

The last inequalities and the ordering convention imply that

$$\operatorname{dist}(\Lambda_1^{j_1}, \Lambda_2^{j_2}) \geqslant (j_1 - 1) \cdot \log^2|\Lambda_1| + (j_2 - 1) \cdot \log^2|\Lambda_2| \tag{5.4}$$

and

$$\frac{|\Lambda_i|}{2\log^2|\Lambda_i|} \leqslant m_i \leqslant \frac{|\Lambda_i|}{\log^2|\Lambda_i|}. \tag{5.5}$$

We now count the number of particles that Ψ_i^U puts in an interval Λ_i^j. Let $\{\Delta_k^i\}_{k=1}^{M_i}$ be the pieces in Λ_i and let Q_k^i be the corresponding occupation numbers. According to the choice of sub-intervals Λ_i^j above, each Λ_i^j is a union of some of the pieces $(\Delta_k^i)_k$. We establish the following natural

Lemma 5.4 *With the above notations, one has*

$$\int_{\Delta_k^i} \rho_{\Psi_i^U}(x)\mathrm{d}x = Q_k^i, \quad i \in \{1, 2\}, \quad k \in \{1, \dots, M_i\}.$$

Proof For convenience, we drop the superscript i in this proof. Recall the decomposition (4.2)

$$\Psi = \sum_{\substack{(n_k)_{1\leqslant k\leqslant M} \\ \forall k,\, n_k\geqslant 1}} a_n \bigwedge_{k=1}^{M} \varphi_{n_k}^k,$$

where $\varphi^k_{n_k}$ are functions of Q_k variables in the piece Δ_k. Keeping the notations, by Theorem 4.2, one has

$$\gamma^{(1)}_{\Psi} = \sum_{k=1}^{M} \sum_{\substack{n_k \geqslant 1 \\ n'_k \geqslant 1}} \sum_{\tilde{n} \in \mathbb{N}^{M-1}} a_{\tilde{n}_k} \overline{a_{\tilde{n}'_k}} \gamma^{(1)}_{n_k, n'_k},$$

where

$$\gamma^{(1)}_{n_k, n'_k}(x, y) = Q_k \int_{(\Delta_k)^{Q_k - 1}} \varphi^k_{n_k}(x, z) \overline{\varphi^k_{n'_k}(y, z)} \mathrm{d}z.$$

The off-diagonal term $\gamma^{(1),o}_{\Psi}$ vanishes because the functions $\Psi_{1,2}$ were chosen of a fixed occupation. This immediately yields

$$\begin{aligned}
\int_{\Delta_k} \rho_{\Psi}(x) \mathrm{d}x &= Q_k \sum_{\substack{n_k \geqslant 1 \\ n'_k \geqslant 1}} \sum_{\tilde{n} \in \mathbb{N}^{M-1}} a_{\tilde{n}_k} \overline{a_{\tilde{n}'_k}} \int_{(\Delta_k)^{Q_k}} \varphi^k_{n_k}(x) \overline{\varphi^k_{n'_k}(x)} \mathrm{d}x \\
&= Q_k \sum_{\tilde{n} \in \mathbb{N}^{M-1}} \int_{(\Delta_k)^{Q_k}} \sum_{n_k \geqslant 1} |a_{\tilde{n}_k}|^2 |\varphi^k_{n_k}(x)|^2 \mathrm{d}x = Q_k,
\end{aligned}$$

where, in the second equality, we used the orthogonality of different Q_k-particle levels in the piece Δ_k and, in the third equality, we used the fact that Ψ is normalized.

This completes the proof of Lemma 5.4. □

Lemma 5.4 immediately entails

Corollary 5.5 *One computes*

$$\int_{\Lambda^j_i} \rho_{\Psi^U_i}(x) \mathrm{d}x = \sum_{k | \Delta^i_k \subset \Lambda^j_i} Q^i_k, \quad i \in \{1, 2\}, \quad j \in \{1, \ldots, m_i\}.$$

Next, we derive a simple bound on the number of particles in Λ^j_i. The total ground state energy is bounded by

$$E^U_\omega(\Lambda_i, n_i) \leqslant C \ell_\rho^{-2} n_i.$$

On the other hand, a system of $q = \sum_{k | \Delta^i_k \subset \Lambda^j_i} Q^i_k$ particles in Λ^j_i has non- interacting energy at least

$$\sum_{s=1}^{q} \frac{\pi^2 s^2}{|\Lambda_i^j|^2} \asymp q^3 |\Lambda_i^j|^{-2}.$$

This implies that

$$q^3 |\Lambda_i^j|^{-2} \leqslant C \ell_\rho^{-2} n_i$$

or, equivalently,

$$\sum_{k|\Delta_k^i \subset \Lambda_i^j} Q_k^i \leqslant C_1 \left(|\Lambda_i^j|/\ell_\rho\right)^{2/3} n_i^{1/3} \leqslant C_2 n_i^{1/3} \log^{4/3} L_i.$$

Let us now estimate the left-hand side of (5.2) using Hölder's inequality ($1/p + 1/q = 1$, $p, q \geqslant 1$) as

$$\begin{aligned} \int_{\Lambda_1 \times \Lambda_2} U(x-y) \rho_{\Psi_1^U}(x) \rho_{\Psi_2^U}(y) \mathrm{d}x \mathrm{d}y &= \sum_{j_1=1}^{m_1} \sum_{j_2=1}^{m_2} \int_{\Lambda_1^{j_1} \times \Lambda_2^{j_2}} U(x-y) \rho_{\Psi_1^U}(x) \rho_{\Psi_2^U}(y) \mathrm{d}x \mathrm{d}y \\ &\leqslant \sum_{j_1=1}^{m_1} \sum_{j_2=1}^{m_2} \|U\|_{p, \Lambda_1^{j_1} \times \Lambda_2^{j_2}} \|\rho_{\Psi_1^U}\|_q \|\rho_{\Psi_2^U}\|_q . \end{aligned} \tag{5.6}$$

where we have set

$$\|U\|_{p, \Lambda_1^{j_1} \times \Lambda_2^{j_2}} := \left(\int_{\Lambda_1^{j_1} \times \Lambda_2^{j_2}} U^p(x-y) dx dy \right)^{1/p} . \tag{5.7}$$

Now, recall that by (6.57), for $i \in \{1, 2\}$, on $\Lambda_i^{j_i}$, one has

$$\|\rho_{\Psi_i^U}\|_{\infty, \Lambda_i^{j_i}} \leqslant 4 \|\Psi_i^U\|_{H^1(\Lambda_i^{j_i})} \|\Psi_i^U\|_{2, \Lambda_i^{j_i}} \leqslant C \left(\langle H_\omega^U(\Lambda_i^{j_i}, n_i) \Psi_i^U, \Psi_i^U \rangle_{\Lambda_i^{j_i}} \right)^{1/2} \|\Psi_i^U\|_2 .$$

Hence, by Corollary 5.5,

$$\|\rho_{\Psi_i^U}\|_q = \left(\int_{\Lambda_i^{j_i}} \rho_{\Psi_i^U}^{q-1} \rho_{\Psi_i^U} \right)^{1/q} \leqslant (Q_i^{j_i})^{1/q} \left(\langle H_\omega^U(\Lambda_i^{j_i}, n_i) \Psi_i^U, \Psi_i^U \rangle_{\Lambda_i^{j_i}} \right)^{(q-1)/2q} \|\Psi_i^U\|_{2, \Lambda_i^{j_i}}^{(q-1)/q} .$$

Recalling (5.6), as $\|\Psi_i^U\|_{2, \Lambda_i^{j_i}} \leqslant 1$ for $i \in \{1, 2\}$, we estimate

$$\int_{\Lambda_1\times\Lambda_2} U(x-y)\rho_{\Psi_1^U}(x)\rho_{\Psi_2^U}(y)\mathrm{d}x\mathrm{d}y$$

$$\leqslant \sum_{j_1=1}^{m_1}\sum_{j_2=1}^{m_2} \|U\|_{p,\Lambda_1^{j_1}\times\Lambda_2^{j_2}} (Q_1^{j_1}Q_2^{j_2})^{1/q}$$

$$\times\left(\langle H_\omega^U(\Lambda_1^{j_1},n_1)\Psi_1^U,\Psi_1^U\rangle_{\Lambda_1^{j_1}}\langle H_\omega^U(\Lambda_2^{j_2},n_2)\Psi_2^U,\Psi_2^U\rangle_{\Lambda_2^{j_2}}\right)^{(q-1)/2q}. \tag{5.8}$$

Now, as $Q_{\Psi_i^U}\lesssim n_i^{1/3}\log^{4/3}L_i\lesssim n^{1/3}\log^{4/3}n$ and as

$$\langle H_\omega^U(\Lambda_i^{j_i},n_i)\Psi_i^U,\Psi_i^U\rangle_{\Lambda_i^{j_i}}\leqslant\langle H_\omega^U(\Lambda_i)\Psi_i^U,\Psi_i^U\rangle\leqslant Cn_i\leqslant Cn,$$

the estimate (5.8) entails

$$\int_{\Lambda_1\times\Lambda_2} U(x-y)\rho_{\Psi_1^U}(x)\rho_{\Psi_2^U}(y)\mathrm{d}x\mathrm{d}y\lesssim n^{(3q-1)/3q}(\log n)^{8/(3q)} \times\sum_{j_1=1}^{m_1}\sum_{j_2=1}^{m_2}\|U\|_{p,\Lambda_1^{j_1}\times\Lambda_2^{j_2}}. \tag{5.9}$$

Hence, to prove (5.1), it suffices to choose q (recall $q\geqslant 1$ and $1/p+1/q=1$) such that

$$\sum_{j_1=1}^{m_1}\sum_{j_2=1}^{m_2}\|U\|_{p,\Lambda_1^{j_1}\times\Lambda_2^{j_2}}=o\left(n^{1/3q}(\log n)^{-8/(3q)}\right). \tag{5.10}$$

Therefore, we recall (5.7) and using the definition of the $(\Lambda_i^{j_i})_{i,j}$, in particular (5.4) and (5.5), we estimate

$$\|U\|_{p,\Lambda_1^{j_1}\times\Lambda_2^{j_2}}\lesssim((j_1+j_2)|\log L|^2)^{-k/p}\left(\int_{\Lambda_1^{j_1}\times\Lambda_2^{j_2}}(x-y)^kU^p(x-y)dxdy\right)^{1/p}. \tag{5.11}$$

Now, by (5.3), as U is even, we have

$$\left(\int_{\Lambda_1^{j_1}\times\Lambda_2^{j_2}}(x-y)^kU^p(x-y)dxdy\right)^{1/p}\lesssim(\log n)^{2/p}\left(\int_{\mathbb{R}+}u^kU^p(u)du\right)^{1/p}. \tag{5.12}$$

On the other hand, if $k/p>1$ and $\max(m_1,m_2)\lesssim L/\log L\lesssim n/\log n$ (with a good probability), one estimates

$$\sum_{\substack{1\leqslant j_1\leqslant m_1\\ 1\leqslant j_2\leqslant m_2}} (j_1+j_2)^{-k/p} \leqslant (\log n)^{k/p-2} n^{2-k/p}.$$

Plugging this (5.12) and (5.11) into the sum in (5.10), we see that (5.10) is a consequence of

$$(\log n)^{2-2/p+8/(3q)} n^{2-k/p-1/(3q)} = (\log n)^{14/3(p-1)/p} n^{5/3-(3k-1)/(3p)} = o(1).$$

as $p^{-1}+q^{-1}=1$.
Thus, it suffices to find $k>0$, $p>1$ such that $u\mapsto u^{k/p}U(u)$ be in $L^p(\mathbb{R}^+)$ and

$$\frac{5}{3}-\frac{3k-1}{3p}<0.$$

Recall that, by assumption $u\mapsto u^\alpha U(u)$ is integrable (for some $\alpha>2$) and $U\in L^r(\mathbb{R}^+)$ for some $r>1$.

We pick $\eta\in(0,1)$ and pick p and k of the form $p=1+\eta(r-1)$ and $k=\dfrac{5p+1}{3}+\eta$. Thus, for $r\in(1,\min(\tilde r,2)]$, setting $\tilde p:=\dfrac{r-p}{r-1}\in(0,1)$, we have

$$\begin{aligned}\frac{5}{3}-\frac{3k-1}{3p}=-\tfrac{\eta}{p}<0,\quad \tfrac{p-\tilde p}{1-\tilde p}=r \quad\text{and}\quad \frac{k}{\tilde p}&=k\frac{r-1}{r-p}\\ &=\left(2+\frac{5}{3}\eta(r-1)\right)\frac{1}{1-\eta}=\alpha\end{aligned}$$

for $\eta\in(0,1)$ well chosen.

For this choice of p, $\tilde p$ and k, using Hölder's inequality, we then estimate

$$\int_{\mathbb{R}^+} u^k U^p(u)du \leqslant \left(\int_{\mathbb{R}^+} u^{k/\tilde p}U(u)du\right)^{\tilde p}\left(\int_{\mathbb{R}^+} U^{(p-\tilde p)/(1-\tilde p)}(u)du\right)^{1-\tilde p} < +\infty$$

This completes the proof of (5.10) and, thus, of Lemma 5.3. □

Lemma 5.4 implies that, under the assumption of Theorem 5.1, in the thermodynamic limit, with probability exponentially close to 1, one has

$$\int_{\Lambda_1\times\Lambda_2} U(x-y)\rho_{\Psi_1^U}(x)\rho_{\Psi_2^U}(y)\mathrm{d}x\mathrm{d}y = o(n_1+n_2).$$

This completes the proof of Theorem 5.1. □

6 Multiple Electrons Interacting in a Fixed Number of Pieces

The main goal of this section is to study a system of two interacting electrons in the interval $[0, \ell]$ for large ℓ and prove Proposition 1.4; this is the purpose of Section 6.1. The two-particle Hamiltonian is given by (1.15). In Section 6.2, we study two electrons in two distinct pieces.

We shall also state and prove one result for more than two interacting electrons in a single piece.

6.1 Two Electrons in the Same Piece

We now study two electrons in a large interval interacting through a pair potential U, that is, the Hamiltonian defined in (1.15). We first prove Proposition 1.4. Next, in Section 6.1.3, we compare the ground state of the interacting system with that of the non-interacting system.

Throughout this section, we will assume U is a repulsive, even pair interaction potential. In the present section, our assumptions on U will be weaker than **(HU)**.

6.1.1 The Proof of Proposition 1.4

Scaling variables to the unit square, the two-particle Hamiltonians $H^U(\ell, 2)$ and $\ell^{-2}H^{U^\ell}(1, 2)$ are unitarily equivalent. Here, we have defined

$$U^\ell(\cdot) := \ell^2 U(\ell \cdot). \tag{6.1}$$

Recall that, for $i \neq j$, $i, j \in \mathbb{N}$, the normalized eigenfunctions of $H^0(1, 2)$ (i.e., of the two-particle free Hamiltonian in a unit square) are given by the determinant

$$\phi_{(i,j)}(x, y) = \sqrt{2}\begin{vmatrix}\sin(\pi i x) & \sin(\pi j x)\\ \sin(\pi i y) & \sin(\pi j y)\end{vmatrix} \quad \text{for } (x, y) \in [0, 1]^2. \tag{6.2}$$

For a two-component index, we will use the shorthand notation $\bar{\imath} = (i, j)$. For the non-interacting ground state $\phi_{(1,2)}$ we will also use the notation ϕ_0. The corresponding ground state energy is $5\pi^2$ and the first excited energy level is at $10\pi^2$.

We decompose $L^2([0, 1]) \wedge L^2([0, 1]) = \mathbb{C}\phi_0 \overset{\perp}{\oplus} \phi_0^\perp$. By the Schur complement formula, E is the ground state energy of $H^{U^\ell}(1, 2)$ if and only if $E < 10\pi^2$ and E satisfies

$$5\pi^2 + U^\ell_{00} - E = U^\ell_{0+}(H_+ + U^\ell_{++} - E)^{-1}U^\ell_{+0}, \tag{6.3}$$

where Π_+ is the orthogonal projector on $\phi_0^\perp$ and

$$\begin{aligned} U^\ell_{00} &= \langle \phi_0, U^\ell \phi_0 \rangle, \quad H_+ = \Pi_+ H^0 \Pi_+, \\ U^\ell_{++} &= \Pi_+ U^\ell \Pi_+, \quad U^\ell_{+0} = \Pi_+ U^\ell \phi_0 \quad U^\ell_{0+} = \left(\Pi_+ U^\ell \phi_0\right)^*. \end{aligned} \tag{6.4}$$

We expand the r.h.s. of (6.3) as

$$\begin{aligned} U^\ell_{0+}(H_+ + U^\ell_{++} - E)^{-1} U^\ell_{+0} &= \langle U^\ell \phi_0, (H_+ - E)^{-1/2} \\ &\quad \times \left(\mathrm{Id} + (H_+ - E)^{-1/2} U^\ell (H_+ - E)^{-1/2}\right)^{-1} \\ &\quad \times (H_+ - E)^{-1/2} U^\ell \phi_0 \rangle \\ &= \frac{1}{\ell} \left\langle \widetilde{\phi}_\ell, A_\ell (\mathrm{Id} + A^*_\ell A_\ell)^{-1} A^*_\ell \widetilde{\phi}_\ell \right\rangle \\ &= \frac{1}{\ell} \left\langle \widetilde{\phi}_\ell, A_\ell A^*_\ell (\mathrm{Id} + A_\ell A^*_\ell)^{-1} \widetilde{\phi}_\ell \right\rangle, \end{aligned} \tag{6.5}$$

where

$$\widetilde{\phi}_\ell = \sqrt{\ell}\sqrt{U^\ell}\phi_0 \quad \text{and} \quad A_\ell = A_\ell(E) = \sqrt{U^\ell}(H_+ - E)^{-1/2}. \tag{6.6}$$

To simplify notations we will drop the reference to the energy E. As $\ell \to +\infty$, the convergence of $\langle \widetilde{\phi}_\ell, A_\ell A^*_\ell (\mathrm{Id} + A_\ell A^*_\ell)^{-1} \widetilde{\phi}_\ell \rangle$ is locally uniform in $(-\infty, 10\pi^2)$. To compute this limit, we shall transform the expression $\langle \widetilde{\phi}_\ell, A_\ell A^*_\ell (\mathrm{Id} + A_\ell A^*_\ell)^{-1} \widetilde{\phi}_\ell \rangle$ once more.

Consider the domain $R_\ell = \{(u, y) \in \mathbb{R} \times [0, 1], \text{ s.t. } y + \ell^{-1} u \in [0, 1]\}$ and the change of variables

$$\begin{aligned} t_\ell : R_\ell &\to [0, 1]^2 \\ (u, y) &\mapsto \left(y + \frac{u}{\ell}, y\right). \end{aligned}$$

Define the partial isometry

$$\begin{aligned} T_\ell : L^2([0, 1]^2) &\to L^2(\mathbb{R} \times [0, 1]) \\ v &\mapsto \ell^{-1/2} \mathbf{1}_{R_\ell} \cdot v \circ t_\ell, \end{aligned}$$

that is, $(T_\ell v)(u, y) = \dfrac{1}{\sqrt{\ell}} \mathbf{1}_{R_\ell}(u, y) v \left(y + \dfrac{u}{\ell}, y\right)$.
One computes its adjoint

$$\begin{aligned} T^*_\ell : L^2(\mathbb{R} \times [0, 1]) &\to L^2([0, 1]^2) \\ v &\mapsto \ell^{1/2} (\mathbf{1}_{R_\ell} v) \circ t_\ell^{-1}, \end{aligned}$$

that is, $(T^*_\ell v)(x, y) = \sqrt{\ell} (\mathbf{1}_{R_\ell} \cdot v)(\ell(x - y), y)$.

One easily checks that

$$T_\ell T_\ell^* = \mathbf{1}_{R_\ell} \quad \text{and} \quad T_\ell^* T_\ell = \mathrm{Id}_{L^2([0,1]^2)} \tag{6.7}$$

where $\mathbf{1}_{R_\ell} : L^2(\mathbb{R} \times [0, 1]) \to L^2(\mathbb{R} \times [0, 1])$ is the orthogonal projector on the functions supported in R_ℓ.

One then computes

$$\left\langle \widetilde{\phi}_\ell, A_\ell A_\ell^* (\mathrm{Id} + A_\ell A_\ell^*)^{-1} \widetilde{\phi}_\ell \right\rangle_{L^2([0,1]^2)} = \left\langle \phi_\ell, K_\ell (\mathrm{Id} + K_\ell)^{-1} \phi_\ell \right\rangle_{L^2(\mathbb{R}\times[0,1])} \tag{6.8}$$

where we have defined

$$\phi_\ell := T_\ell \widetilde{\phi}_\ell \quad \text{and} \quad K_\ell := K_\ell(E) := T_\ell A_\ell A_\ell^* T_\ell^*. \tag{6.9}$$

Define

- the following functions
  - $\phi(u) := u\sqrt{U(u)}$ for $u \in \mathbb{R}$,
  - $\chi_0(y) := \pi\sqrt{2}\,(\sin(3\pi y) - 3\sin(\pi y))$ for $y \in [0, 1]$.
- the non-negative (see (6.47)) operator K on $L^2(\mathbb{R})$ by the kernel

$$K(u, u') = \frac{1}{2}\sqrt{U(u)}(|u + u'| - |u - u'|)\sqrt{U(u')}. \tag{6.10}$$

Define also

$$\tilde{\phi} = \phi \otimes \chi_0 \quad \text{and} \quad \tilde{K} = K \otimes \mathrm{Id}. \tag{6.11}$$

We prove

Lemma 6.1 *Assume that U is non-negative, even, such that $U \in L^p(\mathbb{R})$ for some $p > 1$ and $x \mapsto x^2 U(x)$ is integrable.*

As $\ell \to +\infty$, one has:

(a) *in $L^2(\mathbb{R} \times [0, 1])$, ϕ_ℓ converges to $\tilde{\phi}$;*
(b) *for $\varphi \in C_0^\infty(\mathbb{R} \times (0, 1))$, as $\ell \to +\infty$, the sequence $(K_\ell \varphi)_\ell$ converges in L^2-norm to $\tilde{K}\varphi$*

Proposition 1.4 follows from this result as we shall see now. First, we prove

Lemma 6.2 *Under the assumptions of Lemma 6.1, all the operators $(K_\ell)_\ell$ and the operator K are bounded, respectively, on $L^2(\mathbb{R} \times [0, 1])$ and $L^2(\mathbb{R})$.*

Note however that, depending on U, one may have

$$\|K_\ell\|_{L^2(\mathbb{R}\times[0,1])\to L^2(\mathbb{R}\times[0,1])} \xrightarrow[\ell\to+\infty]{} +\infty.$$

Proof By (6.9), to show the boundedness of K_ℓ, it suffices to show that $\tilde{K}_\ell := \sqrt{U_\ell}(H_+ - E)^{-1}\sqrt{U_\ell}$ is bounded. Note that, by our assumption on U, U_ℓ is in $L^p([0,1]^2)$. Using the eigenfunction expansion of $-\Delta$ on $L^2_-([0,1]^2)$, we write

$$\tilde{K}_\ell = \sum_{\bar{j}\neq(2,1)} \frac{1}{\pi^2|\bar{j}|^2 - E}\sqrt{U_\ell}\phi_{\bar{j}} \otimes \phi_{\bar{j}}\sqrt{U_\ell} \tag{6.12}$$

where the sum is over $\bar{j} = (i, j)$ where $(i, j) \in \mathbb{N}$ such that $i > j$.

For $u \in L^2_-([0,1]^2)$, as $u\sqrt{U_\ell} \in L^{2p/(1+p)}_-([0,1]^2)$ and as the functions $(\phi_{\bar{j}})_{\bar{j}}$ are uniformly bounded, by the Hausdorff-Young inequality (see, e.g., [19]), one has

$$\sum_{\bar{j}} \left|\left\langle\sqrt{U_\ell}\phi_{\bar{j}}, u\right\rangle\right|^{p/(p-1)} \leqslant C_\ell \|u\|_2^{p/(p-1)}. \tag{6.13}$$

Moreover, for some C_ℓ, one has $\|\sqrt{U_\ell}\phi_{\bar{j}}\|_2 \leqslant C_\ell$. Thus, by (6.12), as $p > 1$, we obtain

$$\|\tilde{K}_\ell u\|_2 \leqslant C_\ell \left(\sum_{\bar{j}\neq(2,1)} \frac{1}{(\pi^2|\bar{j}|^2 - E)^p}\right)^{1/p} \|u\|_2 \leqslant C_\ell\|u\|_2.$$

Using the explicit kernel for K given in (6.10), for $u \in L^2(\mathbb{R})$, we compute

$$(Ku)(x)=2\sqrt{U(x)}\int_{-x}^{x} x'\sqrt{U(x')}u(x')dx'+2\sqrt{U(x)}x\int_{-\infty}^{-x}\sqrt{U(x')}(u(x')-u(-x'))dx'$$

Thus,

$$\|K\|_{\mathcal{L}(L^2(\mathbb{R}))} \leqslant 4\sqrt{\|U\|_1\|(\cdot)^2U(\cdot)\|_1}.$$

This completes the proof of Lemma 6.2. □

By Lemma 6.2, $C_0^\infty(\mathbb{R}\times(0,1))$ is a common core for all K_ℓ and $K\otimes\mathrm{Id}$. Thus, by [18, Theorem VIII.25], we know that $K_\ell \xrightarrow[\ell\to+\infty]{} K\otimes\mathrm{Id}$ in the strong resolvent sense. Hence, by [18, Theorem VIII.20], the sequence $(K_\ell(\mathrm{Id}+K_\ell)^{-1})_\ell$ converges to $K(\mathrm{Id}+K)^{-1}\otimes\mathrm{Id}$ strongly. These operators are all bounded uniformly by 1 (as K_ℓ and K are non-negative). Thus, by point (a) of Lemma 6.1 and (6.8), we obtain

$$\left\langle \tilde{\phi}_\ell, A_\ell A_\ell^\star (\mathrm{Id} + A_\ell A_\ell^\star)^{-1} \tilde{\phi}_\ell \right\rangle = \langle \phi \otimes \chi_0, \left[K(\mathrm{Id}+K)^{-1} \otimes \mathrm{Id} \right] \phi \otimes \chi_0 \rangle + o(1)$$
$$= \left\langle \phi, K(\mathrm{Id}+K)^{-1}\phi \right\rangle \cdot \int_0^1 \chi_0^2(y)\mathrm{d}y + o(1)$$
$$= \pi^2 \cdot \left\langle \phi, K(\mathrm{Id}+K)^{-1}\phi \right\rangle + o(1). \tag{6.14}$$

By point (a) of Lemma 6.1, one also computes

$$\ell\, U_{00}^\ell = \|\phi \otimes \chi_0\|^2 + o(1) = \int_{\mathbb{R}} u^2 U(u)\mathrm{d}u \int_0^1 \chi_0^2(y)\mathrm{d}y + o(1)$$
$$= \frac{5}{2}\pi^2 \int_{\mathbb{R}} u^2 U(u)\mathrm{d}u + o(1) \tag{6.15}$$

By (6.15), the eigenvalue equation (6.3) yields that, under the assumptions of Lemma 6.2, the ground state energy of $H^{U^\ell}(1, 2)$ satisfies

$$E^{U^\ell}([0,1],2) = 5\pi^2 + \frac{\gamma(U)}{\ell} + o\left(\frac{1}{\ell}\right) \tag{6.16}$$

where

$$\gamma(U) = 10\pi^2 \left[\|\phi\|^2 - \left\langle \phi, K(\mathrm{Id}+K)^{-1}\phi \right\rangle \right] = 10\pi^2 \left\langle \phi, (\mathrm{Id}+K)^{-1}\phi \right\rangle. \tag{6.17}$$

By Lemma 5.2 and assumption **(HU)**, we know that the assumptions of Lemma 6.2 are satisfied. This proves the asymptotic expansion announced in Proposition 1.4. To complete the proof of this proposition, we simply note that, as K is non-negative and bounded by Lemma 6.2, by (6.17), we know that $\gamma(U) = 0$ if and only if $\phi \equiv 0$, i.e., if and only if $U \equiv 0$. □

Remark 6.3 If one assumes $x \mapsto x^4 U(x)$ to be integrable and U to be in some $L^p(\mathbb{R})$ ($p > 1$) (which is clearly stronger than **(HU)**), one obtains that, $E^U([0,\ell],2)$, the ground state energy of the Hamiltonian defined in (1.15) admits the following more precise expansion

$$E^{U^\ell}([0,1],2) = 5\pi^2 + \frac{\gamma(U)}{\ell} + O\left(\ell^{-2}\right). \tag{6.18}$$

6.1.2 The Proof of Lemma 6.1

We start with a lemma, the result of a computation, that will be used in several parts of the proof.

Lemma 6.4 *For $\bar{j} = (j_1, j_2)$, $j_1 > j_2$, recall that $\phi_{\bar{j}}$, the $\bar{j}$-th normalized eigenvector of H_0, is given by (6.2).*

One has

$$\phi_{\bar{j}}\left(y + \frac{u}{\ell}, y\right) = \phi^0_{\bar{j}}\left(\frac{u}{\ell}, y\right) + \phi^+_{\bar{j}}\left(\frac{u}{\ell}, y\right) + \phi^-_{\bar{j}}\left(\frac{u}{\ell}, y\right) \tag{6.19}$$

where

$$\begin{aligned}
\phi^0_{\bar{j}}(2x, y) &:= 2\sqrt{2}\sin(\pi(j_1 + j_2)x)\sin(\pi(j_2 - j_1)x)\sin(\pi j_1 y)\sin(\pi j_2 y),\\
\phi^+_{\bar{j}}(2x, y) &:= \sqrt{2}\cos(\pi(j_2 - j_1)x)\sin(\pi(j_2 + j_1)x)\sin(\pi(j_1 - j_2)y)\\
\phi^-_{\bar{j}}(2x, y) &:= \sqrt{2}\cos(\pi(j_2 + j_1)x)\sin(\pi(j_2 - j_1)x)\sin(\pi(j_1 + j_2)y)
\end{aligned} \tag{6.20}$$

Proof Using standard sum and product formulas for the sine and cosine, we compute

$$\begin{aligned}
\frac{1}{\sqrt{2}}\phi_{\bar{j}}\left(y + \frac{u}{\ell}, y\right) &= \begin{vmatrix} \sin\left(\pi j_1\left(y + \frac{u}{\ell}\right)\right) & \sin\left(\pi j_1 y\right) \\ \sin\left(\pi j_2\left(y + \frac{u}{\ell}\right)\right) & \sin\left(\pi j_2 y\right) \end{vmatrix}\\
&= \sin\left(\pi j_1 \frac{u}{\ell}\right)\cos\left(\pi j_1 y\right)\sin\left(\pi j_2 y\right) - \sin\left(\pi j_2 \frac{u}{\ell}\right)\cos\left(\pi j_2 y\right)\sin\left(\pi j_1 y\right)\\
&\qquad + \left(\cos\left(\pi j_1 \frac{u}{\ell}\right) - \cos\left(\pi j_2 \frac{u}{\ell}\right)\right)\sin\left(\pi j_1 y\right)\sin\left(\pi j_2 y\right)\\
&= \frac{1}{2}\sin\left(\pi j_1 \frac{u}{\ell}\right)\left(\sin\left(\pi\left(j_1 + j_2\right)y\right) - \sin\left(\pi\left(j_1 - j_2\right)y\right)\right)\\
&\qquad - \frac{1}{2}\sin\left(\pi j_2 \frac{u}{\ell}\right)\left(\sin\left(\pi\left(j_1 + j_2\right)y\right) + \sin\left(\pi\left(j_1 - j_2\right)y\right)\right)\\
&\qquad + \left(\cos\left(\pi j_1 \frac{u}{\ell}\right) - \cos\left(\pi j_2 \frac{u}{\ell}\right)\right)\sin\left(\pi j_1 y\right)\sin\left(\pi j_2 y\right)\\
&= \frac{1}{2}\left(\sin\left(\pi j_1 \frac{u}{\ell}\right) - \sin\left(\pi j_2 \frac{u}{\ell}\right)\right)\sin\left(\pi\left(j_1 + j_2\right)y\right)\\
&\qquad - \frac{1}{2}\left(\sin\left(\pi j_1 \frac{u}{\ell}\right) + \sin\left(\pi j_2 \frac{u}{\ell}\right)\right)\sin\left(\pi\left(j_1 - j_2\right)y\right)\\
&\qquad + \left(\cos\left(\pi j_1 \frac{u}{\ell}\right) - \cos\left(\pi j_2 \frac{u}{\ell}\right)\right)\sin\left(\pi j_1 y\right)\sin\left(\pi j_2 y\right).
\end{aligned}$$

Thus,

$$
\begin{aligned}
\frac{1}{\sqrt{2}}\phi_{\bar{j}}\left(y+\frac{u}{\ell},y\right) &= \sin\left(\pi\frac{j_1-j_2}{2}\frac{u}{\ell}\right)\cos\left(\pi\frac{j_1+j_2}{2}\frac{u}{\ell}\right)\sin\left(\pi\left(j_1+j_2\right)y\right)\\
&\quad -\sin\left(\pi\frac{j_1+j_2}{2}\frac{u}{\ell}\right)\cos\left(\pi\frac{j_1-j_2}{2}\frac{u}{\ell}\right)\sin\left(\pi\left(j_1-j_2\right)y\right)\\
&\quad -2\sin\left(\pi\frac{j_1-j_2}{2}\frac{u}{\ell}\right)\sin\left(\pi\frac{j_1+j_2}{2}\frac{u}{\ell}\right)\sin\left(\pi j_1 y\right)\\
&\quad \sin\left(\pi j_2 y\right).
\end{aligned}
$$

This completes the proof of Lemma 6.4. □

We start with the proof of point (a) of Lemma 6.1. As $\phi_0 = \phi_{(2,1)}$, by (6.19) and (6.20), using the Taylor expansion of the sine and cosine near 0, we compute

$$
\begin{aligned}
(T_\ell\tilde{\phi}_\ell)(u,y) &= \ell\sqrt{U(u)}\mathbf{1}_{R_\ell}(u,y)\phi_{(2,1)}\left(y+\frac{u}{\ell},y\right)\\
&= u\sqrt{U(u)}\chi_0(y)\mathbf{1}_{R_\ell}(u,y)+\frac{u^2}{\ell}\sqrt{U(u)}\chi_1\left(\frac{u}{\ell},y\right)\mathbf{1}_{R_\ell}(u,y)
\end{aligned}
$$

where χ_0 is defined in Lemma 6.1 and χ_1 is continuous and bounded on $\mathbb{R}\times[0,1]$.

We estimate

$$
\left\|\frac{(\cdot)^2}{\ell}\sqrt{U(\cdot)}\chi_1\left(\frac{\cdot}{\ell},\cdot\right)\mathbf{1}_{R_\ell}\right\|^2_{L^2(\mathbb{R}\times[0,1])} \lesssim \int_{R_\ell}\frac{u^2}{\ell^2}u^2U(u)du \leqslant \int_{\mathbb{R}}\frac{u^2\mathbf{1}_{|u|\leqslant\ell}}{\ell^2}u^2U(u)du.
$$

The last integral tends to 0 by the dominated convergence theorem as $u\mapsto u^2U(u)$ is integrable.

This completes the proof of point (a) of Lemma 6.1.

Let us now turn to the analysis of the operator family $(K_\ell)_\ell$. It is easily seen that its kernel (we use the same notations for the operator and its kernel) is given by

$$
K_\ell(E;u,y,u',y') = \ell\mathbf{1}_{R_\ell\times R_\ell}\sqrt{U(u)U(u')}\cdot\tilde{K}\left(E;y+\frac{u}{\ell},y,y'+\frac{u'}{\ell},y'\right)
$$

where $\tilde{K}(E;x,y,x',y')$ is the kernel of $(H_+-E)^{-1}$. The kernel $\tilde{K}(E)$ is easily expressed in terms of the eigenfunctions of H. Using this and the representation yielded by Lemma 6.4 leads to the following representation for the kernel K_ℓ

$$
\begin{aligned}
K_\ell\left(E;u,y,u',y'\right) &= \ell\mathbf{1}_{R_\ell\times R_\ell}\sum_{\bar{j}\neq(2,1)}\frac{\sqrt{U(u)\,U(u')}}{\pi^2|\bar{j}|^2-E}\phi_{\bar{j}}\left(y+\frac{u}{\ell},y\right)\phi_{\bar{j}}\left(y'+\frac{u'}{\cdot\ell},y'\right)\\
&= K_\ell^-\left(E;u,y,u',y'\right)+K_\ell^+\left(E;u,y,u',y'\right)+K_\ell^0\left(E;u,y,u',y'\right)
\end{aligned}
\tag{6.21}
$$

where, for $\bullet\in\{0,+,-\}$, we have set

$$K_\ell^\bullet\left(E;u,y,u',y'\right)=\ell\mathbf{1}_{R_\ell\times R_\ell}\sum_{\bar{j}\neq(2,1)}\frac{\sqrt{U\left(u\right)U\left(u'\right)}}{\pi^2|\bar{j}|^2-E}\phi_{\bar{j}}\left(y+\frac{u}{\ell},y\right)\phi_{\bar{j}}^\bullet\left(\frac{u'}{\ell},y'\right).$$

To prove point (b) of Lemma 6.1, if suffices to prove that, for $v\in C_0^\infty(\mathbb{R}\times(0,1))$, one has $K_l v\to\tilde{K}v$ in $L^2(\mathbb{R}\times[0,1])$. We first prove

Lemma 6.5 *For* $v\in C_0^\infty(\mathbb{R}\times(0,1))$, *one has*

(a) $\|K_\ell^- v\|_2\to 0$ *as* $\ell\to+\infty$,
(b) $\|K_\ell^0 v\|_2\to 0$ *as* $\ell\to+\infty$.

Proof We first study the sequence $K_\ell^+ v$. We compute

$$(K_\ell^- v)(u,y)=\sqrt{U(u)}\sum_{\substack{j\geqslant1,k\geqslant1\\(j,k)\neq(1,1)}}\frac{C_{j,k}(v)}{\pi^2((j+k)^2+j^2)-E}\mathbf{1}_{R_\ell}(u,y)\phi_{(j+k,j)}\left(y+\frac{u}{\ell},y\right)\tag{6.22}$$

where

$$C_{j,k}(v):=\ell\int_{-\ell}^{\ell}\sqrt{U(u')}\sin\left(\pi\frac{(2j+k)u'}{2\ell}\right)\cos\left(\pi\frac{ku'}{\ell}\right)c_{2j+k}(u')du'\tag{6.23}$$

and

$$\begin{aligned}c_j(u')&:=\int_0^1(\mathbf{1}_{R_\ell}v)(u',y')\sin(\pi jy')dy'=\int_{\max(0,-u'/\ell)}^{\min(1,1-u'/\ell)}v(u',y')\sin(\pi jy')dy'\\&=\int_0^1 v(u',y')\sin(\pi jy')dy'\end{aligned}\tag{6.24}$$

for ℓ sufficiently large as $v\in C_0^\infty(\mathbb{R}\times(0,1))$.

Integrating the last integral in (6.24) by parts, we obtain

$$\|c_j\|_{L^2(\mathbb{R})}=O\left(j^{-\infty}\right).\tag{6.25}$$

By (6.24) and (6.23), as $u\mapsto u^2U(u)$ is summable, we obtain

$$\begin{aligned}|C_{j,k}(v)|&\leqslant O\left((2j+k)^{-\infty}\right)\ell\sqrt{\int_{\mathbb{R}}U(u')\sin^2\left(\pi\frac{(2j+k)u'}{2\ell}\right)du'}\\&\leqslant O\left((2j+k)^{-\infty}\right)\min(\ell,2j+k)\end{aligned}\tag{6.26}$$

Estimating $\|K_\ell^- v\|$ using (6.22) and the triangular inequality, as

$$\begin{aligned}&\int_{\mathbb{R}\times[0,1]} U(u)\mathbf{1}_{R_\ell}(u,y)\phi^2_{(j+k,j)}\left(y+\frac{u}{\ell},y\right)dudy\\&\lesssim \int_{\mathbb{R}} U(u)\sin^2\left(\pi k\frac{u}{\ell}\right)du + \int_{\mathbb{R}} U(u)\sin^2\left(\pi(2j+k)\frac{u}{\ell}\right)du\\&\lesssim \frac{\min^2(2j+k,\ell)+\min^2(k,\ell)}{\ell^2},\end{aligned} \tag{6.27}$$

for $p \geqslant 4$, we get

$$\left\|K_\ell^- v\right\| \lesssim \frac{1}{\ell}\sum_{\substack{j\geqslant 1,k\geqslant 1\\(j,k)\neq(1,1)}}\frac{1}{(j+k)^p}.$$

Thus, one gets that $\left\|K_\ell^- v\right\| \to 0$ as $\ell \to +\infty$. This completes the proof of point (a) of Lemma 6.5.

To prove point (b), as $2\sin a\ \sin b = \cos(a-b) - \cos(a+b)$, we compute

$$(K_\ell^0 v)(u,y) = \sqrt{U(u)}\sum_{\substack{j\geqslant 1,k\geqslant 1\\(j,k)\neq(1,1)}}\frac{A^-_{j,k}(v)-A^+_{j,k}(v)}{\pi^2((j+k)^2+j^2)-E}\mathbf{1}_{R_\ell}(u,y)\phi_{(j+k,j)}\left(y+\frac{u}{\ell},y\right)$$

where

$$A^+_{j,k}(v) := \ell\int_{-\ell}^{\ell}\sqrt{U(u')}\sin\left(\pi\frac{(2j+k)u'}{2\ell}\right)\sin\left(\pi\frac{ku'}{\ell}\right)a_{2j+k}(u')du',$$

$$A^-_{j,k}(v) := \ell\int_{-\ell}^{\ell}\sqrt{U(u')}\sin\left(\pi\frac{(2j+k)u'}{2\ell}\right)\sin\left(\pi\frac{ku'}{\ell}\right)a_k(u')du'$$

and

$$a_k(u') := \int_0^1(\mathbf{1}_{R_\ell}v)(u',y')\cos(\pi k y')dy'.$$

As in (6.24), we obtain

$$\|a_k\|_{L^2(\mathbb{R})} = O\left(k^{-\infty}\right).$$

As in (6.26), we obtain

$$|A^\pm_{j,k}(v)| \leqslant O\left(k^{-\infty}\right)\min(\ell,k).$$

By (6.27), for $p \geqslant 2$, we then get

$$\left\|K_\ell^0 v\right\| \lesssim \frac{1}{\ell} \sum_{\substack{j\geqslant 1, k\geqslant 1\\(j,k)\neq(1,1)}} \frac{\min(\ell,k)(\min(\ell,k)+\min(\ell,j+k))}{k^{-p}((j+k)^2+j^2)} \lesssim \frac{1}{\ell}+\sum_{j\geqslant 1}\frac{\min(1,j/\ell)}{j^2} \tag{6.28}$$

The last term converges to 0 by the dominated convergence theorem. This completes the proof of point (b) of Lemma 6.5, thus, of Lemma 6.5. □

Next, we decompose K_ℓ^+ expanding $\phi_{\bar{j}}(y+u/\ell, y)$ according to (6.19). This gives

$$K_\ell^+ = K_\ell^{+,+} + K_\ell^{+,-} + K_\ell^{+,0},$$

where, for $\bullet \in \{0,+,-\}$, we have set

$$K_\ell^{+,\bullet}\left(E;u,y,u',y'\right) = \ell \mathbf{1}_{R_\ell\times R_\ell} \sum_{\bar{j}\neq(2,1)} \frac{\sqrt{U(u)U(u')}}{\pi^2|\bar{j}|^2-E}\phi_{\bar{j}}^\bullet\left(y+\frac{u}{\ell},y\right)\times\phi_{\bar{j}}^+\left(\frac{u'}{\ell},y\right).$$

We now prove

Lemma 6.6 *For* $v\in C_0^\infty(\mathbb{R}\times(0,1))$, *one has*

(a) $\|K_\ell^{-,+}v\| \to 0$ *as* $\ell\to+\infty$,
(b) $\|K_\ell^{0,+}v\| \to 0$ *as* $\ell\to+\infty$.

Proof As in the proof of Lemma 6.5, the two points in Lemma 6.6 are proved in very similar ways. We will only detail the proof of point (a).

We compute

$$(K_\ell^{-,+}v)(u,y) = \sqrt{U(u)} \sum_{\substack{j\geqslant 1,k\geqslant 1\\(j,k)\neq(1,1)}} \frac{C_{j,k}(v)}{\pi^2((j+k)^2+j^2)-E}\mathbf{1}_{R_\ell}(u,y)\phi^-_{(j+k,j)}\left(y+\frac{u}{\ell},y\right) \tag{6.29}$$

where

$$C_{j,k}(v) := \ell\int_{-\ell}^{\ell}\sqrt{U(u')}\sin\left(\pi\frac{(2j+k)u'}{2\ell}\right)\cos\left(\pi\frac{ku'}{2\ell}\right)c_k(u')du' \tag{6.30}$$

and

$$c_k(u') := \int_0^1(\mathbf{1}_{R_\ell}v)(u',y')\sin(\pi k y')dy' = \int_0^1 v(u',y')\sin(\pi k y')dy'$$

for ℓ sufficiently large as $v\in C_0^\infty(\mathbb{R}\times(0,1))$.

Integrating the last integral in (6.24) by parts, we obtain

$$\|c_k\|_{L^2(\mathbb{R})} = O\left(k^{-\infty}\right). \tag{6.31}$$

As in (6.26), we obtain

$$|C_{j,k}(v)| \leqslant O\left(k^{-\infty}\right)\min(\ell, 2j+k). \tag{6.32}$$

Using (6.20), one estimates

$$\sqrt{\int_{\mathbb{R}\times[0,1]} U(u)\mathbf{1}_{R_\ell}(u,y)\left|\phi^-_{(j+k,j)}\left(y+\frac{u}{\ell}, y\right)\right|^2 dudy} \lesssim \frac{\min(k,\ell)}{\ell}. \tag{6.33}$$

Thus, for $p \geqslant 2$, we get

$$\left\|K_\ell^{-,+}v\right\| \lesssim \sum_{\substack{j\geqslant 1, k\geqslant 1\\ (j,k)\neq(1,1)}} \frac{\min(k,\ell)}{\ell}\frac{\min(2j+k,\ell)}{k^p((j+k)^2+j^2)}. \tag{6.34}$$

Thus, by the dominated convergence theorem, as in (6.28), one gets that $\left\|K_\ell^{-,+}v\right\| \to 0$ as $\ell \to +\infty$. This completes the proof of point (a) of Lemma 6.6.

Point (b) is proved similarly except that estimate (6.33) is replaced with

$$\sqrt{\int_{\mathbb{R}\times[0,1]} U(u)\mathbf{1}_{R_\ell}(u,y)\left|\phi^0_{(j+k,j)}\left(y+\frac{u}{\ell}, y\right)\right|^2 dudy} \lesssim \frac{\min(k,\ell)\min(2j+k,\ell)}{\ell^2}.$$

Thus, taking $p > 3$, estimate (6.34) in this case becomes

$$\begin{aligned}\left\|K_\ell^{0,+}v\right\| &\lesssim \sum_{\substack{j\geqslant 1, k\geqslant 1\\ (j,k)\neq(1,1)}} \frac{\min(k,\ell)}{k^p}\frac{\min^2(2j+k,\ell)}{\ell^2((j+k)^2+j^2)} \lesssim \sum_{\substack{j\geqslant 1, k\geqslant 1\\ (j,k)\neq(1,1)}} \frac{1}{k^{p-2}}\frac{\min^2(j,\ell)}{\ell^2 j^2}\\ &\lesssim \sum_{j\geqslant 1}\frac{\min^2(j,\ell)}{\ell^2}\frac{1}{j^2}\end{aligned}$$

which converges to 0 as $\ell \to +\infty$.

This completes the proof of Lemma 6.6. □

We are now left with computing the limit of $K_\ell^{+,+}$ where

$$K_\ell^{+,+}(u,y,u',y') = \sum_{\substack{j\geqslant 1, k\geqslant 1\\ (j,k)\neq(1,1)}} \frac{\ell\sqrt{U(u)U(u')}}{\pi^2((j+k)^2+j^2)-E}$$

$$\times \phi^+_{(j+k,j)}\left(y+\frac{u}{\ell},y\right)\phi^+_{(j,j+k)}\left(y'+\frac{u'}{\ell},y'\right). \tag{6.35}$$

We prove

Lemma 6.7 *In the strong topology, one has*

$$K_\ell^{+,+} \to K \otimes \mathrm{Id} \quad as \quad \ell \to +\infty. \tag{6.36}$$

where K *is defined in* (6.10).

Proof To simplify the computations, we note that it suffices to show the convergence of $K_\ell^{+,+}v$ for $v \in \mathcal{C}_0^\infty(\mathbb{R}\times(0,1))$.For ℓ sufficiently large, compute

$$(K_\ell^{+,+}v)(u,y) = \sum_{k\geqslant 1} \sin(\pi k y)\mathbf{1}_{R_\ell}(u,y)c\left(K_k^\ell,u\right)$$

where

$$c\left(K_k^\ell,u\right) := \frac{1}{2}\sqrt{U(u)}\int_{\mathbb{R}} K_k^\ell(u,u')\sqrt{U(u')}c_k(u')du', \tag{6.37}$$

$u \mapsto c_k(u)$ being defined by (6.24), and

$$K_k^\ell(u,u') := \ell \sum_{\substack{j\in\mathbb{N}\\ (j,k)\neq(1,1)}} \frac{\sin\left(\pi\frac{2j+k}{2\ell}u\right)\sin\left(\pi\frac{2j+k}{2\ell}u'\right)\cos\left(\pi\frac{k}{2\ell}u\right)\cos\left(\pi\frac{k}{2\ell}u'\right)}{\pi^2(j+k/2)^2+(\pi k/2)^2-E} \tag{6.38}$$

Define

$$\begin{aligned}
L_k^\ell(u,u') &:= \ell \sum_{\substack{j\in\mathbb{N}\\ (j,k)\neq(1,1)}} \frac{\sin\left(\pi\frac{2j+k}{2\ell}u\right)\sin\left(\pi\frac{2j+k}{2\ell}u'\right)}{\pi^2(j+k/2)^2},\\
M_k^\ell(u,u') &:= K_k^\ell(u,u') - L_k^\ell(u,u'),\\
(L_\ell^{+,+}v)(u,y) &:= \sum_{k\geqslant 1}\sin(\pi k y)\mathbf{1}_{R_\ell}(u,y)c\left(L_k^\ell,u\right),\\
(M_\ell^{+,+}v)(u,y) &:= \sum_{k\geqslant 1}\sin(\pi k y)\mathbf{1}_{R_\ell}(u,y)c\left(M_k^\ell,u\right).
\end{aligned} \tag{6.39}$$

Here and in the sequel, $c\left(L_k^\ell,u\right)$ and $c\left(M_k^\ell,u\right)$ are defined as $c\left(K_k^\ell,u\right)$ in (6.37) with K_k^ℓ Replaced, respectively, by L_k^ℓ and M_k^ℓ

Note that

$$\left\| L_\ell^{+,+} v \right\|_{L^2(\mathbb{R}\times[0,1])}^2 := \frac{1}{2} \sum_{k\geqslant 1} \int_0^1 \left\| \mathbf{1}_{R_\ell}(y,\cdot) c\left(L_k^\ell,\cdot\right) \right\|_{L^2(\mathbb{R})}^2 dy \leqslant \frac{1}{2} \sum_{k\geqslant 1} \left\| c\left(L_k^\ell,\cdot\right) \right\|_{L^2(\mathbb{R})}^2 \tag{6.40}$$

We prove

Lemma 6.8 *As* $\ell \to +\infty$, $\left\| M_\ell^{+,+} v \right\|_{L^2(\mathbb{R}\times[0,1])} \to 0$

Proof The proof is similar to those of Lemmas 6.5 and 6.6. We write

$$M_k^\ell(u,u') = M_k^{1,\ell}(u,u') + M_k^{2,\ell}(u,u') + M_k^{3,\ell}(u,u')$$

where

$$M_k^{1,\ell}(u,u') = \ell \sum_{\substack{j\in\mathbb{N} \\ (j,k)\neq(1,1)}} \frac{\sin\left(\pi\frac{2j+k}{2\ell}u\right)\sin\left(\pi\frac{2j+k}{2\ell}u'\right)\cos\left(\pi\frac{k}{2\ell}u\right)\cos\left(\pi\frac{k}{2\ell}u'\right)\left((\pi k/2)^2 - E\right)}{\pi^4 (j+k/2)^2\left((j+k/2)^2 + (\pi k/2)^2 - E\right)}$$

$$M_k^{2,\ell}(u,u') := \ell \sum_{\substack{j\in\mathbb{N} \\ (j,k)\neq(1,1)}} \frac{\sin\left(\pi\frac{2j+k}{2\ell}u\right)\sin\left(\pi\frac{2j+k}{2\ell}u'\right)\cos\left(\pi\frac{k}{2\ell}u\right)\left(\cos\left(\pi\frac{k}{2\ell}u'\right) - 1\right)}{\pi^2 (j+k/2)^2}$$

$$M_k^{3,\ell}(u,u') := \ell \sum_{\substack{j\in\mathbb{N} \\ (j,k)\neq(1,1)}} \frac{\sin\left(\pi\frac{2j+k}{2\ell}u\right)\sin\left(\pi\frac{2j+k}{2\ell}u'\right)\left(\cos\left(\pi\frac{k}{2\ell}u\right) - 1\right)}{\pi^2 (j+k/2)^2}.$$

Following the definitions (6.38) and using (6.40), we estimate

$$\begin{aligned}
\left\| M_\ell^{1,+,+} v \right\|_{L^2(\mathbb{R}\times[0,1])}^2 &\leqslant \frac{1}{2} \sum_{k\geqslant 1} \left\| c\left(M_k^{1,\ell},\cdot\right) \right\|_{L^2(\mathbb{R})}^2 \\
&\lesssim \sum_{k\geqslant 1} k^2 \|c_k\|_{L^2(\mathbb{R})}^2 \sum_{\substack{j\geqslant 1 \\ (j,k)\neq(1,1)}} \frac{(\min(2j+k,\ell))^2}{\ell(2j+k)^4} \\
&\lesssim \frac{1}{\ell} \sum_{k\geqslant 1} k^2 \|c_k\|_{L^2(\mathbb{R})}^2
\end{aligned}$$

which, by (6.31), converges to 0 as ℓ goes to $+\infty$.

That the term coming from $M_\ell^{2,+,+}$ (resp. $M_\ell^{3,+,+}$) also vanishes as $\ell \to +\infty$ follows from computations similar to those done in Lemma 6.5 (resp. Lemma 6.6). This completes the proof of Lemma 6.8. □

Note that

$$L_k^\ell(u,u') = \frac{1}{\ell} \sum_{\substack{j\in\mathbb{N} \\ (j,k)\neq(1,1)}} \frac{\sin\left(\pi \frac{2j+k}{2\ell} u\right)\sin\left(\pi \frac{2j+k}{2\ell} u'\right)}{\pi^2 \left(\frac{2j+k}{2\ell}\right)^2} \tag{6.41}$$

Define

$$a(L^+, u) := \frac{1}{2}\sqrt{U(u)} \int_{\mathbb{R}} L^+(u,u')\sqrt{U(u')}c_k(u')du' \tag{6.42}$$

where

$$L^+(u,u') = \int_0^{+\infty} \frac{\sin(\pi x u)\sin(\pi x u')}{\pi^2 x^2} dx. \tag{6.43}$$

We prove

Lemma 6.9 *For any $k \geqslant 1$, one has*

$$\sup_{(u,u')\in[-\ell,\ell]^2} \frac{\left|L_k^\ell(u,u') - L^+(u,u')\right|}{|u||u'|} \lesssim \frac{k}{\ell}. \tag{6.44}$$

Proof Define

$$l(x,u,u') := \frac{\sin(\pi x u)\sin(\pi x u')}{\pi^2 x^2}.$$

Assume first $k \neq 1$. As l is an even function of x, write

$$L_k^\ell(u,u') = \frac{1}{2\ell}\sum_{j\in\mathbb{Z}} l\left(\frac{j+k/2}{\ell}, u, u'\right) - \frac{1}{2\ell}\sum_{j=-k}^{0} l\left(\frac{j+k/2}{\ell}, u, u'\right). \tag{6.45}$$

Using the Poisson formula, one computes

$$\frac{1}{2\ell}\sum_{j\in\mathbb{Z}} l\left(\frac{j+k/2}{\ell}, u, u'\right) = \frac{1}{2}\sum_{j\in\mathbb{Z}} e^{i\pi k j}\cdot \hat{l}\left(j,u,u'\right) \tag{6.46}$$

where $\hat{l}(\cdot,u,u')$ is the Fourier transform of $x \mapsto l(x,u,u')$.

By the Paley-Wiener Theorem (or by a direct computation of the Fourier transform), one checks that $\hat{l}(\cdot,u,u')$ is supported in $[-\pi(|u|+|u'|), \pi(|u|+|u'|)]$. Thus, for $-\ell \leqslant u, u' \leqslant l$, all the terms in right-hand side of (6.46) vanish except the term for $j = 0$. That is, for $-\ell \leqslant u, u' \leqslant l$, one has

$$\frac{1}{2\ell}\sum_{j\in\mathbb{Z}} l\left(\frac{j+k/2}{\ell},u,u'\right)=\frac{1}{2}\hat{l}\left(0,u,u'\right)=L^+(u,u').$$

This and (6.46) then yield that, for $-\ell\leqslant u,u'\leqslant l$,

$$L_k^\ell(u,u')-L^+(u,u')=-\frac{u\,u'}{2\ell}\sum_{j=-k}^{0}\frac{l\left(\frac{j+k/2}{\ell},u,u'\right)}{u\,u'}.$$

Now, as

$$\sup_{(x,u,u')\in\mathbb{R}^3}\left|\frac{l(x,u,u')}{u\,u'}\right|<+\infty,$$

we immediately obtain (6.44) and complete the proof of Lemma 6.9 when $k\neq 1$.

When $k=1$, the proof is done in the same way up to a shift in the index j. This completes the proof of Lemma 6.9. □

As $v\in\mathcal{C}_0^\infty(\mathbb{R}\times(0,1))$, one has

$$\forall N\geqslant 0,\quad \exists C_N>0,\quad \forall k\in\mathbb{Z},\quad \|c_k\|_{L^2(\mathbb{R})}\leqslant C_N\frac{1}{1+|k|^N}.$$

Thus, as $x\mapsto x\sqrt{U(x)}$ is square integrable, the bound (6.44) yields that, for some $C_2>0$, one has

$$\forall k\in\mathbb{Z},\quad \left\|c\left(K_k^\ell,\cdot\right)-c\left(L^+,\cdot\right)\right\|_{L^2([-\ell,\ell])}\leqslant\frac{1}{\ell}\frac{C_2}{1+|k|^2}.$$

Thus, taking into account the following computation

$$\begin{aligned}L^+(u,u')&=\int_{\mathbb{R}}\frac{\sin(\pi x u)\sin(\pi x u')}{\pi^2x^2}\mathrm{d}x\\&=\frac{1}{2\pi^2}\left[\int_{\mathbb{R}}\frac{\cos(\pi x(u-u'))-1}{x^2}\mathrm{d}x+\int_{\mathbb{R}}\frac{1-\cos(\pi x(u+u'))}{x^2}\mathrm{d}x\right]\\&=\frac{1}{2\pi^2}\left[|u-u'|\int_{\mathbb{R}}\frac{\cos(\pi x)-1}{x^2}\mathrm{d}x+|u+u'|\int_{\mathbb{R}}\frac{1-\cos(\pi x)}{x^2}\mathrm{d}x\right]\\&=\frac{1}{2}(|u+u'|-|u-u'|),\end{aligned}\tag{6.47}$$

the definition of K, (6.10) and (6.40), we obtain that

$$\left\|L_\ell^{+,+}v-(K\otimes\mathbf{1})v\right\|_{L^2(\mathbb{R}\times[0,1])}\xrightarrow[\ell\to+\infty]{}0$$

Thus, Lemma 6.7 is proved. □

Clearly, the proof of Lemma 6.1 generalizes to arbitrary $\phi_{(i,j)}$, a normalized eigenfunction of $H^0(1,2)$; one thus proves

Corollary 6.10 *Consider two particles on the i-th and j-th energy levels in an interval of length ℓ. Their interaction amplitude is given by*

$$\langle U\phi_{(i,j)}, \phi_{(i,j)}\rangle = 2\pi^2(i^2+j^2)\cdot \int_{\mathbb{R}} u^2 U(u)\mathrm{d}u \cdot \ell^{-3}(1+O(\ell^{-1})). \tag{6.48}$$

6.1.3 The Ground State of Two Interacting Electrons and Its Density Matrices

Recall that $\varphi^j_{[0,\ell]}$ denotes the j-th normalized eigenvector of $-\Delta^D_{|[0,\ell]}$ and $\zeta^j_{[0,\ell]}$ the j-th normalized eigenvector of (1.15). In the sequel, we drop the subscript $[0,\ell]$ as we always work on the interval $[0,\ell]$.

We remark that, when the interactions are absent, one has

$$\zeta^{1,0} = \varphi^1 \wedge \varphi^2. \tag{6.49}$$

The next proposition estimates the difference $\zeta^{1,U} - \zeta^{1,0}$ induced by the presence of interactions.

Proposition 6.11 *For $\ell \geqslant 1$, one has*

$$\left\| \zeta^{1,U} - \zeta^{1,0} \right\|_{L^2([0,\ell]^2)} \lesssim \ell^{-1/2}. \tag{6.50}$$

Proof Scaling the variables to the unit square (see Section 6.1.1), it suffices to show that the normalized ground state of $H^{U^\ell}(1,2)$ (see (6.1)), say, $\phi_0^{U^\ell}$ satisfies

$$\left\| \phi_0^{U^\ell} - \phi_0 \right\|_{L^2([0,1]^2)} \lesssim \ell^{-1/2}. \tag{6.51}$$

where we recall that $\phi_0 = \phi_{(1,2)}$ (see (6.2)).

Decomposing $L^2([0,1]) \wedge L^2([0,1]) = \mathbb{C}\phi_0 \overset{\perp}{\oplus} \phi_0^{\perp}$ and defining $E_0^{U^\ell}$ to be the ground state energy of $H^{U^\ell}(1,2)$, we rewrite $\phi_0^{U^\ell}$ as

$$\phi_0^{U^\ell} = \alpha\phi_0 + \widetilde{\phi}, \quad \widetilde{\phi} \perp \phi_0, \quad \alpha \in \mathbb{R}^+$$

and the eigenvalue equation it satisfies as

$$\begin{pmatrix} 5\pi^2 + U^\ell_{00} - E_0^{U^\ell} & U^\ell_{0+} \\ U^\ell_{+0} & H_+ + U^\ell_{++} - E_0^{U^\ell} \end{pmatrix} \begin{pmatrix} \alpha \\ \widetilde{\phi} \end{pmatrix} = 0. \tag{6.52}$$

where the terms in the matrix are defined in (6.4).

Thus, to prove (6.51) it suffices to prove that

$$\|\widetilde{\phi}\|_{L^2([0,1])\wedge L^2([0,1])} \leqslant C\ell^{-1/2}.$$

By (6.52), as $\phi_0^{U\ell}$ is normalized, as $10\pi^2 \leqslant H_+ + U_{++}^\ell$ and as $E_0^{U\ell} \xrightarrow[\ell\to+\infty]{} 5\pi^2$, using (6.4) and (6.8), one computes

$$\begin{aligned}\|\widetilde{\phi}\|^2_{L^2([0,1])\wedge L^2([0,1])} &\leqslant U_{0+}^\ell \left(H_+ + U_{++}^\ell - E_0^{U^\ell}\right)^{-2} U_{+0}^\ell \\ &\leqslant \frac{C}{\ell}\left\langle \phi_\ell, K_\ell(\mathrm{Id}+K_\ell)^{-1}\phi_\ell\right\rangle_{L^2(\mathbb{R}\times[0,1])}.\end{aligned}$$

Thus, (6.51) is an immediate consequence of Lemma 6.1. This completes the proof of Proposition 6.11. □

We obtain the following corollary for the one-particle density matrices of $\zeta^{1,U}$.

Corollary 6.12 *Under assumptions of Proposition* 6.11, *one has*

$$\left\|\gamma_{\zeta^{1,U}} - \gamma_{\varphi^1} - \gamma_{\varphi^2}\right\|_1 = O\left(\ell^{-1}\right).$$

Corollary 6.12 is an immediate consequence of (6.50) and

Lemma 6.13 *Let* $\psi, \phi \in L^2([0,\ell]) \wedge L^2([0,\ell])$ *be two normalized two-particle states. Then*

$$\|\gamma_\psi - \gamma_\phi\|_1 \leqslant 4\|\psi - \phi\|.$$

Proof of Lemma 6.13 For $\varphi \in L^2([0,\ell]) \wedge L^2([0,\ell])$, consider the operator A_φ defined as

$$(A_\varphi f)(x) = \int_0^\ell \varphi(x,y) f(y)\mathrm{d}y.$$

Note that A_φ is a Hilbert-Schmidt operator and $\|A_\varphi\|_2 = \|\varphi\|$ and the one-particle density matrix of φ satisfies $\gamma_\varphi = 2A_\varphi^* A_\varphi$. Thus, for ψ, ϕ as in Lemma 6.13, we obtain

$$\begin{aligned}&\|\gamma_\psi - \gamma_\phi\|_1 = 2\|A_\psi^* A_\psi - A_\phi^* A_\phi\|_1 \leqslant 2 \\ &\left(\|A_\psi^*\|_2 \|A_\psi - A_\phi\|_2 + \|A_\psi^* - A_\phi^*\|_2 \|A_\phi\|_2\right) \quad \leqslant 4\|\psi - \phi\|.\end{aligned}$$

This completes the proof of Lemma 6.13. □

6.2 Electrons in Distinct Pieces

In the present section, we assume that U satisfies **(HU)** (see Section 1.1); thus, it decreases sufficiently fast at infinity (roughly better than x^{-4}) and is in L^p for some $p > 1$.

Let the first piece be $\Delta_1 = [-\ell_1, 0]$ and the second be $\Delta_2 = [a, a + \ell_2]$; so, the pieces' lengths are ℓ_1 and ℓ_2, while the distance between them is denoted by a. As for the one-particle systems living in each of these pieces, we will primarily be interested in the following three cases:

(a) the interaction of two eigenstates of the one-particle Hamiltonian on each piece, i.e., following the notations of Section 6.1, of $\varphi^i_{\Delta_1}$ and $\varphi^j_{\Delta_2}$,
(b) the interaction of a one-particle eigenstate with a one-particle reduced density matrix of a two-particle ground state, i.e., $\varphi^i_{\Delta_1}$ with $\gamma_{\zeta^1_{\Delta_2}}$,
(c) the interaction of two one-particle density matrices, i.e., $\gamma_{\zeta^1_{\Delta_1}}$ and $\gamma_{\zeta^1_{\Delta_2}}$.

We observe that for a one-particle eigenstate in a piece of length ℓ, the following uniform pointwise bound holds true:

$$\|\varphi^i_{[0,\ell]}\|_{L^\infty} \leqslant \sqrt{\frac{2}{\ell}}. \tag{6.53}$$

For the one-particle reduced density matrix we establish the following estimates.

Lemma 6.14 *Let $\zeta \in L^2([0,\ell]) \wedge L^2([0,\ell])$ be a two-particle state and $\gamma_\zeta(x,y)$ the kernel of the corresponding one-particle density matrix. Let $p \in \mathbb{N}$. Then, $\zeta \in H^p([0,\ell]^2)$ implies $\gamma_\zeta \in H^p([0,\ell]^2)$ and*

$$\|\gamma_\zeta\|_{H^p} \leqslant 4\|\zeta\|_{H^p}. \tag{6.54}$$

In particular, unconditionally $\|\gamma_\zeta\|_{L^2} \leqslant 4$.

Proof First recall that

$$\gamma_\zeta(x,y) = 2\int_0^\ell \zeta(x,z)\zeta^*(y,z)\mathrm{d}z.$$

Then, one differentiates under the integration sign to get

$$\frac{\partial^p}{\partial x^p}\gamma_\zeta(x,y) = 2\int_0^\ell \partial_x^p\zeta(x,z)\zeta^*(y,z)\mathrm{d}z.$$

This in turn implies by the Cauchy-Schwarz inequality that

$$\left\| \frac{\partial^p}{\partial x^p} \gamma_\zeta \right\|_{L^2}^2 = 4 \int_{[0,\ell]^2} \left| \int_0^\ell \partial_x^p \zeta(x,z) \zeta^*(y,z) \mathrm{d}z \right|^2 \mathrm{d}x\mathrm{d}y$$

$$\leqslant 4 \int_{[0,\ell]^4} \left| \partial_x^p \zeta(x,z) \right|^2 \cdot \left| \zeta(y,z') \right|^2 \mathrm{d}x\mathrm{d}y\mathrm{d}z\mathrm{d}z' = 4 \left\| \partial_x^p \zeta \right\|_{L^2}^2,$$

which proves (6.54). □

Lemma 6.15 *Let $\zeta = \zeta_{[0,\ell]}^{1,U}$ be the ground state of a system of two interacting electrons in $[0,\ell]$. Then, $\zeta \in H^1([0,\ell]^2)$ and there exists a constant $C > 0$ independent of ℓ such that*

$$\|\zeta\|_{H^1} \leqslant C/\sqrt{\ell}. \tag{6.55}$$

Proof We use the construction of the proof of Proposition 6.11. Then, employing the same notations, for the problem scaled to the unit square one has

$$\phi_0^{U^\ell} = \alpha\phi_0 + \widetilde{\phi},$$

where ϕ_0 is the ground state for a system of two non-interacting electrons, $|\alpha| \leqslant 1$ and $\widetilde{\phi} \perp \phi_0$. Obviously, $\phi_0 \in H^p$ for all $p \in \mathbb{N}$. Moreover, according to (6.52),

$$\|\widetilde{\phi}\|_{H^1} = \left\| (H_+ + U_{++}^\ell - E_0^{U^\ell})^{-1} U_{+0}^\ell \alpha\phi_0 \right\|_{H^1}$$

$$\leqslant \left\| (H_+ + U_{++}^\ell - E_0^{U^\ell})^{-1} \right\|_{L^2 \to H^1} \cdot \left\| U_{+0}^\ell \phi_0 \right\|_{L^2}$$

$$\leqslant \left\| (H_+ - E_0^{U^\ell})^{-1} \right\|_{L^2 \to H^1} \cdot \left\| U_{+0}^\ell \phi_0 \right\|_{L^2}.$$

Arguing as in Section 6.1, one can prove that

$$\left\| U_{+0}^\ell \phi_0 \right\|_{L^2} \leqslant \left\| U^\ell \phi_0 \right\|_{L^2} \leqslant C\sqrt{\ell}$$

and $(H_+ - E_0^{U^\ell})^{-1}$ is a bounded operator from $L^2([0,1]^2)$ to $H^1([0,1]^2)$ because H_+ is just a part of $-\Delta_2$ acting in a subspace of functions orthogonal to ϕ_0 and the bottom of its spectrum is separated from $E_0^{U^\ell}$. Thus, we proved that

$$\|\widetilde{\phi}\|_{H^1} \leqslant C\sqrt{\ell}$$

which immediately implies

$$\|\phi_0^{U^\ell}\|_{H^1} \leqslant C\sqrt{\ell}.$$

Scaling back to the original domain $[0, \ell]^2$ yields (6.55) and completes the proof of Lemma 6.15. □

Corollary 6.16 *Restricted to the diagonal, the kernel of the ground state one-particle density matrix* $x \in [0, \ell] \mapsto \gamma_\zeta(x, x)$ *is a bounded function; more precisely, there exists a constant* $C > 0$ *such that*

$$\|\gamma_\zeta\|_{L^\infty([0,\ell])} \leqslant C/\ell. \tag{6.56}$$

Proof Remark first that, as ζ satisfies Dirichlet boundary conditions, so does the kernel $(x, y) \mapsto \gamma_\zeta(x, y)$. Using anti-symmetry, we compute

$$\begin{aligned} |\gamma_\zeta(x,x)| &= 2\left|\int_0^x \frac{d}{dt}\left[\gamma_\zeta(t,t)\right]\right| = 4\left|\mathrm{Im}\left(\int_0^x \int_0^\ell \partial_t\zeta(t,x)\overline{\zeta(t,x)}dxdt\right)\right| \\ &\leqslant 4\|\partial_t\zeta\|_{L^2}\cdot\|\zeta\|_{L^2} \leqslant 4\|\zeta\|_{H^1}^2 \end{aligned} \tag{6.57}$$

Combining this with (6.55) gives (6.56) and completes the proof of Corollary 6.16. □

Having now pointwise bounds (6.53) and (6.56), we estimate the interactions in each of the three cases described in the beginning of the current section. We will also obtain different bounds for close enough and distant pieces $\Delta_1 = [-\ell_1, 0]$ and $\Delta_2 = [a, a + \ell_2]$, i.e., we will discuss different bounds depending on whether a is large or small.

For the case (a) of two interacting one-particle eigenstates we prove the following two estimates. For long distance interactions, i.e., when a is large, we will use

Lemma 6.17 *Suppose* U *satisfies* **(HU)**. *Then, for* $\Delta_1 = [-\ell_1, 0]$ *and* $\Delta_2 = [a, a+\ell_2]$, *one has*

$$\sup_{i,j} \int_{\Delta_1\times\Delta_2} U(x-y)|\varphi^i_{\Delta_1}(x)|^2 \cdot |\varphi^j_{\Delta_2}(y)|^2 \mathrm{d}x\mathrm{d}y \leqslant \frac{2a^{-3}Z(a)}{\max(\ell_1,\ell_2)} \tag{6.58}$$

where Z *is defined in* (1.26).

Proof Let us suppose without loss of generality that Δ_1 is the larger piece, i.e., $\ell_1 \geqslant \ell_2$. Then, using (6.53) and the fact that the functions $(\varphi^i_{\Delta_j})_{i,j}$ are normalized, we compute

$$\begin{aligned} &\int_0^{\ell_1}\int_0^{\ell_2} U(x+y+a)|\varphi^i_{\Delta_1}(x)|^2 \cdot |\varphi^j_{\Delta_2}(y)|^2 \mathrm{d}x\mathrm{d}y \\ &\qquad \leqslant \frac{2}{\ell_1}\int_0^{\ell_1}\int_0^{\ell_2} U(x+y+a)|\varphi^j_{\Delta_2}(y)|^2 \mathrm{d}x\mathrm{d}y \end{aligned}$$

$$\leqslant \frac{2}{\ell_1} \sup_{y\in[0,\ell_2]} \int_0^{\ell_1} U(x+y+a)\mathrm{d}x$$

$$\leqslant \frac{2}{\ell_1} \int_0^{+\infty} U(x+a)\mathrm{d}x$$

$$= \frac{2}{\ell_1} a^{-3} Z(a), \quad a \to +\infty.$$

This completes the proof of Lemma 6.17. □

On the other hand, for close by interactions, i.e., a small and low-lying one-particle energy levels the following lemma gives a more precise estimate.

Lemma 6.18 *Suppose U satisfies* **(HU)**. *Let $(i, j) \in \{1, 2\}^2$. Then, for any $\varepsilon \in (0, 2)$ and $\Delta_1 = [-\ell_1, 0]$ and $\Delta_2 = [a, a + \ell_2]$, one has*

$$\int_{\Delta_1\times\Delta_2} U(x-y)|\varphi^i_{\Delta_1}(x)|^2 \cdot |\varphi^j_{\Delta_2}(y)|^2 \mathrm{d}x\mathrm{d}y = O\left(\frac{a^{-\varepsilon}Z(a)}{\max(\ell_1,\ell_2)^2 \min(\ell_1,\ell_2)^{2-\varepsilon}}\right). \tag{6.59}$$

If $Z(a) = O(a^{-0})$, $a \to +\infty$, then ε can be taken to zero.

Proof As in the proof of the previous lemma we suppose that $\ell_1 \geqslant \ell_2$. If $j \in \{1, 2\}$, then

$$|\varphi^j_{\Delta_1}(x)| = \left|\sqrt{\frac{2}{\ell_1}} \sin\left(\frac{\pi i x}{\ell_1}\right)\right| \leqslant \sqrt{\frac{2}{\ell_1}} \frac{\pi|x|}{\ell_1} \tag{6.60}$$

and the same type inequality holds for $\varphi^j_{\Delta_2}(y)$. Then, using (6.60) and (6.53), we compute

$$\int_0^{\ell_1}\int_0^{\ell_2} U(x+y+a)|\varphi^i_{\Delta_1}(x)|^2 \cdot |\varphi^j_{\Delta_2}(y)|^2 \mathrm{d}x\mathrm{d}y$$

$$\leqslant \frac{C_1}{\ell_1^2\ell_2^{2-\varepsilon}} \int_0^{\ell_1}\int_0^{\ell_2} U(x+y+a)xy^{1-\varepsilon}\mathrm{d}x\mathrm{d}y$$

$$\leqslant \frac{C_1}{\ell_1^2\ell_2^{2-\varepsilon}} \int_{\mathbb{R}^2_+} U(x+y+a)xy^{1-\varepsilon}\mathrm{d}x\mathrm{d}y$$

$$= \frac{C_2}{\ell_1^2\ell_2^{2-\varepsilon}} \int_0^{+\infty}\int_{-s}^{s} U(s+a)(s+t)(s-t)^{1-\varepsilon}\mathrm{d}t\mathrm{d}s$$

$$\leqslant \frac{C_3}{\ell_1^2\ell_2^{2-\varepsilon}} \int_a^{+\infty} U(s)s^{3-\varepsilon}\mathrm{d}s.$$

It is now only left to prove that **(HU)** and (1.26) imply that the last integral converges and is $O(a^{-\varepsilon}Z(a))$. Therefore, we note that

$$\begin{aligned}\int_a^{+\infty} U(s)s^{3-\varepsilon}\mathrm{d}s &= \sum_{n=0}^{+\infty}\int_{2^n a}^{2^{n+1}a} U(s)s^{3-\varepsilon}\mathrm{d}s \leqslant \sum_{n=0}^{+\infty}\left(2^{n+1}a\right)^{3-\varepsilon}\int_{2^n a}^{2^{n+1}a} U(s)\mathrm{d}s\\ &\leqslant 2^{3-\varepsilon}a^{-\varepsilon}\sum_{n=0}^{+\infty}2^{-\varepsilon n}\left(2^n a\right)^3\int_{2^n a}^{+\infty} U(s)\mathrm{d}s\\ &= 2^{3-\varepsilon}a^{-\varepsilon}\sum_{n=0}^{+\infty}2^{-\varepsilon n}Z\left(2^n a\right) \leqslant Ca^{-\varepsilon}Z(a).\end{aligned} \tag{6.61}$$

If $Z(a) = O(a^{-0})$, i.e., if there exists $\delta > 0$ s.t. $Z(a) = O(a^{-\delta})$ for $a \to +\infty$, then the sum in the second line of (6.61) converges for $\varepsilon = 0$.

This concludes the proof of (6.59). □

Let us now pass to the case (b) of one-particle eigenstate interacting with a one-particle density matrix of a two-particle eigenstate. For large a, we prove

Lemma 6.19 *Suppose U satisfies* **(HU)**. *Then, for a sufficiently large, one has*

$$\sup_{i,j}\int_{\Delta_1\times\Delta_2} U(x-y)|\varphi^i_{\Delta_1}(x)|^2\cdot\gamma_{\zeta^j_{\Delta_2}}(y,y)\mathrm{d}x\mathrm{d}y \leqslant \frac{4a^{-3}Z(a)}{\ell_1}. \tag{6.62}$$

Proof The proof follows that of Lemma 6.17. The only change concerns the replacement of the fact that $\varphi^j_{\Delta_2}$ is normalized, $\int_{\Delta_2}|\varphi^j_{\Delta_2}(y)|^2\mathrm{d}y = 1$, by the fact that the trace of $\gamma_{\zeta^j_{\Delta_2}}$ is equal to 2. □

For a small, we prove

Lemma 6.20 *Suppose U satisfies* **(HU)**. *Let $i \in \{1,2\}$. Then, for any $\varepsilon \in (0,2)$,*

$$\int_{\Delta_1\times\Delta_2} U(x-y)|\varphi^i_{\Delta_1}(x)|^2\cdot\gamma_{\zeta^j_{\Delta_2}}(y,y)\mathrm{d}x\mathrm{d}y = O\left(\ell_1^{-3+\varepsilon}\ell_2^{-1/2}a^{-\varepsilon}Z(a)\right). \tag{6.63}$$

If $Z(a) = O(a^{-0})$ as $a \to +\infty$, one can choose $\varepsilon = 0$.

Proof As in the proof of Lemma 6.18 mixing once more (6.53), (6.56), and (6.60), we obtain

$$\begin{aligned}&\int_0^{\ell_1}\int_0^{\ell_2} U(x+y+a)|\varphi^i_{\Delta_1}(x)|^2\gamma_{\zeta^j_{\Delta_2}}(y,y)\mathrm{d}x\mathrm{d}y\\ &\quad\leqslant \frac{C_1}{\ell_1^{3-\varepsilon}\ell_2^{1/2}}\int_0^{\ell_1}\int_0^{\ell_2} U(x+y+a)x^{2-\varepsilon}\mathrm{d}x\mathrm{d}y\end{aligned}$$

$$\leqslant \frac{C_1}{\ell_1^{3-\varepsilon}\ell_2^{1/2}} \int_0^{+\infty} \int_a^{+\infty} U(x+y)x^{2-\varepsilon} \mathrm{d}x\mathrm{d}y$$
$$= \frac{C_2}{\ell_1^{3-\varepsilon}\ell_2^{1/2}} \int_a^{+\infty} \int_{-s}^{s} U(s)(s+t)^{2-\varepsilon} \mathrm{d}t\mathrm{d}s$$
$$\leqslant \frac{C_3}{\ell_1^{3-\varepsilon}\ell_2^{1/2}} \int_a^{+\infty} U(s)s^{3-\varepsilon} \mathrm{d}s$$
$$\leqslant \frac{C_4 a^{-\varepsilon} Z(a)}{\ell_1^{3-\varepsilon}\ell_2^{1/2}}.$$

This completes the proof of Lemma 6.20. □

We are left with the case (c) of two interacting reduced density matrices. We do not make the difference between close and far away pieces in this case.

Lemma 6.21 *Suppose* U *satisfies* **(HU)**. *Then, there exists* $C > 0$ *such that*

$$\sup_{i,j} \int_{\Delta_1\times\Delta_2} U(x-y)\gamma_{\zeta^i_{\Delta_1}}(x,x)\cdot\gamma_{\zeta^j_{\Delta_2}}(y,y)\mathrm{d}x\mathrm{d}y \leqslant C\ell_1^{-1/2}\ell_2^{-1/2}\min(1, a^{-2}Z(a)) \tag{6.64}$$

Proof Using (6.56) one obtains

$$\int_0^{\ell_1}\int_0^{\ell_2} U(x+y+a)\gamma_{\zeta^i_{\Delta_1}}(x,x)\gamma_{\zeta^j_{\Delta_2}}(y,y)\mathrm{d}x\mathrm{d}y$$
$$\leqslant \frac{C_1}{\sqrt{\ell_1\ell_2}} \int_{\mathbb{R}^2_+} U(x+y+a)\mathrm{d}x\mathrm{d}y$$
$$\leqslant \frac{C_2}{\sqrt{\ell_1\ell_2}} \int_0^{+\infty} U(s+a)s\mathrm{d}s$$
$$\leqslant \frac{C_2}{\sqrt{\ell_1\ell_2}} \int_a^{+\infty} \left(\int_t^{+\infty} U(s)ds\right) dt$$

Thus,

$$\int_0^{\ell_1}\int_0^{\ell_2} U(x+y+a)\gamma_{\zeta^i_{\Delta_1}}(x,x)\gamma_{\zeta^j_{\Delta_2}}(y,y)\mathrm{d}x\mathrm{d}y \leqslant \frac{C_3\min(C, a^{-2}Z(a))}{\sqrt{\ell_1\ell_2}}$$

where the last equality is just (6.61) for $\varepsilon = 2$ and $C := \int_0^{+\infty}\left(\int_t^{+\infty} U(s)ds\right) dt < +\infty$. This completes the proof of Lemma 6.21. □

Finally, we give estimates for the case of compactly supported interaction potential U. We prove

Lemma 6.22 *Assume that U has a compact support. Then, there exists $C > 0$ such that, for $i \geqslant 1$ and $j \geqslant 1$, one has*

$$\langle U\phi_{(i,j)}, \phi_{(i,j)}\rangle \leqslant C \cdot \frac{[\min(i,\ell_1)\min(j,\ell_2)]^2}{\ell_1^3\ell_2^3}.$$

Proof Due to the anti-symmetry of the functions $(\phi_{(i,j)})_{i,j,}$, it suffices to compute the scalar product on $[-\ell_1, 0] \times [a, a+\ell_2]$. Thus,

$$\begin{aligned}\langle U\phi_{(i,j)}, \phi_{(i,j)}\rangle &\leqslant \sup_{|a|\leqslant \mathrm{diam}(\mathrm{supp}(U))} \frac{1}{2\ell_1\ell_2}\int_{[0,\ell_1]\times[0,\ell_2]} U(x+y+a) \\ &\qquad \times \sin^2\left(\frac{i\pi x}{\ell_1}\right)\sin^2\left(\frac{j\pi y}{\ell_2}\right)dxdy \\ &\leqslant C(U)\frac{[\min(i,\ell_1)\min(j,\ell_2)]^2}{\ell_1^3\ell_2^3}\end{aligned}$$

where

$$C(U) := \frac{1}{2}\sup_{0\leqslant a\leqslant \mathrm{diam}(\mathrm{supp}(U))}\int_{\mathbb{R}^+\times\mathbb{R}^+} U(x+y+a)(1+x^2)(1+y^2)dxdy.$$

This completes the proof of Lemma 6.22. □

Proposition 6.23 *Consider a system of two interacting electrons, one in $[0, \ell_1]$, another in $[\ell_1 + r, \ell_1 + r + \ell_2]$ with $r \leqslant R_0$. Then, the ground state energy of this system has the following asymptotic expansion*

$$E((\ell_1, r, \ell_2), (1,1)) = \frac{\pi^2}{\ell_1^2} + \frac{\pi^2}{\ell_2^2} + O(\ell_1^{-6} + \ell_2^{-6}). \tag{6.65}$$

Proof Obviously, the energy of this system is greater than the energy of the system without interactions that is given by the main term of (6.65). Taking the ground state of a non-interacting system as a test function and using Lemma 6.22 to estimate the quadratic form of the interaction potential gives the upper bound and, thus, completes the proof. □

6.3 The Proof of Lemma 4.11

Recall that $E^U_{q,n}$ denotes the n-th eigenvalue of $-\sum_{l=1}^{q}\frac{d^2}{dx_l^2} + \sum_{1\leqslant k<l\leqslant q} U(x_k - x_l)$ acting on $\bigwedge_{l=1}^{q} L^2([0,\ell])$. Rescaling as in Section 6.1.1, we need to study the case

$\ell = 1$ and prove that, in this case, there exists $C > 1$ such that, for $n \geqslant 2$ and U^ℓ given by (6.1), one has

$$E_{q,n}^{U^\ell} \geqslant E_{q,1}^{U^\ell} + \frac{1}{C}. \tag{6.66}$$

Indeed in Lemma 4.11, the length ℓ is assumed to be less than $3\ell_\rho$.

As $q \leqslant 3$, the same computations as in the beginning of Section 6.1.1 show that $E_{q,1}^{U^\ell}$ satisfies, for some $C > 1$, for ℓ large,

$$E_{q,1}^{U^\ell} \leqslant E_{q,1}^{0} + \langle \phi_0, U^\ell \phi_0 \rangle \leqslant E_{q,1}^{0} + \frac{C}{\ell}. \tag{6.67}$$

On the other hand, for some $C > 1$, one has

$$E_{q,n}^{U^\ell} \geqslant E_{q,n}^{0} \geqslant E_{q,1}^{0} + \frac{2}{C}.$$

Plugging (6.67) into this immediately yields (6.66) and completes the proof of Lemma 4.11. □

Appendix A The Statistics of the Pieces

In this appendix, we prove most of the results on the statistics of the pieces stated in Section 2.2.

A.1 Facts on the Poisson Process

Let Π be the support of $d\mu(\omega)$, the Poisson process of intensity 1 on $\mathbb{R}_+$ (see Section 1). Let $\Pi \cap [0, L] = \{x_i;\ 1 \leqslant i \leqslant m(\omega) - 1\}$ (where $x_i < x_{i+1}$). Then,

$$\mathbb{P}(\#(\Pi \cap [0, L]) = k) = e^{-L}\frac{L^k}{k!}, \quad k \in \mathbb{N}. \tag{A.1}$$

The following large deviation principle is well known (and easily checked): for any $\beta \in (1/2, 1)$, one has

$$\mathbb{P}(|\#(\Pi \cap [0, L]) - L| \geqslant L^\beta) = O(L^{-\infty}). \tag{A.2}$$

The points $(x_i)_{1 \leqslant i \leqslant m(\omega)-1}$ partition the interval $[0, L]$ in $m(\omega)$ pieces of lengths Δ_i.

For $L > e^{e^2}$, one has

$$\begin{aligned}\mathbb{P}(\exists i;\ |\Delta_i| \geqslant \log L \log\log L) &\leqslant \mathbb{P}(\exists n \in [0, L] \cap \mathbb{N};\\ &\#[\Pi \cap (n + [0, \log L \log\log L/2])] = 0)\\ &\leqslant Le^{-\log L \log\log L/2} = O(L^{-\infty}).\end{aligned}$$

This proves Proposition 2.1.

A.2 *The Proof of Proposition 2.2*

Consider the partition of $[0, L]$ into pieces (see Section 1). For a, b both non-negative, let now $X_{[0,L]}$ to be the number of pieces of length in $[a, a+b]$. We first compute the expectation of $X_{[0,L]}/L$, that is, prove

Proposition A.1 *For $L \geqslant a + b$, one has*

$$\mathbb{E}\left[\frac{X_{[0,L]}}{L}\right] = e^{-a}(1 - e^{-b}) + \frac{e^{-a}((a+b)e^{-b} - a)}{L} = e^{-a}\left(1 - \frac{a}{L}\right) - e^{-a-b}\left(1 - \frac{a+b}{L}\right).$$

Proof Let Π be the support of the support of $d\mu(\omega)$, the Poisson process of intensity 1 on $\mathbb{R}_+$ (see Section 1). Then, one has

$$X_{[0,L]} = \sum_{X \in \Pi} G(\Pi \cap [0, X))$$

where the set-function G is defined as

$$G(\Pi \cap [0, X)) = \begin{cases} 1 & \text{if the distance from } X \text{ to the right most point} \\ & \text{in } \{0\} \cup (\Pi \cap [0, X)) \text{ belongs to } [a, a+b], \\ 0 & \text{if not.} \end{cases} \tag{A.3}$$

The Palm formula (see, e.g., [2, Lemma 2.3]) yields

$$\mathbb{E}(X_{[0,L]}) = \int_{0 \leqslant x \leqslant L} \mathbb{E}\left[G(\Pi \cap [0, x))\right] dx.$$

Now, let $\mathcal{E}$ be an exponential random variable with parameter 1. As the Poisson point process has independent increments, one easily checks that

$$\mathbb{E}\left[G(\Pi \cap [0,x))\right] = \mathbb{P}\left(\min(x,\mathcal{E}) \in [a,a+b]\right) = \begin{cases} e^{-a}\left(1-e^{-b}\right) & \text{if } x \geqslant a+b, \\ e^{-a} & \text{if } x \in [a,a+b], \\ 0 & \text{if } x \leqslant a, \end{cases} \tag{A.4}$$

Hence,

$$\mathbb{E}(X_{[0,L]}) = e^{-a}\left(1-e^{-b}\right)\int_{0\leqslant x\leqslant L} dx + e^{-a-b}\int_a^{a+b} dx - e^{-a}\left(1-e^{-b}\right)$$
$$\int_0^a dx = e^{-a}(1-e^{-b})L - R$$

where

$$R = e^{-a}((a+b)e^{-b} - a). \tag{A.5}$$

This completes the proof of Proposition A.1. □

Let us now prove Proposition 2.2. Therefore, set $M := e^{-a}(1-e^{-b})$ and partition $[0, L] = \cup_{j=1}^{J}[j\ell, (j+1)\ell]$ so that $J \asymp L^{\nu}$ and $\ell \asymp L^{1-\nu}$ for some $\nu \in (0, 1)$ to be fixed. As $(a, b) = (a_L, b_L) \in [0, \log L \cdot \log\log L]^2$, one then has

$$\left| X_{[0,L]} - \sum_{j=1}^{J} X_{[j\ell,(j+1)\ell]} \right| \leqslant 2J. \tag{A.6}$$

Moreover, the random variables $(\ell^{-1}X_{[j\ell,(j+1)\ell]})_{1\leqslant j\leqslant J}$ are independent sub-exponential random variables. Indeed, $X_{[0,L]}$ is clearly bounded by $\#(\Pi \cap [0, L])$, the number of points the Poisson process puts in $[0, L]$ and $L^{-1}\#(\Pi \cap [0, L])$ has a Poisson law with parameter 1. We want to use the Bernstein inequality (see e.g. [22, Proposition 5.16]). To estimate $\left\|\ell^{-1}X_{[j\ell,(j+1)\ell]}\right\|_{\Psi_1}$ (see e.g. [22, Definition 5.13]), we use this bound and the Stirling formula to get, for $p \geqslant 1$,

$$\begin{aligned}
\mathbb{E}\left(\left|X_{[j\ell,(j+1)\ell]}\right|^p\right) &\leqslant e^{-\ell}\sum_{k\geqslant 1}\frac{k^p\,\ell^k}{k!} \leqslant e^{-\ell}\sum_{k=1}^{2p-1}\frac{k^p\,\ell^k}{k!} + e^{-\ell}\sum_{k\geqslant 2p}\frac{k^p\,\ell^k}{k!} \\
&\leqslant (2p)^p + e^{-\ell}\sum_{k\geqslant 2p}\frac{k^p\,\ell^p}{k\cdots(k-p+1)}\frac{\ell^{k-p}}{(k-p)!} \\
&\leqslant (2p)^p + \ell^p\max_{k\geqslant p}\frac{(k+p)^p\,k!}{(k+p)!} \leqslant (2p)^p + (e\ell)^p.
\end{aligned}$$

Hence, for $\ell \geqslant 1$,

$$\left\|\ell^{-1}X_{[j\ell,(j+1)\ell]}\right\|_{\Psi_1}=\frac{1}{\ell}\left\|X_{[j\ell,(j+1)\ell]}\right\|_{\Psi_1}=\frac{1}{\ell}\sup_{p\geqslant 1}\frac{1}{p}\sqrt[p]{\mathbb{E}\left(\left|X_{[j\ell,(j+1)\ell]}\right|^p\right)}$$
$$\leqslant\sup_{p\geqslant 1}\sqrt[p]{\frac{2^p}{\ell^p}+\frac{e^p}{p^p}}\leqslant\frac{2}{\ell}+e\leqslant 2e.$$

Thus, the Bernstein inequality, estimate (A.6), and Proposition A.2 yield that there exists $\kappa>0$ (independent of a,b) such that, for $\alpha=\alpha(L)\geqslant 2(R+2)/\ell$ (here, R is given by (A.5)), one has

$$\mathbb{P}\left(\left|\frac{X_{[0,L]}}{L}-M\right|\geqslant\alpha\right)\leqslant\mathbb{P}\left(\left|\sum_{j=1}^{J}\frac{X_{[j\ell,(j+1)\ell]}-\mathbb{E}[X_{[j\ell,(j+1)\ell]}]}{\ell}\right|\geqslant J\left(\alpha-\frac{R+2}{\ell}\right)\right)$$
$$\leqslant 2e^{-\kappa\alpha^2 J}.$$

To obtain Proposition 2.3, it now suffices to take $\alpha=L^{\beta-1}$ and $(\beta,\nu)\in(0,1)$ such that $1-\beta<1-\nu$ and $2(\beta-1)+\nu>0$; this requires $\beta>2/3$.

The proof of Proposition 2.2 is complete. □

A.3 The Proof of Propositions 2.3 and 2.4

For any a,b,c,d,f,g all positive, define now $X_{[0,L]}$ to be the number of pairs of pieces such that

- the length of the left most piece is contained in $[a,a+b]$,
- the length of the right most piece is contained in $[c,c+d]$,
- the distance between the two pieces belongs to $[g,g+f]$.

Again, we first compute the expectation of $X_{[0,L]}/L$, that is, prove

Proposition A.2 *For $L\geqslant a+b+c+d+f+g$, one has*

$$\mathbb{E}\left[\frac{X_{[0,L]}}{L}\right]=f\,e^{-a-c}(1-e^{-b})(1-e^{-d})+\frac{R_L}{L}\quad\text{where } |R_L|\leqslant fe^{-a-c}. \tag{A.7}$$

Proof Recall that Π denotes the support of $d\mu(\omega)$, the Poisson process of intensity 1 on $\mathbb{R}_+$. Then, one can rewrite

$$X_{[0,L]}=\sum_{\substack{(X,Y)\in\Pi^2\\ X<Y}}\mathbf{1}_{g\leqslant Y-X\leqslant g+f}\,G(\Pi\cap[0,X))\,H(\Pi\cap(Y,L])$$

where the set-functions G and H have been defined, respectively, by (A.3) and

$$H(\Pi \cap (Y, L]) = \begin{cases} 1 & \text{if the distance from } Y \text{ to the left most point} \\ & \text{in } \{L\} \cup (\Pi \cap (Y, L]) \text{ belongs to } [c, c+d], \\ 0 & \text{if not.} \end{cases} \tag{A.8}$$

The Palm formula, thus, yields

$$\begin{aligned} \mathbb{E}(X_{[0,L]}) &= \int_{\substack{0 \leqslant x, y \leqslant L \\ g \leqslant y - x \leqslant g + f}} \mathbb{E}\left[G(\Pi \cap [0, x)) H(\Pi \cap (y, L])\right] dx dy \\ &= \int_{\substack{0 \leqslant x, y \leqslant L \\ g \leqslant y - x \leqslant g + f}} \mathbb{E}\left[G(\Pi \cap [0, x))\right] \mathbb{E}\left[H(\Pi \cap (y, L])\right] dx dy \end{aligned}$$

as the random sets $\Pi \cap [0, x))$ and $\Pi \cap (y, L])$ are independent.

As in (A.4), one checks that

$$\begin{aligned} \mathbb{E}\left[H(\Pi \cap (y, L])\right] &= \mathbb{P}\left(\min(L - y, \mathcal{E}) \in [c, c+d]\right) \\ &= \begin{cases} e^{-c}\left(1 - e^{-d}\right) & \text{if } y \leqslant L - c - d, \\ e^{-c} & \text{if } y \in L - [c, c+d], \\ 0 & \text{if } y \geqslant L - c. \end{cases} \end{aligned}$$

Hence,

$$\begin{aligned} \mathbb{E}(X_{[0,L]}) &= e^{-a-c}\left(1 - e^{-d}\right)\left(1 - e^{-b}\right) \int_{\substack{0 \leqslant x, y \leqslant L \\ g \leqslant y - x \leqslant g + f}} dx dy + R_1 \\ &= f\, e^{-a-c}(1 - e^{-b})(1 - e^{-d}) L + R_2 \end{aligned}$$

where, respectively, $R_1 \leqslant e^{-a-c}$ and

$$R_2 \leqslant R := e^{-a-c}(1 + f^2 + fg). \tag{A.9}$$

This completes the proof of Proposition A.2. □

Let us now prove Proposition 2.3. We want to go along the same lines as in the proof of Proposition 2.2. Therefore, we set $M := f\, e^{-a-c}(1 - e^{-b})(1 - e^{-d})$ and partition $[0, L] = \cup_{j=0}^{J} [j\ell, (j+1)\ell]$ so that $J \asymp L^{\nu}$ and $\ell \asymp L^{1-\nu}$ for some $\nu \in (0, 1)$ to be fixed. For the same reasons as before, the random variables $(\ell^{-1} X_{[j\ell,(j+1)\ell]})_{1 \leqslant j \leqslant J}$ are independent sub-exponential random variables.

We now need a replacement for (A.6). Therefore, we set

$$r := 1 + a + b + c + d + f + g \tag{A.10}$$

and, for $0 \leqslant j \leqslant J$, we let

- Y_j be the number of pieces in the interval $(j+1)\ell+[-r, 0]$ of length in $[a, a+b]$,
- Z_j be the number of pieces in the interval $j\ell + [0, r]$ of length in $[c, c + d]$.

Then, we have

$$- K_a \sum_{j=0}^{J} Y_j - K_c \sum_{j=0}^{J} Z_j \leqslant X_{[0,L]} - \sum_{j=0}^{J} X_{[j\ell,(j+1)\ell]} \leqslant K_a \sum_{j=0}^{J} Y_j + K_c \sum_{j=0}^{J} Z_j \tag{A.11}$$

where we have set

$$K_a := 1 + \frac{f+g}{a} \quad \text{and} \quad K_c = 1 + \frac{f+g}{c}. \tag{A.12}$$

Indeed, if a pair of pieces counted by $X_{[0,L]}$ does not have any of its intervals in any of the $(j\ell + [-r, r])_{1\leqslant j\leqslant J}$, then the convex closure of the pair is inside some $j\ell + [0, \ell]$, thus, the pair is counted by $X_{[j\ell,(j+1)\ell]}$. This yields the upper bound in (A.11) as, any given interval is the left (resp. right) most interval for at most $1 + (f + g)/c$ (resp. $1 + (f + g)/a$) pairs satisfying both the requirements on lengths and distance. The lower bound is obtained in the same way.

For L sufficiently large, the random variables $(Y_j)_{1\leqslant j\leqslant J}$ and $(Z_j)_{1\leqslant j\leqslant J}$ are i.i.d. sub-exponential. Thus, applying the Bernstein inequality as in the proof of Proposition 2.2 yields that, for some constant $\kappa > 0$ (independent of (a, b, c, d, f, g)) and $\beta \in (2/3, 1)$, with probability $1 - O(J^{-\infty}) = 1 - O(L^{-\infty})$, one has

$$\sum_{j=1}^{J} Y_j \leqslant \kappa J(e^{-a} + J^{\beta-1})r \quad \text{and} \quad \sum_{j=0}^{J-1} Z_j \leqslant \kappa J(e^{-c} + J^{\beta-1})r; \tag{A.13}$$

Now, we can estimate $\left\|\ell^{-1} X_{[j\ell,(j+1)\ell]}\right\|_{\Psi_1}$ as in the proof of Proposition 2.2. Thus, the Bernstein inequality and Proposition A.2 yield that, for some κ (independent of (a, b, c, d, f, g)), for $\nu \in (2/3, 1)$ and $\ell \asymp L^{1-\nu}$, with probability $1 - O(L^{-\infty})$, one has

$$\left|\sum_{j=0}^{J} \frac{X_{[j\ell,(j+1)\ell]}}{\ell} - J\,M\right| \leqslant \kappa \frac{R_L\, J}{\ell}.$$

Taking (A.11) and (A.13) into account, we get that, for some $\kappa > 0$ (independent of (a, b, c, d, f, g)), with probability $1 - O(L^{-\infty})$, one has

$$\left|\frac{X_{[0,L]}}{L} - M\right| \leqslant \kappa \frac{R + (K_a e^{-a} + K_c e^{-c} + (K_a + K_c)J^{\beta-1})r}{\ell}.$$

This proves (2.5) where the constants are given by

$$R(a,b,c,d,f,g) = \kappa r \left(R + K_a e^{-a} + K_c e^{-c}\right) \quad \text{and} \quad K(a,c,f,g) = (K_a + K_c) r \tag{A.14}$$

(see (A.9), (A.10), and (A.12).)

The proof of Proposition 2.3 is complete. □

The proof of Proposition 2.4 is identical to that of Proposition 2.3: it suffices to take $b = d = +\infty$.

A.4 The Proof of Proposition 2.5

This proof is essentially identical to that of Proposition 2.3. Let us just say a word about the differences.

For $\ell, \ell', \ell'', d > 0$, let now $X_{[0,L]}$ to be the number of triplets of pieces at most at a distance d from each other such that

- the left most piece is longer than ℓ,
- the middle piece is longer than ℓ',
- the right most piece is longer than ℓ''.

Then, one has

$$X_{[0,L]} = \sum_{\substack{(X,Y,W,Z)\in\Pi^4 \\ X<Y<W<Z}} \mathbf{1}_{\substack{0<Y-X\leqslant d \\ l'\leqslant W-Y \\ 0<Z-W\leqslant d}} G(\Pi\cap[0,X))\, K(\Pi\cap(Y,W))\, H(\Pi\cap(Z,L])$$

where the set-functions G and H have been defined as

$$G(\Pi\cap[0,X)) = \begin{cases} 1 & \text{if the distance from } X \text{ to the right most point} \\ & \text{in } \{0\}\cup(\Pi\cap[0,X)) \text{ belongs to } [l,+\infty), \\ 0 & \text{if not,} \end{cases}$$

$$K(\Pi\cap(Y,W)) = \begin{cases} 1 & \text{if } \Pi\cap(Y,W)=\emptyset, \\ 0 & \text{if not,} \end{cases}$$

$$H(\Pi\cap(Z,L]) = \begin{cases} 1 & \text{if the distance from } Z \text{ to the left most point} \\ & \text{in } \{L\}\cup(\Pi\cap(Z,L]) \text{ belongs to } [l'',+\infty), \\ 0 & \text{if not.} \end{cases}$$

Following the proof of Proposition A.2, one proves

Proposition A.3 *For L sufficiently large, one has*

$$\mathbb{E}\left[\frac{X_{[0,L]}}{L}\right] = d^2 e^{-\ell-\ell'-\ell''} + \frac{R_L}{L} \quad \textit{where } |R_L| \leqslant d^2 e^{-\ell-\ell'-\ell''}.$$

One then derives Proposition 2.2 from Proposition A.3 in the same way as Proposition 2.3 was derived from Proposition A.2.

A.5 *The Proof of Proposition 2.6*

First of all, let us note that a piece of length l in $[k\ell_E, (k+1)\ell_E)$ generates exactly k energy levels that do not exceed E. To count the energies less than E, we are only interested in intervals of length l larger than ℓ_E. Other intervals do not generate any energy levels we are interested in. Thus, by Proposition 2.2, for $\beta \in (2/3, 1)$, we obtain that with probability $1 - O(L^{-\infty})$, the number of intervals generating k energy levels below energy E is

$$L(e^{-k\ell_E} - e^{-(k+1)\ell_E}) + L^\beta R_L \quad \text{where } |R_L| \leqslant 3 \tag{A.15}$$

where $O(\cdot)$ is uniform in k.

Let $m_L = \log L \cdot \log\log L$. By Proposition 2.1, with probability $1 - O(L^{-\infty})$, for L large, one computes

$$\begin{aligned}
N_L^D(E) &= L^{-1} \sum_{k=1}^{[m_L/\ell_E]} k \cdot L(e^{-k\ell_E} - e^{-(k+1)\ell_E}) + m_L L^{-1+\beta} R_L \quad \text{where } |R_L| \leqslant \frac{1}{\ell_E} \\
&= \sum_{k=1}^{[m_L/\ell_E]} e^{-k\ell_E} - \frac{[m_L/\ell_E]}{e^{([m_L/\ell_E]+1)\ell_E}} + m_L L^{-1+\beta} R_L \\
&= \sum_{k=1}^{+\infty} e^{-k\ell_E} + m_L L^{-1+\beta}(R_L + 1) = \frac{e^{-\ell_E}}{1 - e^{-\ell_E}} + m_L L^{-1+\beta}(R_L + 2).
\end{aligned}$$

Thus, decreasing β above somewhat, with probability $1 - O(L^{-\infty})$, for L sufficiently large, one has

$$\left| N_L^D(E) - \frac{e^{-\ell_E}}{1 - e^{-\ell_E}} \right| \leqslant L^{-1+\beta}. \tag{A.16}$$

This proves (2.6). Using the fact that $E \mapsto N_L^D(E)$ is monotonous and the Lipschitz continuity of $E \mapsto N(E)$, (A.16) yields that, for $E_0 > 0$, with probability $1 - O(L^{-\infty})$, for L sufficiently large, one has

$$\sup_{E\in[0,E_0]} \left| N_L^D(E) - \frac{e^{-\ell_E}}{1-e^{-\ell_E}} \right| \leqslant L^{-1+\beta}. \tag{A.17}$$

The formulas (2.8) and (2.9) for the Fermi energy and the Fermi length follow trivially. This completes the proof of Proposition 2.6.

Appendix B A Simple Lemma on Trace Class Operators

The purpose of the present section is to prove

Lemma B.1 *Pick $(\mathcal{H}, \langle\cdot,\cdot\rangle)$ a separable Hilbert space and (Z, μ) a measured space with μ a positive measure. Consider a weakly measurable mapping $z \in Z \to T(z) \in \mathfrak{S}_1(\mathcal{H})$. Here, $\mathfrak{S}_1(\mathcal{H})$ denotes the trace class operators in $\mathcal{H}$, the trace class norm being denoted by $\|\cdot\|_{tr}$.*

Assume

$$\int_Z \|T(z)\|_{tr} d\mu(z) < +\infty. \tag{B.1}$$

Then, the integral $T := \int_Z T(z)d\mu(z)$ converges weakly and defines a trace class operator that satisfies

$$\|T\|_{tr} = \left\| \int_Z T(z)d\mu(z) \right\|_{tr} \leqslant \int_Z \|T(z)\|_{tr} d\mu(z). \tag{B.2}$$

Proof By assumption, for $(\varphi, \psi) \in \mathcal{H}^2$, one has $z \to \langle T(z)\varphi, \psi\rangle$ is measurable and bounded by $z \to \|T(z)\|_{\mathrm{tr}}\|\varphi\|\|\psi\|$ which by (B.1) is integrable. It, thus, is integrable and one has

$$\left| \int_Z \langle T(z)\varphi, \psi\rangle d\mu(z) \right| \leqslant \int_Z |\langle T(z)\varphi, \psi\rangle| \, d\mu(z) \leqslant \int_Z \|T(z)\|_{\mathrm{tr}} d\mu(z) \, \|\varphi\| \, \|\psi\|.$$

Thus, the operator $T := \int_Z T(z)d\mu(z)$ is well defined by

$$\langle T\varphi, \psi\rangle := \int_Z \langle T(z)\varphi, \psi\rangle d\mu(z).$$

and bounded.

Let us prove that it is trace class and satisfies (B.2). Let $(\varphi_n)_{n\geqslant 1}$ be an orthonormal basis of $\mathcal{H}$. Then,

$$|\langle T\varphi_n, \varphi_n\rangle| \leqslant \int_Z |\langle T(z)\varphi_n, \varphi_n\rangle| \, d\mu(z).$$

Thus,

$$\sum_{n=1}^{N} |\langle T\varphi_n, \varphi_n\rangle| \leqslant \int_Z \left(\sum_{n=1}^{N} |\langle T(z)\varphi_n, \varphi_n\rangle|\right) d\mu(z) \leqslant \int_Z \|T(z)\|_{\mathrm{tr}} d\mu(z).$$

Taking $N \to +\infty$ proves that, for any orthonormal basis of $\mathcal{H}$, say, $(\varphi_n)_{n\geqslant 1}$, one has

$$\sum_{n=1}^{+\infty} |\langle T\varphi_n, \varphi_n\rangle| \leqslant \int_Z \|T(z)\|_{\mathrm{tr}} d\mu(z) < +\infty.$$

Thus, T is trace class (see, e.g., [18]) and satisfies (B.2). This completes the proof of Lemma B.1. □

Appendix C Anti-symmetric Tensors: The Projector on Anti-symmetric Functions

Pick $\Psi \in L^2(\Lambda^n)$ and let $\Pi_n^\wedge : L^2(\Lambda^n) \to \bigwedge^n L^2(\Lambda)$ be the orthogonal projector on totally anti-symmetric functions. Then,

$$(\Pi_n^\wedge \Psi)(x) = \frac{1}{n!} \sum_{\substack{\sigma \text{ permutation} \\ \text{of } \{1,\cdots,n\}}} \operatorname{sgn} \sigma \cdot \Psi(\sigma x)$$

where, for $x = (x_1, \cdots, x_n)$, $\sigma x = (x_{\sigma(1)}, \cdots, x_{\sigma(n)})$ and $\operatorname{sgn} \sigma$ is the signature of the permutation σ.

Hence, if $n = Q_1 + \cdots + Q_m$ and, for $1 \leqslant j \leqslant m$, $\varphi_j \in \bigwedge^{Q_j} L^2(\Delta_j)$, we define

$$\left(\prod_{j=1}^{m} \left\|\varphi^j\right\|\right)^{-1} \bigwedge_{j=1}^{m} \varphi^j := \left\|\Pi_n^\wedge \left(\bigotimes_{j=1}^{m} \varphi^j\right)\right\|^{-1} \Pi_n^\wedge \left(\bigotimes_{j=1}^{m} \varphi^j\right) \tag{C.1}$$

and compute

$$\Pi_n^{\wedge}\left(\bigotimes_{j=1}^{m}\varphi^j\right) = \frac{1}{n!}\sum_{\substack{\sigma \text{ permutation} \\ \text{of } \{1,\cdots,n\}}} \operatorname{sgn}\sigma \left(\bigotimes_{j=1}^{m}\varphi^j\right)(\sigma x)$$

$$= \frac{1}{n!}\sum_{\substack{\sigma \text{ permutation} \\ \text{of } \{1,\cdots,n\}}} \operatorname{sgn}\sigma \left(\prod_{j=1}^{m}\varphi^j(x_{\sigma(\mathcal{Q}_j)})\right)$$

where

$$x_{\sigma(\mathcal{Q}_j)} = (x_{\sigma(Q_1+\cdots+Q_{j-1}+1)}, \cdots, x_{\sigma(Q_1+\cdots+Q_{j-1}+Q_j)}),$$

$$\mathcal{Q}_j = \{Q_1 + \cdots + Q_{j-1} + 1, \cdots, Q_1 + \cdots + Q_{j-1} + Q_j\}.$$

Thus,

$$n!\cdot\Pi_n^{\wedge}\left(\bigotimes_{j=1}^{m}\varphi^j\right) = \sum_{\substack{|A_j|=Q_j,\ \forall 1\leqslant j\leqslant m \\ A_1\cup\cdots\cup A_m=\{1,\cdots,n\} \\ A_j\cap A_{j'}=\emptyset \text{ if } j\neq j'}} \sum_{\substack{\sigma \text{ permutation} \\ \text{of } \{1,\cdots,n\} \\ \text{s.t. } \forall j,\ \sigma(\mathcal{Q}_j)=A_j}} \operatorname{sgn}\sigma \left(\prod_{j=1}^{m}\varphi^j(x_{\sigma(\mathcal{Q}_j)})\right)$$

$$= \sum_{\substack{|A_j|=Q_j,\ \forall 1\leqslant j\leqslant m \\ A_1\cup\cdots\cup A_m=\{1,\cdots,n\} \\ A_j\cap A_{j'}=\emptyset \text{ if } j\neq j'}} \left(\sum_{\substack{\sigma \text{ permutation} \\ \text{of } \{1,\cdots,n\} \\ \text{s.t. } \forall j,\ \sigma(\mathcal{Q}_j)=A_j}} \left(\operatorname{sgn}\sigma \prod_{j=1}^{m}\operatorname{sgn}\sigma_{|\mathcal{Q}_j}\right)\left(\prod_{j=1}^{m}\varphi^j(x_{A_j})\right)\right)$$

$$= \prod_{j=1}^{m} Q_j! \sum_{\substack{|A_j|=Q_j,\ \forall 1\leqslant j\leqslant m \\ A_1\cup\cdots\cup A_m=\{1,\cdots,n\} \\ A_j\cap A_{j'}=\emptyset \text{ if } j\neq j'}} \varepsilon(A_1,\cdots,A_m)\left(\prod_{j=1}^{m}\varphi^j(x_{A_j})\right)$$

where we recall that $\varepsilon(A_1, \cdots, A_m)$ is the signature of $\sigma(A_1, \cdots, A_m)$, the unique permutation of $\{1, \cdots, n\}$ such that, if $A_j = \{a_{ij};\ 1 \leqslant i \leqslant Q_j,\ a_{i_1 j} < a_{i_2 j} \text{ for } i_1 < i_2\}$ for $1 \leqslant j \leqslant m$ then $\sigma(a_{ij}) = Q_1 + \cdots + Q_{j-1} + i$.

As $\Delta_j \cap \Delta_k = \emptyset$ if $j \neq k$, one has

$$\left\| \sum_{\substack{|A_j|=Q_j,\ \forall 1\leqslant j\leqslant m \\ A_1\cup\cdots\cup A_m=\{1,\cdots,n\} \\ A_j\cap A_{j'}=\emptyset \text{ if } j\neq j'}} \varepsilon(A_1,\cdots,A_m)\left(\prod_{j=1}^{m}\varphi^j(x_{A_j})\right)\right\|^2 = \prod_{j=1}^{m}\left\|\varphi^j\right\|^2 \sum_{\substack{|A_j|=Q_j,\ \forall 1\leqslant j\leqslant m \\ A_1\cup\cdots\cup A_m=\{1,\cdots,n\} \\ A_j\cap A_{j'}=\emptyset \text{ if } j\neq j'}} 1$$

$$= \frac{n!}{\prod_{j=1}^{m} Q_j!}\prod_{j=1}^{m}\left\|\varphi^j\right\|^2.$$

Hence, by (C.1), we get

$$\left(\bigwedge_{j=1}^{m}\varphi^j\right)(x)=\sqrt{\frac{\prod_{j=1}^{m}Q_j!}{n!}}\sum_{\substack{|A_j|=Q_j,\ \forall 1\leqslant j\leqslant m\\ A_1\cup\cdots\cup A_m=\{1,\cdots,n\}\\ A_j\cap A_{j'}=\emptyset \text{ if } j\neq j'}}\varepsilon(A_1,\cdots,A_m)\left(\prod_{j=1}^{m}\varphi^j(x_{A_j})\right). \tag{C.2}$$

Appendix D The Proofs of the Particle Density Matrix Reduction Theorems

We shall now prove Theorem 4.2 and Theorem 4.4. They will follow from direct computations.

D.1 Proof of the One-Particle Density Matrix Reduction, Theorem 4.2

First, by the bilinearity of formula (1.19), one has

$$\gamma_\Psi^{(1)}=n\sum_{\substack{Q \text{ occ.}\\ \overline{n}\in\mathbb{N}^m}}\sum_{\substack{Q' \text{ occ.}\\ \overline{n}'\in\mathbb{N}^m}}a_{\overline{n}}^{Q}\overline{a_{\overline{n}'}^{Q'}}\,\gamma^{(1)}_{\substack{Q,\overline{n}\\ Q',\overline{n}'}} \tag{D.1}$$

where the trace class operator $\gamma^{(1)}_{Q,n,Q',n'}$ acts on $L^2([0,L])$ and has the kernel

$$\gamma^{(1)}_{\substack{Q,\overline{n}\\ Q',\overline{n}'}}(x,y):=\int_{[0,L]^{n-1}}\left[\bigwedge_{j=1}^{m}\varphi^j_{Q_j,n_j}\right](x,z)\overline{\left[\bigwedge_{j=1}^{m}\varphi^j_{Q'_j,n'_j}\right](y,z)}dz.$$

Recall (C.2), that is, in the present case

$$\left[\bigwedge_{j=1}^{m}\varphi^j_{Q_j,n_j}\right](z_1,z_2,\cdots,z_n)$$
$$=c(Q)\cdot\sum_{\substack{|A_j|=Q_j,\ \forall 1\leqslant j\leqslant m\\ A_1\cup\cdots\cup A_m=\{1,\cdots,n\}\\ A_j\cap A_{j'}=\emptyset \text{ if } j\neq j'}}\varepsilon(A_1,\cdots,A_m)\prod_{j=1}^{m}\varphi^j_{Q_j,n_j}((z_l)_{l\in A_j}) \tag{D.2}$$

where

- $\varepsilon(A_1, \cdots, A_m)$ is the signature of $\sigma(A_1, \cdots, A_m)$, the unique permutation of $\{1, \cdots, n\}$ such that, if $A_j = \{a_{ij};\ 1 \leqslant i \leqslant Q_j\}$ for $1 \leqslant j \leqslant m$ then $\sigma(a_{ij}) = Q_1 + \cdots + Q_{j-1} + i$,
- and $c(Q)$ is such that $\| \wedge_j \varphi^j_{Q_j,n_j} \| = 1$, i.e.

$$c(Q) = \sqrt{\frac{\prod_{j=1}^m Q_j!}{n!}}. \tag{D.3}$$

Thus, by (1.19), one has

$$\frac{\gamma^{(1)}_{\substack{Q,\bar{n}\\ Q',\bar{n}'}}(x, y)}{c(Q)c(Q')} = \sum_{\substack{|A_j|=Q_j,\ \forall 1\leqslant j\leqslant m\\ A_1\cup\cdots\cup A_m=\{1,\cdots,n\}\\ A_j\cap A_{j'}=\emptyset \text{ if } j\neq j'}} \sum_{\substack{|A'_j|=Q'_j,\ \forall 1\leqslant j\leqslant m\\ A'_1\cup\cdots\cup A'_m=\{1,\cdots,n\}\\ A'_j\cap A'_{j'}=\emptyset \text{ if } j\neq j'}} (-1)^{\varepsilon((A_j))+\varepsilon((A'_j))} I((A_j)_j, (A'_j)_j) \tag{D.4}$$

where

$$I(A, A') := I((A_j)_j, (A'_j)_j)$$
$$= \int_{[0,L]^{n-1}} \left[\prod_{j=1}^m \varphi^j_{Q_j,n_j}((x_l)_{l\in A_j}) \overline{\varphi^j_{Q'_j,n'_j}((y_l)_{l\in A'_j})} \right]_{\substack{x_1=x\\ y_1=y\\ x_j=y_j \text{ if } j\geqslant 2}} dx_2 \cdots dx_n. \tag{D.5}$$

To evaluate this last integral, we note that, for any pair of partitions $(A_j)_j$ and $(A'_j)_j$ (as in the indices of the sum in (D.4)), if there exists $j \neq j'$ such that $A_j \cap A'_{j'} \cap \{2, \cdots, n\} \neq \emptyset$, then the integral $I(A, A')$ vanishes.

Now, note that, if $d_1(Q, Q') > 2$, then, for any pair of partitions $(A_j)_j$ and $(A'_j)_j$, there exists $j \neq j'$ such that $A_j \cap A'_{j'} \cap \{2, \cdots, n\} \neq \emptyset$; thus, the integral $I(A, A')$ above always vanishes and, summing this, one has

$$\gamma^{(1)}_{\substack{Q,\bar{n}\\ Q',\bar{n}'}} = 0 \quad \text{if} \quad d_1(Q, Q') > 2.$$

So we are left with the case $Q = Q'$ or $d_1(Q, Q') = 2$.

Assume first $Q = Q'$. Consider the sums in (D.4). If $1 \in A_{j_0}$ and $1 \notin A'_{j_0}$, then, as $\forall j$, $|A'_j| = |A_j|$, there exists $\alpha \in A'_{j_0} = A'_{j_0} \cap \{2, \cdots, n\}$ and $j \neq j_0$ such that $\alpha \in A_j$. That is, there exists $j \neq j'$ such that $A_j \cap A'_{j'} \cap \{2, \cdots, n\} \neq \emptyset$, thus, the integral $I(A, A')$ vanishes. Thus, we rewrite

$$\frac{\gamma^{(1)}_{\substack{Q,\overline{n}\\Q,\overline{n}'}}(x,y)}{c^2(Q)} = \sum_{\substack{j_0=1\\Q_{j_0}\geqslant 1}}^{m} \sum_{\substack{1\in A_{j_0}\\|A_j|=Q_j,\ \forall 1\leqslant j\leqslant m\\A_1\cup\cdots\cup A_m=\{1,\cdots,n\}\\A_j\cap A_{j'}=\emptyset \text{ if } j\neq j'}} \sum_{\substack{1\in A'_{j_0}\\|A'_{j_0}|=Q_{j_0}\\A'_j=A_j \text{ if } j\neq j_0}} (-1)^{\varepsilon((A_j))+\varepsilon((A'_j))} I(A,A')$$

$$= \sum_{\substack{j_0=1\\Q_{j_0}\geqslant 1}}^{m} \sum_{\substack{1\in A_{j_0}\\|A_j|=Q_j,\ \forall 1\leqslant j\leqslant m\\A_1\cup\cdots\cup A_m=\{1,\cdots,n\}\\A_j\cap A_{j'}=\emptyset \text{ if } j\neq j'}} I(A) \tag{D.6}$$

where, using the support and orthonormality properties of the functions $(\varphi^j_{q,n})_{1\leqslant n}$, one computes

$$I(A) := \left(\int_{\Delta_{j_0}^{Q_{j_0}-1}} \varphi^{j_0}_{Q_{j_0},n_{j_0}}(x,z)\,\overline{\varphi^{j_0}_{Q_{j_0},n'_{j_0}}(y,z)}\,dz\right) \prod_{\substack{j=1\\j\neq j_0}}^{m} \int_{\Delta_j^{Q_j}} \varphi^j_{Q_j,n_j}(z)\overline{\varphi^j_{Q_j,n'_j}(z)}\,dz$$

$$= \prod_{j\neq j_0} \delta_{n_j=n'_j} \cdot \left(\int_{\Delta_{j_0}^{Q_{j_0}-1}} \varphi^{j_0}_{Q_{j_0},n_{j_0}}(x,z)\overline{\varphi^{j_0}_{Q_{j_0},n'_{j_0}}(y,z)}\,dz\right).$$

As

$$\#\{(A_j)_j;\ 1\in A_{j_0},\ \forall j,\ |A_j|=Q_j\} = \frac{(n-1)!Q_{j_0}}{\prod_{j=1}^m Q_j!}$$

by (D.3) and (D.6), one computes

$$\gamma^{(1)}_{\substack{Q,\overline{n}\\Q,\overline{n}'}}(x,y) = \sum_{\substack{j=1\\Q_j\geqslant 1}}^{m} \frac{Q_j}{n} \int_{\Delta_j^{Q_j-1}} \varphi^j_{Q_j,n_j}(x,z)\,\overline{\varphi^j_{Q_j,n'_j}(y,z)}\,dz \prod_{j\neq j_0} \delta_{n_j=n'_j}$$

$$= \frac{1}{n}\sum_{\substack{j=1\\Q_j\geqslant 1}}^{m} \gamma^{(1)}_{\substack{Q_j\\n_j,n'_j}}(x,y).$$

We now assume that $d_1(Q,Q')=2$. Thus, there exist $1\leqslant i_0\neq j_0\leqslant m$ such that $Q_{j_0}\geqslant 1$, $Q'_{i_0}=Q_{i_0}+1$, $Q_{j_0}=Q'_{j_0}+1$ and $Q_k=Q'_k$ for $k\notin\{i_0,j_0\}$. Consider the sums in (D.4). If $1\notin A_{j_0}$ (or $1\notin A'_{i_0}$), then as $|A'_{j_0}|=Q'_{j_0}=Q_{j_0}-1$, there exists $\alpha\in A_{j_0}=A_{j_0}\cap\{2,\cdots,n\}$ and $i\neq j_0$ such that $\alpha\in A'_i$. That is, there exists $j\neq j'$ such that $A_j\cap A'_{j'}\cap\{2,\cdots,n\}\neq\emptyset$, thus, the integral $I(A,A')$ vanishes. The reasoning is the same if $1\notin A'_{i_0}$. Moreover, if $1\in A_{j_0}$ and $1\in A'_{i_0}$, then, as in the derivation of (D.6), we see that $I(A, A') = 0$ except if $A_j = A'_j$ for all $j \notin \{i_0, j_0\}$. Therefore, if $d_1(Q, Q') = 2$, we rewrite

$$\frac{\gamma^{(1)}_{\substack{Q,\overline{n}\\Q',\overline{n}'}}(x,y)}{c^2(Q)} = \sum_{\substack{j_0,i_0=1\\i_0\neq j_0\\Q_{j_0}\geqslant 1}}^{m} \sum_{\substack{1\in A_{j_0}\\|A_l|=Q_l,\ \forall 1\leqslant j\leqslant m\\A_1\cup\cdots\cup A_m=\{1,\cdots,n\}\\A_j\cap A_{j'}=\emptyset \text{ if } j\neq j'}} \sum_{\substack{A'_{i_0}=\{1\}\cup A_{i_0}\\A'_{j_0}=A_{j_0}\setminus\{1\}\\A'_j=A_j \text{ if } j\notin\{i_0,j_0\}}} (-1)^{\varepsilon((A_j))+\varepsilon((A'_j))} I(A, A'). \tag{D.7}$$

For such $(A_j)_j$ and $(A'_j)_j$, one has $(-1)^{\varepsilon((A_j))+\varepsilon((A'_j))} = 1$ and we compute

$$I(A, A') = \int_{\Delta^{Q_{j_0}-1}_{j_0}} \varphi^{j_0}_{Q_{j_0},n_{j_0}}(x,z)\overline{\varphi^{j_0}_{Q_{j_0}-1,n'_{j_0}}(z)}dz$$
$$\int_{\Delta^{Q_{i_0}}_{i_0}} \varphi^{i_0}_{Q_{i_0},n_{i_0}}(z)\overline{\varphi^{i_0}_{Q_{i_0}+1,n'_{i_0}}(y,z)}dz \prod_{j\notin\{i_0,j_0\}} \delta_{n_j=n'_j} \tag{D.8}$$

with the convention described in Remark 4.3.

The number of partitions coming up in (D.7) is given by

$$\sum_{\substack{1\in A_{j_0}\\|A_l|=Q_l,\ \forall 1\leqslant j\leqslant m\\A_1\cup\cdots\cup A_m=\{1,\cdots,n\}\\A_j\cap A_{j'}=\emptyset \text{ if } j\neq j'}} \sum_{\substack{A'_{i_0}=\{1\}\cup A_{i_0}\\A'_{j_0}=A_{j_0}\setminus\{1\}\\A'_j=A_j \text{ if } j\notin\{i_0,j_0\}}} 1 = \frac{(n-Q_{j_0}-Q_{i_0}-1)!Q_{i_0}!Q_{j_0}!}{Q_1!\cdots Q_m!}.$$

Plugging this and (D.8) into (D.7), we obtain (4.5). This completes the proof of Theorem 4.2.

D.2 *Proof of the Two-Particle Density Matrix Reduction, Theorem 4.4*

Theorem 4.4 follows from a direct computation that we now perform. First, by the bilinearity of formula (1.20), one has

$$\gamma^{(2)}_{\Psi} = \frac{n(n-1)}{2} \sum_{\substack{Q \text{ occ.}\\ \overline{n}\in\mathbb{N}^m}} \sum_{\substack{Q' \text{ occ.}\\ \overline{n}'\in\mathbb{N}^m}} a^{Q}_{\overline{n}}\overline{a^{Q'}_{\overline{n}'}}\gamma^{(2)}_{\substack{Q,\overline{n}\\Q',\overline{n}'}} \tag{D.9}$$

where the trace class operator $\gamma^{(2)}_{\substack{Q,\overline{n}\\Q',\overline{n}'}}$ acts on $L^2([0,L]) \bigwedge L^2([0,L])$ and has the kernel

$$\gamma^{(2)}_{\substack{Q,\overline{n}\\ Q',\overline{n}'}}(x,x',y,y') := \int_{[0,L]^{n-2}} \left[\bigwedge_{j=1}^{m} \varphi^{j}_{Q_j,n_j}\right](x,x',z_3,\cdots,z_n) \overline{\left[\bigwedge_{j=1}^{m} \varphi^{j}_{Q'_j,n'_j}\right](y,y',z_3,\cdots,z_n)} dz_3\cdots dz_n. \tag{D.10}$$

By (D.2), one has

$$\frac{\gamma^{(2)}_{\substack{Q,\overline{n}\\ Q',\overline{n}'}}(x,x',y,y')}{c(Q)c(Q')} = \sum_{\substack{|A_j|=Q_j,\ \forall 1\leqslant j\leqslant m\\ A_1\cup\cdots\cup A_m=\{1,\cdots,n\}\\ A_j\cap A_{j'}=\emptyset \text{ if } j\neq j'}} \sum_{\substack{|A'_j|=Q'_j,\ \forall 1\leqslant j\leqslant m\\ A'_1\cup\cdots\cup A'_m=\{1,\cdots,n\}\\ A'_j\cap A'_{j'}=\emptyset \text{ if } j\neq j'}} (-1)^{\varepsilon((A_j))+\varepsilon((A'_j))} I(A,A') \tag{D.11}$$

where

$$I(A,A') := \int_{[0,L]^{n-2}} \left[\prod_{j=1}^{m} \varphi^{j}_{Q_j,n_j}((x_l)_{l\in A_j}) \overline{\varphi^{j}_{Q'_j,n'_j}((y_l)_{l\in A'_j})}\right]_{\substack{x_1=x,\ x_2=x'\\ y_1=y,\ y_2=y'\\ x_j=y_j \text{ if } j\geqslant 3}} dx_3\cdots dx_n. \tag{D.12}$$

To evaluate this last integral, we note that, for any pair of partitions $(A_j)_j$ and $(A'_j)_j$ (as in the indices of the above sum), if there exists $j\neq j'$ such that $A_j\cap A'_{j'}\cap\{3,\cdots,n\}\neq\emptyset$, then the integral $I(A,A')$ vanishes.

Now, note that, if $d_1(Q,Q')>4$, then, for any pair of partitions $(A_j)_j$ and $(A'_j)_j$, there exists $j\neq j'$ such that $A_j\cap A'_{j'}\cap\{3,\cdots,n\}\neq\emptyset$; thus, the integral $I(A,A')$ above always vanishes and, summing this, one has

$$\gamma^{(2)}_{\substack{Q,\overline{n}\\ Q',\overline{n}'}} = 0 \quad \text{if} \quad d_1(Q,Q')>4.$$

So we are left with the cases $Q=Q'$, $d_1(Q,Q')=2$ or $d_1(Q,Q')=4$.

Assume first $Q=Q'$. Consider the sums in (D.11). If $\{1,2\}\subset A_{i_0}\cup A_{j_0}$ and $\{1,2\}\not\subset A'_{i_0}\cup A'_{j_0}$ then, as $\forall j$, $|A'_j|=|A_j|$, there exists $\alpha\in(A'_{i_0}\cup A'_{j_0})\cap\{3,\cdots,n\}$ and $j\notin\{i_0,j_0\}$ such that $\alpha\in A_j$. That is, there exists $j\neq j'$ such that $A_j\cap A'_{j'}\cap\{3,\cdots,n\}\neq\emptyset$, thus, the integral $I(A,A')$ vanishes. Moreover, if $\{1,2\}\subset A_{j_0}$ and $\{1,2\}\not\subset A'_{j_0}$, then there exists $\alpha\in A'_{j_0}\cap\{3,\cdots,n\}$ and $j\neq j_0$ such that $\alpha\in A_j$, thus, the integral $I(A,A')$ vanishes. Thus, we rewrite

$$\frac{\gamma^{(2)}_{\substack{Q,\overline{n}\\Q,\overline{n}'}}(x,x',y,y')}{c^2(Q)} = \sum_{i_0,j_0=1}^{m} \sum_{\substack{\{1,2\}\subset A_{i_0}\cup A_{j_0}\\ |A_l|=Q_l,\ \forall 1\leqslant j\leqslant m\\ A_1\cup\cdots\cup A_m=\{1,\cdots,n\}\\ A_j\cap A_{j'}=\emptyset \text{ if } j\neq j'}} \sum_{\substack{\{1,2\}\subset A'_{i_0}\cup A'_{j_0}\\ |A'_{i_0}|=Q_{i_0}\\ |A'_{j_0}|=Q_{j_0}\\ A'_j=A_j \text{ if } j\notin\{i_0,j_0\}}} (-1)^{\varepsilon((A_j))+\varepsilon((A'_j))} I(A,A')$$

$$= \sum_{\substack{j_0=1\\ Q_{j_0}\geqslant 2}}^{m} \sum_{\substack{\{1,2\}\subset A_{j_0}\\ |A_j|=Q_j,\ \forall 1\leqslant j\leqslant m\\ A_1\cup\cdots\cup A_m=\{1,\cdots,n\}\\ A_j\cap A_{j'}=\emptyset \text{ if } j\neq j'}} I(A) + \sum_{\substack{i_0\neq j_0\\ Q_{i_0}\geqslant 1\\ Q_{j_0}\geqslant 1}} \sum_{\substack{1\in A_{i_0},\ 2\in A_{j_0}\\ |A_j|=Q_j,\ \forall 1\leqslant j\leqslant m\\ A_1\cup\cdots\cup A_m=\{1,\cdots,n\}\\ A_j\cap A_{j'}=\emptyset \text{ if } j\neq j'}} J(A) \tag{D.13}$$

where

$$\begin{aligned}
J(A) := \prod_{j\notin\{i_0,j_0\}} \delta_{n_j=n'_j} \Bigg(&\int_{\Delta_{i_0}^{Q_{i_0}-1}} \varphi^{i_0}_{Q_{i_0},n_{i_0}}(x,z)\overline{\varphi^{i_0}_{Q_{i_0},n'_{i_0}}(y,z)}dz \\
&\cdot \int_{\Delta_{j_0}^{Q_{j_0}-1}} \varphi^{j_0}_{Q_{j_0},n_{j_0}}(x',z')\overline{\varphi^{j_0}_{Q_{j_0},n'_{j_0}}(y',z')}dz' \\
&- \int_{\Delta_{j_0}^{Q_{j_0}-1}} \varphi^{j_0}_{Q_{j_0},n_{j_0}}(x,z)\overline{\varphi^{j_0}_{Q_{j_0},n'_{j_0}}(y',z)}dz \\
&\cdot \int_{\Delta_{i_0}^{Q_{i_0}-1}} \varphi^{i_0}_{Q_{i_0},n_{i_0}}(x',z')\overline{\varphi^{i_0}_{Q_{i_0},n'_{i_0}}(y,z')}dz' \\
&- \int_{\Delta_{j_0}^{Q_{j_0}-1}} \varphi^{j_0}_{Q_{j_0},n_{j_0}}(x',z)\overline{\varphi^{j_0}_{Q_{j_0},n'_{j_0}}(y,z)}dz \\
&\cdot \int_{\Delta_{i_0}^{Q_{i_0}-1}} \varphi^{i_0}_{Q_{i_0},n_{i_0}}(x,z')\overline{\varphi^{i_0}_{Q_{i_0},n'_{i_0}}(y',z')}dz' \\
&+ \int_{\Delta_{j_0}^{Q_{j_0}-1}} \varphi^{j_0}_{Q_{j_0},n_{j_0}}(x,z)\overline{\varphi^{j_0}_{Q_{j_0},n'_{j_0}}(y,z)}dz \\
&\cdot \int_{\Delta_{i_0}^{Q_{i_0}-1}} \varphi^{i_0}_{Q_{i_0},n_{i_0}}(x',z')\overline{\varphi^{i_0}_{Q_{i_0},n'_{i_0}}(y',z')}dz' \Bigg)
\end{aligned}$$

and

$$I(A) := \prod_{j\neq j_0} \delta_{n_j=n'_j} \int_{\Delta_{j_0}^{Q_{j_0}-2}} \varphi^{j_0}_{Q_{j_0},n_{j_0}}(x,x',z)\overline{\varphi^{j_0}_{Q_{j_0},n'_{j_0}}(y,y',z)}dz.$$

As

$$\#\{(A_j)_j;\ \{1,2\}\subset A_{j_0},\ \forall j,\ |A_j|=Q_j\}=\frac{(n-2)!Q_{j_0}(Q_{j_0}-1)}{\prod_{j=1}^m Q_j!}$$

and

$$\#\{(A_j)_j;\ 1\in A_{i_0},\ 2\in A_{j_0},\ \forall j,\ |A_j|=Q_j\}=\frac{(n-2)!Q_{i_0}Q_{j_0}}{\prod_{j=1}^m Q_j!}\quad\text{if}\quad i_0\neq j_0$$

by (D.3) and (D.13), one obtains

$$\sum_{\substack{Q \text{ occ.}\\ \overline{n}\in\mathbb{N}^m\\ \overline{n}'\in\mathbb{N}^m}} a_{\overline{n}}^{Q}\overline{a_{\overline{n}'}^{Q}}\gamma^{(2)}_{\substack{Q,\overline{n}\\ Q,\overline{n}'}}=\gamma_\Psi^{(2),d,d}+\gamma_\Psi^{(2),d,o} \tag{D.14}$$

where $\gamma_\Psi^{(2),d,d}$ and $\gamma_\Psi^{(2),d,o}$ are defined in Theorem 4.4.

Let us now assume $d_1(Q,Q')=2$. Thus, there exists $1\leqslant i_0\neq j_0\leqslant m$ such that $Q_{j_0}\geqslant 1$, $Q'_{i_0}=Q_{i_0}+1$, $Q_{j_0}=Q'_{j_0}+1$ and $Q_k=Q'_k$ for $k\notin\{i_0,j_0\}$.

Consider now the sums in (D.11). If $\{1,2\}\cap A_{j_0}=\emptyset$, then as $|A'_{j_0}|=Q'_{j_0}=Q_{j_0}-1$, there exists $\alpha\in A_{j_0}=A_{j_0}\cap\{3,\cdots,n\}$ and $i\neq j_0$ such that $\alpha\in A'_i$. Thus, the integral $I(A,A')$ vanishes. If $A_{j_0}=\{1\}\cup B$ (resp. $A_{j_0}=\{2\}\cup B$) with $B\subset\{3,\cdots,n\}$, either $A'_{j_0}=B$ (and $\{1,2\}\subset A'_{i_0}$) or the integral $I(A,A')$ vanishes. Finally, if $A_{j_0}=\{1,2\}\cup B$ with $B\subset\{3,\cdots,n\}$, then $A'_{j_0}=\{1\}\cup B$ or $A'_{j_0}=\{2\}\cup B$ or $I(A,A')=0$. The same holds true for A_{j_0} replaced with A'_{i_0}.

Therefore, using the definition of $\varepsilon((A_j))$, if $d_1(Q,Q')=2$, we rewrite

$$\frac{\gamma^{(2)}_{\substack{Q,\overline{n}\\ Q',\overline{n}'}}(x,y)}{c^2(Q)}=\sum_{\substack{i_0\neq j_0\\ Q_{j_0}\geqslant 2}}\Sigma_1(i_0,j_0)-\Sigma_2(i_0,j_0)+\sum_{\substack{i_0\neq j_0\\ Q_{i_0}\geqslant 1}}\Sigma_3(i_0,j_0)-\Sigma_4(i_0,j_0) \tag{D.15}$$

where

$$\Sigma_1(i_0,j_0):=\sum_{\substack{\{1,2\}\subset A_{j_0}\\ |A_j|=Q_j,\ \forall 1\leqslant j\leqslant m\\ A_1\cup\cdots\cup A_m=\{1,\cdots,n\}\\ A_j\cap A_{j'}=\emptyset \text{ if } j\neq j'}}\ \sum_{\substack{A'_{i_0}=\{1\}\cup A_{i_0}\\ A'_{j_0}=A_{j_0}\setminus\{1\}\\ A'_j=A_j \text{ if } j\notin\{i_0,j_0\}}} I(A,A'), \tag{D.16}$$

$$\Sigma_2(i_0, j_0) := \sum_{\substack{\{1,2\}\subset A_{j_0} \\ |A_j|=Q_j,\ \forall 1\leqslant j\leqslant m \\ A_1\cup\cdots\cup A_m=\{1,\cdots,n\} \\ A_j\cap A_{j'}=\emptyset \text{ if } j\neq j'}} \sum_{\substack{A'_{i_0}=\{2\}\cup A_{i_0} \\ A'_{j_0}=A_{j_0}\setminus\{2\} \\ A'_j=A_j \text{ if } j\notin\{i_0,j_0\}}} I(A, A'), \tag{D.17}$$

$$\Sigma_3(i_0, j_0) := \sum_{\substack{\{1,2\}\subset A'_{i_0} \\ |A_j|=Q_j,\ \forall 1\leqslant j\leqslant m \\ A_1\cup\cdots\cup A_m=\{1,\cdots,n\} \\ A_j\cap A_{j'}=\emptyset \text{ if } j\neq j'}} \sum_{\substack{A_{i_0}=A'_{i_0}\setminus\{1\} \\ A_{j_0}=A'_{j_0}\cup\{1\} \\ A'_j=A_j \text{ if } j\notin\{i_0,j_0\}}} I(A, A'), \tag{D.18}$$

$$\Sigma_4(i_0, j_0) := \sum_{\substack{\{1,2\}\subset A'_{i_0} \\ |A_j|=Q_j,\ \forall 1\leqslant j\leqslant m \\ A_1\cup\cdots\cup A_m=\{1,\cdots,n\} \\ A_j\cap A_{j'}=\emptyset \text{ if } j\neq j'}} \sum_{\substack{A_{i_0}=A'_{i_0}\setminus\{2\} \\ A_{j_0}=A'_{j_0}\cup\{2\} \\ A'_j=A_j \text{ if } j\notin\{i_0,j_0\}}} I(A, A') \tag{D.19}$$

and

- for the summands in $\Sigma_1(i_0, j_0)$:

$$I(A, A') = \int_{\Delta_{j_0}^{Q_{j_0}-2}} \varphi^{j_0}_{Q_{j_0},n_{j_0}}(x, x', z)\overline{\varphi^{j_0}_{Q_{j_0}-1,n'_{j_0}}(y', z)}dz$$
$$\int_{\Delta_{i_0}^{Q_{i_0}}} \varphi^{i_0}_{Q_{i_0},n_{i_0}}(z')\overline{\varphi^{i_0}_{Q_{i_0}+1,n'_{i_0}}(y, z')}dz' \prod_{j\notin\{i_0,j_0\}} \delta_{n_j=n'_j}$$

- for the summands in $\Sigma_2(i_0, j_0)$:

$$I(A, A') = \int_{\Delta_{j_0}^{Q_{j_0}-2}} \varphi^{j_0}_{Q_{j_0},n_{j_0}}(x, x', z)\overline{\varphi^{j_0}_{Q_{j_0}-1,n'_{j_0}}(y, z)}dz$$
$$\int_{\Delta_{i_0}^{Q_{i_0}}} \varphi^{i_0}_{Q_{i_0},n_{i_0}}(z')\overline{\varphi^{i_0}_{Q_{i_0}+1,n'_{i_0}}(y', z')}dz' \prod_{j\notin\{i_0,j_0\}} \delta_{n_j=n'_j}$$

- for the summands in $\Sigma_3(i_0, j_0)$:

$$I(A, A') = \int_{\Delta_{j_0}^{Q_{j_0}-1}} \varphi^{j_0}_{Q_{j_0},n_{j_0}}(x', z)\overline{\varphi^{j_0}_{Q_{j_0}-1,n'_{j_0}}(z)}dz$$
$$\int_{\Delta_{i_0}^{Q_{i_0}-1}} \varphi^{i_0}_{Q_{i_0},n_{i_0}}(x, z')\overline{\varphi^{i_0}_{Q_{i_0}+1,n'_{i_0}}(y, y', z')}dz' \prod_{j\notin\{i_0,j_0\}} \delta_{n_j=n'_j}$$

- for the summands in $\Sigma_4(i_0, j_0)$:

$$I(A, A') = \int_{\Delta_{j_0}^{Q_{j_0}-1}} \varphi^{j_0}_{Q_{j_0},n_{j_0}}(x, z)\overline{\varphi^{j_0}_{Q_{j_0}-1,n'_{j_0}}(z)}dz$$

$$\int_{\Delta_{i_0}^{Q_{i_0}-1}} \varphi^{i_0}_{Q_{i_0},n_{i_0}}(x', z')\overline{\varphi^{i_0}_{Q_{i_0}+1,n'_{i_0}}(y, y', z')}dz' \prod_{j\notin\{i_0,j_0\}} \delta_{n_j=n'_j}$$

with the convention described in Remark 4.1.

The number of partitions coming up in (D.16), (D.17), (D.18), and (D.19) is the same: indeed, it suffices to invert the roles of 1 and 2 and i_0 and j_0. We compute

$$\sum_{\substack{\{1,2\}\subset A_{j_0} \\ |A_j|=Q_j,\ \forall 1\leqslant j\leqslant m \\ A_1\cup\cdots\cup A_m=\{1,\cdots,n\} \\ A_j\cap A_{j'}=\emptyset \text{ if } j\neq j'}} \sum_{\substack{A'_{i_0}=\{1\}\cup A_{i_0} \\ A'_{j_0}=A_{j_0}\setminus\{1\} \\ A'_j=A_j \text{ if } j\notin\{i_0,j_0\}}} 1 = \frac{(n - Q_{j_0} - Q_{i_0} - 2)!Q_{i_0}!Q_{j_0}!}{Q_1!\cdots Q_m!}.$$

Hence, we get that

$$\frac{n(n-1)}{2} \sum_{\substack{Q,\ Q' \text{ occ.} \\ d_1(Q,Q')=2 \\ \overline{n},\ \overline{n}'\in\mathbb{N}^m}} a^{Q}_{\overline{n}} \overline{a^{Q'}_{\overline{n}'}} \gamma^{(2)}_{\substack{Q,\overline{n} \\ Q',\overline{n}'}} = \sum_{i\neq j} \sum_{\tilde{n}\in\mathbb{N}^{m-2}} \sum_{\substack{Q \text{ occ.} \\ Q_j\geqslant 1 \\ Q':\ Q'_k=Q_k \text{ if } k\notin\{i,j\} \\ Q'_i=Q_i+1 \\ Q'_j=Q_j-1}} \sum_{\substack{n_j,n'_j\geqslant 1 \\ n_i,n'_i\geqslant 1}} a^{Q}_{\tilde{n}_{i,j}} \overline{a^{Q'}_{\tilde{n}'_{i,j}}} \gamma^{(2),2}_{\substack{Q_i,Q_j \\ n_i,n_j \\ n'_i,n'_j}} \tag{D.20}$$

where $\gamma^{(2),2}_{\substack{Q_i,Q_j \\ n_i,n_j \\ n'_i,n'_j}}$ is defined in (4.19).

Let us now assume $d_1(Q, Q') = 4$. Thus,

(a) either there exist $1 \leqslant i_0 \neq j_0 \leqslant m$ such that $Q_{j_0} \geqslant 2$, $Q'_{i_0} = Q_{i_0} + 2$, $Q_{j_0} = Q'_{j_0} + 2$ and $Q_k = Q'_k$ for $k \notin \{i_0, j_0\}$.

In this case, either $A_{j_0} = \{1, 2\}\cup A'_{j_0}$ and $A'_{i_0} = \{1, 2\}\cup A_{i_0}$ with $A_{i_0}, A'_{j_0}, \subset \{3, \cdots, n\}$ or $I(A, A') = 0$ vanishes. Thus,

$$\frac{\gamma^{(2)}_{\substack{Q,\overline{n} \\ Q',\overline{n}'}}(x, y)}{c^2(Q)} = \sum_{\substack{\{1,2\}\subset A_{j_0} \\ |A_j|=Q_j,\ \forall 1\leqslant j\leqslant m \\ A_1\cup\cdots\cup A_m=\{1,\cdots,n\} \\ A_j\cap A_{j'}=\emptyset \text{ if } j\neq j'}} \sum_{\substack{A'_{i_0}=\{1,2\}\cup A_{i_0} \\ A'_{j_0}=A_{j_0}\setminus\{1,2\} \\ A'_j=A_j \text{ if } j\notin\{i_0,j_0\}}} (-1)^{\varepsilon((A_j))+\varepsilon((A'_j))} I(A, A'), \tag{D.21}$$

and

$$I(A,A') = \int_{\Delta_{j_0}^{Q_{j_0}-2}} \varphi^{j_0}_{Q_{j_0},n_{j_0}}(x,x',z)\overline{\varphi^{j_0}_{Q_{j_0}-2,n'_{j_0}}(z)}dz$$
$$\int_{\Delta_{i_0}^{Q_{i_0}}} \varphi^{i_0}_{Q_{i_0},n_{i_0}}(z')\overline{\varphi^{i_0}_{Q_{i_0}+2,n'_{i_0}}(y,y',z')}dz' \prod_{j\notin\{i_0,j_0\}} \delta_{n_j=n'_j}.$$

Hence, taking (4.20) into account, we get

$$\begin{aligned}\frac{n(n-1)}{2} & \sum_{\substack{Q,\,Q' \text{ occ.}\\ \exists i\neq j,\ Q_j\geqslant 2\\ Q':\ Q'_k=Q_k \text{ if } k\notin\{i,j\}\\ Q'_i=Q_i+2\\ Q'_j=Q_j-2}} \sum_{\substack{\overline{n}\in\mathbb{N}^m\\ \overline{n}'\in\mathbb{N}^m}} a^Q_{\overline{n}}\overline{a^{Q'}_{\overline{n}'}}\gamma^{(2)}_{\substack{Q,\overline{n}\\ Q',\overline{n}'}} \\ &= \sum_{i\neq j}\sum_{\tilde{n}\in\mathbb{N}^{m-2}} \sum_{\substack{Q \text{ occ.}\\ Q_j\geqslant 2\\ Q':\ Q'_k=Q_k \text{ if } k\notin\{i,j\}\\ Q'_i=Q_i+2\\ Q'_j=Q_j-2}} C_2(Q,i,j) \sum_{\substack{n_j,n'_j\geqslant 1\\ n_i,n'_i\geqslant 1}} a^Q_{\tilde{n}_{i,j}}\overline{a^{Q'}_{\tilde{n}'_{i,j}}}\gamma^{(2),4,2}_{\substack{Q_i,Q_j\\ n_i,n_j\\ n'_i,n'_j}}\end{aligned} \tag{D.22}$$

as

$$\sum_{\substack{\{1,2\}\subset A_j\\ |A_l|=Q_l,\ \forall 1\leqslant l\leqslant m\\ A_1\cup\cdots\cup A_m=\{1,\cdots,n\}\\ A_l\cap A_{l'}=\emptyset \text{ if } l\neq l'}} \sum_{\substack{A'_i=\{1,2\}\cup A_i\\ A'_j=A_j\setminus\{1,2\}\\ A'_l=A_l \text{ if } j\notin\{i,j\}}} 1 = \frac{(n-Q_j-Q_i-2)!Q_i!Q_j!}{Q_1!\cdots Q_m!} = \frac{2\,C_2(Q,i,j)}{n(n-1)\,c(Q)^2}.$$

(b) or there exist $1\leqslant i_0, j_0, k_0\leqslant m$ distinct such that $Q_{j_0}\geqslant 2$, $Q'_{j_0}=Q_{j_0}-2$, $Q_{i_0}=Q'_{i_0}+1$, $Q_{k_0}=Q'_{k_0}+1$, and $Q_k=Q'_k$ for $k\notin\{i_0,j_0,k_0\}$.
In this case, either $A_{j_0}=\{1,2\}\cup A'_{j_0}$ and $((A'_{i_0}=\{1\}\cup A_{i_0}$ and $A'_{k_0}=\{2\}\cup A_{k_0})$ or $(A'_{i_0}=\{2\}\cup A_{i_0}$ and $A'_{k_0}=\{1\}\cup A_{k_0})$) with $A_{j_0}, A'_{i_0}, A'_{k_0}\subset\{3,\cdots,n\}$ or $I(A,A')=0$ vanishes. Thus,

$$\frac{\gamma^{(2)}_{\substack{Q,\overline{n}\\ Q',\overline{n}'}}(x,y)}{c^2(Q)} = \sum_{\substack{\{1,2\}\subset A_{j_0}\\ |A_j|=Q_j,\ \forall 1\leqslant j\leqslant m\\ A_1\cup\cdots\cup A_m=\{1,\cdots,n\}\\ A_j\cap A_{j'}=\emptyset \text{ if } j\neq j'}} \left(\sum_{\substack{A'_{i_0}=\{1\}\cup A_{i_0}\\ A'_{k_0}=\{2\}\cup A_{k_0}\\ A'_{j_0}=A_{j_0}\setminus\{1,2\}\\ A'_j=A_j \text{ if } j\notin\{i_0,j_0,k_0\}}} I(A,A') - \sum_{\substack{A'_{i_0}=\{2\}\cup A_{i_0}\\ A'_{k_0}=\{1\}\cup A_{k_0}\\ A'_{j_0}=A_{j_0}\setminus\{1,2\}\\ A'_j=A_j \text{ if } j\notin\{i_0,j_0,k_0\}}} I(A,A') \right) \tag{D.23}$$

and, if $A'_{i_0}=\{1\}\cup A_{i_0}$ and $A'_{k_0}=\{2\}\cup A_{k_0}$, one has

$$\begin{aligned} I(A,A') = &\int_{\Delta_{j_0}^{Q_{j_0}-2}} \varphi^{j_0}_{Q_{j_0},n_{j_0}}(x,x',z)\overline{\varphi^{j_0}_{Q_{j_0}-2,n'_{j_0}}(z)}dz \\ &\int_{\Delta_{i_0}^{Q_{i_0}}} \varphi^{i_0}_{Q_{i_0},n_{i_0}}(z')\overline{\varphi^{i_0}_{Q_{i_0}+1,n'_{i_0}}(y,z')}dz' \\ &\quad\int_{\Delta_{k_0}^{Q_{k_0}}} \varphi^{i_0}_{Q_{k_0},n_{k_0}}(z'')\overline{\varphi^{k_0}_{Q_{k_0}+1,n'_{k_0}}(y',z'')}dz'' \prod_{j\notin\{i_0,j_0,k_0\}} \delta_{n_j=n'_j} \end{aligned}$$

and, if $A'_{i_0}=\{2\}\cup A_{i_0}$ and $A'_{k_0}=\{1\}\cup A_{k_0}$, one has

$$\begin{aligned} I(A,A') = &\int_{\Delta_{j_0}^{Q_{j_0}-2}} \varphi^{j_0}_{Q_{j_0},n_{j_0}}(x,x',z)\overline{\varphi^{j_0}_{Q_{j_0}-2,n'_{j_0}}(z)}dz \\ &\int_{\Delta_{i_0}^{Q_{i_0}}} \varphi^{i_0}_{Q_{i_0},n_{i_0}}(z')\overline{\varphi^{i_0}_{Q_{i_0}+1,n'_{i_0}}(y',z')}dz' \\ &\quad\int_{\Delta_{k_0}^{Q_{k_0}}} \varphi^{i_0}_{Q_{k_0},n_{k_0}}(z'')\overline{\varphi^{k_0}_{Q_{k_0}+1,n'_{k_0}}(y,z'')}dz'' \prod_{j\notin\{i_0,j_0,k_0\}} \delta_{n_j=n'_j}. \end{aligned}$$

For i_0, j_0, k_0 distinct, one has

$$\sum_{\substack{\{1,2\}\subset A_{j_0}\\ |A_j|=Q_j,\ \forall 1\leqslant j\leqslant m\\ A_1\cup\cdots\cup A_m=\{1,\cdots,n\}\\ A_j\cap A_{j'}=\emptyset \text{ if } j\neq j'}}\ \sum_{\substack{A'_{i_0}=\{1\}\cup A_{i_0}\\ A'_{k_0}=\{2\}\cup A_{k_0}\\ A'_{j_0}=A_{j_0}\setminus\{1,2\}\\ A'_j=A_j \text{ if } j\notin\{i_0,j_0,k_0\}}} 1 = \frac{(n-Q_{j_0}-Q_{i_0}-Q_{k_0}-2)!\,Q_{i_0}!\,Q_{j_0}!\,Q_{k_0}!}{Q_1!\cdots Q_m!}$$

$$= \frac{2\,C_3(Q,i_0,j_0,k_0)}{n(n-1)\,c(Q)^2}.$$

Inverting the roles of 1 and 2 we see that the number of partitions coming up in the second sum in (D.23) is the same. Thus, taking (4.20) into account, we get

$$\frac{n(n-1)}{2} \sum_{\substack{Q,\ Q' \text{ occ.}\\ \exists i,j,k \text{ distinct}\\ Q_j\geqslant 2\\ Q':\ Q'_l=Q_l \text{ if } l\notin\{i,j,k\}\\ Q'_j=Q_j-2\\ Q'_i=Q_i+1,\ Q'_k=Q_k+1}} \sum_{\substack{\overline{n}\in\mathbb{N}^m\\ \overline{n}'\in\mathbb{N}^m}} a^Q_{\overline{n}}\, \overline{a^{Q'}_{\overline{n}'}}\, \gamma^{(2)}_{\substack{Q,\overline{n}\\ Q',\overline{n}'}}$$

$$= \sum_{\substack{i,j,k\\ \text{distinct}}} \sum_{\tilde{n}\in\mathbb{N}^{m-3}} \sum_{\substack{Q \text{ occ.}\\ Q_j\geqslant 2\\ Q':\ Q'_l=Q_l \text{ if } l\notin\{i,j,k\}\\ Q'_j=Q_j-2\\ Q'_i=Q_i+1,\ Q'_k=Q_k+1}} C_3(Q,i,j,k) \sum_{\substack{n_i,n_j,n_k\geqslant 1\\ n'_i,n'_j,n'_k\geqslant 1}} a^Q_{\tilde{n}_{i,j,k}}\, \overline{a^{Q'}_{\tilde{n}'_{i,j,k}}}\, \gamma^{(2),4,3}_{\substack{Q_i,Q_j,Q_k\\ n_i,n_j,n_k\\ n'_i,n'_j,n'_k}}. \tag{D.24}$$

(c) or there exist $1 \leqslant i_0, j_0, k_0 \leqslant m$ distinct such that $Q_{i_0} \geqslant 1$, $Q_{k_0} \geqslant 1$, $Q'_{j_0} = Q_{j_0} + 2$, $Q_{i_0} = Q'_{i_0} - 1$, $Q_{k_0} = Q'_{k_0} - 1$, and $Q_k = Q'_k$ for $k \notin \{i_0, j_0, k_0\}$.
We see that we are back to case (b) if we invert the roles of Q and Q'. Thus, we get

$$\frac{n(n-1)}{2} \sum_{\substack{Q,\ Q' \text{ occ.}\\ \exists i,j,k \text{ distinct}\\ Q_i\geqslant 1,\ Q_k\geqslant 1\\ Q':\ Q'_l=Q_l \text{ if } l\notin\{i,j,k\}\\ Q'_j=Q_j+2\\ Q'_i=Q_i-1,\ Q'_k=Q_k-1}} \sum_{\substack{\overline{n}\in\mathbb{N}^m\\ \overline{n}'\in\mathbb{N}^m}} a^Q_{\overline{n}}\, \overline{a^{Q'}_{\overline{n}'}}\, \gamma^{(2)}_{\substack{Q,\overline{n}\\ Q',\overline{n}'}}$$

$$= \sum_{\substack{i,j,k\\ \text{distinct}}} \sum_{\tilde{n}\in\mathbb{N}^{m-3}} \sum_{\substack{Q \text{ occ.}\\ Q_i\geqslant 1,\ Q_k\geqslant 1\\ Q':\ Q'_l=Q_l \text{ if } l\notin\{i,j,k\}\\ Q'_j=Q_j+2\\ Q'_i=Q_i-1,\ Q'_k=Q_k-1}} C_3(Q,i,j,k) \sum_{\substack{n_i,n_j,n_k\geqslant 1\\ n'_i,n'_j,n'_k\geqslant 1}} a^Q_{\tilde{n}_{i,j,k}}\, \overline{a^{Q'}_{\tilde{n}'_{i,j,k}}}\, \gamma^{(2),4,3'}_{\substack{Q_i,Q_j,Q_k\\ n_i,n_j,n_k\\ n'_i,n'_j,n'_k}}. \tag{D.25}$$

(d) or there exist $1 \leqslant i_0, j_0, k_0, l_0 \leqslant m$ distinct such that $Q_{j_0} \geqslant 1$, $Q_{l_0} \geqslant 1$, $Q'_{i_0} = Q_{i_0} - 1$, $Q_{j_0} = Q'_{j_0} - 1$, $Q'_{k_0} = Q_{k_0} + 1$, $Q_{l_0} = Q'_{l_0} + 1$ and $Q_k = Q'_k$ for $k \notin \{i_0, j_0, k_0, l_0\}$.

Then, either $I(A, A') = 0$ or

(i) either $A_{i_0} = \{1\} \cup A'_{i_0}$ and $A_{j_0} = \{2\} \cup A'_{j_0}$ and $A'_{i_0}, A'_{j_0} \subset \{3, \cdots, n\}$,
(ii) or $A_{i_0} = \{2\} \cup A'_{i_0}$ and $A_{j_0} = \{1\} \cup A'_{j_0}$ and $A'_{i_0}, A'_{j_0} \subset \{3, \cdots, n\}$.

Moreover, in each of the cases (i) and (ii), either $I(A, A') = 0$ or

(i) either $A'_{k_0} = \{1\} \cup A_{k_0}$ and $A'_{l_0} = \{2\} \cup A_{l_0}$ and $A_{k_0}, A_{l_0} \subset \{3, \cdots, n\}$,
(ii) or $A'_{k_0} = \{2\} \cup A_{k_0}$ and $A'_{l_0} = \{1\} \cup A_{l_0}$ and $A_{k_0}, A_{l_0} \subset \{3, \cdots, n\}$.

In the 4 cases when $I(A, A')$ does not vanish, one computes

- $I(A, A') = \alpha(x, x', y, y')$ in case (i.i),
- $I(A, A') = \alpha(x', x, y, y')$ in case (ii.i),
- $I(A, A') = \alpha(x, x', y, y')$ in case (i.ii),
- $I(A, A') = \alpha(x', x, y', y)$ in case (ii.ii),

where

$$\alpha(x, x', y, y') := \int_{\Delta_i^{Q_i-1}} \varphi^i_{Q_i,n_i}(x, z)\overline{\varphi^i_{Q_i-1,n'_i}(z)}dz \int_{\Delta_j^{Q_j-1}} \varphi^j_{Q_j,n_j}(x', z)\overline{\varphi^j_{Q_j-1,n'_j}(z)}dz$$
$$\times \int_{\Delta_k^{Q_k}} \varphi^k_{Q_k,n_k}(z)\overline{\varphi^k_{Q_k+1,n'_k}(y, z)}dz \int_{\Delta_l^{Q_l}} \varphi^l_{Q_l,n_l}(z)\overline{\varphi^l_{Q_l+1,n'_l}(y', z)}dz.$$

Hence, if $d_1(Q, Q') = 4$, we obtain

$$\frac{\gamma^{(2)}_{\substack{Q,\overline{n}\\ Q',\overline{n}'}}(x, y)}{c^2(Q)}$$
$$= \sum_{\substack{1\in A_{i_0},\ 2\in A_{j_0}\\ |A_j|=Q_j,\ \forall 1\leqslant j\leqslant m\\ A_1\cup\cdots\cup A_m=\{1,\cdots,n\}\\ A_j\cap A_{j'}=\emptyset \text{ if } j\neq j'}} \left(\sum_{\substack{A'_{k_0}=\{1\}\cup A_{k_0},\ A'_{l_0}=\{2\}\cup A_{l_0}\\ A'_{i_0}=A_{i_0}\setminus\{1\},\ A'_{j_0}=A_{j_0}\setminus\{2\}\\ A'_j=A_j \text{ if } j\notin\{i_0,j_0,k_0,l_0\}}} I(A, A') - \sum_{\substack{A'_{k_0}=\{2\}\cup A_{k_0},\ A'_{l_0}=\{1\}\cup A_{l_0}\\ A'_{i_0}=A_{i_0}\setminus\{1\},\ A'_{j_0}=A_{j_0}\setminus\{2\}\\ A'_j=A_j \text{ if } j\notin\{i_0,j_0,k_0,l_0\}}} I(A, A') \right)$$
$$- \sum_{\substack{2\in A_{i_0},\ 1\in A_{j_0}\\ |A_j|=Q_j,\ \forall 1\leqslant j\leqslant m\\ A_1\cup\cdots\cup A_m=\{1,\cdots,n\}\\ A_j\cap A_{j'}=\emptyset \text{ if } j\neq j'}} \left(\sum_{\substack{A'_{k_0}=\{1\}\cup A_{k_0},\ A'_{l_0}=\{2\}\cup A_{l_0}\\ A'_{i_0}=A_{i_0}\setminus\{2\},\ A'_{j_0}=A_{j_0}\setminus\{1\}\\ A'_j=A_j \text{ if } j\notin\{i_0,j_0,k_0,l_0\}}} I(A, A') - \sum_{\substack{A'_{k_0}=\{2\}\cup A_{k_0},\ A'_{l_0}=\{1\}\cup A_{l_0}\\ A'_{i_0}=A_{i_0}\setminus\{2\},\ A'_{j_0}=A_{j_0}\setminus\{1\}\\ A'_j=A_j \text{ if } j\notin\{i_0,j_0,k_0,l_0\}}} I(A, A') \right). \tag{D.26}$$

For i_0, j_0, k_0, l_0 distinct, the number of partitions coming up in the first sum in (D.26) is given by

$$\sum_{\substack{1\in A_{i_0},\ 2\in A_{j_0}\\ |A_j|=Q_j,\ \forall 1\leqslant j\leqslant m\\ A_1\cup\cdots\cup A_m=\{1,\cdots,n\}\\ A_j\cap A_{j'}=\emptyset \text{ if } j\neq j'}} \sum_{\substack{A'_{k_0}=\{1\}\cup A_{k_0},\ A'_{l_0}=\{2\}\cup A_{l_0}\\ A'_{i_0}=A_{i_0}\setminus\{1\},\ A'_{j_0}=A_{j_0}\setminus\{2\}\\ A'_j=A_j \text{ if } j\notin\{i_0,j_0,k_0,l_0\}}} 1$$

$$= \frac{(n-Q_{j_0}-Q_{i_0}-Q_{k_0}-Q_{l_0}-2)!Q_{i_0}!Q_{j_0}!Q_{k_0}!Q_{l_0}!}{Q_1!\cdots Q_m!}$$

$$= \frac{2\,C_4(Q,i_0,j_0,k_0,l_0)}{n(n-1)\,c(Q)^2}.$$

Inverting the roles of i_0, j_0, k_0, l_0, we see that the number of partitions involved is the same in the three remaining sums of (D.26).

Thus, taking (4.20) into account, we get

$$\frac{n(n-1)}{2} \sum_{\substack{Q,\ Q' \text{ occ.}\\ \exists i,j,k,l \text{ distinct}\\ Q_i\geqslant 1,\ Q_j\geqslant 1\\ Q':\ Q'_p=Q_p \text{ if } p\notin\{i,j,k,l\}\\ Q'_i=Q_i-1,\ Q'_j=Q_j-1\\ Q'_k=Q_k+1,\ Q'_l=Q_l+1}} \sum_{\substack{\overline{n}\in\mathbb{N}^m\\ \overline{n}'\in\mathbb{N}^m}} a^Q_{\overline{n}} \overline{a^{Q'}_{\overline{n}'}}\, \gamma^{(2)}_{\substack{Q,\overline{n}\\ Q',\overline{n}'}}$$

$$= \sum_{\substack{i,j,k,l\\ \text{distinct}}} \sum_{\tilde{n}\in\mathbb{N}^{m-4}} \sum_{\substack{Q \text{ occ.}\\ Q_i\geqslant 1,\ Q_j\geqslant 1\\ Q':\ Q'_p=Q_p \text{ if } p\notin\{i,j,k,l\}\\ Q'_i=Q_i-1,\ Q'_j=Q_j-1\\ Q'_k=Q_k+1,\ Q'_l=Q_l+1}} C_4(Q,i,j,k) \sum_{\substack{n_i,n_j,n_k,n_l\geqslant 1\\ n'_i,n'_j,n'_k,n'_l\geqslant 1}} a^Q_{\tilde{n}_{i,j,k,l}} \overline{a^{Q'}_{\tilde{n}'_{i,j,k,l}}}\, \gamma^{(2),4,4}_{\substack{Q_i,Q_j,Q_k,Q_l\\ n_i,n_j,n_k,n_l\\ n'_i,n'_j,n'_k,n'_l}}. \tag{D.27}$$

Plugging this (D.22) and (D.20) into (D.9), we obtain (4.9). This completes the proof of Theorem 4.4.

References

1. M. Aizenman, S. Warzel, Localization bounds for multiparticle systems. Commun. Math. Phys. **290**(3), 903–934 (2009)
2. J. Bertoin, *Random Fragmentation and Coagulation Processes*, vol. 102. Cambridge Studies in Advanced Mathematics (Cambridge University Press, Cambridge, 2006)
3. X. Blanc, M. Lewin, Existence of the thermodynamic limit for disordered quantum Coulomb systems. J. Math. Phys. **53**(9), 095209, 32 (2012)

4. V. Chulaevsky, Y. Suhov, Multi-particle Anderson localisation: induction on the number of particles. Math. Phys. Anal. Geom. **12**(2), 117–139 (2009)
5. H.L. Cycon, R.G. Froese, W. Kirsch, B. Simon, *Schrödinger Operators with Application to Quantum Mechanics and Global Geometry*, study edition. Texts and Monographs in Physics (Springer, Berlin, 1987)
6. E.B. Davies, *Heat Kernels and Spectral Theory*, vol. 92. Cambridge Tracts in Mathematics (Cambridge University Press, Cambridge, 1990)
7. F. Germinet, F. Klopp, Spectral statistics for random Schrödinger operators in the localized regime. JEMS. (2010). ArXiv http://arxiv.org/abs/1011.1832
8. F. Germinet, F. Klopp, Enhanced Wegner and Minami estimates and eigenvalue statistics of random Anderson models at spectral edges. Ann. Henri Poincaré **14**(5), 1263–1285 (2013)
9. L.N. Grenkova, S.A. Molčanov, Ju.N. Sudarev, On the basic states of one-dimensional disordered structures. Commun. Math. Phys. **90**(1), 101–123 (1983)
10. W. Kirsch, An invitation to random Schrödinger operators, in *Random Schrödinger Operators*, vol. 25. Panor. Synthèses (Société Mathématique de France, Paris, 2008), pp. 1–119. With an appendix by Frédéric Klopp
11. W. Kirsch, A Wegner estimate for multi-particle random Hamiltonians. Zh. Mat. Fiz. Anal. Geom. **4**(1), 121–127, 203 (2008)
12. F. Klopp, The low lying states of the Poisson-Anderson model in one-dimension (2020, in preparation)
13. I.M. Lifshits, S.A. Gredeskul, L.A. Pastur, *Introduction to the Theory of Disordered Systems*. A Wiley-Interscience Publication (Wiley, New York, 1988). Translated from the Russian by Eugene Yankovsky [E. M. Yankovskiĭ]
14. J.M. Luttinger, H.K. Sy, Low-lying energy spectrum of a one-dimensional disordered system. Phys. Rev. A **7**, 701–712 (1973)
15. N. Minami, Local fluctuation of the spectrum of a multidimensional Anderson tight binding model. Commun. Math. Phys. **177**(3), 709–725 (1996)
16. L. Pastur, A. Figotin, *Spectra of Random and Almost-Periodic Operators*, vol. 297. Grundlehren der Mathematischen Wissenschaften [Fundamental Principles of Mathematical Sciences] (Springer, Berlin, 1992)
17. M. Reed, B. Simon, *Methods of Modern Mathematical Physics. IV. Analysis of Operators* (Academic [Harcourt Brace Jovanovich Publishers], New York, 1978)
18. M. Reed, B. Simon, *Methods of Modern Mathematical Physics. I*, 2nd edn. (Academic [Harcourt Brace Jovanovich Publishers], New York), 1980. Functional analysis
19. W. Rudin, *Real and Complex Analysis*, 3rd edn. (McGraw-Hill Book Co., New York, 1987)
20. N.A. Veniaminov, Limite thermodynamique pour un système de particules quantiques en interaction dans un milieu aléatoire. PhD thesis, Université Paris 13, 2012
21. N.A. Veniaminov, The existence of the thermodynamic limit for the system of interacting quantum particles in random media. Ann. Henri Poincaré **14**(1), 63–94 (2013)
22. R. Vershynin, Introduction to the non-asymptotic analysis of random matrices, in *Compressed Sensing* (Cambridge University Press, Cambridge, 2012), pp. 210–268

Entropies for Negatively Curved Manifolds

François Ledrappier and Lin Shu

2010 Mathematics Subject Classification 37D40, 58J65

This is a survey of several notions of entropy related to a compact manifold of negative curvature and of some relations between them. Namely, let (M, g) be a C^∞ compact boundaryless Riemannian connected manifold with negative curvature. After recalling the basic definitions, we will define and state the first properties of

(1) the volume entropy V,
(2) the dynamical entropies of the geodesic flow, in particular the entropy H of the Liouville measure and the topological entropy (which coincides with V),
(3) the stochastic entropy h_ρ of a family of (biased) diffusions related to the stable foliation of the geodesic flow,
(4) the relative dynamical entropy of natural stochastic flows representing the (biased) diffusions.

Most of the material in this survey are not new, some are classical, and we apologize in advance for any inaccuracy in the attributions. New observations are Theorems 2.5 and 4.9, but the main goal of this survey is to present together related notions that are spread out in the literature. In particular, we are interested in the different so-called rigidity results and problems that (aim to) characterize locally

The second author was partially supported by NSFC (No.11331007 and No.11422104).

F. Ledrappier (✉)
Sorbonne Université, UMR 8001, LPSM, Boîte Courrier 158, 4, Place Jussieu, 75252 Paris Cedex 05, France
e-mail: fledrapp@nd.edu

L. Shu
LMAM, School of Mathematical Sciences, Peking University, 100871 Beijing, People's Republic of China
e-mail: lshu@math.pku.edu.cn

N. Anantharaman et al. (eds.), *Frontiers in Analysis and Probability*,
https://doi.org/10.1007/978-3-030-56409-4_6

symmetric spaces among negatively curved manifolds by equalities in general entropy inequalities.

These notes grew out from lectures delivered by the second author in the workshop *Probabilistic methods in negative curvature* in ICTS, Bengaluru, India, and we thank Riddhipratim Basu, Anish Ghosh, and Mahan Mj for giving us this opportunity. We also thank Nalini Anantharaman, Ashkan Nikeghbali for organizing the 2nd Strasbourg/Zurich Meeting on *Frontiers in Analysis and Probability* and Michail Rassias for allowing us to publish these notes that have only a loose connection with the talk of the first author there.

1 Local Symmetry and Volume Growth

Let (M, g) be a C^∞ compact boundaryless connected d-dimensional Riemannian manifold and for u, v vector fields on M we denote $\nabla_u v$ the covariant derivative of v in the direction of u. Given $u, v \in T_x M$, the *curvature tensor* R associates with a vector $w \in T_x M$ the vector $R(u, v)w$ given by

$$R(u, v)w \;=\; \nabla_u \nabla_v w - \nabla_v \nabla_u w - \nabla_{[u,v]} w.$$

The space (M, g) is called *locally symmetric* if $\nabla R = 0$.

Consider the case (M, g) has *negative sectional curvature*; i.e., for all non-colinear $u, v \in T_x M$, $x \in \widetilde{M}$, the *sectional curvature* $K(u, v) := \dfrac{< R(u, v)v, u >}{|u \wedge v|^2}$ is negative. Simply connected locally symmetric spaces of negative sectional curvature are non-compact. They have been classified and are one of the hyperbolic spaces $\mathbb{H}^n_{\mathbb{R}}, \mathbb{H}^n_{\mathbb{C}}, \mathbb{H}^n_{\mathbb{H}}, \mathbb{H}^2_{\mathbb{O}}$, respectively of dimension respectively $n, 2n, 4n, 16$. Hyperbolic spaces are obtained as quotients of semisimple Lie groups of real rank one (respectively $SO(n, 1)$, $SU(n, 1)$, $Sp(n, 1)$, $F_{4(-20)}$), endowed with the metrics coming from the Killing forms, by maximal compact subgroups. By general results of Borel [6] and Selberg [51], these spaces admit compact boundaryless quotient manifolds and those locally symmetric (M, g_0) are the basic examples of our objects of study. Clearly, C^2 small C^∞ perturbations of g_0 on the same space M yield other examples of compact negatively curved manifolds. Different examples of non-locally symmetric, compact, negatively curved manifolds have been constructed (see [16, 18, 22, 45]). They are supposed to be abundant, even if constructing explicit ones is often delicate.

It is natural to ask if we can recognize locally symmetric spaces through global properties or quantities. One supportive example is the volume entropy. Let $\widetilde{M}$ be the universal cover space of M such that $M = \widetilde{M}/\Gamma$, where $\Gamma := \Pi_1(M)$ is the fundamental group of M, and endow $\widetilde{M}$ with metric $\widetilde{g}$, which is the Γ-invariant extension of g. The volumes on (M, g) and $(\widetilde{M}, \widetilde{g})$ are denoted Vol_g and $\mathrm{Vol}_{\widetilde{g}}$, respectively. (We will fix a connected fundamental domain M_0 for the action of Γ on $\widetilde{M}$. The restriction of $\mathrm{Vol}_{\widetilde{g}}$ on M_0 is also denoted Vol_g.) For $x \in \widetilde{M}$, let

$B_{\widetilde{M}}(x,r), r > 0$, denote the ball centered at x with radius r. The following limit exists (independent of $x \in \widetilde{M}$) and defines the *volume entropy* (Manning, [43]):

$$V(g) := \lim_{r\to\infty} \frac{1}{r} \log \mathrm{Vol}_{\widetilde{g}} B_{\widetilde{M}}(x,r).$$

Since (M,g) is negatively curved, by Bishop comparison theorem, $V(g) > 0$. The following rigidity result is shown by Besson–Courtois–Gallot [5]:

Theorem 1.1 ([5]) *Let (M, g_0) be closed locally symmetric space of negative curvature, and consider another metric g on M with negative curvature and such that* $\mathrm{Vol}_g(M) = \mathrm{Vol}_{g_0}(M)$. *Then,*

$$V(g) \geq V(g_0).$$

If $d = dim(M) > 2$, one has equality only if (M,g) is isometric to (M, g_0).

If $d = 2$, equality holds if, and only if, the curvature is constant (Katok, [30]). In the case $d > 2$, Katok [30] proved Theorem 1.1 under the hypothesis that g is conformally equivalent to g_0.

Remark 1.2 The theorem holds even if g' is a metric on another manifold M', homotopically equivalent to M.

The locally symmetric property can also be interpreted as geodesic symmetry. A *geodesic* in M is a curve $t \mapsto \gamma(t), t \in \mathbb{R}$, such that, if $\dot\gamma(t) := \frac{d}{dt}\gamma(s)\big|_{s=t}$, satisfies $\nabla_{\dot\gamma(t)}\dot\gamma(t) = 0$ for all t. For all $v \in TM$, there is a unique geodesic γ_v such that $\dot\gamma_v(0) = v$. The *exponential map* $\exp_x : T_xM \to M$ is given by $\exp_x v = \gamma_v(1)$. By compactness, there exists $\iota > 0$ such that, for all $x \in M$, $\exp_x$ is a diffeomorphism between the ball of radius ι in (T_xM, g_x) and the ball of radius ι about x in M. The Cartan–Ambrose–Hicks Theorem implies that the space is locally symmetric if, and only if, for any $x \in M$, the geodesic symmetry about x defined by $y \mapsto \exp_x(-\exp_x^{-1} y)$ is a local isometry.

One natural dynamics related to geodesics is the geodesic flow. Let $SM := \{v, v \in TM : \|v\| = 1\}$ be the unit tangent bundle. The *geodesic flow* φ_t on SM is such that $\varphi_t(v) = \dot\gamma_v(t)$ for $t \in \mathbb{R}$. Denote $\overline{X}(v) \in T_vSM$ the vector field on SM generating the geodesic flow. The derivative $D_v\varphi_t$ is described using *Jacobi fields.* Let $s \mapsto v(s)$ be a curve in SM with $v(0) = v$, $\dot v(0) = w \in T_vSM$. Then, $s \mapsto \gamma_{v(s)}(t)$ is a curve with tangent vector $J(t)$ at $\gamma_v(t)$. $J(t)$ satisfies the *Jacobi equation:*

$$\nabla_{\dot\gamma}\nabla_{\dot\gamma}J(t) + R(J(t), \dot\gamma(t))\dot\gamma(t) = 0. \tag{1.1}$$

Proof By definition,

$$R(J(t), \dot\gamma(t))\dot\gamma(t) = \nabla_{J(t)}\nabla_{\dot\gamma(t)}\dot\gamma(t) - \nabla_{\dot\gamma(t)}\nabla_{J(t)}\dot\gamma(t) - \nabla_{[J(t),\dot\gamma(t)]}\dot\gamma(t).$$

We have $\nabla_{\dot\gamma(t)}\dot\gamma(t) = 0$ by definition, $[J(t), \dot\gamma(t)] = [\frac{\partial}{\partial s}, \frac{\partial}{\partial t}] = 0$ and so $\nabla_{J(t)}\dot\gamma(t) = \nabla_{\dot\gamma(t)}J(t)$ (we use the fact that $\nabla_u v - \nabla_v u = [u, v]$). □

We will consider C^∞ compact boundaryless connected Riemannian manifolds with negative sectional curvature. It follows from (1.1) that $t \mapsto \|J(t)\|^2$ is a strictly convex function (by a direct computation). In particular, $\exp_x$ is a diffeomorphism from $T_x M$ to the universal cover $\widetilde{M}$. Two geodesic rays γ_1, γ_2 in $\widetilde{M}$ are said to be equivalent if $\sup_{t\geq 0} d(\gamma_1(t), \gamma_2(t)) < \infty$. The space of equivalence classes $\partial\widetilde{M} := \{[\gamma_v(t), t \geq 0], v \in T\widetilde{M}\}$ is the *geometric boundary at infinity*. For $x \in \widetilde{M}, \pi_x : S_x\widetilde{M} \to \partial\widetilde{M}, \pi_x(v) = [\gamma_v(t), t \geq 0]$ is one-to-one (π_x is injective by convexity (of $t \mapsto d(\gamma_v(t), \gamma_w(t))$ for $w \in S_x\widetilde{M}$ with $w \neq v$) and for any geodesic ray γ, any $t > 0$, one can find $v_t \in S_x\widetilde{M}$ such that $\gamma(t) \in \gamma_{v_t}(s), s \geq 0$; any limit point v of $v_t, t \to +\infty$, is such that γ_v is equivalent to γ). Thus, the unit tangent bundle $S\widetilde{M}$ is identified with $\widetilde{M} \times \partial\widetilde{M}$. For any two points ξ, η in $\partial\widetilde{M}$, there is a unique geodesic $\gamma_{\eta,\xi}$ (up to time translation) such that $\gamma_{\eta,\xi}(+\infty) := \lim_{t\to+\infty}\gamma_{\eta,\xi}(t) = \xi$ and $\gamma_{\eta,\xi}(-\infty) := \lim_{t\to-\infty}\gamma_{\eta,\xi}(t) = \eta$. The topology on $\widetilde{M} \times \partial\widetilde{M}$ is such that two pairs (x, ξ) and (y, η) are close if x and y are close and the distance from x to the geodesic $\gamma_{\eta,\xi}$ is large. The group Γ acts discretely and cocompactly on $\widetilde{M}$. The action of Γ extends continuously to $\partial\widetilde{M}$ and the diagonal action of Γ on $\widetilde{M} \times \partial\widetilde{M}$ is again discrete and cocompact. The quotient $(\widetilde{M} \times \partial\widetilde{M})/\Gamma = S\widetilde{M}/\Gamma$ is identified with SM.

We continue to use φ_t to denote the geodesic flow on $S\widetilde{M}$. It has the *Anosov property* [3]: each $\varphi_t, t \neq 0$, has no fixed point and there is a continuous decomposition $\{T_vS\widetilde{M} = E^{ss}(v) \oplus \overline{X}(v) \oplus E^{su}(v), v \in S\widetilde{M}\}$ with $\overline{X}(v)$ being the geodesic spray tangent to the flow direction and constants $C, C > 0, \lambda, \lambda \in (0, 1)$, such that, for $t > 0$,

$$\|D_v\varphi_t w_s\| \leq C\lambda^t\|w_s\|,\ \forall w_s \in E^{ss}(v),\ \|D_v\varphi_{-t}w_u\| \leq C\lambda^t\|w_u\|,\ \forall w_u \in E^{su}(v).$$

For $v = (x, \xi) \in S\widetilde{M}$, the *stable manifold at* v of the geodesic flow,

$$\widetilde{W}^s(v) := \left\{w : \sup_{t\geq 0} d(\varphi_t w, \varphi_t v) < +\infty\right\}$$

is tangent to $E^{ss}(v) \oplus \overline{X}(v)$. The $\widetilde{W}^s(v)$ can be identified with $\widetilde{M} \times \{\xi\}$ and hence is endowed naturally with the metric $\widetilde{g}$. The quotient $(\widetilde{M} \times \{\xi\})/\Gamma$ is the *stable manifold* $W^s(v)$. As ξ varies, they form a Hölder continuous lamination $\mathcal{W}^s$ of SM into C^∞ manifolds of dimension d which is called the *stable foliation*. Therefore, the metric on each individual stable manifold comes from the local identification with $\widetilde{M}$. The *strong stable manifold at* v,

$$\widetilde{W}^{ss}(v) := \left\{(y, \xi) : \lim_{t\to+\infty} d(\gamma_{x,\xi}(t), \gamma_{y,\xi}(t)) = 0\right\}$$

has tangent $E^{ss}(v)$. Let $\underline{v}$ be the projection of v on SM; then, $\widetilde{W}^{ss}(v)$ projects onto

$$W^{ss}(\underline{v}) := \left\{ w \in SM : \lim_{t \to +\infty} d(\gamma_w(t), \gamma_{\underline{v}}(t)) = 0 \right\}.$$

The collection of $\{W^{ss}(\underline{v}), \underline{v} \in SM\}$ forms a Hölder continuous lamination $\mathcal{W}^{ss}$ of SM into C^∞ manifolds of dimension $d - 1$ which is called the *strong stable foliation*.

For $v = (x, \xi) \in S\widetilde{M}$, define the *Busemann function*

$$b_{x,\xi}(y) = b_{x,\xi}(y, \xi) := \lim_{z \to \xi} (d(y, z) - d(x, z)), \ \forall y \in \widetilde{M}.$$

The level set $\{(y, \xi) : b_{x,\xi}(y, \xi) = 0\}$ coincides with $\widetilde{W}^{ss}(x, \xi)$ and the set of its foot points is the horosphere of (x, ξ). Denote Div^s, ∇^s the divergence and gradient along $\widetilde{W}^s$ (and W^s) induced by the metric $\widetilde{g}$ on $\widetilde{M} \times \{\xi\}$, $\Delta^s = \mathrm{Div}^s \nabla^s$. Then,

$$\nabla_y b_{x,\xi}(y)|_{y=x} = -(x, \xi) \text{ or } \nabla^s_w b_v(w)|_{w=v} = -\overline{X}(v).$$

Set

$$B(x, \xi) := \Delta_y b_{x,\xi}(y)|_{y=x} = -\mathrm{Div}^s \overline{X}(v).$$

Geometrically, the $B(x, \xi)$ is the mean curvature at x of the horosphere of (x, ξ). The function B is a Γ-invariant function on $S\widetilde{M}$. We still denote B the function on the quotient SM. From the definition follows:

$$B(v) = -\frac{d}{dt} \log \mathrm{Det} D_v \varphi_t|_{W^{ss}(v)}\Big|_{t=0}. \tag{1.2}$$

So, dynamically, $-B$ tells the exponential growth rate of the volume on W^{ss} under the geodesic flow φ_t, $t > 0$. It follows from (1.2) that the function B is Hölder continuous on SM. The main property of the function B is the following, whose proof combines the works of Benoist–Foulon–Labourie [4], Foulon–Labourie [20], and Besson–Courtois–Gallot [5].

Theorem 1.3 ([4, 5, 20]) *The function B is constant if, and only if, the space (M, g) is locally symmetric.*

Remark 1.4 There is a positive operator U on the orthogonal space to v in $T_x M$ satisfying the Riccati equation $\dot{U} + U^2 + R(\cdot, \dot{\gamma}(t))\dot{\gamma}(t) = 0$ and such that $B = \mathrm{Tr} U$. If $d = 2$, the equation reduces to $\dot{B} + B^2 + K = 0$. Clearly, if B is constant, then the curvature K is the constant $-B^2$. If $d = 3$, one can also conclude from the Riccati equation and some matrix calculations that B is constant if, and only if, the sectional curvature is constant (see Knieper [33]).

2 Dynamical Entropy and an Application of Thermodynamical Formalism

More quantities related to V, B can be introduced through a dynamical point of view.

2.1 Dynamical Entropy

Let T be a continuous transformation of a compact metric space $\mathbf{X}$. For $x \in \mathbf{X}, \varepsilon > 0, n \in \mathbb{N}$, define the *Bowen ball* $B(x, \varepsilon, n)$

$$B(x, \varepsilon, n) := \{y \in \mathbf{X} : d(T^j y, T^j x) < \varepsilon \text{ for } 0 \leq j \leq n\}$$

and the *entropy* $h_m(T)$ of a T-invariant probability measure m

$$h_m(T) := \sup_{\varepsilon} \int \left(\limsup_{n} -\frac{1}{n} \log m(B(x, \varepsilon, n)) \right) dm(x).$$

It is easy to see that for $j \in \mathbb{Z}, h_m(T^j) = |j| h_m(T)$. A useful upper bound of $h_m(T)$ is given by Ruelle inequality [50] using the average maximal exponential growth rate of all the parallelograms under the iteration of the tangent map DT.

Theorem 2.1 (Ruelle, [50]) *Assume* $\mathbf{X}$ *is a compact manifold and* T *a* C^1 *mapping of* $\mathbf{X}$. *Then, for any* T*-invariant probability measure* m,

$$h_m(T) \leq \int \left(\sup_k \limsup_n \frac{1}{n} \log \| \wedge^k D_x T^n \| \right) dm(x),$$

where $\wedge^k D_x T^n$ *denotes the k-th exterior power of* $D_x T^n$.

Corollary 2.2 *If* $\mathbf{X} = SM$, *where* (M, g) *is a compact, boundaryless,* C^2 *Riemannian manifold with negative sectional curvature and dimension* d, m *a geodesic flow invariant probability measure, and* $t \in \mathbb{R}$,

$$h_m(\varphi_t) \leq |t| \int_{SM} B\, dm.$$

Proof For $v \in SM, t < 0, |t|$ large, the highest value of $\| \wedge^k D_v \varphi_t \|$ is obtained for $k = d - 1$ and is the Jacobian of $D_v \varphi_t$ restricted to $T_v W^{ss}(v)$. By (1.1), this is $e^{\int_t^0 B(\varphi_s v)\, ds}$. By the ergodic theorem,

$$\lim_{n \to +\infty} \frac{1}{n} \log \| \wedge^{d-1} D_v \varphi_{nt}|_{W^{ss}} \| = \lim_{n \to +\infty} \frac{1}{n} \int_{nt}^0 B(\varphi_s v)\, ds$$

exists and has integral $|t| \int B\, dm$. The conclusion follows by Ruelle inequality. □

Another general inequality is given by

Theorem 2.3 (Manning, [43]) *Let (M, g) be a compact, boundaryless, C^2 Riemannian manifold with negative sectional curvature and dimension d, m a geodesic flow invariant probability measure, and $t \in \mathbb{R}$,*

$$h_m(\varphi_t) \leq |t| V.$$

Remark 2.4 The proof of Theorem 2.3 is based on the following consequence of nonpositive curvature ([43], Lemma page 571). For any $v, w \in SM$, any $r \geq 1$,

$$\begin{aligned}\max\{\sup_{0\leq s\leq 1} d(\varphi_s v, \varphi_s w), \sup_{r-1\leq s\leq r} d(\varphi_s v, \varphi_s w)\} &\leq \sup_{0\leq s\leq r} d(\varphi_s v, \varphi_s w)\\ &\leq \sup_{0\leq s\leq 1} d(\varphi_s v, \varphi_s w)\\ &\quad + \sup_{r-1\leq s\leq r} d(\varphi_s v, \varphi_s w).\end{aligned}$$

This observation can also be used to give a direct proof of Corollary 2.2.

2.2 Thermodynamical Formalism

For simplicity, we introduce the notion of pressure by the classical variational principle. Let $(\mathbf{X}, T)$ be a continuous mapping of a compact metric space. The *Pressure* $P(F)$ of a continuous function $F : \mathbf{X} \to \mathbb{R}$ is defined by

$$P(F) := \sup_m \left\{ h_m(T) + \int F\, dm \right\},$$

where m runs over all T-invariant probability measures. Let $\mathbf{X} = SM$, where M is closed negatively curved and $T = \varphi_1$. From Ruelle and Manning inequalities follow

$$P(-B) \leq 0 \quad \text{and} \quad P(0) \leq V.$$

We will construct later the *Liouville measure* m_L with the property (Theorem 2.6)

$$h_{m_L}(\varphi_1) = \int B\, dm_L =: H \tag{2.1}$$

and the *Bowen–Margulis measure* m_{BM} such that (Theorem 3.3)

$$h_{m_{BM}}(\varphi_1) = V. \tag{2.2}$$

This will show that $P(-B) = 0$ and $P(0) = V$. Using these properties, we can prove:

Theorem 2.5 *Let (SM, φ_t) be the geodesic flow on a closed manifold of negative curvature. Let $\mathcal{M}$ be the set of φ_t-invariant probability measures, H and V as defined above. Then,*

$$\inf_{m\in\mathcal{M}} \int B\,dm \leq H \leq V \leq \sup_{m\in\mathcal{M}} \int B\,dm, \tag{2.3}$$

with equality in one of the inequalities if, and only if, $m_L = m_{BM}$. Moreover, in that case, $\int B\,dm = V$ for all $m \in \mathcal{M}$.

Proof Since the function B is Hölder continuous on SM, for each $s \in \mathbb{R}$, there exists a unique invariant probability measure m_s (equilibrium measure for sB) such that P$(s) := P(sB) = h_{m_s}(\varphi_1) + s\int B\,dm_s$ [46, Proposition 4.10].[1] For example, by (2.1), (2.2), m_L, m_{BM} are equilibrium measures for $-B$ and 0, respectively. Together with Corollary 2.2, we obtain

$$\inf_{m\in\mathcal{M}} \int B\,dm \leq \int B\,dm_L = H \leq \sup_{m\in\mathcal{M}} \{h_m(\varphi_1)\} = V \leq \int B\,dm_{BM} \leq \sup_{m\in\mathcal{M}} \int B\,dm,$$

which gives (2.3).

Clearly, using the uniqueness of m_s, we have that $H = V$ if, and only if $m_L = m_{BM}$. To show any equality in the other inequalities of (2.3) holds if, and only if, $m_L = m_{BM}$, we use properties of the Pressure function, in particular of the convex function $s \mapsto$ P(s). We already know that P$(-1) = 0$ and that P$(0) = V$. From the definition follows that $\inf_{m\in\mathcal{M}} \int B\,dm$ and $\sup_{m\in\mathcal{M}} \int B\,dm$ are the slopes of the asymptotes of the function P(s) as $s \to -\infty$ and $+\infty$, respectively. Since the function B is Hölder continuous on SM, the function $s \mapsto$ P(s) is real analytic [46, Proposition 4.8]. Moreover, the slope at s is given by $\int B\,dm_s$ [46, Proposition 4.10]. Now, if $H = \inf_{m\in\mathcal{M}} \int B\,dm$, the function $s \mapsto$ P(s) is affine on $[-\infty, -1]$ and thus everywhere. Since the slopes of P(s) at -1 and 0 are $\int B\,dm_L = H$ and $\int B\,dm_{BM}$, respectively, and $H \leq V \leq \int B\,dm_{BM}$, hence we must have $V = \int B\,dm_{BM}$, which implies that m_{BM} coincides with m_L and $V = H$. Finally, if $V = \sup_{m\in\mathcal{M}} \int B\,dm$, the measure m_{BM} is the equilibrium measure for $-B$, which must coincide with m_L.

Assume m_{BM} and m_L coincide, then by [46, Proposition 4.9], there exists a continuous function F on SM, C^1 along the trajectories of the geodesic flow, such that

[1] Chapter 4 in [46] is only concerned with subshifts of finite type. The extension of [46] Propositions 4.8, 4.9, 4.10 to suspended flows is direct (see [46], Chapter 6) and the application to geodesic flows on compact negatively curved manifolds is standard (cf. [46], Appendix 3).

$$-B = \mathrm{P}(-1) - \mathrm{P}(0) + \frac{\partial}{\partial t} F \circ \varphi_t \big|_{t=0}.$$

In particular, $\int B\, dm = P(0) = V$ for all $m \in \mathcal{M}$. □

2.3 Liouville Measure

For $x \in \widetilde{M}$, let λ_x denote the pull back measure on $\partial \widetilde{M}$ of the Lebesgue probability measure on $S_x\widetilde{M}$ through the mapping $\pi_x^{-1} : \partial \widetilde{M} \mapsto S_x \widetilde{M},\ \xi \mapsto (x, \xi)$. Define a measure $\widetilde{m}_L$ on $\widetilde{M} \times \partial \widetilde{M}$ by setting

$$\int F(x,\xi)\, d\widetilde{m}_L = \int_{\widetilde{M}} \left(\int_{\partial \widetilde{M}} F(x,\xi)\, d\lambda_x(\xi) \right) \frac{d\mathrm{Vol}_{\widetilde{g}}(x)}{\mathrm{Vol}_g(M)}.$$

It is clear from the definition that the measure $\widetilde{m}_L$ is Γ-invariant. There is a $D\varphi_t$-invariant 2-form on $\overline{X}^{\perp}$ in TSM defined by the Wronskian $\mathcal{W}$

$$\mathcal{W}\big((J_1, J_1'), (J_2, J_2')\big) \ := < J_1(t), J_2'(t) > - < J_1'(t), J_2(t) > .$$

Assume M is orientable. The $(2d-1)$-form $\wedge^{d-1}\mathcal{W} \wedge dt$ is nondegenerate and invariant. For $v \in SM$, take a positively oriented orthonormal basis $\{e_0, \cdots, e_{n-1}\}$ in T_xM such that $e_0 = v$. By computing $\wedge^{d-1}\mathcal{W} \wedge dt$ on the $(2d-1)$-vector $\big((e_1, 0), (0, e_1), \cdots, (e_{n-1}, 0), (0, e_{n-1}), \overline{X}\big)$, one sees that the measure associated with this volume form is the one we defined. So the measure $\widetilde{m}_L$ is invariant under the geodesic flow. We do the same computation on a double cover of M if M is not orientable.

The measure m_L on SM that extends to $\widetilde{m}_L$ is a φ_t-invariant probability measure which is called the *Liouville probability measure*. It satisfies

Theorem 2.6 *For all* $t \in \mathbb{R}$, $h_{m_L}(\varphi_t) = |t| \int B\, dm_L$.

Proof (Sketch) It suffices to prove the theorem for $t = -1$. In the definition of entropy, we can use the flow Bowen balls $\mathbf{B}(v, \varepsilon, r)$, $\varepsilon, r > 0$,

$$\mathbf{B}(v, \varepsilon, r) := \left\{ w : \sup_{-r \le s \le 0} d(\varphi_s v, \varphi_s w) < \varepsilon \right\}.$$

By Remark 2.4,

$$\mathbf{B}(v, \varepsilon/2, 1) \cap \varphi_{r-1}\mathbf{B}(\varphi_{-r+1}v, \varepsilon/2, 1) \subset \mathbf{B}(v, \varepsilon, r) \subset \mathbf{B}(v, \varepsilon, 1) \cap \varphi_{r-1}\mathbf{B}(\varphi_{-r+1}v, \varepsilon, 1).$$

Estimating the Liouville measure of $\mathbf{B}(v, \varepsilon, 1) \cap \varphi_{r-1}\mathbf{B}(\varphi_{-r+1}v, \varepsilon, 1)$ reduces to estimating the d-dimensional measure of $B^s(v, \varepsilon) \cap \varphi_{r-1}B^s(\varphi_{-r+1}v, \varepsilon)$, where $B^s(v, a)$ is the ball of radius a and center v in $W^s(v)$. It follows from (1.2) that this measure is, up to error terms that depend on ε small enough, but not on r, equal to

$$\mathrm{Det} D_{\varphi_{-r+1}v}\varphi_r|_{W^s(\varphi_{-r+1}v)} = e^{-\int_{-r+1}^0 B(\varphi_s v)\,ds}.$$

It follows that, if one takes ε small enough,

$$h_{m_L}(\varphi_{-1}) = \lim_{r\to+\infty}\frac{1}{r}\int_{SM}\left(\int_{-r+1}^0 B(\varphi_s v)\,ds\right) dm_L(v) = \int_{SM} B\,dm_L.$$

□

Observe that, since m_L is a measure realizing the maximum in $P(-B)$, it is ergodic.

Remark 2.7 Basic facts about ergodic theory and thermodynamic formalism are in Bowen [8]; see also Parry–Pollicott [46]. The definition of the entropy given here is due to Brin–Katok [11]. The ergodicity of m_L with respect to the geodesic flow is a landmark result of Anosov [3].

3 Patterson–Sullivan, Bowen–Margulis, Burger–Roblin

In analogy to the construction of the measure m_L, one can obtain the Bowen–Margulis measure m_{BM} using a class of measures (Patterson–Sullivan measures) on the boundary at infinity.

3.1 Patterson–Sullivan

Theorem 3.1 *There exists a family of measures on $\partial\widetilde{M}$, $x \mapsto \nu_x$, $x \in \widetilde{M}$, such that*

$$\nu_{\beta x} = \beta_*\nu_x, \text{ for } \beta\in\Gamma, \text{ and } \frac{d\nu_y}{d\nu_x}(\xi) = e^{-V b_{x,\xi}(y)}. \tag{3.1}$$

The family is unique if normalized by $\int_M \nu_x(\partial\widetilde{M})\,d\mathrm{Vol}_g(x) = 1$. *Moreover, the measures ν_x are continuous.*

Proof We first show the existence of such a family. Fix $x_0 \in \widetilde{M}$. It suffices to construct the family $\nu_{\beta x_0}$, $\beta\in\Gamma$, such that

$$\text{for all } \beta\in\Gamma,\ \nu_{\beta x_0} = \beta_*\nu_{x_0} \text{ and } \frac{d\nu_{\beta x_0}}{d\nu_{x_0}}(\xi) = e^{-V b_{x_0,\xi}(\beta x_0)}. \tag{3.2}$$

Indeed, assume such a family $\nu_{\beta x_0}, \beta \in \Gamma$, is constructed, we then set $\nu_y := e^{-Vb_{x_0,\xi}(y)}\nu_{x_0}$ for all $y \in \widetilde{M}$. Using the cocycle property of the Busemann function:

$$b_{x,\xi}(\beta y) = b_{x,\beta^{-1}\xi}(y) + b_{x,\xi}(\beta x), \ \forall x, y \in \widetilde{M}, \xi \in \partial\widetilde{M},$$

one can easily check that the class of measures $\{\nu_y\}$ satisfies the requirement of (3.1).

Recall $V = \lim_{R\to+\infty} \frac{1}{R} \log \mathrm{Vol}_{\widetilde{g}} B_{\widetilde{M}}(x_0, R)$. Set, for $s > V$, a family $\nu^s_{\beta x_0}, \beta \in \Gamma$, with $d\nu^s_{\beta x_0}(y) := \dfrac{e^{-sd(\beta x_0,y)}\, d\mathrm{Vol}_{\widetilde{g}}(y)}{\int_{\widetilde{M}} e^{-sd(x_0,y)}\, d\mathrm{Vol}_{\widetilde{g}}(y)}$. We have

$$\begin{aligned} \beta_* d\nu^s_{x_0}(y) &= d\nu^s_{x_0}(\beta^{-1}y) = \frac{e^{-sd(x_0,\beta^{-1}y)}\, d\mathrm{Vol}_{\widetilde{g}}(y)}{\int_{\widetilde{M}} e^{-sd(x_0,y)}\, d\mathrm{Vol}_{\widetilde{g}}(y)} \\ &= \frac{e^{-sd(\beta x_0,y)}\, d\mathrm{Vol}_{\widetilde{g}}(y)}{\int_{\widetilde{M}} e^{-sd(x_0,y)}\, d\mathrm{Vol}_{\widetilde{g}}(y)} = d\nu^s_{\beta x_0}(y). \end{aligned}$$

Recall that $\widetilde{M} \cup \partial\widetilde{M}$ is compact and assume that $\int_{\widetilde{M}} e^{-sd(x_0,y)}\, d\mathrm{Vol}_{\widetilde{g}}(y) \to \infty$ as $s \searrow V$. Choose $s_n \searrow V$ such that $\nu^{s_n}_{x_0}$ weak* converge to ν_{x_0}. Then, ν_{x_0} is supported by $\partial\widetilde{M}$. Moreover, for any $\beta \in \Gamma$, $\nu^{s_n}_{\beta x_0}$ weak* converge as well and call $\nu_{\beta x_0} := \lim_{s_n\searrow V} \nu^{s_n}_{\beta x_0}$. The family $\nu_{\beta x_0}, \beta \in \Gamma$, satisfies (3.2). Indeed, $\nu_{\beta x_0} = \beta_*\nu_{x_0}$. Moreover, consider an open cone C based on x_0. We have, for any $\beta \in \Gamma$,

$$\nu_{\beta x_0}(C) = \lim_{s_n\searrow V} \nu^{s_n}_{\beta x_0}(C) = \lim_{s_n\searrow V} \int_C e^{-s_n(d(\beta x_0,y)-d(x_0,y))}\, d\nu^{s_n}_{x_0}(y).$$

As $s_n \searrow V$, most of the $\nu^{s_n}_{x_0}$ measure is supported by a neighborhood of $\partial\widetilde{M}$ and, for y close to $\xi \in \partial\widetilde{M}$, $d(\beta x_0, y) - d(x_0, y)$ is close to $b_{x_0,\xi}(\beta x_0)$. The density property follows.

If $\int_{\widetilde{M}} e^{-sd(x_0,y)}\, d\mathrm{Vol}_{\widetilde{g}}(y)$ is bounded, use Patterson's trick [47, Lemma 3.1]: one can find a real function L on $\mathbb{R}_+$ such that

$$\lim_{s\searrow V} \int_{\widetilde{M}} L(d(x_0, y))e^{-sd(x_0,y)}\, d\mathrm{Vol}_{\widetilde{g}}(y) = \infty \quad \text{and} \quad \forall a \in \mathbb{R}, \ \lim_{t\to+\infty} \frac{L(t+a)}{L(t)} = 1.$$

We can then replace the previous family $\nu^s_{\beta x_0}, \beta \in \Gamma$, by $\nu'^s_{\beta x_0}\beta \in \Gamma$, with $d\nu'^s_{\beta x_0}(y) := \frac{L(d(\beta x_0,y))e^{-sd(\beta x_0,y)}\, d\mathrm{Vol}_{\widetilde{g}}(y)}{\int_{\widetilde{M}} L(d(x_0,y))e^{-sd(x_0,y)}\, d\mathrm{Vol}_{\widetilde{g}}(y)}$.

The function $x \mapsto \nu_x(\partial\widetilde{M})$ is Γ-invariant and continuous; in particular, it is bounded. This implies that the measure ν_{x_0} is continuous since otherwise, there is $\xi \in \partial\widetilde{M}$ with $\nu_{x_0}(\{\xi\}) = a > 0$. When $\{y_n\}_{n\in\mathbb{N}} \in \widetilde{M}$ converge to ξ, $\nu_{y_n}(\{\xi\}) = e^{-Vb_{x_0,\xi}(y_n)}a \to +\infty$, a contradiction.

We will see later (Remark 3.5) that such a family is unique, up to multiplication by a constant factor. □

The family $\nu_x, x \in \widetilde{M}$, is called the family of *Patterson–Sullivan measures.*

3.2 Bowen–Margulis

Define, for $x \in \widetilde{M}, \xi, \eta \in \partial\widetilde{M}$, the *Gromov product*

$$(\xi, \eta)_x := \frac{1}{2} \lim_{y \to \xi, z \to \eta} \left(d(x, y) + d(x, z) - d(y, z) \right).$$

The Gromov product is a nonnegative number (by the triangle inequality) and because of pinched negative curvature, the Gromov product is finite; actually it is (exercise) uniformly bounded away from the distance from x to the geodesic $\gamma_{\eta,\xi}$. Moreover, the Gromov product satisfies the cocycle relation

$$(\xi, \eta)_{x'} - (\xi, \eta)_x = \frac{1}{2}(b_{x,\xi}(x') + b_{x,\eta}(x')). \tag{3.3}$$

Let $\widetilde{M}^{(2)} := \{(\xi, \eta) \in \partial\widetilde{M} \times \partial\widetilde{M}, \xi \neq \eta\}$. Then, $S\widetilde{M}$ is identified with $\widetilde{M}^{(2)} \times \mathbb{R}$ by the *Hopf coordinates:*

$$v \mapsto (\gamma_v(+\infty), \gamma_v(-\infty), b_v(x_0)).$$

Proposition 3.2 *Let $\nu_x, x \in \widetilde{M}$, be the family of Patterson–Sullivan measures. The measure ν with $d\nu(\xi, \eta) := \frac{d\nu_x(\xi) \times d\nu_x(\eta)}{e^{-2V(\xi,\eta)_x}}$ does not depend on x. The measure $\nu \times dt$ on $\widetilde{M}^{(2)} \times \mathbb{R}$ is Γ-invariant and invariant by the geodesic flow.*

Proof The first affirmation follows directly from the cocycle relation (3.3). In particular, the measure ν is Γ-invariant on $\partial\widetilde{M} \times \partial\widetilde{M}$. The measure ν is supported by $\widetilde{M}^{(2)}$ because ν_x is continuous. The actions of Γ and of φ_s in Hopf coordinates are given by:

$$\beta(\xi, \eta, t) = (\beta\xi, \beta\eta, t + b_{x_0,\xi}(\beta^{-1}x_0)), \text{ for } \beta \in \Gamma,$$
$$\varphi_s(\xi, \eta, t) = (\xi, \eta, t + s).$$

The invariance of $\nu \times dt$ under the actions of Γ and of φ_s follows. □

We call *Bowen–Margulis measure* m_{BM} the unique probability measure on SM such that its Γ-invariant extension is proportional to $\nu \times dt$. It satisfies

Theorem 3.3 $h_{m_{BM}}(\varphi_t) = |t|V.$

Proof (Sketch) We follow the sketch of the proof of Theorem 2.6. We have to estimate $m_{BM}\left(\mathbf{B}(v, \varepsilon, 1) \cap \varphi_{r-1}\mathbf{B}(\varphi_{-r+1}v, \varepsilon, 1)\right)$. Choose ε small enough that this set lifts to $S\widetilde{M}$ into a set of the same form. In Hopf coordinates, this is, up to some constant A, of the form:

$\mathbf{B}(v, \varepsilon, 1)$

$$\asymp \left\{(\xi, \eta, t) : \xi \in C(\varphi_{1/2}v, A^{\pm 1}\varepsilon), \eta \in C(-\varphi_{1/2}v, A^{\pm 1}\varepsilon), b_v(x_0) \leq t \leq b_v(x_0)+1\right\},$$

where, for $w \in S\widetilde{M}$ and $0 < \delta < \pi$, $C(w, \delta)$ is the cone of geodesics starting from w with an angle smaller than δ. Our set $\mathbf{B}(v, \varepsilon, 1) \cap \varphi_{r-1}\mathbf{B}(\varphi_{-r+1}v, \varepsilon, 1)$ is

$$\left\{(\xi, \eta, t) : \xi \in C(\varphi_{1/2}v, A^{\pm 1}\varepsilon), \eta \in C(-\varphi_{-r+3/2}v, A^{\pm 1}\varepsilon), b_v(x_0) \leq t \leq b_v(x_0)+1\right\}.$$

The $\nu \times dt$ measure of this set is within $A^{\pm 2}e^{(-r+3/2)V} m_{BM}(\mathbf{B}(v, \varepsilon, 1))$. □

Corollary 3.4 *$P(0) = V$ and m_{BM} is the measure of maximal entropy for the geodesic flow φ_t. In particular, m_{BM} is ergodic.*

Remark 3.5 It also follows from this construction that the Patterson–Sullivan family ν_x is unique. Indeed, let ν'_x be another Patterson–Sullivan family. One can construct as above a family ν', $d\nu'(\xi, \eta) := \frac{d\nu_x(\xi) \times d\nu'_x(\eta)}{e^{-2V(\xi,\eta)_x}}$. By the same reasoning, the measure $\nu' \times dt$ is proportional to an invariant probability measure with entropy V. It follows that ν' is proportional to ν; i.e., ν'_x is proportional to ν_x for all x.

3.3 Burger–Roblin

Define a measure $\widetilde{m}_{BR}$ on $\widetilde{M} \times \partial \widetilde{M}$ by setting, for all continuous function F with compact support on $S\widetilde{M}$,

$$\int F(x, \xi)\, d\widetilde{m}_{BR} = \int_{\widetilde{M}} \left(\int_{\partial \widetilde{M}} F(x, \xi)\, d\nu_x(\xi) \right) d\mathrm{Vol}_{\widetilde{g}}(x). \tag{3.4}$$

It follows from the definition that the measure $\widetilde{m}_{BR}$ is Γ-invariant. Call m_{BR} the induced measure on SM; by our normalization, we have $m_{BR}(SM) = 1$. The measure m_{BR} is called the *Burger–Roblin measure.* Many of its properties follow from

Theorem 3.6 *For any vector field Z on SM such that $Z(v)$ is tangent to $W^s(v)$ for all $v \in SM$, we have*

$$\int_{SM} \mathrm{Div}^s Z(v) + V < Z(v), \overline{X}(v) > dm_{BR}(v) \;=\; 0. \tag{3.5}$$

Proof Using a partition of unity, we may assume that Z has compact support inside a flow-box for the foliation. Choosing a reference point x_0, we can write $dm_{BR}(x, \xi) = e^{-Vb_{x_0,\xi}(y)} d\nu_{x_0}(y) d\mathrm{Vol}_{\widetilde{g}}(y)$. Since Z has compact support on each local stable leaf $W^s_{loc}(x, \xi)$, we have

$$\int_{W^s_{loc}(x,\xi)} \mathrm{Div}^s_y \left(e^{-Vb_{x_0,\xi}(y)} Z(y,\xi) \right) \Big|_{y=z} d\mathrm{Vol}_{\widetilde{g}}(z) = 0$$

for all $(x,\xi) \in S\widetilde{M}$. Then, (3.5) follows by developing

$$\mathrm{Div}^s_y \left(e^{-Vb_{x_0,\xi}(y)} Z(y,\xi) \right) \Big|_{y=z}$$
$$= \left(\mathrm{Div}^s_y Z(y,\xi) \Big|_{y=z} + V < Z(z,\xi), \overline{X}(z,\xi) > \right) e^{-Vb_{x_0,\xi}(z)}.$$

□

Corollary 3.7 $\int B \, dm_{BR} = V$.

Proof Apply (3.5) to $Z = \overline{X}$. □

Corollary 3.8 *The operator $\Delta^s + V\overline{X}$ is symmetric for m_{BR}: for $F_1, F_2 \in C^\infty(SM)$, the set of smooth functions on SM,*

$$\int_{SM} F_1(\Delta^s + V\overline{X})F_2 \, dm_{BR} = \int_{SM} F_2(\Delta^s + V\overline{X})F_1 \, dm_{BR}.$$

Hence, m_{BR} is also stationary for the operator $\Delta^s + V\overline{X}$, i.e., for all $F \in C^\infty(SM)$, $\int_{SM}(\Delta^s + V\overline{X})F \, dm_{BR} = 0$.

Proof Apply (3.5) to $Z = F_1 \nabla^s F_2$ to get

$$\int_{SM} F_1(\Delta^s + V\overline{X})F_2 \, dm_{BR} = -\int_{SM} < \nabla^s F_1, \nabla^s F_2 > dm_{BR}.$$

The Right Hand Side is invariant when switching F_1 and F_2. □

Corollary 3.9 *The measure m_{BR} is symmetric for the Laplacian Δ^{ss} along the strong stable foliation $\mathcal{W}^{ss}$: for $F_1, F_2 \in C^\infty(SM)$,*

$$\int_{SM} F_1 \Delta^{ss} F_2 \, dm_{BR} = \int_{SM} F_2 \Delta^{ss} F_1 \, dm_{BR}.$$

So, m_{BR} is also stationary for the operator Δ^{ss}, i.e., for all $F \in C^\infty(SM)$, $\int_{SM} \Delta^{ss} F \, dm_{BR} = 0$.

Proof Apply (3.5) to $Z = F_1 \frac{d}{dt} F_2 \circ \varphi_t \big|_{t=0} \overline{X}$ to obtain that

$$\int_{SM} F_1 \left(\frac{d^2}{dt^2} F_2 \circ \varphi_t \big|_{t=0} - B \frac{d}{dt} F_2 \circ \varphi_t \big|_{t=0} + V \frac{d}{dt} F_2 \circ \varphi_t \big|_{t=0} \right) dm_{BR}$$
$$= -\int_{SM} \overline{X} F_1 \overline{X} F_2 \, dm_{BR}.$$

Recall that in horospherical coordinates, Δ^s can be written as

$$\Delta^s F = \frac{d^2}{dt^2} F \circ \varphi_t \big|_{t=0} - B \frac{d}{dt} F \circ \varphi_t \big|_{t=0} + \Delta^{ss} F.$$

Replacing in the formula above, we get that

$$-\int_{SM} F_1 \Delta^{ss} F_2 \, dm_{BR} + \int_{SM} F_1 (\Delta^s + V\overline{X}) F_2 \, dm_{BR} \;=\; -\int_{SM} \overline{X} F_1 \overline{X} F_2 \, dm_{BR}.$$

The conclusion follows from Corollary 3.8. □

Remark 3.10 We observe that m_{BR} is ergodic. Indeed, strong stable manifolds have polynomial volume growth[2], so a symmetric measure for the Laplacian Δ^{ss} along the strong stable foliation $\mathcal{W}^{ss}$ is given locally by the product of the Lebesgue measure along the W^{ss} leaves and some family of measures on the transversals (Kaimanovich, [28]). This family has to be *invariant* under the holonomy map of the W^{ss} leaves. By Bowen–Marcus [9], there exists only one holonomy-invariant family on the transversals to the $\mathcal{W}^{ss}$ foliation, up to a multiplication by a constant factor.

Remark 3.11 The family of measures in this section has a long history. The invariant measures for the $\mathcal{W}^{ss}$ foliation were first constructed by Margulis [44] and used to construct the invariant measure m_{BM}. Margulis' construction (in the strong unstable case) amounts to taking the limit of the normalized Lebesgue measure on $\varphi_T S_x M$ (see also Knieper [33]). Margulis did not state that the measure m_{BM} has maximal entropy, and the measure of maximal entropy was constructed by Bowen (cf. Bowen [8], Bowen–Ruelle [10]) as the limit as $T \to +\infty$ of equidistributed measures on closed geodesics of length smaller than T. Bowen also showed that the measure of maximal entropy is unique, so that the two constructions give the same measure m_{BM}. Independently, Patterson [47] constructed the measures ν_x in the case of hyperbolic surfaces, not necessarily compact; Sullivan [52] extended the construction to a general hyperbolic space, observed that it is, up to normalization, the Hausdorff measure on the limit set of the discrete group in its Hausdorff dimension for the angle metric, that it is also the conformal measure for the action of the group on its limit set and moreover, the exit measure of the Brownian motion with suitable drift. He also made its connection with the measure of maximal entropy (in the constant curvature case). Hamenstädt [24] connected m_{BM} with the Patterson–Sullivan construction and then many authors extended the Patterson–Sullivan construction to many circumstances (see Paulin–Pollicott–Schapira [48] for a detailed recent survey). Again in the hyperbolic geometrically finite case, Burger [12] considered m_{BR} as the measure invariant by the horocycle action; finally,

[2] There are constants C, k such that the volume of the balls of radius r for the induced metric on strong stable manifolds is bounded by Cr^k.

Roblin [49] considered the general case of a group acting discretely on a $CAT(-1)$ space. What is remarkable is that in all these constructions, these measures were introduced as tools, and not, like here, as objects interesting in their own right. A posteriori, their interest comes from all these applications.

4 A Family of Stable Diffusions; Probabilistic Rigidity

Recall (Corollary 3.8) that the Burger–Roblin measure m_{BR} is a stationary measure for $\Delta^s + V\overline{X}$. In this section, we study the stationary measures for $\Delta^s + \rho\overline{X}$, $\rho < V$, characterize them in analogy to m_{BR}, and state a rigidity result concerning these measures.

4.1 Foliated Diffusions

A differential operator $\mathcal{L}$ on SM is called *subordinate to the stable foliation* $\mathcal{W}^s$ if, for any $F \in C^\infty(SM)$, $\mathcal{L}F(v)$ depends only on the values of F along $W^s(v)$. It is given by a Γ-equivariant family $\mathcal{L}_\xi$ on $\widetilde{M} \times \{\xi\}$. A probability measure m is called *stationary* for $\mathcal{L}$ (or *$\mathcal{L}$-stationary, $\mathcal{L}$-harmonic*) if, for all $F \in C^\infty(SM)$,

$$\int \mathcal{L}F(v)\, dm(v) \;=\; 0.$$

Theorem 4.1 (Garnett, [21]) *Assume $\mathcal{L}$-stationary is an operator which is subordinate to $\mathcal{W}^s$, has continuous coefficients, and is elliptic on W^s leaves. Then, the set of $\mathcal{L}$-stationary probability measures is a non-empty convex compact set. Extremal points are called ergodic.*

We will consider the operators $\mathcal{L}^\rho := \Delta^s + \rho\overline{X}$ for $\rho \in \mathbb{R}$. Clearly, each $\mathcal{L}^\rho$ is subordinate to $\mathcal{W}^s$ and for all $F \in C^\infty(SM)$,

$$\mathcal{L}^\rho_\xi F(x,\xi) = \Delta^s_y F(y,\xi)|_{y=x} + \rho < \overline{X}, \nabla^s_y F(y,\xi)|_{y=x} >_{x,\xi} .$$

For a fixed ξ, $\mathcal{L}^\rho_\xi$ is elliptic on $\widetilde{M}$ and Markovian ($\mathcal{L}^\rho_\xi 1 \;=\; 0$). Hence, by Theorem 4.1, there is always some $\mathcal{L}^\rho$-stationary measure. Let m_ρ be a $\mathcal{L}^\rho$-stationary measure. Then, locally [21], on a local flow-box of the lamination the measure m_ρ has conditional measures along the leaves that are absolutely continuous with respect to Lebesgue, and the density K^ρ satisfies $\mathcal{L}^{\rho *}K^\rho \;=\; 0$, where $\mathcal{L}^{\rho *}$ is the formal adjoint of $\mathcal{L}^\rho$ with respect to Lebesgue measure on the leaf, i.e.,

$$\mathcal{L}^{\rho *}F \;=\; \Delta^s F - \rho \mathrm{Div}^s(F\overline{X}). \tag{4.1}$$

Globally, there exists a Γ-equivariant family of measures ν_x^ρ such that the Γ-invariant extension $\widetilde{m}_\rho$ of m_ρ is given by a formula analogous to (3.4):

$$\int F(x,\xi)\, d\widetilde{m}_\rho = \int_{\widetilde{M}} \left(\int_{\partial \widetilde{M}} F(x,\xi)\, d\nu_x^\rho(\xi) \right) d\mathrm{Vol}_{\widetilde{g}}(x).$$

Indeed, choose a transversal to the foliation $\mathcal{W}^s$, say the sphere $S_{x_0}M$ and write SM as $M_0 \times S_{x_0}M$. A stationary measure m_ρ is given by an integral for some measure $d\nu(\xi)$ of measures of the form $\mathsf{K}^\rho(x,\xi)\, d\mathrm{Vol}_g(x)$, where Vol_g is the volume on M_0. We can arrange that $\mathsf{K}^\rho(x_0,\xi) = 1$, ν-a.e.. For a lift $\widetilde{x}_0 =: x$, set $\nu_x^\rho = (\pi_x)_*\nu$. The family $\nu_{\beta x}^\rho$, $\beta \in \Gamma$, is Γ-equivariant by construction. Starting from a different point $y_0 \in M_0$, the same construction gives a Γ-equivariant family $\nu_{\beta y}^\rho$, $\beta \in \Gamma$, for the lifts y of y_0. By construction also,

$$\frac{d\nu_y^\rho}{d\nu_x^\rho}(\xi) = \frac{\mathsf{K}^\rho(y,\xi)}{\mathsf{K}^\rho(x,\xi)}.$$

The same proof as for the relation (3.5) yields, for any vector field Z on SM such that $Z(v)$ is tangent to $W^s(v)$ for all $v \in SM$,

$$\int_{SM} \mathrm{Div}^s Z + < Z, \nabla_y^s \log \mathsf{K}^\rho(y,\xi)\big|_{y=x} > dm_\rho(v) = 0. \tag{4.2}$$

For each $\mathcal{L}^\rho$, there is a *diffusion*, i.e., a Γ-equivariant family of probability measures $\widetilde{\mathbb{P}}_{x,\xi}^\rho$ on $C(\mathbb{R}_+, S\widetilde{M})$ such that $t \mapsto \widetilde{\omega}(t)$ is a Markov process with generator $\mathcal{L}_\xi^\rho$, $\widetilde{\mathbb{P}}_{x,\xi}^\rho$-a.s. $\widetilde{\omega}(0) = (x,\xi)$ and $\widetilde{\omega}(t) \in \widetilde{M} \times \{\xi\}$, $\forall t > 0$. The distribution of $\widetilde{\omega}(t)$ under $\widetilde{\mathbb{P}}_{x,\xi}^\rho$ is $p_\xi^\rho(t,x,y) d\mathrm{Vol}_{\widetilde{g}}(y)\delta_\xi(\eta)$, where $p_\xi^\rho(t,x,y)$ is the fundamental solution of the equation $\frac{\partial F}{\partial t} = \mathcal{L}_\xi^\rho F$. The quotient $\mathbb{P}_v^\rho$ defines a Markov process on SM such that for all $t \geq 0$, $\omega(t) \in W^s(\omega(0))$. For any $\mathcal{L}^\rho$-stationary measure m_ρ, the probability measure $\mathbb{P}_{m_\rho}^\rho := \int \mathbb{P}_v^\rho \, dm_\rho(v)$ is invariant under the shift on $C(\mathbb{R}_+, SM)$ (cf. [21, 26]). If the measure m_ρ is an extremal point of the set of stationary measures for $\mathcal{L}^\rho$, then the probability measure $\mathbb{P}_{m_\rho}^\rho$ is invariant ergodic under the shift on $C(\mathbb{R}_+, SM)$.

Proposition 4.2 *Let m_ρ be a stationary ergodic measure for $\mathcal{L}^\rho$. Then, for $\mathbb{P}_{m_\rho}^\rho$ a.e. ω and any lift $\widetilde{\omega}$ of ω to $S\widetilde{M}$,*

$$\lim_{t\to+\infty} \frac{1}{t} b_{\widetilde{\omega}(0)}(\widetilde{\omega}(t)) = -\rho + \int B\, dm_\rho =: \ell_\rho(m_\rho). \tag{4.3}$$

In particular, for $\rho = V, m_\rho = m_{BR}$, *we have* $\ell_V(m_{BR}) = V - \int B\, dm_{BR} = 0$.

By Remark 3.10, the measure m_{BR} is ergodic.

Proof Let $\sigma_t, t \in \mathbb{R}_+$, be the shift transformation on $C(\mathbb{R}_+, S\widetilde{M})$. For any $\widetilde{\omega} \in C(\mathbb{R}_+, S\widetilde{M})$, $t, s \in \mathbb{R}_+$, $b_{\widetilde{\omega}(0)}(\widetilde{\omega}(t+s)) = b_{\widetilde{\omega}(0)}(\widetilde{\omega}(t)) + b_{\sigma_t\widetilde{\omega}(0)}(\sigma_t\widetilde{\omega}(s))$. By Γ-equivariance, $b_{\widetilde{\omega}(0)}(\widetilde{\omega}(t))$ takes the same value for all $\widetilde{\omega}$ with the same projection in $C(\mathbb{R}_+, SM)$ and defines an additive functional on $C(\mathbb{R}_+, SM)$. Moreover, $\sup_{0\le t\le 1} b_{\widetilde{\omega}(0)}(\widetilde{\omega}(t)) \le \sup_{0\le t\le 1} d(\widetilde{\omega}(0), \widetilde{\omega}(t))$, so that the convergence in (4.3) holds $\mathbb{P}^\rho_{m_\rho}$-a.e. and in $L^1(\mathbb{P}^\rho_{m_\rho})$. By ergodicity of the process and additivity of the functional $b_{\widetilde{\omega}(0)}(\widetilde{\omega}(t))$, the limit is $\frac{1}{t}\mathbb{E}^\rho_{m_\rho}\left(b_{\widetilde{\omega}(0)}(\widetilde{\omega}(t))\right)$, for all $t > 0$. In particular,

$$\begin{aligned}\lim_{t\to+\infty} \frac{1}{t} b_{\widetilde{\omega}(0)}(\widetilde{\omega}(t)) &= \lim_{t\to 0^+} \frac{1}{t}\mathbb{E}^\rho_{m_\rho}\left(b_{\widetilde{\omega}(0)}(\widetilde{\omega}(t))\right) \\ &= \int_{SM} \Delta^s_y b_{x,\xi}(y)\big|_{y=x} + \rho < \overline{X}, \nabla^s_y b_{x,\xi}(y)\big|_{y=x} >_{x,\xi}\, dm_\rho(x,\xi).\end{aligned}$$

Equation (4.3) follows. □

Following Ancona [1] and Hamenstädt [26], we call our operator $\mathcal{L}^\rho$ *weakly coercive* if there is some $\varepsilon > 0$ such that for all $\xi \in \partial\widetilde{M}$, there exists a positive superharmonic function for the operator $\mathcal{L}^\rho_\xi + \varepsilon$ (i.e., a positive F such that $\mathcal{L}^\rho_\xi F + \varepsilon F \le 0$). As a corollary of Proposition 4.2, we see that if m_ρ is a $\mathcal{L}^\rho$-stationary measure with $\ell_\rho(m_\rho) > 0$, then for $\widetilde{m}_\rho$ almost all $\widetilde{\omega}(0)$ and $\widetilde{\mathbb{P}}^\rho_{x,\xi}$ almost all $\widetilde{\omega}$, $\widetilde{\omega}(+\infty) = \lim_{t\to+\infty}\widetilde{\omega}(t) \in (\partial\widetilde{M}\setminus\{\xi\}) \times \{\xi\}$. This, together with the negative curvature and the cocompact assumption of the underlying space, implies that

Corollary 4.3 [26, Corollary 3.10] *Assume the operator* $\mathcal{L}^\rho$ *is such that there exists some* $\mathcal{L}^\rho$*-stationary ergodic measure* m_ρ *with* $\ell_\rho(m_\rho) > 0$. *Then,* $\mathcal{L}^\rho$ *is weakly coercive.*

4.2 Stable Diffusions

For a weakly coercive $\mathcal{L}^\rho$, we want to understand more about its diffusions. Hamenstädt developed in [26] many tools for the study of the foliated diffusions subordinate to the stable foliation $\mathcal{W}^s$, using dynamics and thermodynamical formalism. We review in this subsection her results when applied for our $\mathcal{L}^\rho$.

For each $\mathcal{L}^\rho$, $\rho \in \mathbb{R}$, recall that $p^\rho_\xi(t,x,y)$ is the fundamental solution of the equation $\frac{\partial F}{\partial t} = \mathcal{L}^\rho_\xi F$. We write $G^\rho_\xi(x,y)$ for the Green function of $\mathcal{L}^\rho$: for $x, y \in \widetilde{M}$,

$$G^\rho_\xi(x,y) := \int_0^\infty p^\rho_\xi(t,x,y)\, dt.$$

For weakly coercive operators on a pinched negatively curved simply connected manifold, Ancona's Martin boundary theory [1] shows the following

Theorem 4.4 ([1]) *Assume that the operator $\Delta^s + \rho\overline{X}$ is weakly coercive and recall that the sectional curvature of $\widetilde{M}$ is between two constants $-a^2$ and $-b^2$. There exists a constant C such that for any $\xi \in \partial\widetilde{M}$, any three points x, y, z in that order on the same geodesic in $\widetilde{M}$ and such that $d(x, y), d(y, z) \geq 1$, we have:*

$$C^{-1}G_\xi^\rho(x, y)G_\xi^\rho(y, z) \leq G_\xi^\rho(x, z) \leq CG_\xi^\rho(x, y)G_\xi^\rho(y, z). \tag{4.4}$$

(In particular, by Corollary 4.3, the inequality (4.4) holds for ρ such that there is an ergodic $\mathcal{L}^\rho$-stationary measure m_ρ with $\ell_\rho(m_\rho) > 0$.)

Ancona [1] deduced from (4.4) that the *Martin boundary* of each weakly coercive operator $\mathcal{L}_\xi^\rho$ is the geometric boundary $\partial\widetilde{M}$. Namely, for any $x, y \in \widetilde{M}, \xi, \eta \in \partial\widetilde{M}$, there exists a function $K_{\xi,\eta}^\rho(x, y)$ such that

$$\lim_{z\to\eta} \frac{G_\xi^\rho(y, z)}{G_\xi^\rho(x, z)} = K_{\xi,\eta}^\rho(x, y).$$

The function $K_{\xi,\eta}^\rho(x, y)$ is $\mathcal{L}_\xi^\rho$-harmonic and therefore smooth in x and y. Moreover, the functions $(x, \eta) \mapsto K_{\xi,\eta}^\rho(x, y)$, $(x, \eta) \mapsto \nabla_y K_{\xi,\eta}^\rho(x, y)\big|_{y=x}$ are Hölder continuous (cf. [26], Appendix B). By uniformity of the constant C in (4.4), the functions $(x, \xi) \mapsto K_{\xi,\eta}^\rho(x, y)$, $\xi \mapsto \nabla_y K_{\xi,\eta}^\rho(x, y)\big|_{y=x}$ are continuous into the space of Hölder continuous functions on SM (see e.g. [37], Proposition 3.9).

Let $\mathcal{L}^{\rho*}$ be the leafwise formal adjoint of $\mathcal{L}^\rho$ (see (4.1)). Then, $\mathcal{L}^{\rho*}$ is subordinate to $\mathcal{W}^s$ and the corresponding Green function $G_\xi^{\rho*}(x, y)$ is given by $G_\xi^{\rho*}(x, y) = G_\xi^\rho(y, x)$. In particular, the Green function $G_\xi^{\rho*}(x, y)$ satisfies (4.4) as well and we find, for $\xi, \eta \in \partial\widetilde{M}, x, y \in \widetilde{M}$, the Martin kernel $K_{\xi,\eta}^{\rho*}(x, y)$ given by:

$$K_{\xi,\eta}^{\rho*}(x, y) = \lim_{z\to\eta} \frac{G_\xi^{\rho*}(y, z)}{G_\xi^{\rho*}(x, z)} = \lim_{z\to\eta} \frac{G_\xi^\rho(z, y)}{G_\xi^\rho(z, x)}.$$

Again, the function $K_{\xi,\eta}^{\rho*}(x, y)$ is $\mathcal{L}_\xi^\rho$-harmonic and therefore smooth in x and y. Moreover, the functions $(x, \eta) \mapsto K_{\xi,\eta}^{\rho*}(x, y)$, $(x, \eta) \mapsto \nabla_y K_{\xi,\eta}^{\rho*}(x, y)\big|_{y=x}$ are Hölder continuous and the functions $(x, \xi) \mapsto K_{\xi,\eta}^{\rho*}(x, y)$, $\xi \mapsto \nabla_y K_{\xi,\eta}^{\rho*}(x, y)\big|_{y=x}$ are continuous into the space of Hölder continuous functions on SM. Observe also that the relation (4.4) is satisfied also by the resolvent $G_\xi^{\lambda,\rho*}(x, y) := \int_0^\infty e^{-\lambda t} p_\xi^\rho(t, y, x)\, dt$, uniformly for $\lambda > 0$ close to 0 and for $\xi \in \partial\widetilde{M}$, so that we also have:

$$K_{\xi,\eta}^{\rho*}(x,y) = \lim_{z\to\eta,\lambda\to 0+} \frac{G_\xi^{\lambda,\rho*}(y,z)}{G_\xi^{\lambda,\rho*}(x,z)}. \tag{4.5}$$

We can use the function $K_{\xi,\eta}^{\rho,*}(x,y)$ to express the function K^ρ in (4.2).

Proposition 4.5 *Assume $\ell_\rho(m_\rho) > 0$ and m_ρ is ergodic. Then, the corresponding K^ρ in (4.2) is given by $\frac{\mathsf{K}^\rho(y,\xi)}{\mathsf{K}^\rho(x,\xi)} = K_{\xi,\xi}^{\rho*}(x,y)$.*

Proof Let ν_x^ρ be the family such that $d\widetilde{m}_\rho(x,\xi) = d\mathrm{Vol}_{\widetilde{g}}(x)d\nu_x^\rho(\xi)$. For $F \in C(SM)$, the set of continuous functions on SM, set $\widetilde{F}$ for the Γ-periodic function on $\widetilde{M}\times\partial\widetilde{M}$ extending F. Since m_ρ is ergodic, we have, for m_ρ-a.e. (x,ξ),

$$\int_{SM} F\, dm_\rho = \lim_{\lambda\to 0+} \lambda \int_0^\infty e^{-\lambda t}\left(\int p_\xi^\rho(t,x,y)\widetilde{F}(y,\xi)\, d\mathrm{Vol}_{\widetilde{g}}(y)\right) dt.$$

The inner integral can be written

$$\sum_{\beta\in\Gamma}\int p_\xi^\rho(t,x,\beta y)\widetilde{F}(\beta y,\xi)\, d\mathrm{Vol}_g(y) = \sum_{\beta\in\Gamma}\int p_{\beta^{-1}\xi}^\rho(t,\beta^{-1}x,y)\widetilde{F}(y,\beta^{-1}\xi)\, d\mathrm{Vol}_g(y),$$

where Vol_g is the restriction of $\mathrm{Vol}_{\widetilde{g}}$ on the fundamental domain M_0, so that we have

$$\int_{SM} F\, dm_\rho = \lim_{\lambda\to 0+}\sum_{\beta\in\Gamma}\lambda\int G_{\beta^{-1}\xi}^{\lambda,\rho*}(y,\beta^{-1}x)F(y,\beta^{-1}\xi)\, d\mathrm{Vol}_g(y).$$

By Harnack inequality, all ratios $\frac{G_{\beta^{-1}\xi}^{\lambda,\rho*}(y,\beta^{-1}x)}{G_{\beta^{-1}\xi}^{\lambda,\rho*}(z,\beta^{-1}x)}$ for $y,z\in M_0$ are of the same order as soon as $d(\beta^{-1}x, M_0)\ge 1$. Choose an open $A\subset\partial\widetilde{M}$ disjoint from $\{\xi\}$. If, for β large enough, $\beta^{-1}\xi\in A$, then $\beta^{-1}x$ is close to A. Then, by (4.5) and Harnack inequality, given $\varepsilon>0$, for all $x\in M_0$, $\xi\in\partial\widetilde{M}$, for all $\beta\in\Gamma$ so that $\beta^{-1}x$ is close enough to $\beta^{-1}\xi$, y' close enough to y, z' close enough to z,

$$\frac{G_{\beta^{-1}\xi}^{\lambda,\rho*}(y',\beta^{-1}x)}{G_{\beta^{-1}\xi}^{\lambda,\rho*}(z',\beta^{-1}x)} \sim^{1+\varepsilon} K_{\beta^{-1}\xi,\beta^{-1}\xi}^{\rho,*}(z,y),$$

where, for $a,b\in\mathbb{R}$, $a\sim^{1+\varepsilon} b$ means $(1+\varepsilon)^{-1}b\le a\le(1+\varepsilon)b$. Consider as functions F_y, F_z the indicator of $\mathcal{U}_y\times A, \mathcal{U}_z\times A$, where $\mathcal{U}_y,\mathcal{U}_z$ are respectively small neighborhoods of y,z. Then

$$\int_{SM} F_y\, dm_\rho = \int_{\mathcal{U}_y}\nu_{y'}^\rho(A)\, d\mathrm{Vol}_g(y') = \lim_{\lambda\to 0+}\sum_{\beta\in\Gamma,\beta^{-1}\xi\in A}\lambda\int_{\mathcal{U}_y} G_{\beta^{-1}\xi}^{\lambda,\rho*}(y',\beta^{-1}x)\, d\mathrm{Vol}_g(y'),$$

$$\int_{SM} F_z \, dm_\rho = \int_{\mathcal{U}_z} \nu_{z'}^\rho(A) \, d\mathrm{Vol}_g(z') = \lim_{\lambda \to 0+} \sum_{\beta \in \Gamma, \beta^{-1}\xi \in A} \lambda \int_{\mathcal{U}_z} G_{\beta^{-1}\xi}^{\lambda,\rho*}(z', \beta^{-1}x) \, d\mathrm{Vol}_g(z').$$

As $\lambda \to 0^+$, the β's involved in the sums are such that the distance $d(y, \beta^{-1}x), d(z, \beta^{-1}x)$ is larger and larger. It follows that, for ν_z^ρ-a.e. η,

$$\frac{d\nu_y^\rho}{d\nu_z^\rho}(\eta) = K_{\eta,\eta}^{\rho,*}(z, y).$$

□

Corollary 4.6 *Assume $\ell_\rho(m_\rho) > 0$ for some ergodic $\mathcal{L}^\rho$-stationary measure m_ρ. Then, m_ρ is the only $\mathcal{L}^\rho$-stationary probability measure.*

Proof By Proposition 4.5, any ergodic $\mathcal{L}^\rho$-stationary measure is described by a Γ-equivariant family of measures at the boundary ν_x that satisfies

$$\frac{d\nu_y}{d\nu_z}(\eta) = K_{\eta,\eta}^{\rho,*}(z, y).$$

Since the cocycle depends Hölder-continuously on η, there is a unique equivariant family with that property (see, e.g., [36, Théorème 1.d], [48, Corollary 5.12]). □

4.3 Stochastic Entropy and Rigidity

Let m_ρ be an ergodic $\mathcal{L}^\rho$-stationary measure, and assume that $\ell_\rho(m_\rho) > 0$. The following theorems are the counterpart of the more familiar random walks properties in our setting.

Theorem 4.7 (Kaimanovich, [27]) *Let m_ρ be an ergodic $\mathcal{L}^\rho$-stationary measure, and assume that $\ell_\rho(m_\rho) > 0$. For $\mathbb{P}_{m_\rho}$-a.e. $\omega \in C(\mathbb{R}_+, SM)$, the following limits exist*

$$\begin{aligned} h_\rho(m_\rho) &= \lim_{t \to +\infty} -\frac{1}{t} \log p_\xi^\rho(t, \widetilde{\omega}(0), \widetilde{\omega}(t)) \\ &= \lim_{t \to +\infty} -\frac{1}{t} \log G_\xi^\rho(\widetilde{\omega}(0), \widetilde{\omega}(t)), \end{aligned}$$

where $\widetilde{\omega}(t), t \geq 0$, is a lift of ω to $S\widetilde{M}$. Moreover,

$$h_\rho(m_\rho) = \int_{SM} \left(\|\nabla^s \log K^\rho(x, \xi)\|^2 - \rho B(x, \xi) \right) dm_\rho.$$

Proof The first part is proven in details in [38], Proposition 2.4. For the final formula, we follow [38], Erratum. Since the notations are not exactly the same, for the sake of clarity, we give the main ideas of the proof. We firstly claim is that, since $\ell_\rho(m_\rho) > 0$, for $\mathbb{P}_{m_\rho}$-a.e. $\omega \in C(\mathbb{R}_+, SM)$,

$$\limsup_{t\to+\infty} \left|\log G_\xi^\rho(\widetilde{\omega}(0), \widetilde{\omega}(t)) - \log K_{\xi,\xi}^{\rho*}(\widetilde{\omega}(0), \widetilde{\omega}(t))\right| < +\infty.$$

Indeed, let z_t be the point on the geodesic ray $\gamma_{\widetilde{\omega}(t),\xi}$ closest to x. Then, as $t \to +\infty$,

$$G_\xi^\rho(\widetilde{\omega}(0), \widetilde{\omega}(t)) \asymp G_\xi^\rho(z_t, \widetilde{\omega}(t)) \asymp \frac{G_\xi^\rho(y, \widetilde{\omega}(t))}{G_\xi^\rho(y, z_t)}$$

for all y on the geodesic going from $\widetilde{\omega}(t)$ to ξ with $d(y, \widetilde{\omega}(t)) \geq d(y, z_t) + 1$, where $\asymp$ means up to some multiplicative constant independent of t. The first $\asymp$ comes from Harnack inequality using the fact that $\sup_t d(x, z_t)$ is finite $\mathbb{P}_{m_\rho}$-almost everywhere. (Since $\ell_\rho(m_\rho) > 0$, for $\mathbb{P}_{m_\rho}$-a.e. $\omega \in C(\mathbb{R}_+, SM)$, $\eta = \lim_{t\to+\infty} \widetilde{\omega}(t)$ differs from ξ and $d(x, z_t)$, as $t \to +\infty$, converge to the distance between x and $\gamma_{\xi,\eta}$.) The second $\asymp$ comes from Ancona inequality (4.4). Replace $\frac{G_\xi^\rho(y,\widetilde{\omega}(t))}{G_\xi^\rho(y,z_t)}$ by its limit as $y \to \xi$, which is $K_{\xi,\xi}^{\rho*}(z_t, \widetilde{\omega}(t))$ by (4.5), which is itself $\asymp K_{\xi,\xi}^{\rho*}(\widetilde{\omega}(0), \widetilde{\omega}(t))$ by Harnack inequality again. It follows that, for $\mathbb{P}_{m_\rho}^\rho$-a.e. $\omega \in C(\mathbb{R}_+, SM)$,

$$h_\rho(m_\rho) = \lim_{t\to+\infty} -\frac{1}{t} \log K_{\xi,\xi}^{\rho*}(\widetilde{\omega}(0), \widetilde{\omega}(t)).$$

By Harnack inequality, there is a constant C such that $|\log K_{\xi,\xi}^{\rho*}(\widetilde{\omega}(0), \widetilde{\omega}(t))| \leq Cd(\widetilde{\omega}(0), \widetilde{\omega}(t))$. Since $\log K_{\xi,\xi}^{\rho*}(\widetilde{\omega}(0), \widetilde{\omega}(t))$ is additive along the trajectories, and $\mathbb{P}_{m_\rho}^\rho$ is shift ergodic, the limit reduces to

$$\begin{aligned} h_\rho(m_\rho) &= \lim_{t\to 0^+} -\frac{1}{t}\mathbb{E}_{m_\rho} \log K_{\xi,\xi}^{\rho*}(\widetilde{\omega}(0), \widetilde{\omega}(t)) \\ &= -\int_{SM} \left(\Delta_y^s \log K_{\xi,\xi}^{\rho*}(x, y)\big|_{y=x} + \rho < \overline{X}, \nabla_y^s \log K_{\xi,\xi}^{\rho*}(x, y)\big|_{y=x} >_{x,\xi}\right) dm_\rho(x, \xi) \\ &= -\int_{SM} \left(\Delta^s \log \mathsf{K}^\rho(x, \xi) + \rho < \overline{X}, \nabla^s \log \mathsf{K}^\rho > (x, \xi)\right) dm_\rho(x, \xi), \end{aligned}$$

where we used Proposition 4.5 to replace $\nabla_y^s \log K_{\xi,\xi}^{\rho*}(x, y)\big|_{y=x}$ by $\nabla^s \log \mathsf{K}^\rho(x, \xi)$. Finally, we use (4.2) applied to $Z = \nabla^s \log \mathsf{K}^\rho(x, \xi)$ to write

$$-\int_{SM} \Delta^s \log \mathsf{K}^\rho(x, \xi)\, dm_\rho(x, \xi) = \int_{SM} \|\nabla^s \log \mathsf{K}^\rho(x, \xi)\|^2\, dm_\rho(x, \xi)$$

and applied to $Z = \overline{X}$ to write

$$\int B\, dm_\rho = \int < \overline{X}, \nabla^s \log \mathsf{K}^\rho > dm_\rho. \tag{4.6}$$

The formula for the entropy follows. □

Theorem 4.8 (Guivarc'h, [23]) *Assume that* $\ell_\rho(m_\rho) > 0$. *Then,* $h_\rho(m_\rho) \leq \ell_\rho(m_\rho)V$.

Proof Fix $(x, \xi) \in S\widetilde{M}$ such that $\frac{1}{t} b_{x,\xi}(\widetilde{\omega}(t)) \to \ell_\rho(m_\rho)$ and $-\frac{1}{t}\log p_\xi^\rho(t, \widetilde{\omega}(0), \widetilde{\omega}(t)) \to h_\rho(m_\rho)$, $\widetilde{\mathbb{P}}^\rho_{x,\xi}$-a.e., as $t \to +\infty$. There is a constant $\widetilde{C}$ depending only on the curvature bounds such that one can find a partition $\mathcal{A} = \{A_k, k \in \mathbb{N}\}$ of $\widetilde{M}$ such that the sets A_k have diameter at most $\widetilde{C}$ and inner diameter at least 1. Set for $k \in \mathbb{N}, t > 0$, $q_k^\rho(t) := \widetilde{\mathbb{P}}^\rho_{x,\xi}(\{\widetilde{\omega} : \widetilde{\omega}(t) \in A_k\})$. The family $\{q_k^\rho(t), k \in \mathbb{N}\}$ is a probability on $\mathbb{N}$ with the property that, with high probability, $q_k^\rho(t) \leq e^{-t(h_\rho(m_\rho)-\varepsilon)}$ and $k \in N_t$, where $N_t := \{k : A_k \subset B(x, t(\ell_\rho(m_\rho) + \varepsilon))\}$. Then,

$$-\sum_{k\in N_t} q_k^\rho(t) \log q_k^\rho(t) \leq \sum_{k \in N_t} q_k^\rho(t) \times \log \# N_t.$$

Since $\# N_t \leq C e^{t(\ell_\rho(m_\rho)+\varepsilon)(V+\varepsilon)}$, for some constant C, Theorem 4.8 follows. □

Theorem 4.9 *Assume that* $\ell_\rho(m_\rho) > 0$. *Then,* $\int B\, dm_\rho \leq V$, *with equality in this inequality only when* (M, g) *is locally symmetric.*

Proof Recall Equation (4.6): $\int B\, dm_\rho = \int < \overline{X}, \nabla^s \log \mathsf{K}^\rho > dm_\rho$, so that, by Schwarz inequality,

$$\left(\int B\, dm_\rho\right)^2 \leq \int_{SM} \|\nabla^s \log \mathsf{K}^\rho_{x,\xi}\|^2\, dm_\rho,$$

with equality only if $\nabla^s \log \mathsf{K}^\rho = \tau(\rho)\overline{X}$ for some real number $\tau(\rho)$. Abbreviate $h_\rho(m_\rho), \ell_\rho(m_\rho)$ as h_ρ, ℓ_ρ. We write

$$\begin{aligned} h_\rho &= \int_{SM} \left(\|\nabla^s \log \mathsf{K}^\rho_{x,\xi}\|^2 - \rho B(x,\xi)\right) dm_\rho \\ &\geq \left(\int B\, dm_\rho\right)^2 - \rho \int B\, dm_\rho = \ell_\rho \int B\, dm_\rho. \end{aligned}$$

We indeed have $\int B\, dm_\rho \leq V$, with equality only if $\nabla^s \log \mathsf{K}^\rho = \tau(\rho)\overline{X}$ for some real number $\tau(\rho)$. Then, Equation (3.5) holds with V replaced by $\tau(\rho)$. The proof of Corollary 3.9 applies and the operator Δ^{ss} is symmetric with respect to the measure m_ρ. By Remark 3.10, $m_\rho = m_{BR}$. Then, $\tau(\rho) = V$ and from $\int B\, dm_\rho = \int B\, dm_{BR} = V$ and $\ell_\rho(m_\rho) > 0$, we have $\rho \neq V$. We have

$0 = \mathcal{L}_y^{\rho *} e^{-Vb_{x,\xi}(y)}\Big|_{y=x} = (V - B(x,\xi))(V-\rho)$. It follows that $B = V$ is constant. By Theorem 1.3, the space (M, g) is locally symmetric. □

The conclusion in Theorem 4.9 actually holds true for all $\rho < V$ due to the following.

Proposition 4.10 *Let $\rho \in \mathbb{R}$. There is some $\mathcal{L}^\rho$-stationary ergodic measure m_ρ such that $\ell_\rho(m_\rho) > 0$ if, and only if, $\rho < V$. Moreover, the measures m_ρ weak* converge to m_{BR} as $\rho \nearrow V$.*

Proof Let ρ_0 be such that there is some $\mathcal{L}^{\rho_0}$-stationary measure m_{ρ_0} with $\ell_{\rho_0}(m_{\rho_0}) \leq 0$, but such that there exist $\{\rho_n\}_{n\in\mathbb{N}}$ with $\lim_{n\to+\infty} \rho_n = \rho_0$ and $\ell_{\rho_n}(m_{\rho_n}) > 0$ (we know that m_{ρ_n} is unique by Corollary 4.6). Observe that by Equation (4.3), $\ell_\rho > 0$ for ρ sufficiently close to $-\infty$. On the other hand, if $\ell_{\rho_n}(m_{\rho_n}) > 0$, we must have $\rho_n < V$ by Equation (4.3) and Theorem 4.9. Therefore one can choose ρ_0 and ρ_n with those properties. Let m be a weak* limit of the measures m_{ρ_n}. We are going to show that $m = m_{BR}$ and that $\rho_0 = V$.

Observe that $\ell_{\rho_0}(m) \leq 0$ since otherwise m is the only stationary measure and we cannot have $\ell_{\rho_0}(m_{\rho_0}) \leq 0$ for some other $\mathcal{L}^{\rho_0}$-stationary measure m_{ρ_0}. On the other hand, $\ell_{\rho_0}(m) \geq 0$ by continuity, so $\ell_{\rho_0}(m) = 0$ and $\lim_{n\to+\infty} \ell_{\rho_n}(m_{\rho_n}) = 0$. By Theorem 4.8, $\lim_{n\to+\infty} h_{\rho_n}(m_{\rho_n}) = 0$ as well. We have

$$\begin{aligned} 0 = \lim_{n\to+\infty} h_{\rho_n}(m_{\rho_n}) &= \lim_{n\to+\infty} \int_{SM} \left(\|\nabla^s \log \mathsf{K}^{\rho_n}(x,\xi)\|^2 - \rho_n B(x,\xi)\right) dm_{\rho_n} \\ &= \lim_{n\to+\infty} \int_{SM} \left(\|\nabla^s \log \mathsf{K}^{\rho_n}(x,\xi)\|^2 - \rho_n < \overline{X}, \nabla^s \log \mathsf{K}^{\rho_n} > \right) dm_{\rho_n}. \end{aligned}$$

Write $Z_n := \nabla^s \log \mathsf{K}^{\rho_n}(x,\xi) - \left(\int_{SM} < \overline{X}, \nabla^s \log \mathsf{K}^{\rho_n} > dm_{\rho_n}\right) \overline{X}$. We have

$$\begin{aligned} \lim_{n\to+\infty} \int_{SM} \|Z_n\|^2 \, dm_{\rho_n} &= \lim_{n\to+\infty} \int_{SM} \left(\|\nabla^s \log \mathsf{K}^{\rho_n}(x,\xi)\|^2 \right) dm_{\rho_n} - \left(\int_{SM} B \, dm_{\rho_n} \right)^2 \\ &= \lim_{n\to+\infty} \left(h_{\rho_n}(m_{\rho_n}) - \ell_{\rho_n}(m_{\rho_n}) \int_{SM} B \, dm_{\rho_n} \right) \end{aligned}$$

and so $\lim_{n\to+\infty} \int_{SM} \|Z_n\|^2 \, dm_{\rho_n} = 0$. In other words, Equation (3.5) holds with V replaced by $\int_{SM} B \, dm_{\rho_n}$ with an error $\int_{SM} < Z, Z_n > dm_{\rho_n}$. The proof of Corollary 3.9 applies and the operator Δ^{ss} is symmetric with respect to the measure m_{ρ_n}, up to an error which goes to 0 as $n \to +\infty$. It follows that the operator Δ^{ss} is symmetric with respect to the limit measure m. By Remark 3.10, $m = m_{BR}$. Since $\ell_{\rho_0}(m) = 0$, $\rho_0 = \int_{SM} B \, dm = \int_{SM} B \, dm_{BR} = V$. □

Remark 4.11 Anderson and Schoen [2] described the Martin boundary for the Laplacian on a simply connected manifold with pinched negative curvature. Regularity of the Martin kernel in the [2] proof yields, in the cocompact case, nice properties of the harmonic measure (i.e., the stationary measure for $\mathcal{L}^0 = \Delta^s$).

This was observed by [25, 29] and [34]. Ancona [1] extended [2]'s results to the general weakly coercive operator and proved the basic inequality (4.4). This allowed Hamenstädt to consider the general case that $\mathcal{L} = \Delta^s + Y$, with Y^*, the dual of Y in the cotangent bundle to the stable foliation over SM, satisfying $dY^* = 0$ leafwisely [26]. The criterion she obtained for the existence of a $\mathcal{L}$-stationary ergodic measure m with $\ell_{\mathcal{L}}(m) := \int_{M_0 \times \partial \widetilde{M}} \left(- < Y, \overline{X} > + B\right) \, dm > 0$ is $P\left(- < \overline{X}, Y >\right) > 0$. Our presentation follows [26], with a few simplifications when $Y = \rho \overline{X}$. Theorem 4.9 was shown by Kaimanovich [27] in the case $\rho = 0$. From [26], Theorem A (2), the measure m_{BR} is the only symmetric measure for $\mathcal{L}^V$. It is not known whether m_{BR} is the only stationary measure for $\mathcal{L}^V$. The second statement in Proposition 4.10 would also follow from such a uniqueness result.

5 Stochastic Flows of Diffeomorphisms and a Relative Entropy

In this section, we introduce a stochastic flow associated with $\mathcal{L}^\rho$. In the case of $\rho = 0$ our object has been considered as a *stochastic (analogue of) the geodesic flow* (cf. [14, 17]). It gives rise to a random walk on the space of homeomorphisms of a bigger compact manifold and the relative entropy of this random walk of homeomorphisms is our fourth entropy. The continuity of this entropy as $\rho \to -\infty$ will be used to prove that the measures m_ρ converge to m_L as $\rho \to -\infty$ (see Theorem 5.5 below).

5.1 Stochastic Flow Adapted to $\mathcal{L}^\rho$

Let $O\widetilde{M}$ be the orthonormal frame bundle (OFB) of $(\widetilde{M}, \widetilde{g})$:

$$O\widetilde{M} := \left\{x \mapsto u(x) : \; u(x) = (u_1, \cdots, u_d) \in O(S_x \widetilde{M})\right\}$$

and consider $O\widetilde{M} \times \{\xi\} =: O^s S\widetilde{M}$, the OFB in $T\widetilde{W}^s$ and $O^s SM := O^s S\widetilde{M}/\Gamma$, the OFB in TW^s. For $v \in S\widetilde{M}$, $u \in O^s_v S\widetilde{M}$, the *horizontal* subspace of $T_u O^s S\widetilde{M}$ is the space of directions w such that $\nabla_u w = 0$.

Denote $D^r(O^s S\widetilde{M})$ ($r \in \mathbb{N}$ or $r = \infty$) the space of homeomorphisms Φ such that

$$\Phi(x, u, \xi) := \left(\phi_\xi(x, u), \xi\right),$$

where ϕ_ξ is a C^r diffeomorphism of $O\widetilde{M}$, which depends continuously on ξ in $\partial\widetilde{M}$. We use stochastic flow theory to define a random walk on $D^\infty(O^s S\widetilde{M})$.

Theorem 5.1 ([17]) *Let $(\Omega, \mathbb{P})$ be a $\mathbb{R}^d$ Brownian motion (with covariance $2t\mathbb{I}$). For $\mathbb{P}$-a.e. $\omega \in \Omega$, all $t > 0$, there exists $\Phi_t^\rho = \left(\phi_{\xi,t}^\rho, \xi\right) \in D^\infty(O^s S\widetilde{M})$ such that for all $u \in O^s S\widetilde{M}$, $(\omega, t) \mapsto u_t = \phi_{\xi,t}^\rho(u)$ solves the Stratonovich Stochastic Differential Equation (SDE)*

$$du_t = \rho \widehat{X}(u_t) + \sum_{i=1}^d \widehat{H}(u_t^i) \circ dB_t^i, \tag{5.1}$$

where $\widehat{X}, \widehat{H}(u^i)$ are the horizontal lifts of $\overline{X}, u^i \in T_v \widetilde{W}^s(v)$ to $T_u O^s S\widetilde{M}$. Moreover,

1) *for $\mathbb{P}$-a.e. $\omega \in \Omega$, all $t, s > 0$, $\rho < V$, $\xi \in \partial\widetilde{M}$,*

$$\phi_{\xi,t+s}^\rho(\omega) = \phi_{\xi,t}^\rho(\sigma_s\omega) \circ \phi_{\xi,s}^\rho(\omega),$$

   where σ_s is the shift on Ω,
2) *for $\mathbb{P}$-a.e. $\omega \in \Omega$, for all $\beta \in \Gamma$, all $t > 0$, $D\beta \circ \phi_{\xi,t}^\rho(\omega) = \phi_{\xi,t}^\rho(\omega) \circ D\beta$, and*
3) *for $\mathbb{P}$-a.e. $\omega \in \Omega$, all $t > 0$, $\rho \mapsto \Phi_t^\rho(\omega)$ is continuous in $D^\infty(O^s S\widetilde{M})$ and the derivatives are solutions to the derivative SDE.*

Relation (5.1) implies that for all $(x, \xi, u), u \in OS_x\widetilde{M}$, the projection of $\phi_{\xi,t}^\rho(\omega)(u)$ on $S\widetilde{M}$ is a realization of the $\mathcal{L}^\rho$ diffusion starting from (x, ξ).

Property 1) and independence of the increments of the Brownian motion give that if $\kappa_{\rho,s}$ is the distribution of $\Phi_{\rho,s}(\omega)$ in $D^\infty(O^s S\widetilde{M})$, we can write

$$\kappa_{\rho,s+t} = \kappa_{\rho,t} * \kappa_{\rho,s},$$

where $*$ denotes the convolution in the group $D^\infty(O^s S\widetilde{M})$. So we have a *stochastic flow*. Property 2) yields a stochastic flow on $D^\infty(O^s SM)$. Property 3) will allow to control derivatives.

Fix $t > 0$. A probability measure $\overline{m}$ on $O^s SM$ is said to be *stationary* for $\kappa_{\rho,t}$, if for any $F \in C(O^s SM)$, the set of continuous functions on $O^s SM$,

$$\int_{O^s SM} F(u)\, d\overline{m}(u) = \int_{D^\infty(O^s SM)} \int_{O^s SM} F(\Phi u)\, d\overline{m}(u)\, d\kappa_{\rho,t}(\Phi).$$

Proposition 5.2 *Fix any $\rho < V, t > 0$. The probability measure $\overline{m}_\rho$ on $O^s SM$ that projects to m_ρ on SM and is the normalized Lebesgue measure on the fibers is stationary for $\kappa_{\rho,t}$. If we identify $O^s SM = \{(x, u, \xi) : x \in M_0, u \in O_x\widetilde{M}, \xi \in \partial\widetilde{M}\}$, then, up to a normalizing constant,*

$$d\overline{m}_\rho(x, u, \xi) = d\nu_x^\rho(\xi) d\mathrm{Vol}(x, u).$$

5.2 Entropy of a Random Transformation

There is a notion of entropy for random transformations with a stationary measure (see [31] for details).

Let $\mathbf{X}$ be a compact metric space and $D^0\mathbf{X}$ the group of homeomorphisms of $\mathbf{X}$. Let κ be a probability measure on $D^0\mathbf{X}$ and let $\overline{m}$ be a stationary measure for κ. Let σ be the shift on $(D^0\mathbf{X})^{\otimes\mathbb{N}}$, $\mathcal{K} = \kappa^{\otimes\mathbb{N}}$ the Bernoulli σ-invariant measure, $\overline{\sigma}$ the skew-product transformation on $(D^0\mathbf{X})^{\otimes\mathbb{N}} \times \mathbf{X}$

$$\overline{\sigma}(\underline{\phi}, x) := (\sigma\underline{\phi}, \phi_0 x), \ \forall\underline{\phi} = (\phi_0, \phi_1, \cdots) \in (D^0\mathbf{X})^{\otimes\mathbb{N}}.$$

Proposition 5.3 *Let $\overline{m}$ be a stationary measure for κ. Then, the measure $\mathcal{K} \times \overline{m}$ is $\overline{\sigma}$-invariant.*

For $\underline{\phi} \in (D^0\mathbf{X})^{\otimes\mathbb{N}}$, $x \in \mathbf{X}$, $\varepsilon > 0$, $n \in \mathbb{N}$, define a *random Bowen ball* by

$$\overline{B}(\underline{\phi}, x, \varepsilon, n) := \{y : \ y \in \mathbf{X}, d(\phi_k \circ \cdots \circ \phi_0 y, \phi_k \circ \cdots \circ \phi_0 x) < \varepsilon, \ \forall 0 \le k < n\}$$

and the *relative entropy* $h_{\overline{m}}(\mathcal{K})$ as the $\mathcal{K}$-a.e. value of

$$\sup_{\varepsilon} \int_{\mathbf{X}} \limsup_{n\to+\infty} -\frac{1}{n} \log \overline{m}(\overline{B}(\underline{\phi}, x, \varepsilon, n))\, d\overline{m}(x).$$

With the preceding notations, take $\mathbf{X} = O^s SM$, $\kappa = \kappa_{\rho,t}$ for some (ρ, t), $\rho < V, t > 0$, and the stationary measure $\overline{m}_\rho$. We want to estimate the relative entropy $h_{\overline{m}_\rho}(\mathcal{K}_{\rho,t})$.

Proposition 5.4 ([39]) *We have*

$$h_{\overline{m}_\rho}(\mathcal{K}_{\rho,t}) \ge \int \log \left|\mathrm{Det} D_u \Phi|_{T_u O^s S\widetilde{M}}\right| d\kappa_{\rho,t}(\Phi)\, d\overline{m}_\rho(u).$$

Recall that $\overline{m}_\rho$ has absolutely continuous conditional measures on the foliation $\overline{\mathcal{W}}^s$ defined by $(O\widetilde{M} \times \{\xi\})/\Gamma$. The proof uses ingredients from the proof of Pesin formula in the non-uniformly hyperbolic case (cf. [42]) and the non-invertible case [40, 41]. Observe that, even if $\Phi_{-1}|_{\overline{\mathcal{W}}^s}$ has only nonnegative exponents, there might be negative exponents for the random walk, and the inequality in Proposition 5.4 might be strict.

5.3 Continuity of the Relative Entropy

We now indicate the main ideas of the proof of the following theorem

Theorem 5.5 ([39]) *For $\rho < V$, let m_ρ be the stationary measure for the diffusion on SM with generator $\mathcal{L}^\rho = \Delta^s + \rho \overline{X}$. Then, as $\rho \to -\infty$, m_ρ weak* converge to the Liouville measure m_L.*

Corollary 5.6 $\lim_{\rho \to -\infty} \int B \, dm_\rho = \int B \, dm_L = H.$

Proof Set $\kappa_\rho = \kappa_{\rho, \frac{-1}{\rho}}$. We first observe that as $\rho \to -\infty$, κ_ρ weak* converge on $D^\infty(O^s SM)$ to the Dirac measure on the reverse frame flow Φ_{-1}. Moreover, for any $r \in \mathbb{N}, r \geq 1$,

$$\limsup_{\rho \to -\infty} C_r(\rho) < +\infty, \text{ where } C_r(\rho) := \int \|\Phi\|_{D^r(O^s SM)} \, d\kappa_\rho(\Phi),$$

where $\|\cdot\|_{D^r(O^s SM)}$ is the supremum of leafwise C^r norm. Indeed, by definition, κ_ρ is the distribution of the time one of the stochastic flow associated with the Stratonovich SDE

$$du_t = -\widehat{X}(u_t) + \frac{-1}{\rho} \sum_{i=1}^{d} \widehat{H}(u_t^i) \circ dB_t^i.$$

When $\rho \to -\infty$, the SDE converge to the ODE on $O^s SM$, $du_t = -\widehat{X}(u_t)$. The convergence, and the control on C_r, follow by continuity of the solutions in $D^\infty(O^s SM)$.

Let then m be a weak* limit of the measures m_ρ as $\rho \to -\infty$, $\overline{m}$ its extension to $O^s SM$ by the Lebesgue measure on the fibers. The measure m is φ_{-1} invariant, $\overline{m}$ is the weak* limit of the measures $\overline{m}_\rho$, and $\overline{m}$ is Φ_{-1} invariant. Moreover, $h_m(\varphi_{-1}) = h_{\overline{m}}(\Phi_{-1})$ (this is a compact isometric extension) and

$$\int \log \left| \mathrm{Det} D_v \varphi_{-1}|_{T_v W^s(v)} \right| dm(v) = \int \log \left| \mathrm{Det} D_u \Phi_{-1}|_{T_u O^s S\widetilde{M}} \right| d\overline{m}(u)$$

$$= \lim_{\rho \to -\infty} \int \log \left| \mathrm{Det} D_u \Phi|_{T_u O^s S\widetilde{M}} \right| d\overline{m}_\rho(u) \, d\kappa_\rho(\Phi).$$

By [10], the Liouville measure is the only φ_{-1} invariant measure with

$$h_m(\varphi_{-1}) = \int \log \left| \mathrm{Det} D_v \varphi_{-1}|_{T_v W^s(v)} \right| dm(v).$$

To conclude the theorem, using Proposition 5.4, it suffices to show

$$h_{\overline{m}}(\Phi_{-1}) \geq \limsup_{\rho \to -\infty} h_{\overline{m}_\rho}(\mathcal{K}_\rho).$$

This will follow from the properties of the *topological relative conditional entropy* in the next subsection. □

5.4 Topological Relative Conditional Entropy

The following definition extends the definition of Bowen [7] to the random case, following Kifer–Yomdin [32] and Cowieson–Young [15].

For $\varepsilon > 0$ and $\underline{\phi} \in (D^0 X)^{\otimes \mathbb{N}}$, $x \in \mathbf{X}$, $\tau > 0^+$, $n \in \mathbb{N}$, set $r(\varepsilon, \underline{\phi}, x, \tau, n)$ for the smallest number of random $\overline{B}(\underline{\phi}, y, \tau, n)$ balls needed to cover $\overline{B}(\underline{\phi}, x, \varepsilon, n)$ and

$$h_{loc}(\varepsilon, \underline{\phi}) := \sup_x \lim_{\tau \to 0^+} \limsup_{n \to +\infty} \frac{1}{n} \log r(\varepsilon, \underline{\phi}, x, \tau, n).$$

The function $\underline{\phi} \mapsto h_{loc}(\varepsilon, \underline{\phi})$ is σ-invariant. For $\mathbf{X} = O^s SM$, write $h_{\rho,loc}(\varepsilon)$ for the $\mathcal{K}_\rho$-essential value of $h_{loc}(\varepsilon, \underline{\phi})$. The conclusion follows from the two following facts (cf. [39], Section 4).

Proposition 5.7 *For all $\varepsilon > 0$,*

$$h_{\overline{m}}(\Phi_{-1}) \geq \limsup_{\rho \to -\infty} h_{\overline{m}_\rho}(\mathcal{K}_\rho) - \limsup_{\rho \to -\infty} h_{\rho,loc}(\varepsilon).$$

Proposition 5.8 *There is a constant C such that, for all $r \in \mathbb{N}$, $r \geq 1$, there is ρ_r such that, for $\rho < \rho_r$,*

$$\lim_{\varepsilon \to 0^+} \sup_{\rho < \rho_r} h_{\rho,loc}(\varepsilon) \leq \frac{C}{r} C_1,$$

where $C_1 = \sup_{\rho < \rho_1} \int \|\Phi\|_{D^1(O^s SM)} d\kappa_\rho(\Phi)$.

Proposition 5.7 in the deterministic case is due to Bowen [7]. Proposition 5.8 in the deterministic case is a famous result of Yomdin [53, 54] and Buzzi [13]. By Proposition 5.8, since r is arbitrary, $\lim_{\varepsilon \to 0^+} \limsup_{\rho \to -\infty} h_{\rho,loc}(\varepsilon) = 0$. Proposition 5.7 then yields the claimed inequality.

5.5 Conclusion. Katok's Conjecture

Let (M, g) be a C^∞ d-dimensional Riemannian manifold with negative curvature. We introduced in Sections 1 and 2 the numbers H, the entropy of the Liouville measure for the geodesic flow, V, the topological entropy of the geodesic flow, and the function B on SM. The function B is constant if, and only if (M, g) is a locally symmetric space (Theorem 1.3). Using thermodynamical formalism, $H \leq V$ and if $H = V$, there exists a continuous function F on SM, C^1 along the trajectories of the flow, such that $B = V - \frac{\partial}{\partial t} F \circ \varphi_t \big|_{t=0}$ (see Theorem 2.5). Katok's conjecture (see [35] and [55] for some history of this topic) is that this can only happen when (M, g) is a locally symmetric space, that is, when B is constant on SM. This was proven by Katok [30] in dimension 2 and more generally if g is conformally equivalent to a locally symmetric g_0. It was also proven by Flaminio [19] in a C^2 neighborhood of a constant curvature metric g_0. Here, we introduced a family of measures $m_\rho, \rho \leq V$, such that $\int B\, dm_V = V$ and for $\rho < V, \int B\, dm_\rho \leq V$ with equality only in the case of locally symmetric spaces (Theorem 4.9). Finally, in the C^∞ case, we also show that $\lim_{\rho \to -\infty} \int B\, dm_\rho = H$ (Corollary 5.6).

References

1. A. Ancona, Negatively curved manifolds, elliptic operators and the Martin boundary. Ann. Math. (2) **125**, 495–536 (1987)
2. M. Anderson, R. Schoen, Positive harmonic functions on complete manifolds of negative curvature. Ann. Math. (2) **121**, 429–461 (1985)
3. D.V. Anosov, Geodesic flow on closed Riemannian manifolds of negative curvature. Trudy. Mat. Inst. Steklov. **90** (1967), 209pp
4. Y. Benoist, P. Foulon, F. Labourie, Flots d'Anosov à distributions stables et instables différentiables. J. Am. Math. Soc. **5**, 33–74 (1992)
5. G. Besson, G. Courtois, S. Gallot, Entropies et rigidités des espaces localement symétriques de courbure strictement négative. Geom. Funct. Anal. **5**, 731–799 (1995)
6. A. Borel, Compact Clifford-Klein forms of symmetric spaces. Topology **2**, 111–122 (1963)
7. R. Bowen, Entropy-expansive maps. Trans. Am. Math. Soc. **164**, 323–331 (1972)
8. R. Bowen, *Equilibrium States and the Ergodic Theory of Anosov Diffeomorphisms*. Lecture Notes in Math., vol. 470 (Springer, Berlin, 1975)
9. R. Bowen, B. Marcus, Unique ergodicity for horocycle foliations. Isr. J. Math. **26**, 43–67 (1977)
10. R. Bowen, D. Ruelle, The ergodic theory of Axiom A flows. Invent. Math. **29**, 181–202 (1975)
11. M. Brin, A. Katok, On local entropy, in *Geometric Dynamics*. Lecture Notes in Math., vol. 1007 (Springer, Berlin, 1983), pp. 30–38
12. M. Burger, Horocycle flow on geometrically finite surfaces. Duke Math. J. **61**, 779–803 (1990)
13. J. Buzzi, Intrinsic ergodicity of smooth interval maps. Isr. J. Math. **100**, 125–161 (1997)
14. A.P. Carverhill, K.D. Elworthy, Lyapunov exponents for a stochastic analogue of the geodesic flow. Trans. Am. Math. Soc. **295**, 85–105 (1986)
15. W. Cowieson, L.-S. Young, SRB measures as zero-noise limits. Ergod. Theory Dyn. Syst. **25**, 1115–1138 (2005)
16. M. Deraux, A negatively curved Kähler threefold not covered by the ball. Invent. Math. **160**, 501–525 (2005)
17. K.D. Elworthy, *Stochastic Differential Equations on Manifolds*. London Mathematical Society Lecture Note Series, vol. 70 (Cambridge University Press, Cambridge, 1982)
18. F.T. Farrell, L.E. Jones, Negatively curved manifolds with exotic smooth structures. J. Am. Math. Soc. **2**, 899–908 (1989)
19. L. Flaminio, Local entropy rigidity for hyperbolic manifolds. Commun. Anal. Geom. **3**, 555–596 (1995)
20. P. Foulon, F. Labourie, Sur les variétés compactes asymptotiquement harmoniques. Invent. Math. **109**, 97–111 (1992)
21. L. Garnett, Foliations, the ergodic theorem and Brownian motion. J. Funct. Anal. **51**, 285–311 (1983)
22. M. Gromov, W. Thurston, Pinching constants for hyperbolic manifolds. Invent. Math. **89**, 1–12 (1987)
23. Y. Guivarc'h, Sur la loi des grands nombres et le rayon spectral d'une marche aléatoire. Astérisque **74**, 47–98 (1980)

24. U. Hamenstädt, A new description of the Bowen-Margulis measure. Ergod. Theory Dyn. Syst. **9**, 455–464 (1989)
25. U. Hamenstädt, An explicit description of harmonic measure. Math. Z. **205**, 287–299 (1990)
26. U. Hamenstädt, Harmonic measures for compact negatively curved manifolds. Acta Math. **178**, 39–107 (1997)
27. V.A. Kaimanovich, Brownian motion and harmonic functions on covering manifolds. An entropic approach. Sov. Math. Dokl. **33**, 812–816 (1986)
28. V.A. Kaimanovich, Brownian motion on foliations: entropy, invariant measures, mixing. Funct. Anal. Appl. **22**, 326–328 (1988)
29. V.A. Kaimanovich, Invariant measures of the geodesic flow and measures at infinity on negatively curved manifolds. Ann. Inst. H. Poincaré Phys. Théor. **53**, 361–393 (1990)
30. A. Katok, Entropy and closed geodesics. Ergod. Theory Dyn. Syst. **2**, 339–365 (1982)
31. Y. Kifer, P.D. Liu, Random dynamics, in *Handbook of Dynamical Ssytems*, vol. 1 (Elsevier, Amsterdam, 2006), pp. 379–499
32. Y. Kifer, Y. Yomdin, Volume growth and topological entropy for random transformations, in *Dynamical Systems*. Lecture Notes in Math., vol. 1342 (Springer, Berlin, 1988), pp. 361–373
33. G. Knieper, Spherical means on compact Riemannian manifolds of negative curvature. Differ. Geom. Appl. **4**, 361–390 (1994)
34. F. Ledrappier, Ergodic properties of Brownian motion on covers of compact negatively-curve manifolds. Bol. Soc. Brasil. Mat. **19**, 115–140 (1988)
35. F. Ledrappier, Applications of dynamics to compact manifolds of negative curvature, in *Proceedings of the International Congress of Mathematicians*, Vol. 1, 2 (Zürich, 1994), 1195–1202 (Birkhäuser, Basel, 1995)
36. F. Ledrappier, Structure au bord des variétés à courbure négative, in *Séminaire de théorie spectrale et géométrie*, Grenoble 1994–1995 (1995), pp. 97–122
37. F. Ledrappier, S. Lim, Local limit Theorem in negative curvature. Preprint, arXiv:1503.04156v5
38. F. Ledrappier, L. Shu, Differentiability of the stochastic entropy for compact negatively curved spaces under conformal changes. Ann. Inst. Fourier **67**, 1115–1183 (2017). Erratum, Ann. Inst. Fourier
39. F. Ledrappier, L. Shu, Robustness of Liouville measure under a family of stable diffusions, to appear in Comm. Pure Appl. Math. https://doi.org/10.1002/cpa.21935
40. P.D. Liu, Pesin's entropy formula for endomorphisms. Nagoya Math. J. **150**, 197–209 (1998)
41. P.D. Liu, L. Shu, Absolute continuity of hyperbolic invariant measures for endomorphisms. Nonlinearity **24**, 1595–1611 (2011)
42. R. Mañé, A proof of Pesin's formula. Ergod. Theory Dyn. Syst. **1**, 95–102 (1981). Errata Ergod. Theory Dyn. Syst. **3**, 159–160 (1983)
43. A. Manning, Topological entropy for geodesic flow. Ann. Math. (2) **110**, 567–573 (1979)
44. G.A. Margulis, Certain measures associated with U-flows. Funct. Anal. Appl. **4**, 55–67 (1970)
45. G.D. Mostow, Y.T. Siu, A compact Kähler surface of negative curvature not covered by the ball. Ann. Math. **112**, 321–360 (1980)
46. W. Parry, M. Pollicott, *Zeta Functions and the Periodic Orbit Structure of Hyperbolic Dynamics*. Astérisque 187–188 (Société Mathématique de France, Montrouge, 1990), 268pp
47. S.J. Patterson, The limit set of a Fuchsian group. Acta Math. **136**, 241–273 (1976)
48. F. Paulin, M. Pollicott, B. Schapira, *Equilibrium States in Negative Curvature*. Astérisque, vol. 373 (Société Mathématique de France, Montrouge, 2015), 289pp
49. T. Roblin, *Ergodicité et équidistribution en courbure négative*. Mém. Soc. Math. France, vol. 95 (Société Mathématique de France, Paris, 2003), vi+99pp
50. D. Ruelle, An inequality for the entropy of differentiable maps. Bul. Braz. Math. Soc. **9**, 83–87 (1978)
51. A. Selberg, On discontinuous groups in higher dimensional symmetric spaces, in *Contributions to Function Theory* Tata IFR, Bombay (1960), pp. 147–164
52. D. Sullivan, The density at infinity of a discrete group of hyperbolic motions. Pub. Math. I.H.É.S. **50**, 171–202 (1979)

53. Y. Yomdin, Volume growth and entropy. Isr. J. Math. **57**, 285–300 (1987)
54. Y. Yomdin, C^k-resolution of semialgebraic mappings. Addendum to: "Volume growth and entropy". Isr. J. Math. **57**, 301–317 (1987)
55. C. Yue, Rigidity and dynamics around manifolds of negative curvature. Math. Res. Lett. **1**, 123–147 (1994)

Two-Dimensional Quantum Yang–Mills Theory and the Makeenko–Migdal Equations

Thierry Lévy

2010 Mathematics Subject Classification 46T12, 60B20, 81T13

Introduction

These notes, echoing a conference given at the Strasbourg–Zurich seminar in October 2017, are written to serve as an introduction to 2-dimensional quantum Yang–Mills theory and to the results obtained in the last five to ten years about its so-called large N limit.

Quantum Yang–Mills theory, at least in the flavour that we will describe, combines differential geometric and probabilistic ideas. We would like to think, and hope to convince the reader, that this is less a complication than a source of beauty and enjoyment.

Some parts of our presentation will rely more distinctly on a probabilistic or a differential geometric background. We will however always try to keep technicalities aside and to favour explanation over demonstration. This is thus not, in the purest sense, a mathematical text: there will be essentially no proof. On the other hand, we will give fairly detailed examples of some computations that, we hope, are typical of the theory and illustrate it.

Slightly different in aim and content, but also introductory, the notes [26] written with four hands with Ambar Sengupta can serve as counterpoint, or complement, to the present text.

T. Lévy (✉)
Laboratoire de Probabilités, Statistique et Modélisation (LPSM), Sorbonne Université, Paris, France
e-mail: thierry.levy@sorbonne-universite.fr

N. Anantharaman et al. (eds.), *Frontiers in Analysis and Probability*,
https://doi.org/10.1007/978-3-030-56409-4_7

These notes are split in three parts. In the first, we explain the nature of the Yang–Mills holonomy process, which is the main object of interest of the theory. We do it from two perspectives, one differential geometric, and one probabilistic. This leads us to the definition of Wilson loop expectations, which are the most important numerical quantities of the theory.

In the second part, we discuss several approaches to the computation of Wilson loop expectations, and illustrate them on several examples. The large N limit of the theory makes a first appearance in this section, and we derive by hand some concrete instances of the Makeenko–Migdal equations which are the subject of the third part. We also included in the second part a discussion of the holonomy process on the sphere, and of the Douglas–Kazakov phase transition.

In the third part, we describe the Makeenko–Migdal equations. In keeping with the style of these notes, we do not offer a proof of these equations, but we describe as carefully as we can Makeenko and Migdal's original derivation of them. Then, we discuss the amount of information carried by these equations and illustrate their power in the computation of the so-called master field, that is the large N limit of Wilson loop functionals.

1 Quantum Yang–Mills Theory on Compact Surfaces

1.1 The Holonomy Process and the Yang–Mills Action

The central object of study of quantum 2-dimensional Yang–Mills theory is a collection of random unitary matrices indexed by the class $\mathscr{L}_m(M)$ of Lipschitz continuous loops based at some point m on a compact surface M. This collection of random variables is called the *Yang–Mills holonomy process* and it is denoted by

$$(H_\ell)_{\ell \in \mathscr{L}_m(M)} \tag{1}$$

The idea of this collection of random variables arose, along a fairly convoluted path, from physical considerations relating to the description of certain kinds of fundamental interactions.[1] It is, fortunately, not necessary to be familiar with the original motivation of Yang and Mills to understand what the Yang–Mills holonomy process is.

In very broad terms, the basic data of the theory is a compact surface M (for example a disk, a sphere, a cylinder, a torus) and a compact matrix group G (for

[1] We will not describe this path, but indicate that it is marked by contributions of Chen Ning Yang and Robert Mills, the classical reference being [48], of Alexander Migdal, who in [32] provided mathematicians with a usable description of a crucial part of Yang–Mills theory, of Leonard Gross who initiated a school of mathematical study of the 2-dimensional Yang–Mills theory [13–15], of Bruce Driver and Ambar Sengupta, who finally gave in [6, 40] the first mathematically rigorous definitions of the Yang–Mills holonomy process. This enumeration is of course much too short not to leave many important contributions aside: a more extensive bibliography can for instance be found in [26].

example U(1), SO(3), U(N)). From this data, an infinite dimensional space of *connections* can be built[2], on which an infinite dimensional symmetry group, the *gauge group* acts[3], with infinite dimensional quotient, and one of the fundamental maps of the theory is the *holonomy map*

$$\{\text{connections}\}/\{\text{gauge group}\} \xrightarrow{\text{holonomy}} \mathrm{Maps}(\mathscr{L}_m(M), G)/G$$

On the right-hand side, the action of G on the space of maps from $\mathscr{L}_m(M)$ to G is by conjugation. Leaving this action aside, note that the distribution of the holonomy process (1) is a probability measure on the space $\mathrm{Maps}(\mathscr{L}_m(M), G)$. We will call this space the space of *holonomies*.

One property that makes the holonomy map so important is that it is injective. It is thus legitimate to say that a connection is well described by its holonomy.

Another fundamental map of the theory is the *Yang–Mills action* S_{YM} which is a non-negative functional traditionally defined on the space of connections, but that can also be defined on the space of holonomies, so that the situation is

$$\begin{array}{ccc} \{\text{connections}\}/\{\text{gauge group}\} & \xrightarrow{\text{holonomy}} & \mathrm{Maps}(\mathscr{L}_m(M), G)/G \\ & {\scriptstyle S_{\mathsf{YM}}} \searrow \quad \swarrow {\scriptstyle S_{\mathsf{YM}}} & \\ & [0, \infty] & \end{array} \tag{2}$$

The *Yang–Mills measure* is heuristically described as the Boltzmann probability measure, on the space of connections or on the space of holonomies, associated with the Yang–Mills action. The typical formula that one finds in the literature is

$$\mathrm{d}\mu_{\mathsf{YM}}(\omega) = \frac{1}{Z} e^{-\frac{1}{2T} S_{\mathsf{YM}}(\omega)} \, \mathrm{d}\omega \tag{3}$$

where T is a positive real parameter called the coupling constant. Here, ω is meant to stand for a connection or for a holonomy, depending on one's preferred point of view. This expression is however plagued with difficulties: on the infinite dimensional spaces where the Yang–Mills measure is supposed to live, there is no Lebesgue-like reference measure that could reasonably play the role of $\mathrm{d}\omega$, and even if there were, one would not expect the Yang–Mills measure to be absolutely

[2]The exact nature of these connections can be ignored for the moment. If $G = \mathrm{U}(1)$, they can be pictured as magnetic potentials on M.

[3]In physical terms, two connections related by a gauge transformation represent two magnetic potentials corresponding to the same magnetic field.

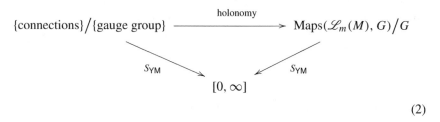

continuous with respect to it; moreover, because of the action of the gauge group, the most sensible value for the normalisation constant would be $Z = +\infty$; and one does finally not expect a typical ω in the sense of the Yang–Mills measure to be regular enough to have a finite Yang–Mills action.

One of the goals of the 2-dimensional quantum Yang–Mills theory is to find a way of sorting out these difficulties and to construct rigorously a probability measure that can honestly be called the Yang–Mills measure. The situation may look rather desperate, but it is uplifting to realise that after replacing the space of connections, or holonomies, by a space of real-valued functions on $[0, 1]$ and the Yang–Mills action by the square of the Sobolev H^1 norm, the analogous problem is almost just as ill-posed but has a very well-known solution, namely the Wiener measure. The main difference between the Wiener and the Yang–Mills cases is the presence in the latter of the gauge symmetry. Symmetry can however be a nuisance or a guide, and it turns out to be possible, in Yang–Mills theory, to make gauge symmetry an ally rather than a foe.

We will now describe more precisely the three maps appearing in the diagram (2). The holonomy map and the Yang–Mills action on the space of connections are differential geometric in nature. We start by describing them, and then turn to the Yang–Mills action on the space of holonomies. It would be unfair to say that the content of Section 1.2 can safely be completely ignored: we will refer to it later, in particular in Section 3.2. However, it is certainly possible to skip it at first reading and to jump to Section 1.3.

1.2 The Yang–Mills Action: Connections

In this section, we assume from the reader some familiarity with the differential geometry of principal bundles. We give brief reminders of the main definitions, but this is of course not the place for a complete exposition. For details, and although some might find it too Bourbakist in style, we recommend the second chapter of the first volume of the classical opus by Kobayashi and Nomizu [21].

1.2.1 The Yang–Mills Action

Although we are concerned in this text with compact surfaces, we will describe the Yang–Mills action in the more general context of compact Riemannian manifolds of arbitrary dimension—this is not more difficult.

Let M be a compact connected Riemannian manifold. Let G be a compact Lie group with Lie algebra $\mathfrak{g}$. Assume that $\mathfrak{g}$ is endowed with a scalar product $\langle \cdot, \cdot \rangle$ that is invariant under the adjoint representation $\mathrm{Ad} : G \to \mathrm{GL}(\mathfrak{g})$.[4] Let $\pi : P \to M$

[4]The typical example that we have in mind is $G = \mathrm{U}(N)$ and, for all $X, Y \in \mathfrak{u}(N)$ skew-Hermitian $N \times N$ matrices, $\langle X, Y \rangle = N\mathrm{Tr}(X^*Y)$.

be a principal G-bundle over M.[5] Let $\mathscr{A}$ denote the space of connections on P. It is an affine subspace of the space of $\mathfrak{g}$-valued differential 1-forms on P. For every connection $\omega \in \mathscr{A}$, the curvature of ω is the form $\Omega = d\omega + \frac{1}{2}[\omega \wedge \omega]$.[6] This $\mathfrak{g}$-valued 2-form on P vanishes on vertical vectors and is G-equivariant. It can thus be seen as a 2-form on M with values in the adjoint bundle $\mathrm{Ad}(P)$. Using the Hodge operator of the Riemannian structure of M, one can form the $(\mathrm{Ad}(P) \otimes \mathrm{Ad}(P))$-valued form of top degree $\Omega \wedge \star\Omega$ on M. Contracting this form with the Euclidean structure of $\mathrm{Ad}(P)$ induced by the invariant scalar product on $\mathfrak{g}$ yields the real-valued differential form of top degree $\langle \Omega \wedge \star\Omega \rangle$. This form can be integrated[7] to

[5] The manifold P is thus acted on, on the right, by G. For small open subsets U of M, the part $\pi^{-1}(U)$ of the manifold P that sits above U is equivariantly diffeomorphic to $U \times G$, with π being the first coordinate map and G acting by translations on the right on the second coordinate. A principal bundle is *trivial* if it is globally isomorphic to $M \times G$.

[6] This definition of the curvature is made slightly ambiguous by the coexistence, in the literature, of two different conventions regarding the definition of the exterior product and the exterior differential of differential forms. Since it took me some time to clarify this elementary point, I want to record it here, to the price of a rather long footnote.

The two conventions could be called 'simplicial' and 'cubical' according to their respective definitions of the exterior product of 1-forms:

$$(\alpha_1 \wedge \ldots \wedge \alpha_k)(X_1, \ldots, X_k) = \begin{cases} \frac{1}{k!} \det\left[(\alpha_i(X_j))_{i,j=1\ldots k}\right] & \text{(simplicial)} \\ \det\left[(\alpha_i(X_j))_{i,j=1\ldots k}\right] & \text{(cubical)} \end{cases}$$

Each convention is supported by illustrious authors, including, for the simplicial one, Kobayashi and Nomizu [21, p. 35] and Morita [33, Eq. (2.14) p. 70], and for the cubical one, Spivak [43, p. 203]. Since everyone agrees on the formula $d(\alpha\wedge\beta) = d\alpha\wedge\beta+(-1)^{\deg(\alpha)\deg(\beta)}\alpha\wedge d\beta$, there must also be two competing definitions of the exterior differential. Specifically, the two definitions are related by the formula $d^{\mathrm{simplicial}}\alpha = \frac{1}{\deg(\alpha)+1} d^{\mathrm{cubical}}\alpha$ (compare, for instance, [21, p. 36] or [33, Thm. 2.9 p. 71] and [43, Thm 13 p. 213]). The formula $d\alpha(X, Y) = X\alpha(Y) - Y\alpha(X) - \alpha([X, Y])$, for instance, belongs to the cubical school.

Returning to the definition of the curvature, it has a different meaning with each convention, but fortunately, the simple relation $\Omega^{\mathrm{simplicial}} = \frac{1}{2}\Omega^{\mathrm{cubical}}$ holds. Let us be more explicit about this definition: the expression $\omega \wedge \omega$ is to be understood as a $\mathfrak{g} \otimes \mathfrak{g}$-valued 2-form, which is then composed by the Lie bracket to yield a $\mathfrak{g}$-valued 2-form. Explicitly, if X and Y are two vector fields defined on an open subset of P, then the curvature of ω is defined on this open set by

$$\Omega^{\mathrm{cubical}}(X, Y) = 2\Omega^{\mathrm{simplicial}}(X, Y) = X\omega(Y) - Y\omega(X) - \omega([X, Y]) + [\omega(X), \omega(Y)]$$

Note that there is universal agreement on what it means for the curvature to vanish.

Finally, since everyone also agrees that Stokes' formula is free of any coefficient, each convention on the definition of the exterior differential entails its own definition of the integral. This is slightly hidden by the fact that everyone agrees on the formula $\int_{[0,1]^n} dx_1 \wedge \ldots \wedge dx_n = 1$ (see [33, Sec. 3.2 (a), p. 104] and [43, Prop. 1 p. 247]), but it must be realised that the differential form that is denoted by $dx_1 \wedge \ldots \wedge dx_n$ is not the same for everyone. Specifically, the relation is $\int^{\mathrm{simplicial}} \alpha = \deg(\alpha)! \int^{\mathrm{cubical}} \alpha$.

Finally, there is agreement on the meaning of the curvature as a linear map from the space of smooth 2-chains in P to $\mathfrak{g}$.

[7] The definition of the Yang–Mills action seems to require an orientation of M. In fact, this orientation is used twice, once to define the Hodge dual $\star\Omega$ of Ω and once to integrate $\langle \Omega, \star\Omega \rangle$

yield the Yang–Mills action of ω:

$$S_{\mathsf{YM}}(\omega) = \frac{1}{2} \int_M \langle \Omega \wedge \star\Omega \rangle \tag{4}$$

In words, the Yang–Mills action of a connection is nothing more than one half of the squared L^2 norm of its curvature.[8]

Let us describe locally, in coordinates, the scalar function that is integrated over M to compute $S_{\mathsf{YM}}(\omega)$. For this, let us consider an open subset U of M on which there exist local coordinates $x_1, \ldots, x_n$ on M and over which P is trivial. Let us choose a section[9] $\sigma : U \to P$ of P over U. Let us define $A = \sigma^*\omega$. Then in the local coordinates on U, the 1-form A writes $A_1 \,\mathrm{d}x_1 + \ldots + A_n \,\mathrm{d}x_n$, where $A_1, \ldots, A_n$ are maps from U to $\mathfrak{g}$. Then $F = \sigma^*\Omega$ writes

$$F = \sum_{1 \leqslant i < j \leqslant n} \big(\partial_i A_j - \partial_j A_i + [A_i, A_j]\big) \,\mathrm{d}x_i \wedge \mathrm{d}x_j$$

and the contribution of U to the Yang–Mills action of ω is

$$\frac{1}{2} \int_U \langle \Omega \wedge \star\Omega \rangle = \frac{1}{2} \sum_{1 \leqslant i < j \leqslant n} \int_U \big\| \partial_i A_j - \partial_j A_i + [A_i, A_j] \big\|^2 \,\mathrm{dvol}(x)$$

where $\mathrm{dvol}(x)$ is the Riemannian volume measure on M, and $\| \cdot \|$ is the Euclidean norm on $\mathfrak{g}$ associated with the invariant scalar product $\langle \cdot, \cdot \rangle$. The analogy with the squared Sobolev H^1 norm should be even more obvious on this expression.

1.2.2 Gauge Transformations

The gauge group, that we denote by $\mathscr{J}$, is the group of G-equivariant diffeomorphisms of P over the identity of M.[10] It acts by pull-back on $\mathscr{A}$ and a

over M. Reversing the orientation changes the Hodge dual and the integral by a sign, so that if M is orientable, the definition of S_{YM} is independent of the choice of orientation of M. Moreover, if M is not orientable, S_{YM} can still be defined using a partition of unity.

[8] Considering that the curvature is a kind of derivative of the connection, the Yang–Mills action stands thus in close analogy with the squared H^1 norm of a real-valued function on $[0, 1]$.

[9] To say that σ is a section of P over U means that $\pi \circ \sigma = \mathrm{id}_U$. The existence of such a section is equivalent to the triviality of the restriction of P over U. In particular, the existence of a global section $\sigma : M \to P$ is equivalent to the triviality of the bundle $\pi : P \to M$. The reader who is more familiar with vector bundles than principal bundles might at first be surprised by this statement, since a vector bundle can admit a global section, even a non-vanishing one, without being trivial. However, the existence of a section for a principal bundle corresponds, for a vector bundle, to the existence of a basis of sections.

[10] An element j of the gauge group is a diffeomorphism $j : P \to P$ that leaves each fibre of P globally stable, and acts on it in a way that commutes with the action of G on the right on P. For the bundle $P = M \times G \to M$, the gauge group can be identified with $\mathscr{J} = C^\infty(M, G)$ acting pointwise on P by multiplication on the left on the second coordinate.

routine verification shows that it leaves S_{YM} invariant. Thus, the Yang–Mills action descends to a function

$$S_{\mathsf{YM}} : \mathscr{A}/\mathscr{J} \to [0, \infty)$$

the study of which is the subject of classical Yang–Mills theory.

Let us display the formulas which give, through a local section of P, the action of the gauge group on a connection and its curvature. These formulas are indeed useful, and ubiquitous in the literature. Let $j : P \to P$ be a gauge transformation. Let $\sigma : U \to P$ be a local section of P over an open subset U of M. Then there exists a unique function $g : U \to G$ such that for every $x \in U$, one has $j(\sigma(x)) = \sigma(x)g(x)$. Then, letting j act on a connection ω yields the new connection $j \cdot \omega = j^*\omega$ and transforms on one hand A into

$$A^g = \sigma^*(j \cdot \omega) = g^{-1}Ag + g^{-1}\,\mathrm{d}g$$

and on the other hand F into

$$F^g = g^{-1}Fg$$

This formula explains the invariance of the Yang–Mills action: without trying to be perfectly precise, one can say that the action of a gauge transformation conjugates the curvature at each point of M by some element of G, and thus leaves its Euclidean norm unchanged.

1.2.3 Some Questions of Classical Yang–Mills Theory

Let us mention, without giving any details, a few examples of the questions that arise in the study of the Yang–Mills action.

- The set $S_{\mathsf{YM}}^{-1}(0)$ is the *moduli space of flat connections*, that is, the quotient of the set of flat connections by the action of the gauge group. It is a finite-dimensional orbifold with a rich geometric structure, the study of which is both an old and an active subject of investigation [11, 12, 20, 28, 29, 45, 46].
- The Yang–Mills action can be understood as arising, through appropriate reformulation and generalisation, from a Lagrangian formulation of Maxwell's equations of the electromagnetic field. The critical points of the Yang–Mills action are thus of special interest: they are, in a sense, the classical physical fields of Yang–Mills theory. They are called *Yang–Mills connections* and a milestone in their study in the 2-dimensional case is [1].
- When M is 4-dimensional, the Yang–Mills action is conformally invariant, in the sense that it depends on the Riemannian metric on M only through its conformal class. There is an extensive literature devoted to Yang–Mills connections on 4-dimensional manifolds [18]. Looking for self-dual Yang–Mills connections on

$\mathbb{R}^4$ that are invariant by translation in two directions, for example, leads to the study of Hitchin equations and Higgs bundles [19].

- From a physical point of view, the Yang–Mills action of a connection is an appropriate measure of its non-triviality. From an analytical point of view, however, it turns out that a natural way of measuring a connection is its Sobolev H^1 norm.[11] The Yang–Mills action is controlled by the H^1 norm, but not conversely. A flat connection, that is, a connection with Yang–Mills action 0, can be given an arbitrarily large H^1 norm by an appropriate gauge transformation. A beautiful theorem of Karen Uhlenbeck states that level sets of the Yang–Mills action, that is, the sets of the form $\{S_{\mathsf{YM}} \leqslant c\}$, $c \in \mathbb{R}_+$, are sequentially weakly compact in H^1 up to gauge transformation: from any sequence of connections with bounded Yang–Mills action, one can extract a subsequence which, after suitable gauge transformation of each term, converges weakly in H^1 [44].
- The Yang–Mills action gives rise to a gradient flow, which formally is the solution of the differential equation $\partial_t \omega_t = -\nabla_{\omega_t} S_{\mathsf{YM}}$. This is the *Yang–Mills flow* [36]. There is currently an active investigation of stochastic perturbations of this flow in cases where M is 2- or 3-dimensional [4, 41].

1.2.4 The Holonomy Map

A fundamental construction associated with a connection is that of the *holonomy*, or *parallel transport*, that it induces. For every continuous and piecewise smooth curve $c : [0, 1] \to M$, the parallel transport along c determined by the connection ω is the G-equivariant mapping $\mathrm{hol}(\omega, c) : P_{c_0} \to P_{c_1}$ which to every point p of P_{c_0} associates the endpoint of the unique continuous curve $\tilde{c} : [0, 1] \to P$ such that $\tilde{c}_0 = p$, $\pi \circ \tilde{c} = c$ and for all $t \in [0, 1]$ at which c is differentiable, $\omega(\dot{\tilde{c}}_t) = 0$.

This parallel transport enjoys the following properties, which are of fundamental importance.

- It is unaffected by a change of parametrisation of the curve.
- If $c : [0, 1] \to M$ is a curve and c^{-1} denotes the same curve traced backwards, that is, $c_t^{-1} = c_{1-t}$, then $\mathrm{hol}(\omega, c^{-1}) = \mathrm{hol}(\omega, c)^{-1}$.
- If c and c' are two curves such that $c_1 = c'_0$, so that the concatenation cc' is well defined, then $\mathrm{hol}(\omega, cc') = \mathrm{hol}(\omega, c') \circ \mathrm{hol}(\omega, c)$.

It will be useful to understand a bit more concretely how this parallel transport can be computed, and how it gives rise to a holonomy in the sense that we gave to this word in Section 1.1.

Assume that the range of the curve c lies in an open subset U of M over which the fibre bundle P is trivial.[12] Let $\sigma : U \to P$ be a section of P over U. Set $A = \sigma^*\omega$.

---

[11] Here, we are talking about connections as elements of $\mathscr{A}$, not of the quotient $\mathscr{A}/\mathscr{J}$.

[12] If c does not lie in such an open subset, it can be split into finitely many pieces which do and the holonomy along c is simply the product of the holonomies along these shorter pieces.

It is a 1-form on U with values in $\mathfrak{g}$. The solution of the differential equation

$$\dot{h}_t = -A(\dot{c}_t)h_t, \ h_0 = 1_G \tag{5}$$

is a curve $h : [0, 1] \to G$ which starts from the unit element 1_G. The endpoint of this curve computes the parallel transport along c determined by ω in the sense that

$$\mathrm{hol}(\omega, c)(\sigma(c_0)) = \sigma(c_1)h_1$$

This relation is illustrated in Figure 1.

Let us introduce the notation

$$\mathrm{hol}_\sigma(\omega, c) = h_1$$

the holonomy of ω along c read in the section σ. This object has the drawback of depending on the choice of a local section of the bundle, but the great advantage of being fairly concrete, namely an element of G, that is, in many situations, a matrix.

If $j \in \mathscr{J}$ is a gauge transformation of P, recall from Section 1.2.2 that $j \cdot \omega = j^*\omega$ is the pull-back of ω by the diffeomorphism j of P. Then the holonomy of $j \cdot \omega$ along c is related to that of ω by the relation

$$\mathrm{hol}(j \cdot \omega, c) = j^{-1}_{|P_{c_1}} \circ \mathrm{hol}(\omega, c) \circ j_{|P_{c_0}}$$

Through the local section $\sigma : U \to M$, and letting $g : U \to G$ be the function such that $j(\sigma(x)) = \sigma(x)g(x)$ for every $x \in U$, this relation takes the more explicit form

$$\mathrm{hol}_\sigma(j \cdot \omega, c) = g_{c_1}^{-1}\mathrm{hol}_\sigma(\omega, c)g_{c_0} \tag{6}$$

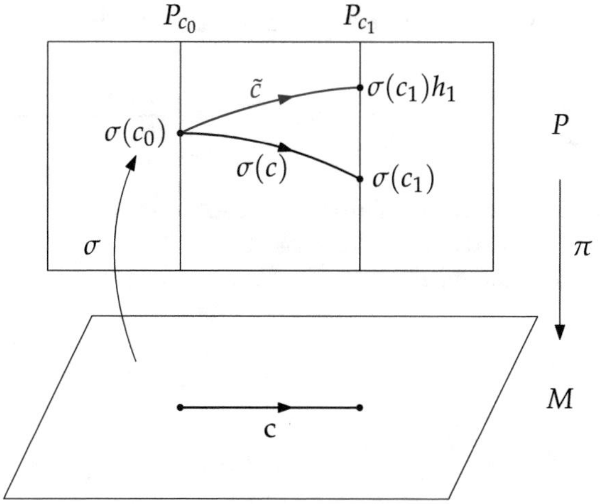

Fig. 1 The difference between the horizontal lift of c starting at $\sigma(c_0)$, denoted in this picture by $\tilde{c}$, and $\sigma(c)$, the image of c by the local section σ, is measured by the function h which solves (5)

It follows from (6) that for all loop ℓ on M, that is, all curve which ends at its starting point, the conjugacy class of $\mathrm{hol}_\sigma(\omega, \ell)$ is not affected[13] by a gauge transformation of ω.

More generally, given a base point m on M, and denoting by $\mathscr{L}_m^\infty(M)$ the class of piecewise smooth loops on M based at m, the orbit of

$$(\mathrm{hol}_\sigma(\omega, \ell) : \ell \in \mathscr{L}_m^\infty(M)) \in \mathrm{Maps}(\mathscr{L}_m^\infty(M), G)$$

under the action of G by simultaneous conjugation is not affected by a gauge transformation of ω. This explains how a connection modulo gauge transformation defines a holonomy modulo conjugation.

The following result makes precise the statement that the horizontal arrow of (2) is injective.

Theorem 1.1 *Let m be a point of M. Let σ be a section of P in a neighbourhood of m. For any two connections ω and ω' on P, the following assertions are equivalent.*

1. *There exists a gauge transformation $j \in \mathscr{J}$ such that $j \cdot \omega = \omega'$.*
2. *There exists $g \in G$ such that for all loop $\ell \in \mathscr{L}_m^\infty(M)$, the equality $hol_\sigma(\omega', l) = g^{-1} hol_\sigma(\omega, l) g$ holds.*

1.3 The Yang–Mills Action: Holonomies

We will now give an alternative of the Yang–Mills action that is less classical and, most importantly, specific to the 2-dimensional case. To give an idea of the nature of this second description, let us pursue the analogy with the Wiener measure and the Sobolev H^1 norm. Consider a smooth function $b : [0, 1] \to \mathbb{R}$ with $b(0) = 0$. The squared H^1 norm of b can be expressed at least in the following two ways:

$$\|b\|_{H^1}^2 = \int_0^1 |\dot{b}(t)|^2 \, \mathrm{d}t = \sup_{0 \leqslant t_0 < t_1 < \ldots < t_n \leqslant 1} \sum_{k=1}^{n} \frac{|b(t_k) - b(t_{k-1})|^2}{t_k - t_{k-1}} \tag{7}$$

The integral expression corresponds to the description of the Yang–Mills action that we gave in the last section and is very similar to (4). We will now give another description, similar to the second, more combinatorial one.

[13] Incidentally, this class does not depend on the local section σ either.

1.3.1 Holonomies

The main algebraic property of the holonomy of a connection, already mentioned in Section 1.2.4, is that it is a *multiplicative* map from $\mathscr{L}_m^\infty(M)$ to G. Let us formulate this in a slightly different way.

Recall that M is a compact Riemannian manifold and G a compact Lie group. Let $\mathscr{P}(M)$ denote the set of all Lipschitz continuous[14] paths on M, two paths being identified if they differ only by an increasing change of parametrisation. Let us call a function $h : \mathscr{P}(M) \to G$ *multiplicative* if it satisfies the following two properties.

- For all path c, letting c^{-1} denote the same path traced backwards, one has $h(c^{-1}) = h(c)^{-1}$.
- For all paths c and c' such that c finishes where c' starts, so that the concatenated path cc' is defined, one has $h(cc') = h(c')h(c)$.

More generally, given a subset P of $\mathscr{P}(M)$, we say that a function $h : P \to G$ is multiplicative if it satisfies the above conditions whenever all the paths involved belong to the subset P.

Let us denote by $\mathrm{Mult}(\mathscr{P}(M), G)$ (resp. by $\mathrm{Mult}(P, G)$) the subset of $\mathrm{Maps}(\mathscr{P}(M), G)$ (resp. of $\mathrm{Maps}(P, G)$) formed by all multiplicative maps.

There is an action of the *gauge group* $\mathrm{Maps}(M, G)$ on $\mathrm{Mult}(\mathscr{P}(M), G)$ defined as follows. Consider $g : M \to G$ and a multiplicative map $h : \mathscr{P}(M) \to G$. For all path c starting at c_0 and finishing at c_1, define

$$(g \cdot h)(c) = g_{c_1}^{-1} h(c) g_{c_0} \tag{8}$$

an equation that should be compared with (6). It is not difficult to check that the map $g \cdot h$ is still multiplicative.

Let m be a point of M. A multiplicative function can be restricted to $\mathscr{L}_m(M)$ and the action of $\mathrm{Maps}(M, G)$ on this restricted map reduces to the action of G by conjugation. The following fact may seem surprising at first sight, but it is not difficult to prove.

Proposition 1.2 *For all $m \in M$, the restriction map*

$$\mathit{Mult}(\mathscr{P}(M), G)\big/\mathit{Maps}(M, G) \longrightarrow \mathrm{Mult}(\mathscr{L}_m(M), G)\big/G$$

is a bijection.

[14] In this text, we consider alternatively paths that are piecewise smooth and paths that are Lipschitz continuous. We do so for reasons of technical convenience, and the reader should not be overly worried by what can safely be regarded as a secondary issue.

We call either side of this bijection the space of *holonomies*. Thanks to the multiplicativity and the gauge symmetry, a holonomy can either be seen as a group-valued function on the set of all paths, or on the set of all loops based at some reference point m on M.

1.3.2 Graphs on Surfaces

We will now assume that M is a 2-dimensional manifold: it is thus a compact surface. We announced an expression of the Yang–Mills action similar to the rightmost term of (7): the role of subdivisions of the interval $[0, 1]$ will be played by *graphs* on M. This will be the occasion of a first encounter with this notion that is central to the construction of the 2-dimensional Yang–Mills measure.

Let us call *edge* an element of $\mathscr{P}(M)$ that is injective — note that this does not depend on the way in which the path is parameterised. A *graph* is a finite set of edges, stable by the reversal map $e \mapsto e^{-1}$, and in which any two edges either form a pair $\{e, e^{-1}\}$, or meet, if at all, at some of their endpoints.

The *vertices* of a graph are the endpoints of its edges. The *faces* of a graph are the connected components of the complement in M of the union of its edges. A graph is conveniently described as a triple $\mathbb{G} = (\mathbb{V}, \mathbb{E}, \mathbb{F})$ consisting of a set of vertices, a set of edges and a set of faces, but it is in fact entirely determined by the set $\mathbb{E}$ of its edges.

A crucial additional assumption is that every face of a graph must be homeomorphic to a disk. This guarantees that the 1-skeleton of the graph correctly represents the topology of the surface, to the extent that a 1-dimensional object can represent a 2-dimensional one.

1.3.3 The Yang–Mills Action

Let $\mathbb{G}$ be a graph on our compact surface M. We will denote by $\mathscr{P}(\mathbb{G})$ the set of paths that can be constructed as concatenations of edges of $\mathbb{G}$. To each face F of $\mathbb{G}$, we can associate in an almost unequivocal way a loop ∂F that winds exactly once around F. To give a perfectly rigourous definition of this loop is less simple than one might expect, but there is nothing counterintuitive in it. It is only *almost* well defined because there is no preferred starting point for this loop. However, if $f : \mathscr{P}(\mathbb{G}) \to G$ is a multiplicative function, then the conjugacy class of the element $h(\partial F)$ of G is well defined. In particular, the Riemannian distance, in G, between the element $h(\partial F)$ and the unit element 1_G, is well defined.[15] This distance is, moreover, not affected by the action of an element of the gauge group $\mathrm{Maps}(M, G)$ on h.

[15] This distance is defined by the bi-invariant Riemannian metric on G associated with the invariant scalar product chosen on its Lie algebra, see the first lines of Section 1.2.1.

We can now define the Yang–Mills action on the space of holonomies by setting, for all $h \in \mathrm{Mult}(\mathscr{P}(M), G)$,

$$S_{\mathsf{YM}}(h) = \sup\left\{ \sum_{F \in \mathbb{F}} \frac{d_G(1_G, h(\partial F))^2}{\mathrm{area}(F)} \;:\; \mathbb{G} \text{ graph on } M \right\} \tag{9}$$

where the area of a face F is computed using the Riemannian structure on M.

It is manifest on this expression that, in the case where M is a surface, the only part of the Riemannian structure on M that is used in the definition of the Yang–Mills action is the Riemannian volume, in this case the Riemannian area. This is of course also true, be it in a slightly less apparent way, of the definition (4).

Proposition 1.3 *Assume that M is 2-dimensional. Then the definitions* (4) *and* (9) *of the Yang–Mills action agree. More precisely, for every connection ω inducing a holonomy h, the equality $S_{\mathsf{YM}}(\omega) = S_{\mathsf{YM}}(h)$ holds.*

1.4 The Yang–Mills Holonomy Process

We will now explain how to construct the Yang–Mills holonomy process. Although the definition of this process is derived, at a heuristic level, from the Yang–Mills action, the process and the action are logically unrelated. We can thus start afresh, from a compact surface M on which we have a Riemannian structure, or at least a measure of area, and a compact Lie group G, on the Lie algebra of which we have an invariant scalar product.

1.4.1 The Configuration Space of Lattice Yang–Mills Theory

One piece of information that we need to retain from the previous sections is the notion of graph on our surface M (see Section 1.3.2). Let us choose a graph $\mathbb{G} = (\mathbb{V}, \mathbb{E}, \mathbb{F})$ on M. The *configuration space* associated with a graph $\mathbb{G}$ on our surface M is the manifold

$$\mathscr{C}_{\mathbb{G}} = \{g = (g_e)_{e \in \mathbb{E}} \in G^{\mathbb{E}} : \forall g \in G, g_{e^{-1}} = g_e^{-1}\} = \mathrm{Mult}(\mathbb{E}, G)$$

of all ways of assigning an element of G to each oriented edge, in a way that is consistent with the orientation reversal.

Recall that we denote by $\mathscr{P}(\mathbb{G})$ the set of paths that can be constructed as concatenations of edges of $\mathbb{G}$. The configuration space $\mathscr{C}_{\mathbb{G}}$ is naturally in one-to-one correspondence with the set $\mathrm{Mult}(\mathscr{P}(\mathbb{G}), G)$ of all multiplicative maps from $\mathscr{P}(\mathbb{G})$ to G.

Choosing an *orientation* of $\mathbb{G}$, that is, a subset $\mathbb{E}^+ \subset \mathbb{E}$ containing exactly one element in each pair $\{e, e^{-1}\}$ allows one to realise the configuration space in the slightly less canonical, but easier to handle, way

$$\mathscr{C}_{\mathbb{G}} = G^{\mathbb{E}^+}$$

This makes it easy, for instance, to endow $\mathscr{C}_{\mathbb{G}}$ with a probability measure, namely the Haar measure on $G^{\mathbb{E}^+}$. The invariance of the Haar measure on the compact group G under the inverse map $x \mapsto x^{-1}$ implies that this measure on $\mathscr{C}_{\mathbb{G}}$ does not depend on the choice of orientation. We denote it by $\mathrm{d}g$.

Every path $c \in \mathscr{P}(\mathbb{G})$ can be uniquely written as a concatenation of edges $c = e_1^{\epsilon_1} \dots e_n^{\epsilon_n}$ with $e_1, \dots, e_n \in \mathbb{E}^+$ and $\epsilon_1, \dots, \epsilon_n \in \{-1, 1\}$. To such a path $c = e_1^{\epsilon_1} \dots e_n^{\epsilon_n}$ we associate a holonomy map

$$\begin{aligned} h_c : \mathscr{C}_{\mathbb{G}} &\longrightarrow G \\ g &\longmapsto g_{e_n}^{\epsilon_n} \cdots g_{e_1}^{\epsilon_1} \end{aligned} \tag{10}$$

Our goal is to endow the configuration space $\mathscr{C}_{\mathbb{G}}$ with an interesting probability measure, so as to make the collection of maps $(h_c)_{c\in\mathscr{P}(\mathbb{G})}$ into a collection of G-valued random variables.

1.4.2 The Driver–Sengupta Formula

In order to define this probability measure, we need to introduce the heat kernel on G, or more accurately the fundamental solution of the heat equation. The invariant scalar product on the Lie algebra $\mathfrak{g}$ determines a bi-invariant Riemannian structure on G, and a Laplace-Beltrami operator Δ. We consider the function $p : \mathbb{R}_+^* \times G \to \mathbb{R}_+^*$ that is the unique positive solution of the heat equation $(\partial_t - \frac{1}{2}\Delta)p = 0$ with initial condition $p(t, x)\,\mathrm{d}x \Rightarrow \delta_{1_G}$ as $t \to 0$. We use the notation $p_t(x) = p(t, x)$. A crucial property of this function is that, for all $t > 0$ and all $x, y \in G$, we have $p_t(yxy^{-1}) = p_t(x)$. We refer to this property as the *invariance under conjugation* of the heat kernel.

We mentioned at the end of Section 1.3.3 that, in the 2-dimensional setting, the Yang–Mills action depends on a Riemannian structure of the surface M only through the Riemannian area that it induces. We will denote by $|F|$ the area of a Borel subset F of M.

Given a face F of our graph, recall that we denote by ∂F a path that goes once around this face in the positive direction. Recall also that this path is ill-defined because there is no preferred vertex on the boundary of F from which to start it. However, this indeterminacy only results in an indeterminacy *up to conjugation* for the holonomy map $h_{\partial F}$. Thanks to the invariance under conjugation of the heat kernel, the function $g \mapsto p_t(h_{\partial F}(g))$ is still well defined on $\mathscr{C}_{\mathbb{G}}$ for every $t > 0$.

We can now write the formula which is the basis of the definition of the 2-dimensional Yang–Mills measure. It is due to Bruce Driver in the case where M is the plane, or a disk, and to Ambar Sengupta when M is an arbitrary compact surface. Recall that T is a positive real parameter of the measure. We define, on $\mathscr{C}_{\mathbb{G}}$, the probability measure

$$\mathrm{d}\mu_{\mathsf{YM}}^{\mathbb{G},T}(g) = \frac{1}{Z(\mathbb{G},T)} \prod_{F \in \mathbb{F}} p_{T|F|}(h_{\partial F}(g))\, \mathrm{d}g \tag{DS}$$

Here, $Z(\mathbb{G},T)$ is the normalisation constant that makes $\mu_{\mathsf{YM}}^{\mathbb{G},T}$ a probability measure on $\mathscr{C}_{\mathbb{G}}$.

The gauge group Maps$(\mathbb{V}, G)$ acts on the configuration space $\mathscr{C}_{\mathbb{G}}$ by a formula analogous to (8), and the measure $\mu_{\mathsf{YM}}^{\mathbb{G},T}$ is invariant under this action. Indeed, this action preserves the reference measure $\mathrm{d}g$ and transforms the holonomy along loops, in this case along boundaries of faces, by conjugation, which leaves the value of the fundamental solution of the heat equation on these holonomies unchanged.[16]

[16]Let us say a word about the way in which the presence of a boundary to the surface M should be taken into account in (DS), and how to treat the case where M is not orientable. The only place where we used the orientability and orientation of M is when we defined the boundary of a face as a loop winding *positively* around M. However, since the heat kernel also enjoys the invariance property $p_t(x) = p_t(x^{-1})$, it does not matter which orientation we choose around each face of the graph. Thus, (DS) is valid without any modification on a non-orientable surface.

In the case where M has a boundary, this boundary is a finite union of circles. Our assumption that each face of a graph is homeomorphic to a disk implies that each of these circles is a path in any graph on M. In this case, (DS) still makes sense and corresponds to free boundary conditions along the boundary of M. Fixed boundary conditions can be imposed: it is possible to insist that the holonomy along each boundary component belongs to a specific conjugacy class in G. If we wish to set the boundary condition for which the holonomy along a boundary component $c = e_1 \dots e_n$ belongs to a conjugacy class C of G, the basic ingredient is the unique probability measure $\nu_{n,C}$ on $\mathcal{O}_{n,C} = \{(x_1,\dots,x_n) \in G^n : x_n \dots x_1 \in C\}$ invariant under the transitive action of G^n given by

$$(y_1,\dots,y_n)\cdot(x_1,\dots,x_n) = (y_1 x_1 y_n^{-1}, y_2 x_2 y_1^{-1}, \dots, y_n x_n y_{n-1}^{-1})$$

This measure is easily described by the formula

$$\int_{\mathcal{O}_{n,C}} f \,\mathrm{d}\nu_{n,C} = \int_{G^n} f(x_1,\dots,x_{n-1}, x_n z x_n^{-1} x_1^{-1} \dots x_{n-1}^{-1})\, \mathrm{d}x_1 \dots \mathrm{d}x_n$$

for an arbitrary $z \in C$. The way in which (DS) should be modified is that the uniform measure on $\mathscr{C}$ should be replaced, for the edges lying on the boundary of M, by the appropriate copy of a measure of the form $\nu_{n,C}$.

1.4.3 Invariance Under Subdivision

Starting from a graph $\mathbb{G}$ on our surface M, we built the configuration space $\mathscr{C}_{\mathbb{G}}$ and endowed, thanks to the Driver–Sengupta formula, this space with a probability measure, the lattice 2-dimensional Yang–Mills measure on $\mathbb{G}$. In doing so, we automatically produced a collection

$$(h_c)_{c\in\mathscr{P}(\mathbb{G})} \text{ or } (h_\ell)_{\ell\in\mathscr{L}_m(\mathbb{G})}$$

of G-valued random variables.[17]

The property of this construction that makes it so extremely pleasant is the fact that it is *invariant under subdivision*.

To articulate this fundamental property, let us say that a graph $\mathbb{G}_2$ is *finer* than a graph $\mathbb{G}_1$ if $\mathbb{G}_2$ can be obtained from $\mathbb{G}_1$ by subdividing and adding edges. More precisely, $\mathbb{G}_2$ is finer than $\mathbb{G}_1$ if $\mathbb{E}_1 \subset \mathscr{P}(\mathbb{G}_2)$: each edge of $\mathbb{G}_1$ is a path in $\mathbb{G}_2$. When this happens, there is a natural map

$$\begin{aligned}\mathscr{C}_{\mathbb{G}_2} &\longrightarrow \mathscr{C}_{\mathbb{G}_1}\\ g^{(2)} &\longmapsto \big(h_e^{(2)}(g^{(2)})\big)_{e\in\mathbb{E}_1}\end{aligned}$$

where each edge e of $\mathbb{G}_1$ is seen as a path in $\mathbb{G}_2$ and thus assigned a holonomy by the configuration $g^{(2)}$.

The main result of 2-dimensional lattice Yang–Mills theory is the following.

Theorem 1.4 *Let $\mathbb{G}_1$ and $\mathbb{G}_2$ be two graphs on M. Assume that $\mathbb{G}_2$ is finer than $\mathbb{G}_1$. Then for all $T > 0$, the equality $Z(\mathbb{G}_1, T) = Z(\mathbb{G}_2, T)$ holds and the push-forward of the measure $\mu_{\mathsf{YM}}^{\mathbb{G}_2,T}$ by the natural map $\mathscr{C}_{\mathbb{G}_2} \to \mathscr{C}_{\mathbb{G}_1}$ is the measure $\mu_{\mathsf{YM}}^{\mathbb{G}_1,T}$.*

This theorem is so important that we are going to give an idea of the mechanism of its proof.

Proof The first observation is that one can always go from a graph to a finer graph by an appropriate succession of elementary operations consisting either in adding a new vertex in the middle of an existing edge or in adding a new edge between two existing vertices. We need to understand why neither of these elementary operations affect the partition function, nor transform essentially the measure.

The subdivision of an edge e into two new edges e' and e'' amounts, in the integral defining the partition function and in the expression defining the discrete Yang–Mills measure, to the replacement of every occurrence of the integration variable g_e by the product of the two new variables $g_{e''}g_{e'}$. The invariance by translation of the Haar measure ensures that this does not affect the result of any computation.

[17] Thanks to the multiplicativity of the holonomy and the gauge invariance of the construction of the lattice Yang–Mills measure, the point of view of a collection of random variables indexed by all paths in $\mathbb{G}$ or by the set of loops based at a specific reference point is equivalent, see Proposition 1.2.

The case of the addition of a new edge is more interesting. This edge e splits a face F into two faces F_1 and F_2, the boundaries of which are of the form ea and be^{-1} for some paths a and b. Observe that ba is a loop going along the boundary of F. In the computation of the partition function of the Yang–Mills measure on the finer graph, or of the integral of any functional on the configuration space of the coarser graph with respect to the image of the discrete Yang–Mills measure on the finer graph, we find an integral of a product of many factors, among which the two factors

$$p_{T|F_1|}\big(h_a(g)g_e\big)\; p_{T|F_2|}\big(g_e^{-1}h_b(g)\big)$$

contain the only two occurrences of the integration variable g_e. We can thus easily integrate with respect to g_e, using the convolution property of the heat kernel, namely the equality $p_t * p_s = p_{t+s}$, to find these two factors replaced by

$$p_{T(|F_1|+|F_2|)}\big(h_a(g)h_b(g)\big) = p_{T|F|}\big(h_{ba}(g)\big) = p_{T|F|}\big(h_{\partial F}(g)\big)$$

We are thus left with the partition function, or the integral of our functional, relative to the coarser graph. □

The partition function $Z(\mathbb{G}, T)$, which is now promoted to a function of T alone, is a very interesting object. Let us give without proof an expression of this function. We use the notation $[a, b] = aba^{-1}b^{-1}$ for the commutator of two elements a and b of G.

Proposition 1.5 *Assume that M is a surface of genus g without boundary. Then for all $T > 0$, the partition function of the 2-dimensional Yang–Mills theory on M is given by*

$$Z_M(T) = \int_{G^{2g}} p_{T|M|}([a_1, b_1]\dots[a_g, b_g])\, \mathrm{d}a_1\mathrm{d}b_1\dots\mathrm{d}a_g\mathrm{d}b_g$$

1.4.4 The Continuum Limit

Up to some conceptually inessential but technically annoying complications, the invariance by subdivision of the discrete theory allows one to take the limit of the discrete measures as the graphs on the surface become infinitely fine. The technical complications have to do with the fact that, because two edges of two distinct graphs can intersect in a rather pathological way, it is not always true that given two graphs, there exists a third graph that is finer than these two graphs. The net effect of this complication is the persistence, in the theorem asserting the existence and uniqueness of the Yang–Mills holonomy process, of a continuity condition. We say that a sequence of paths $(c_n)_{n\geqslant 1}$ on M *converges* to a path c *with fixed endpoints* if all paths $c, c_1, c_2, \dots$ start at the same point and finish at the same (possibly different) point, and if the sequence of the paths $(c_n)_{n\geqslant 1}$ parameterised at unit speed converges uniformly to c.

Theorem 1.6 (The Yang–Mills holonomy process, [23, 40]) *Let M be a compact surface endowed with a smooth*[18] *measure of area. Let G be a compact Lie group, the Lie algebra of which is endowed with an invariant scalar product. There exists a collection of G-valued random variables $(H_c)_{c\in\mathscr{P}(M)}$ such that*

- *for every graph $\mathbb{G} = (\mathbb{V}, \mathbb{E}, \mathbb{F})$, the distribution of $(H_e)_{e\in\mathbb{E}}$ is the measure $\mu_{YM}^{\mathbb{G},T}$,*
- *whenever a sequence $(c_n)_{n\geqslant 1}$ of paths converges with fixed endpoints to a path c, the sequence of random variables $(H_{c_n})_{n\geqslant 1}$ converges in probability to H_c.*

Moreover, any two collections of G-valued random variables with these properties have the same distribution.

The Yang–Mills holonomy process $(H_c)_{c\in\mathscr{P}(M)}$ is invariant in distribution under the action of the gauge group. This means that for every function $g : M \to G$, the following equality in distribution holds:

$$\left(g(\overline{c})^{-1} H_c g(\underline{c})\right)_{c\in\mathscr{P}(M)} \overset{(d)}{=} (H_c)_{c\in\mathscr{P}(M)} \tag{11}$$

where $\underline{c}$ and $\overline{c}$ denote respectively the starting and finishing point of a path c. In particular, the distribution of H_c is uniform on G for every path c that is not a loop. Of course, this huge collection of uniform random variables is correlated in a complicated way, in particular to allow the random variables associated with loops to have non-uniform distributions.

The holonomy process also enjoys a property of invariance under area-preserving maps of M: if $\phi : M \to M$ is an area-preserving diffeomorphism, then ϕ preserves the class $\mathscr{P}(M)$ and the family $(H_{\phi(c)})_{c\in\mathscr{P}(M)}$ has the same distribution as the family $(H_c)_{c\in\mathscr{P}(M)}$. This is because the Driver–Sengupta formula depends only on the combinatorial structure of the graph under consideration, and on the areas of its faces. This is consistent with the fact that the Yang–Mills action, which we originally defined on a Riemannian manifold by (4), depends, if the manifold is 2-dimensional, on the Riemannian structure only through the Riemannian area. We already mentioned this important point in relation with the expression (9) of the Yang–Mills action.

[18] By a smooth measure, we mean a measure that admits a smooth positive density with respect to the Lebesgue measure in any coordinate chart.

1.4.5 The Structure of the Holonomy Process

The structure of the Yang–Mills holonomy process can be described fairly concretely provided one understands the structure of the set of loops on a graph.

Let us consider a graph $\mathbb{G}$ on M and a vertex m of this graph. We denote naturally by $\mathscr{L}_m(\mathbb{G})$ the set of loops in $\mathbb{G}$ based at m. The operation of concatenation makes $\mathscr{L}_m(\mathbb{G})$ a monoid, with unit element the constant loop at m. Each element ℓ of this monoid has an 'inverse' ℓ^{-1}, but it is not true, unless ℓ is already the constant loop, that $\ell\ell^{-1}$ is the constant loop. In order to make $\mathscr{L}_m(\mathbb{G})$ a group, into which ℓ^{-1} is truly the inverse of ℓ, it is natural to introduce on it the *backtracking equivalence* relation, for which two loops are equivalent if one can go from one to the other by successively erasing or inserting sub-loops of the form ee^{-1}, where e is an edge of the graph.

Each equivalence class of loops contains a unique loop of shortest length, which is also the unique *reduced* loop in this class, where by a reduced loop we mean one without any sub-loop of the form ee^{-1}.

Moreover, concatenation is compatible with this equivalence relation and the quotient monoid is a group. This quotient monoid can be more concretely described as the set $\mathscr{L}_m^{\mathrm{red}}(\mathbb{G})$ of reduced loops endowed with the operation of concatenation-followed-by-reduction.

With this group of reduced loops in hand, we can make several observations.

- Each element g of the configuration space $\mathscr{C}_{\mathbb{G}}$ induces, by the holonomy map, a map $\mathscr{L}_m^{\mathrm{red}}(\mathbb{G}) \to G$, which sends a loop ℓ to $h_\ell(g)$. This map is a group homomorphism, and the map

$$\mathscr{C}_{\mathbb{G}} \longrightarrow \mathrm{Hom}(\mathscr{L}_m^{\mathrm{red}}(\mathbb{G}), G)$$

  is onto. Moreover, it descends to a bijection

$$\mathscr{C}_{\mathbb{G}}/\mathrm{Maps}(\mathbb{V}, G) \overset{\sim}{\longrightarrow} \mathrm{Hom}(\mathscr{L}_m^{\mathrm{red}}(\mathbb{G}), G)/G$$

  where the action on the left is that of the gauge group, and on the action on the right is that of G by conjugation.
- Let Γ denote the 1-skeleton of the graph, that is, the union of the ranges of its edges. The map $\mathscr{L}_m^{\mathrm{red}}(\mathbb{G}) \to \pi_1(\Gamma, m)$ which simply sends a reduced loop to its homotopy class is an isomorphism.
- The group $\mathscr{L}_m^{\mathrm{red}}(\mathbb{G})$, being isomorphic to the fundamental group of a graph, or of a 1-dimensional complex, is a free group. The rank of this group is equal to $|\mathbb{E}| - |\mathbb{V}| + 1 = |\mathbb{F}| - \chi(M) + 1 = |\mathbb{F}| + 2g - 1$, where $\chi(M)$ is the Euler characteristic of M and g its genus.

It is useful to recognise that the free group $\mathscr{L}_m^{\mathrm{red}}(\mathbb{G})$ admits nice bases.[19] Let us call *lasso* around a face F of $\mathbb{G}$ any loop of the form $c.\partial F.c^{-1}$, where c is a path from m to a vertex on the boundary of F, and ∂F is a loop going once around F.

It is now quite easy to describe the holonomy process. Let us begin with the case of the plane, or the disk.

Proposition 1.7 *Assume that M is a disk or the plane. Let $\mathbb{G}$ be a graph on M. The free group $\mathscr{L}_m^{\mathrm{red}}(\mathbb{G})$ admits a basis $\{\lambda_F : F \in \mathbb{F}\}$ such that*

- *for each face F, the loop λ_F is a lasso around F,*
- *under the lattice Yang–Mills measure $\mu_{\mathsf{YM}}^{\mathbb{G},T}$, the random variables $(H_{\lambda_F} : F \in \mathbb{F})$ are independent, each H_{λ_F} being distributed according to the measure $p_{T|F|}(g)\,\mathrm{d}g$.*

In a sense, the holonomy process has independent increments distributed according to the fundamental solution of the heat equation: it can be described as a 'Brownian motion on G indexed by loops' on the disk, or on the plane. The role of time is played by area, and increments occur along faces of the graph, or lassos, instead of intervals of time.

In the case of a closed surface, the situation is slightly different. In this case, the most natural presentation of the group $\mathscr{L}_m^{\mathrm{red}}(\mathbb{G})$ is not as a free group (which it is), but with one generator too many, and one relation.

Proposition 1.8 *Assume that M is a closed surface of genus g. Let $\mathbb{G}$ be a graph on M. Set $r = |\mathbb{F}|$. The free group $\mathscr{L}_m^{\mathrm{red}}(\mathbb{G})$ admits a presentation*

$$\mathscr{L}_m^{\mathrm{red}}(\mathbb{G}) = \left\langle \lambda_{F_1}, \ldots, \lambda_{F_r}, a_1, b_1, \ldots, a_g, g_b \,\middle|\, [a_1, b_1] \ldots [a_g, b_g] = \lambda_{F_1} \ldots \lambda_{F_r} \right\rangle$$

where

- *the loops $\lambda_{F_1}, \ldots, \lambda_{F_r}$ are lassos around the r faces of $\mathbb{G}$,*
- *the homotopy classes of the loops $a_1, b_1, \ldots, a_g, b_g$ generate $\pi_1(M, m)$,*
- *for every test function $f : G^{2g+r} \to \mathbb{C}$, one has*

$$\int_{\mathscr{C}} f(H_{\lambda_1}, \ldots, H_{\lambda_r}, H_{a_1}, H_{b_1}, \ldots, H_{a_g}, H_{b_g})\, \mathrm{d}\mu_{\mathsf{YM}}^{\mathbb{G},T} \tag{12}$$

$$= Z_M(T)^{-1} \int_{G^{2g+r-1}} f(z_1, \ldots, z_{r-1}, z_r, x_1, y_1, \ldots, x_g, y_g) p_{T|F_1|}(z_1) \ldots$$

$$p_{T|F_r|}(z_r)\, \mathrm{d}z_1 \ldots \mathrm{d}z_{r-1}\, \mathrm{d}x_1\, \mathrm{d}y_1 \ldots \mathrm{d}x_g\, \mathrm{d}y_g$$

where in the last integral, z_r stands for

[19] Recall that a free group admits bases, that is, subsets by which it is freely generated. Any two bases have the same cardinality, called the rank of the group. Any subgroup of a free group is itself a free group, but the rank of a subgroup can be larger than the rank of the group. In fact, the free group of rank 2 contains subgroups of arbitrary finite or (countably) infinite rank.

$$z_r = (z_{r-1} \dots z_1 [a_g, b_g] \dots [a_1, b_1])^{-1}$$

Let us try to spell out the probabilistic content of this result. The presentation of the group $\mathscr{L}_m^{\mathrm{red}}(\mathbb{G})$ that we chose splits it into a homotopically trivial part, giving rise to the random variables $H_{\lambda_1}, \dots, H_{\lambda_r}$, and a system of generators of the fundamental group of M, associated with the random variables $H_{a_1}, H_{b_1}, \dots, H_{a_g}, H_{b_g}$. A particular role is played by the homotopically trivial loop $C = [a_1, b_1] \dots [a_g, b_g]$.

- The distribution of the random variable H_C is such that for every continuous test function $\tilde{f} : G \to \mathbb{C}$,

$$\int_{\mathscr{C}_G} \tilde{f}(H_C) \, \mathrm{d}\mu_{\mathsf{YM}}^{\mathbb{G},T} = Z_M(T)^{-1} \int_{G^{2g}} (\tilde{f} p_{T|M|})([a_1, b_1] \dots [a_g, b_g]) \, \mathrm{d}a_1 \, \mathrm{d}b_1 \dots \, \mathrm{d}a_g \, \mathrm{d}b_g$$

  This does not seem to be a particularly well-known distribution. It needs not have a density with respect to the Haar measure: for instance if $G = \mathrm{U}(N)$, it is supported by the Haar-negligible subgroup $\mathrm{SU}(N)$. However, it is, by definition, absolutely continuous with respect to the distribution of the product of g independent commutators of independent uniformly distributed random variables, and this distribution, for example if $G = \mathrm{SU}(N)$ and provided $g \geqslant 2$, is absolutely continuous with respect to the Haar measure. It is also possible to write a Fourier series for this distribution, but it involves Littlewood–Richardson coefficients, or more generally an understanding of the tensor product of irreducible representations of G.
- Conditional on H_C, the families $(H_{\lambda_1}, \dots, H_{\lambda_r})$ and $(H_{a_1}, H_{b_1}, \dots, H_{a_g}, H_{b_g})$ are independent. It is also true that the random variables

$$(H_{\lambda_1}, \dots, H_{\lambda_r}) \bmod G \quad \text{and} \quad (H_{a_1}, H_{b_1}, \dots, H_{a_g}, H_{b_g}) \bmod G$$

  with values in G^r/G and G^{2g}/G, where G acts by conjugation, are independent conditional on H_C mod G, that is, conditional on the conjugacy class of H_C.

On a surface of genus g, the probabilistic backbone of the holonomy process can thus be described as consisting of a segment of a Brownian motion on G of length $T|M|$ and $2g$ independent Haar distributed random variables on G, jointly conditioned on the final point of the Brownian motion being equal to the products of the g commutators of the uniform random variables taken in pairs.

The case where M is a sphere is special, in the sense that it involves no uniform random variables, but only a Brownian bridge on G going from 1_G to 1_G in a time equal to T times the total area of the sphere.

1.5 Wilson Loop Expectations

A different approach to the description of the distribution of the Yang–Mills holonomy process consists in identifying a natural class of scalar, gauge-invariant, functionals of this process, the distribution of which is hoped to contain as much information as possible. The most natural class of such functionals is that of *Wilson loop functionals*, which are indeed the most important scalar observables of the theory. A Wilson loop functional is constructed by choosing a certain number of loops $\ell_1, \dots, \ell_n$ on M, then the same number of conjugation-invariant functions $\chi_1, \dots, \chi_n : G \to \mathbb{C}$ and by forming the product

$$\chi_1(H_{\ell_1}) \dots \chi_n(H_{\ell_n}) \tag{13}$$

When G is a group of matrices, the simplest choice of conjugation-invariant function is the trace. The *Wilson loop expectations*, which play in this theory the role of n-point functions, are the numbers

$$\mathbb{E}[\mathrm{Tr}(H_{\ell_1}) \dots \mathrm{Tr}(H_{\ell_n})] \tag{14}$$

the computation of which is a seemingly endless subject of reflection. We will discuss in the next section a few concrete examples of computation of such numbers. For the time being, let us say a word about the amount of information that they carry.

Suppose we know the collection of all the numbers (14), or more generally the expectation of all functionals of the form (13). Then we know the joint distribution of all random variables of the form $\chi(H_\ell)$ where ℓ is a loop and $\chi : G \to \mathbb{C}$ is an invariant function. Since G is compact, invariant functions separate conjugacy classes and we know, in fact, the joint distribution of the conjugacy classes of all variables H_ℓ. This is certainly an important piece of information. However, the form of the action of the group of gauge transformations on the collection of holonomies, as given by (11), indicates that this action preserves more than just the individual conjugacy classes of the holonomies. Indeed, if $\ell_1, \dots, \ell_n$ are based at the same point, then it is the orbit of $(H_{\ell_1}, \dots, H_{\ell_n})$ under the operation of *simultaneous conjugation*

$$(h_1, \dots, h_n) \mapsto (gh_1g^{-1}, \dots, gh_ng^{-1})$$

that is gauge-invariant. To grasp the geometric meaning of this invariance, it is useful to take a concrete example for G, say $G = \mathrm{SU}(N)$ or even $G = \mathrm{SO}(3)$. In these groups, knowing the individual conjugacy classes of a collection of elements amounts to knowing their eigenvalues, that is, in the case of SO(3), the angles of the rotations. On the other hand, to know the orbit of these elements under simultaneous conjugation requires the additional knowledge of the relative positions of their eigenspaces, or for rotations, the relative positions of their axes.

The main question is then the following. Is it the case that the Wilson loop expectations describe not only the individual conjugacy classes of the G-valued random variables that constitute the Yang–Mills process, but also the simultaneous conjugacy class of all variables associated with the loops based at some point m of M? In more precise terms, is it true that the algebra of functions on $\mathscr{A}/\mathscr{J}$ generated by Wilson loop functionals separates points? If not, it cannot be said that the Wilson loop functionals constitute a complete set of gauge-invariant scalar observables.

The answer turns out to depend entirely on the group G, and it does not seem to be known in all cases, even for compact Lie groups.[20] The property that G must have for the answer to be positive is the following.[21]

Definition 1 *(Property W)* We say that a group G has the property W if for any $n \geqslant 2$ and any two collections $x_1, \ldots, x_n$ and $x'_1, \ldots, x'_n$ of elements of G, the assumption that every word in $x_1, \ldots, x_n$ and their inverses is conjugated to the same word in $x'_1, \ldots, x'_n$ and their inverses implies the existence of an element y of G such that $x'_1 = yx_1y^{-1}, \ldots, x'_n = yx_ny^{-1}$.

Since this long definition is maybe not very pleasant to read, let us word it differently. We are comparing two relations between n-tuples $(x_1, \ldots, x_n)$ and $(x'_1, \ldots, x'_n)$ of elements of G. The first is the relation of simultaneous conjugation

$$\exists y \in G,\ x'_1 = yx_1y^{-1}, \ldots, x'_n = yx_ny^{-1} \tag{SC}$$

The second could be called *lexical conjugation* and holds exactly when

$$\text{every word in } x_1, \ldots, x_n \text{ is conjugated to the same word in } x'_1, \ldots, x'_n \tag{LC}$$

where a word in a certain set of letters can involve these letters and their inverses. We also considered a third property of individual conjugation

$$\exists y_1, \ldots, y_n \in G,\ x'_1 = y_1x_1y_1^{-1}, \ldots, x'_n = y_nx_ny_n^{-1} \tag{IC}$$

In any group, one has the chain of implications

$$(\text{SC}) \Rightarrow (\text{LC}) \Rightarrow (\text{IC})$$

Unless the group G has very special properties (for instance that of being abelian), the second implication is not an equivalence, and the property (IC) is much weaker than the property (LC). For the group G to have the property W means that the properties (SC) and (LC) are equivalent. The proof of the following result can be found in [22], see also [10, 39].

[20]It would be more prudent to say that it is not known to the author.

[21]The name of Property W is by no means standard.

Theorem 1.9 *Any Cartesian product of special orthogonal, orthogonal, special unitary, unitary and symplectic groups has the property W.*

It is known that some non-compact groups fail to have the property W. However, it seems not be known whether this equivalence holds, for instance, for spin groups.

2 Computation of Wilson Loop Expectations

In this section, we will give a few concrete examples of computations with the Yang–Mills holonomy process, with an eye to its so-called *large N limit*, that is, its behaviour when the group G is taken to be $\mathrm{U}(N)$ with an appropriately scaled invariant product on its Lie algebra, and N tends to infinity.[22]

The basis of virtually any computation in 2-dimensional Yang–Mills theory is the Driver–Sengupta formula (DS). This formula can be combined with an expression of the heat kernel on G, for example its Fourier expansion, and lead to very concrete calculations. It is also possible to use a more dynamical, either analytic or probabilistic approach to the heat kernel, by seeing it as the solution of the heat equation or, almost equivalently, as the density of the distribution of the Brownian motion on G. We will illustrate these possibilities on a few examples in the simplest case where M is the plane, and then turn to the much more complicated case where M is the 2-dimensional sphere. For the sake of simplicity, we will assume in this section that the coupling constant T that appears in (DS) is equal to 1.

2.1 The Brownian Motion on the Unitary Group

In order to be as concrete as possible, and because we are interested in the large N limit, we will in this section choose $G = \mathrm{U}(N)$, the unitary group of rank N. As indicated earlier (see Footnote 4), we endow the Lie algebra of $\mathrm{U}(N)$, which is the space $\mathfrak{u}(N)$ of $N \times N$ skew-Hermitian matrices, with the scalar product $\langle X, Y\rangle = N\mathrm{Tr}(X^*Y)$. In the Euclidian space $(\mathfrak{u}(N), \langle\cdot,\cdot\rangle)$, we consider a linear Brownian motion $(K_t)_{t\geqslant 0}$, use it to form the stochastic differential equation

$$dU_t = U_t\,\mathrm{d}K_t - \frac{1}{2}U_t\,\mathrm{d}t\ , \quad U_0 = I_N \tag{15}$$

and call the unique solution to this equation the Brownian motion on $\mathrm{U}(N)$.

---

[22]The notion of large N limit also applies to the cases where $G = \mathrm{SO}(N)$ and $G = \mathrm{Sp}(N)$, the real and quaternionic analogues of $\mathrm{U}(N)$ or $\mathrm{SU}(N)$. As far as we understand today, there is no essential difference between the three cases. More precisely, the computations for finite N are similar in the three cases, if generally a bit more complicated in the orthogonal case and even more so in the symplectic case, and the large N limits are identical.

Using the notation Tr for the usual trace of a $N \times N$ matrix and $\mathrm{tr} = \frac{1}{N}\mathrm{Tr}$ for its normalised trace, the usual rules of stochastic calculus take, in this matricial context, the following nice form: for all $N \times N$ matrix A, measurable with respect to $\sigma(K_s : s \leqslant t)$, we have

$$\mathrm{d}K_t A \,\mathrm{d}K_t = -\mathrm{tr}(A)\,\mathrm{d}t \quad \text{and} \quad \mathrm{d}K_t \,\mathrm{tr}(A\,\mathrm{d}K_t) = -\frac{1}{N^2} A \,\mathrm{d}t \tag{16}$$

This relation can be used to check that $\mathrm{d}(U_t U_t^*) = 0$, so that the trajectories of the process B stay almost surely, as expected, in $\mathrm{U}(N)$.

The density of the distribution of U_t with respect to the normalised Haar measure on $\mathrm{U}(N)$ is the function p_t appearing in the Driver–Sengupta formula, and that we described in Section 1.4.2.

It will be useful to know the Fourier series of this function $p_t : \mathrm{U}(N) \to \mathbb{R}$. To describe it, let us introduce the set $\widehat{\mathrm{U}}(N)$ of equivalence classes of irreducible representations (or *irreps*) of $\mathrm{U}(N)$. For every $\alpha \in \widehat{\mathrm{U}}(N)$, let us denote by d_α the degree of α, that is, the dimension of the space on which $\mathrm{U}(N)$ acts through α. Let us also denote by $\chi_\alpha : \mathrm{U}(N) \to \mathbb{C}$ the character of α, and by $c_2(\alpha)$ the quadratic Casimir number of α, that is, the non-negative real number such that

$$\Delta \chi_\alpha = -c_2(\alpha)\chi_\alpha$$

The Fourier series of the heat kernel is then

$$p_t = \sum_{\alpha \in \widehat{\mathrm{U}}(N)} e^{-\frac{c_2(\alpha)t}{2}} d_\alpha \chi_\alpha \tag{17}$$

and there is nothing specific to $\mathrm{U}(N)$ in this formula.

It is however possible, in the case of $\mathrm{U}(N)$, to write explicitly each of its ingredients. Indeed, the set of irreps of $\mathrm{U}(N)$ is conveniently labelled by non-increasing sequences of N relative integers $\lambda = (\lambda_1 \geqslant \ldots \geqslant \lambda_N)$, called dominant weights. The dimension and quadratic Casimir number of the irrep with highest weight λ are given by the formulas

$$d_\lambda = \prod_{1 \leqslant i < j \leqslant N} \frac{\lambda_i - \lambda_j + j - i}{j - i} \quad \text{and} \quad N c_2(\lambda) = \sum_{1 \leqslant i \leqslant N} \lambda_i^2 + \sum_{1 \leqslant i < j \leqslant N} (\lambda_i - \lambda_j) \tag{18}$$

The character of this representation is given, up to a power of the determinant, by a Schur function, but we will not need its explicit formula.

We are now equipped to make some computations with the Yang–Mills holonomy process.

2.2 *The Simple Loop on the Plane*

2.2.1 Using Harmonic Analysis

Let us consider, on the plane, a loop ℓ that is a simple loop going once around a domain of area t (see, if needed, Figure 2). The partition function of the Yang–Mills model on the plane is equal to 1 and the Driver–Sengupta formula (DS) tells us that for every continuous test function $f : \mathrm{U}(N) \to \mathbb{C}$, we have

$$\mathbb{E}[f(H_\ell)] = \int_{\mathrm{U}(N)} f(x)p_t(x)\,\mathrm{d}x$$

In other words, H_ℓ has the same distribution as U_t, the value at time t of the Brownian motion on $\mathrm{U}(N)$ defined in the previous section.

Using the Fourier expansion (17) and the classical orthogonality relations between characters, we find, for every irrep α of $\mathrm{U}(N)$ acting on the vector space V_α, the equality

$$\mathbb{E}[\alpha(H_\ell)] = e^{-\frac{c_2(\alpha)t}{2}}\,\mathrm{id}_{V_\alpha}$$

which holds in $\mathrm{End}(V_\alpha)$. In particular, since the usual trace is, on $\mathrm{U}(N)$, the character of the natural representation, which has highest weight $(1, 0, \ldots, 0)$, dimension N and quadratic Casimir 1, we find

$$\mathbb{E}[H_\ell] = e^{-\frac{t}{2}} I_N \quad \text{and} \quad \mathbb{E}[\mathrm{tr}(H_\ell)] = e^{-\frac{t}{2}} \tag{19}$$

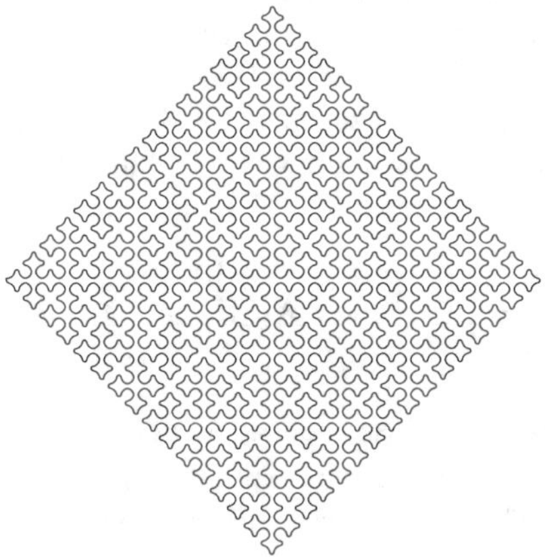

Fig. 2 A simple loop on the plane

Suppose now that we want to compute the expectation of $\mathrm{tr}(H_\ell^2)$, which is also the expectation of $\mathrm{tr}(H_{\ell^2})$, where ℓ^2 is the loop ℓ gone along twice. From the Driver–Sengupta formula and the Fourier expansion of the heat kernel, we get the expression

$$\mathbb{E}[\mathrm{tr}(H_\ell^2)] = \sum_{\lambda \in \widehat{\mathrm{U}}(N)} e^{-\frac{c_2(\lambda)t}{2}} d_\lambda \int_{\mathrm{U}(N)} \mathrm{tr}(x^2)\chi_\lambda(x)\, \mathrm{d}x$$

In order to go further, we need to know that, at least when $N \geqslant 2$,

$$\mathrm{tr}(x^2) = \chi_{(2,0,\ldots,0)}(x) - \chi_{(1,1,0\ldots,0)}(x)$$

Using again the orthogonality of characters, we find, after some reordering of the terms,

$$\mathbb{E}[\mathrm{tr}(H_\ell^2)] = e^{-t}\Big(\cosh\frac{t}{N} - N\sinh\frac{t}{N}\Big) \tag{20}$$

It is possible to go further down this road, by systematically writing the function $x \mapsto \mathrm{tr}(x^n)$ as a linear combination of characters. This is what Philippe Biane did to determine the large N limit of the non-commutative distribution of the Brownian motion on the unitary group. The simplest non-trivial case is the large N limit of (20):

$$\lim_{N\to\infty} \mathbb{E}[\mathrm{tr}(H_\ell^2)] = e^{-t}(1-t) \tag{21}$$

The general formula is nice enough, at least in the limit when N tends to infinity, to be quoted explicitly. It was discovered independently by Philippe Biane and Eric Rains, who formulated it in terms of the Brownian motion on $\mathrm{U}(N)$ rather than the Yang–Mills holonomy process.

Theorem 2.1 (Biane [2], Rains [37]) *With the current notation, and for every integer $n \geqslant 1$,*

$$\lim_{N\to\infty} \mathbb{E}[\mathrm{tr}(H_\ell^n)] = e^{-\frac{nt}{2}} \sum_{k=0}^{n-1} \frac{(-t)^k}{k!} n^{k-1} \binom{n}{k+1} \tag{22}$$

It must be said that this result already appeared, without proof, in Isadore Singer's seminal paper on the large N limit of the Yang–Mills holonomy field [42].[23]

[23] Singer and Rains recognise, in the right-hand side of (22), modified Laguerre polynomials of the first kind. As far as I know, a structural explanation for the appearance of these polynomials in this context has yet to be given.

One of Biane's aims in [2] was to prove the following theorem concerning the limit as N tends to infinity of the Brownian motion on $\mathrm{U}(N)$ as a stochastic process. This convergence result is stated in the language of free probability, a theory presented in detail in the book of Alexandru Nica and Roland Speicher [34].

Theorem 2.2 (Biane [2]) *As N tends to infinity, the Brownian motion on* $\mathrm{U}(N)$ *converges in non-commutative distribution, as a process, towards a unitary non-commutative process $(u_t)_{t\geqslant 0}$ with free stationary multiplicative increments such that for all integer $n \geqslant 0$ and all real $t \geqslant 0$, the expectation of u_t^n and that of $(u_t^*)^n$ are given by the right-hand side of* (22).

2.2.2 Using Stochastic Calculus

Let us illustrate, on the same example of a simple loop on the plane, the dynamical approach to the same computations, based on the use of Itō's formula. The general principle of these computations is to see the quantities such as the left-hand sides of (19) and (20) as functions of t, and to write a differential equation that they satisfy. Recall that t, in our current notation, is the area of the disk enclosed by the simple loop ℓ. A variation of t can thus be described, in geometrical terms, as a variation of the area of the unique face enclosed by ℓ.

As a first example, let us use (15) and Itō's formula to find

$$\frac{d}{dt}\mathbb{E}[\mathrm{tr}(H_\ell)] = \frac{d}{dt}\mathbb{E}[\mathrm{tr}(U_t)] = -\frac{1}{2}\mathbb{E}[\mathrm{tr}(U_t)]$$

which, together with the information $\mathbb{E}[\mathrm{tr}(U_0)] = 1$, yield immediately (19).

Let us apply the same strategy to the computation of $\mathbb{E}[\mathrm{tr}(H_\ell^2)] = \mathbb{E}[\mathrm{tr}(U_t^2)]$. The computation is more interesting and involves the first of the two rules (16). We find

$$\frac{d}{dt}\mathbb{E}[\mathrm{tr}(U_t^2)] = -\mathbb{E}[\mathrm{tr}(U_t^2)] - \mathbb{E}[\mathrm{tr}(U_t)^2] \tag{23}$$

and see a function of t pop up that we were initially not interested in, namely $\mathbb{E}[\mathrm{tr}(U_t)^2]$. The only way out left to us is retreat forwards and we compute the derivative with respect to t of this new function, using now the second rule of (16):

$$\frac{d}{dt}\mathbb{E}[\mathrm{tr}(U_t)^2] = -\frac{1}{N^2}\mathbb{E}[\mathrm{tr}(U_t^2)] - \mathbb{E}[\mathrm{tr}(U_t)^2] \tag{24}$$

All's well that ends well: (23) and (24) form a closed system of ordinary differential equations that is easily solved and from which we recover, in particular, (20). As a bonus, we get

$$\mathbb{E}[\mathrm{tr}(H_\ell)^2] = e^{-t}\left(\cosh\frac{t}{N} - \frac{1}{N}\sinh\frac{t}{N}\right) \tag{25}$$

The only change with respect to (20) is the change from N to $\frac{1}{N}$ in front of the hyperbolic sine, with the effect that

$$\lim_{N\to\infty} \mathbb{E}[\mathrm{tr}(H_\ell)^2] = e^{-t} = \lim_{N\to\infty} \mathbb{E}[\mathrm{tr}(H_\ell)]^2 \tag{26}$$

This is an instance of a general factorisation property which was observed, among others, by Feng Xu [47], and which is a consequence of the concentration, in the limit where N tends to infinity, of the spectra of the random matrices that we are considering.

2.3 Yin . . .

Let us consider a slightly more complicated loop depicted on Figure 3. This loop goes once around a domain of area $s + t$ and then once around a smaller domain of area t contained in the first one.

Let us apply the Driver–Sengupta formula in this case. We denote a generic element of the configuration space $\mathrm{U}(N)^2$ by (x_a, x_b), in relation with our labelling by a and b of the two edges of the graph formed by ℓ. Thus, for every continuous test function $f : \mathrm{U}(N) \to \mathbb{C}$, we have

$$\mathbb{E}[f(H_\ell)] = \int_{\mathrm{U}(N)^2} f(x_b x_a) p_s(x_b^{-1} x_a) p_t(x_b)\, \mathrm{d}x_a\, \mathrm{d}x_b$$

Note that, according to (10), the discrete holonomy map is order-reversing, so that the loop $\ell = ab$ gives rise to the map $h_\ell(x_a, x_b) = x_b x_a$.

The change of variables $(y, z) = (x_b^{-1} x_a, x_b)$ preserves the Haar measure on $\mathrm{U}(N)^2$ and we have

$$\mathbb{E}[f(H_\ell)] = \int_{\mathrm{U}(N)^2} f(z^2 y) p_s(y) p_t(z)\, \mathrm{d}y\, \mathrm{d}z \tag{27}$$

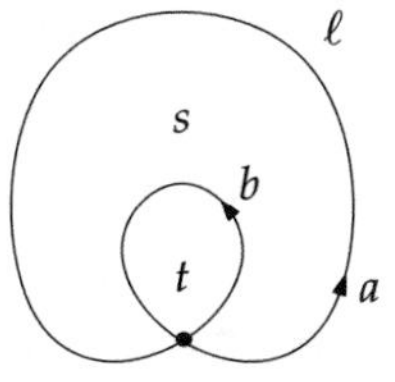

Fig. 3 The loop ℓ goes first once along the larger circle (the edge a) and then once along the smaller circle (the edge b). The loop ab is equivalent to the concatenation of ab^{-1}, b and b. The loops ab^{-1} and b are essentially simple loops surrounding disjoint domains

This corresponds to the fact, explained in the caption of Figure 3, that the loop ℓ can be written as $\ell_1\ell_2\ell_2$, where ℓ_1 goes around the moon-shaped domain sitting between the two disks, and ℓ_2 goes around the small circle of area t. These loops enclose disjoint domains, and although ℓ_1 is not strictly speaking self-intersection free, they are essentially simple, in the sense that they can be approximated by simple loops.

From this graphical decomposition of ℓ, or from (27), we infer that H_ℓ has the distribution of $V_t^2 U_s$, where U and V are independent Brownian motions on $\mathrm{U}(N)$.[24] Using the independence, the fact that the expectation of U_s is $e^{-\frac{s}{2}} I_N$ (see (19)), and (20), we find

$$\mathbb{E}[\mathrm{tr}(H_\ell)] = e^{-\frac{s}{2}-t}\Big(\cosh\frac{t}{N} - N\sinh\frac{t}{N}\Big) \tag{28}$$

and, letting N tend to infinity,

$$\lim_{N\to\infty} \mathbb{E}[\mathrm{tr}(H_\ell)] = e^{-\frac{s}{2}-t}(1-t) \tag{29}$$

We succeeded in computing the expectation of $\mathrm{tr}(H_\ell)$, but we did so by taking advantage of the favourable circumstances, namely the fact that the word $V_t^2 U_s$ is a very simple one, with two independent Brownian motions appearing one after the other (and not, for example, as $U_s V_t U_s V_t$), and the fact that the expectation of U_s is a very simple matrix.

A more systematic approach is possible, by looking at $\mathbb{E}[\mathrm{tr}(V_t^2 U_s)]$ as a function of s and t and by using Itō's formula to compute its partial derivatives. One finds

$$\begin{aligned}
\partial_s \mathbb{E}[\mathrm{tr}(V_t^2 U_s)] &= -\frac{1}{2}\mathbb{E}[\mathrm{tr}(V_t^2 U_s)]\\
\partial_t \mathbb{E}[\mathrm{tr}(V_t^2 U_s)] &= -\mathbb{E}[\mathrm{tr}(V_t^2 U_s)] - \mathbb{E}[\mathrm{tr}(V_t)\mathrm{tr}(V_t U_s)]
\end{aligned}$$

Once again, a function appears that we were not considering at first. Let us apply the same treatment to this new function:

$$\begin{aligned}
\partial_s \mathbb{E}[\mathrm{tr}(V_t)\mathrm{tr}(V_t U_s)] &= -\frac{1}{2}\mathbb{E}[\mathrm{tr}(V_t)\mathrm{tr}(V_t U_s)]\\
\partial_t \mathbb{E}[\mathrm{tr}(V_t)\mathrm{tr}(V_t U_s)] &= -\mathbb{E}[\mathrm{tr}(V_t)\mathrm{tr}(V_t U_s)] - \frac{1}{N^2}\mathbb{E}[\mathrm{tr}(V_t^2 U_s)]
\end{aligned}$$

It is possible to solve this system and to recover (28).

[24] Thanks to the independence of the multiplicative increments of the Brownian motion, this distribution is of course also that of $V_t^2(V_t^{-1}V_{t+s}) = V_t V_{s+t}$. Reasoning in this way amounts to undo the change of variables that we did to obtain (27).

An interesting observation is the fact that the linear combination $2\partial_s - \partial_t$ of partial derivatives is particularly simple:

$$(2\partial_s - \partial_t)\mathbb{E}[\mathrm{tr}(V_t^2 U_s)] = \mathbb{E}[\mathrm{tr}(V_t)\mathrm{tr}(V_t U_s)] \tag{30}$$

$$\text{and } (2\partial_s - \partial_t)\mathbb{E}[\mathrm{tr}(V_t)\mathrm{tr}(V_t U_s)] = \frac{1}{N^2}\mathbb{E}[\mathrm{tr}(V_t^2 U_s)] \tag{31}$$

These are instances of the Makeenko–Migdal equations that we will discuss in greater detail in the next section. Before that, let us study another example.

2.4 *. . . And Yang*

Let us now consider the eight-shaped loop drawn on Figure 4. The Driver–Sengupta formula yields, with the by now usual notation, and taking the inversion of the order into account,

$$\mathbb{E}[f(H_\ell)] = \int_{\mathrm{U}(N)^6} f(x_f x_e x_d x_c x_b x_a) p_s(x_a x_c x_e) p_t(x_f x_b x_d) p_u(x_c^{-1} x_f) p_v(x_a^{-1} x_d)\, \mathrm{d}x$$

The appropriate change of variables is dictated by the geometry of the loop, more precisely by a decomposition in product of lassos, one of which is given in the caption of Figure 4. Accordingly, let us set

$$(y, z, g, h, e, f) = (x_c x_e x_a, x_f x_b x_d, x_c x_f^{-1}, x_d^{-1} x_a, x_e, x_f)$$

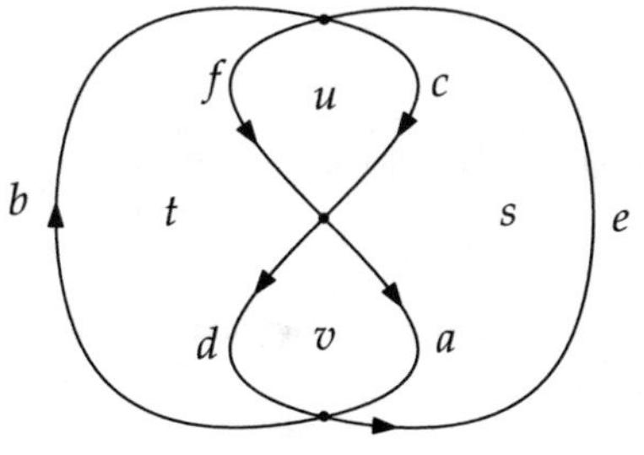

Fig. 4 An eight-shaped loop on the plane. The letters s, t, u, v in the faces indicate the areas of the faces. The other letters label the edges of the graph. The loop can be decomposed, as we did for the heart-shaped loop, as a product of lassos enclosing pairwise disjoint domains: $abcdef = (ad^{-1})(dbf)(f^{-1}c)(da^{-1})(aec)(c^{-1}f)$. Here, by a lasso, we mean a loop of the form clc^{-1}, where c is a path starting from the starting point of our loop and l is a simple loop. In this particular case, the path c is always the constant path

This change of variables preserves the Haar measure on $\mathrm{U}(N)^6$.[25] Thus, we find

$$\mathbb{E}[f(H_\ell)] = \int_{\mathrm{U}(N)^4} f(g^{-1}yh^{-1}gzh)p_s(y)p_t(z)p_u(g)p_v(h)\, \mathrm{d}g\, \mathrm{d}h\, \mathrm{d}y\, \mathrm{d}z$$

after integrating with respect to e and f which do not appear in the integrand. Thus, considering four independent Brownian motions G, H, Z, Y on $\mathrm{U}(N)$, we find the equality in distribution

$$H_\ell \overset{\text{dist.}}{=} G_u^{-1} Y_s H_v^{-1} G_u Z_t H_v \tag{32}$$

The quantity $\mathbb{E}[\mathrm{tr}(H_\ell)]$ appears now as a function of the four real parameters s, t, u, v and we can use stochastic calculus to differentiate it with respect to each of them. In fact, using the first assertion of (19), which in the language of Brownian motion reads $\mathbb{E}[Y_s] = e^{-\frac{s}{2}} I_N$ and $\mathbb{E}[Z_t] = e^{-\frac{t}{2}} I_N$, we can simplify the problem to

$$\mathbb{E}[\mathrm{tr}(H_\ell)] = e^{-\frac{s+t}{2}} \mathbb{E}[\mathrm{tr}(G_u^{-1} H_v^{-1} G_u H_v)]$$

The expectation in the right-hand side of this equality is a symmetric function of u and v. Using stochastic calculus, we find

$$\partial_u \mathbb{E}[\mathrm{tr}(G_u^{-1} H_v^{-1} G_u H_v)] = -\mathbb{E}[\mathrm{tr}(G_u^{-1} H_v^{-1} G_u H_v)] + \mathbb{E}[\mathrm{tr}(H_v^{-1})\mathrm{tr}(H_v)] \tag{33}$$

The new function $\mathbb{E}[\mathrm{tr}(H_v^{-1})\mathrm{tr}(H_v)]$ of v can in turn be computed using Itō's formula, since it is equal to 1 when $v = 0$ and satisfies the differential equation

$$\partial_v \mathbb{E}[\mathrm{tr}(H_v^{-1})\mathrm{tr}(H_v)] = -\mathbb{E}[\mathrm{tr}(H_v^{-1})\mathrm{tr}(H_v)] + \frac{1}{N^2}$$

---

[25] This is because the normalised Haar measure on $\mathrm{U}(N)^n$, or on G^n for any compact topological group G, is pushed forward onto itself by each of the elementary maps

- $(x_1, x_2, \dots, x_n) \mapsto (x_1^{-1}, x_2, \dots, x_n)$
- $(x_1, x_2, \dots, x_n) \mapsto (x_1 x_2, x_2, \dots, x_n)$
- $(x_1, \dots, x_n) \mapsto (x_{\sigma(1)}, \dots, x_{\sigma(n)})$, where σ is any permutation of $\{1, \dots, n\}$

and it is not difficult to check that our change of variables can be obtained as a composition of these maps.

Interestingly, these elementary operations are exactly the Nielsen transformations, which generate the group of automorphisms of the free group of rank n (see [30]). Thus, the random homomorphism from the free group $\mathbb{F}_n$ to a compact topological group G constructed by picking a basis of $\mathbb{F}_n$ and sending this basis to a uniformly chosen element of G^n does not depend, in distribution, on the basis of $\mathbb{F}_n$ used to construct it. In particular, the distribution of the image of every element of the free group is intrinsically defined, and one may for instance wonder, for specific or for general G, which elements of $\mathbb{F}_n$ are sent to a uniformly distributed element of G. I am grateful to the referee for pointing out to me that this problem was solved for finite groups in [35].

which is solved in

$$\mathbb{E}[\mathrm{tr}(H_v^{-1})\mathrm{tr}(H_v)] = \frac{1}{N^2}(1-e^{-v}) + e^{-v} \tag{34}$$

Replacing in (33) and solving, we find finally

$$\mathbb{E}[\mathrm{tr}(H_\ell)] = e^{-\frac{s+t}{2}}\left(e^{-u} + e^{-v} - e^{-(u+v)} + \frac{1}{N^2}(1-e^{-u})(1-e^{-v})\right) \tag{35}$$

and, letting N tend to infinity,

$$\lim_{N\to\infty} \mathbb{E}[\mathrm{tr}(H_\ell)] = e^{-\frac{s+t}{2}}\left(e^{-u} + e^{-v} - e^{-(u+v)}\right) \tag{36}$$

We did these computations without taking great care of a possible geometric meaning of the successive steps. Anticipating our discussion of the Makeenko–Migdal equations, it is interesting to check that

$$(\partial_u + \partial_v - \partial_s - \partial_t)\mathbb{E}[\mathrm{tr}(H_\ell)] = e^{-\frac{s+t}{2}}\left(e^{-(u+v)} + \frac{1}{N^2}(1-e^{-(u+v)})\right) = \mathbb{E}[\mathrm{tr}(H_{\ell'})\mathrm{tr}(H_{\ell''})] \tag{37}$$

where ℓ' and ℓ'' are the loops drawn on Figure 5.

Perhaps even more interesting than the fact that (37) holds, which after all is a consequence of Theorem 3.1, is the observation that (37) does not seem to be easily guessed from (32) and Itō's formula. More precisely, Itō's formula allows us to give an expression of the left-hand side of (37) and it is not obvious that this expression coincides with the right-hand side of (37). We take this as a sign that the Makeenko–Migdal equations give an information that is practically non-trivial.

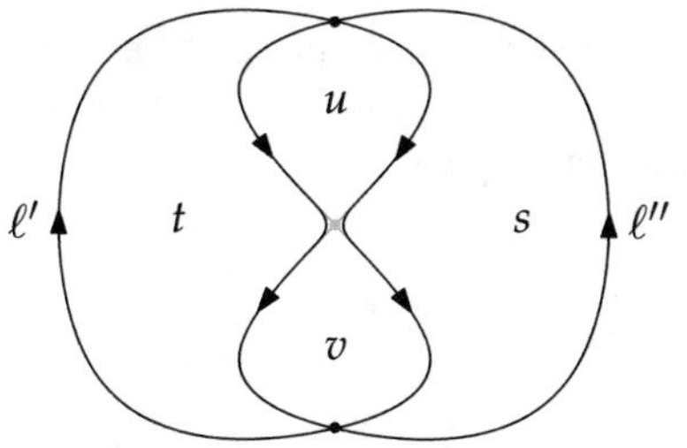

Fig. 5 The loops ℓ' and ℓ'' are obtained from ℓ by an operation that will feature prominently in Section 3

2.5 The Case of the Sphere: A Not So Simple Loop

Computations involving the Yang–Mills holonomy process on the sphere, although in principle based on the same formulas as in the case of the plane, are in general much more complicated. This can be explained by the fact that, as we indicated in Section 1.4.5, the stochastic core of the Yang–Mills holonomy process on a sphere is a Brownian *bridge* on $\mathrm{U}(N)$, or on the compact Lie group G, instead of a Brownian motion.

In this section, we are going to illustrate some of the difficulties that one meets when working on a sphere. The first is that the partition function is not equal to 1 anymore. Instead, according to (1.5), it is given, on a sphere of total area T, by

$$Z_{S^2}(T) = p_T(I_N) = \|p_{\frac{T}{2}}\|^2_{L^2(\mathrm{U}(N))} = \sum_{\alpha \in \widehat{\mathrm{U}}(N)} e^{-\frac{T}{2}c_2(\alpha)} d_\alpha^2$$

This is also an expression in which nothing is specific to $\mathrm{U}(N)$: it is valid for any compact Lie group.[26]

The most basic question about the Yang–Mills holonomy process on the sphere is the analogue to the question that we treated in Section 2.2, namely to compute the expectation of the normalised trace of the holonomy along a simple loop ℓ enclosing a domain of area t. The Driver–Sengupta formula yields the following expression for this expectation:

$$\mathbb{E}[\mathrm{tr}(H_\ell)] = \frac{1}{Z_{S^2}(T)} \int_{\mathrm{U}(N)} \mathrm{tr}(x) p_t(x) p_{T-t}(x^{-1}) \, \mathrm{d}x \tag{38}$$

Using the Fourier expansion of the heat kernel, one finds

$$\mathbb{E}[\mathrm{tr}(H_\ell)] = \frac{1}{Z_{S^2}(T)} \sum_{\lambda,\mu \in \widehat{\mathrm{U}}(N)} e^{-c_2(\lambda)\frac{t}{2} - c_2(\mu)\frac{T-t}{2}} d_\lambda d_\mu \int_{\mathrm{U}(N)} \mathrm{tr}(x) \chi_\lambda(x) \chi_\mu(x^{-1}) \, \mathrm{d}x$$

The integral can be computed thanks to Pieri's rule: it is equal to 0 unless μ is obtained from λ by adding 1 to exactly one component, in which case it is equal to 1. We write $\lambda \nearrow \mu$ when this happens. Thus,

$$\mathbb{E}[\mathrm{tr}(H_\ell)] = \frac{1}{Z_{S^2}(T)} \sum_{\lambda \in \widehat{\mathrm{U}}(N)} e^{-c_2(\lambda)\frac{T}{2}} d_\lambda^2 \underbrace{\left[\sum_{\substack{\mu \in \widehat{\mathrm{U}}(N) \\ \lambda \nearrow \mu}} e^{-(c_2(\mu) - c_2(\lambda))\frac{T-t}{2}} \frac{d_\mu}{d_\lambda} \right]}_{f_1(\lambda)} \tag{39}$$

[26]Note that T, which used to denote the coupling constant in (1.5), now denotes the total area of our surface. This is not a problem because the only meaningful quantity is the product of the coupling constant by the total area of the surface.

It seems difficult to give an expression of $\mathbb{E}[\mathrm{tr}(H_\ell)]$ much simpler than (38) or (39) which, as is hardly necessary to emphasise, is much more complicated than the one that we obtained in the case of the plane.[27]

It is, however, possible to analyse the limit of this quantity as N tends to infinity. A first step in this direction is based on the realisation that Pieri's rule is simple, and the quantity between square brackets, which we denote by $f_1(\lambda)$ is a finite sum and can be written explicitly using (18):

$$f_1(\lambda) = e^{-\frac{T-t}{2}} \sum_{i=1}^{N} \mathbb{1}_{\{i=1 \text{ or } \lambda_{i-1}>\lambda_i\}} e^{-(T-t)\left(\lambda_i + \frac{N-2i+1}{2}\right)} \prod_{\substack{1 \leqslant j \leqslant N \\ j \neq i}} \left(1 + \frac{1}{\lambda_i - \lambda_j + j - i}\right)$$

This suggests to associate with the highest weight λ the decreasing sequence $l = (l_1 > \ldots > l_N)$ of half-integers defined by

$$l_i = \lambda_i + \frac{N - 2i + 1}{2}$$

so that

$$f_1(\lambda) = e^{-\frac{T-t}{2}} \sum_{i=1}^{N} \mathbb{1}_{\{i=1 \text{ or } \lambda_{i-1}>\lambda_i\}} e^{-\frac{T-t}{N} l_i} \prod_{\substack{1 \leqslant j \leqslant N \\ j \neq i}} \left(1 + \frac{1}{l_i - l_j}\right)$$

Let us now introduce the probability measure $\pi_{N,T}$ on $\widehat{\mathrm{U}}(N)$ such that for every highest weight λ, one has

$$\pi_{N,T}(\{\lambda\}) \propto e^{-c_2(\lambda)\frac{T}{2}} d_\lambda^2$$

Then (39) can be written more compactly as

$$\mathbb{E}[\mathrm{tr}(H_\ell)] = \int_{\widehat{\mathrm{U}}(N)} f_1(\lambda)\, \mathrm{d}\pi_{N,T}(\lambda) \tag{40}$$

Moreover, there exists for each integer $n \geqslant 2$ a function f_n on $\widehat{\mathrm{U}}(N)$, not very different from f_1, and the integral of which against $\pi_{N,T}$ yields $\mathbb{E}[\mathrm{tr}(H_\ell^n)]$.

We would like to express that, as N tends to infinity, the measure $\pi_{N,T}$ concentrates on a few highest weights, characterised by a certain limiting shape. One unpleasant feature of (40) in this respect is that the set on which the integral is taken, namely $\widehat{\mathrm{U}}(N)$, depends on N. It is thus uneasy to formulate a concentration

[27] Let us drive the point home: (39), once made fully explicit using (18), is the exact analogue of the $e^{-\frac{t}{2}}$ that we see in the second assertion of (19).

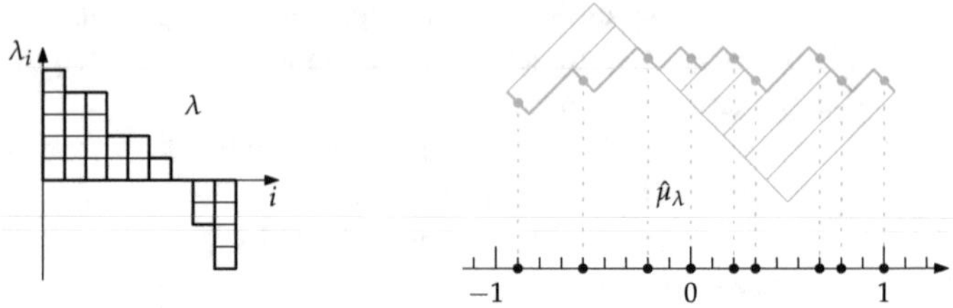

Fig. 6 With $N = 9$, the highest weight $\lambda = (5, 4, 4, 2, 2, 1, 0, -2, -4)$ drawn in the style of a Young diagram, and its empirical measure. Each dot represents $\frac{1}{9}$ of mass and any two dots are distant by a multiple of $\frac{1}{9}$

result. One classical and efficient way around this problem is to associate with each highest weight λ its *empirical measure* (Figure 6)

$$\hat{\mu}_\lambda = \frac{1}{N}\sum_{i=1}^{N} \delta_{\frac{l_i}{N}} = \frac{1}{N}\sum_{i=1}^{N} \delta_{\frac{1}{N}(\lambda_i + \frac{N-2i+1}{2})}$$

Pushing the probability measure $\pi_{N,T}$ forward by the map $\lambda \mapsto \hat{\mu}_\lambda$ yields a probability measure, which we denote by $\Pi_{N,T}$, on the set of probability measures on the real line. It is possible to predict the behaviour of this probability as N tends to infinity by writing $c_2(\lambda)$ and d_λ in terms of the empirical measure of λ. Up to some inessential terms (see [25, Eq. (24)] for complete expressions), one finds

$$c_2(\lambda) \simeq N^2 \int_{\mathbb{R}} x^2 \, \mathrm{d}\hat{\mu}_\lambda(x) \quad \text{and}$$

$$d_\lambda^2 \simeq \exp\Big[-N^2 \int_{\{(x,y)\in\mathbb{R}^2, x\neq y\}} -\log|x-y| \, \mathrm{d}\hat{\mu}_\lambda(x) \, \mathrm{d}\hat{\mu}_\lambda(y) \Big]$$

Introducing, for every probability measure μ, the quantity

$$\mathcal{J}_T(\mu) = \int_{\{(x,y)\in\mathbb{R}^2, x\neq y\}} -\log|x-y| \, \mathrm{d}\mu(x) \, \mathrm{d}\mu(y) + \frac{T}{2}\int_{\mathbb{R}} x^2 \, \mathrm{d}\mu(x)$$

we see that the probability measure $\Pi_{N,T}$ assigns to any probability measure μ that is the empirical measure of a highest weight a mass proportional to

$$\Pi_{N,T}(\{\mu\}) \propto \exp(-N^2 \mathcal{J}_T(\mu))$$

In the large N limit, it seems plausible that $\Pi_{N,T}$ will concentrate on the minimisers, or even better, on the unique minimiser of the functional $\mathcal{J}_T$. This turns out to be true, with a little twist that we will explain and contributes to making the story much more interesting than it already is. Let us summarise the main results on which one can ground a rigorous analysis of the situation.

- Minimising the functional $\mathcal{J}_T$ on the space of all probability measures on $\mathbb{R}$ is one of the simplest examples of a rich and well-developed theory which is, for example, exposed in the book of Edward Saff and Vilmos Totik [38]. This is also a very common problem in random matrix theory. Indeed, the unique minimiser of $\mathcal{J}_T$ is Wigner's semi-circular distribution with variance $\frac{1}{T}$:

$$\mathrm{d}\sigma_{1/T}(x) = \frac{T}{2\pi}\sqrt{\frac{4}{T} - x^2}\, \mathbb{1}_{\left[-\frac{2}{\sqrt{T}}, \frac{2}{\sqrt{T}}\right]}(t)\, \mathrm{d}t \tag{41}$$

- The fact that the measure $\Pi_{N,T}$ concentrates, as N tends to infinity, to the minimiser of $\mathcal{J}_T$ is a special case of a principle of large deviations proved by Alice Guionnet and Mylène Maïda in [16]. However, the minimiser of $\mathcal{J}_T$ that one must consider is not the absolute minimiser on the set of all probability measures on $\mathbb{R}$. Indeed, for all $N \geqslant 1$, the measure $\Pi_{N,T}$ is supported by the set of empirical measures of highest weights of $\mathrm{U}(N)$, which form a rather special set of probability measures. A distinctive feature of these measures is that they are atomic, with atoms of mass $\frac{1}{N}$ spaced by integer multiples of $\frac{1}{N}$. Weak limits, as N tends to infinity, of such measures can only be absolutely continuous with respect to the Lebesgue measure on $\mathbb{R}$, with a density not exceeding 1: a class of probability measures that we will denote by $\mathcal{L}(\mathbb{R})$. The result of Guionnet and Maïda asserts that the measure $\Pi_{N,T}$ concentrates exponentially fast, as N tends to infinity, around the unique minimiser μ_T^* of $\mathcal{J}_T$ on the closed set $\mathcal{L}(\mathbb{R})$.
- The problem of minimising $\mathcal{J}_T$ under the constraint of having a density not exceeding 1 is a problem that is, in principle, just as well understood as the unconstrained problem. The book [38] contains results ensuring the existence and uniqueness of the minimiser, and others allowing one to determine its support. In fact, the measure $\sigma_{1/T}$ given by (41), and which is the absolute minimiser of $\mathcal{J}_T$, is absolutely continuous with a maximal density of $\sqrt{T}/\pi$, so that it belongs to $\mathcal{L}(\mathbb{R})$ provided $T \leqslant \pi^2$. For $T > \pi^2$, the constraint becomes truly restrictive, and one must make do with a probability measure which is, in $\mathcal{L}(\mathbb{R})$, the best available substitute for $\sigma_{1/T}$. The actual determination of this minimiser μ_T^* is, depending on one's background, a more or less elementary exercise in Riemann–Hilbert theory, and involves manipulating elliptic functions. The density of μ_T^* for $T > \pi^2$ is represented on Figure 7. An exact expression of this density can be found in [25, Eq. (37)].

Having established the exponential concentration, as N tends to infinity, of the measure $\Pi_{N,T}$ around μ_T^*, it is possible to come back to our initial problem of computing $\mathbb{E}[\mathrm{tr}(H_\ell)]$. After noticing that $f_1(\lambda)$ can be expressed as a functional $F_1(\hat{\mu}_\lambda)$ of the empirical measure of λ, it can be guessed that $\mathbb{E}[\mathrm{tr}(H_\ell)]$ is related to $F_1(\mu_T^*)$. Antoine Dahlqvist and James Norris were the first to rigorously and

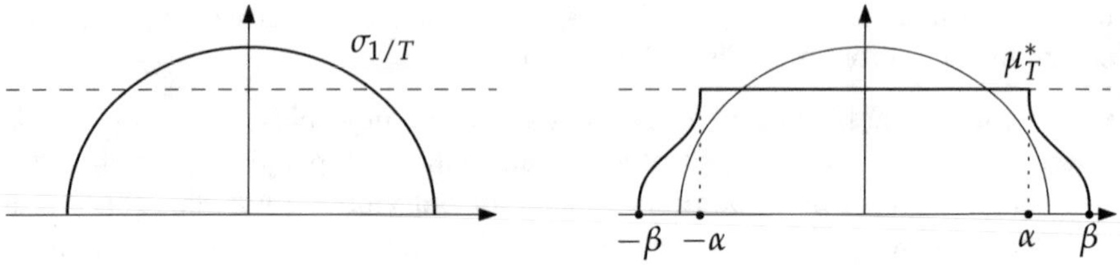

Fig. 7 For $T > \pi^2$, the absolute minimiser of the functional $\mathcal{J}_T$ does not belong to the class of probabilities on $\mathbb{R}$ with a density not exceeding 1. The minimiser within this class is represented on the right. Its density is identically equal to 1 on an interval in the middle of its support, and given by elliptic functions outside this interval

successfully pursue this line of reasoning, and to obtain the following remarkably elegant result.

Theorem 2.3 (Dahlqvist–Norris [5]) *Let ρ_T denote the density of the minimiser μ_T^*. Then, for all integer $n \geqslant 0$, one has*

$$\lim_{N\to\infty} \mathbb{E}[\mathrm{tr}(H_\ell^n)] = \lim_{N\to\infty} \mathbb{E}[\mathrm{tr}(H_\ell^{-n})] = \frac{1}{n\pi}\int_{\mathbb{R}} \cosh\left(\frac{nx}{2}(T-2t)\right)\sin(n\pi\rho_T(x))\,dx \tag{42}$$

To conclude this long discussion of the simple loop on the sphere, let us mention another result for the statement of which we have all the concepts at hand. Our description of the behaviour of the measure $\Pi_{N,T}$ suggests that the partition function itself is dominated by the contribution of the highest weights that have an empirical measure close to μ_T^*. This is indeed true, and the fact that the shape of μ_T^* changes suddenly when T crosses the critical value π^2 gives rise to a phase transition, in this case of third order, first discovered by Douglas and Kazakov, and named after them. It was first proved rigorously, in a slightly different but equivalent language, by Karl Liechty and Dong Wang in [27], and by Mylène Maïda and the author in [25].

Theorem 2.4 (Douglas–Kazakov phase transition) *The free energy of the Yang–Mills model on a sphere of total area T is given by*

$$F(T) = \lim_{N\to\infty}\frac{1}{N^2}\log Z_{S^2}(T) = \frac{T}{24} + \frac{3}{2} - \mathcal{J}_T(\mu_T^*)$$

The function F is of class C^2 on $(0,\infty)$ and smooth on $(0,\infty)\setminus\{\pi^2\}$. The third derivative of F admits a jump discontinuity at π^2.

This phase transition is not one that is easily detected numerically, as Figure 8 shows.

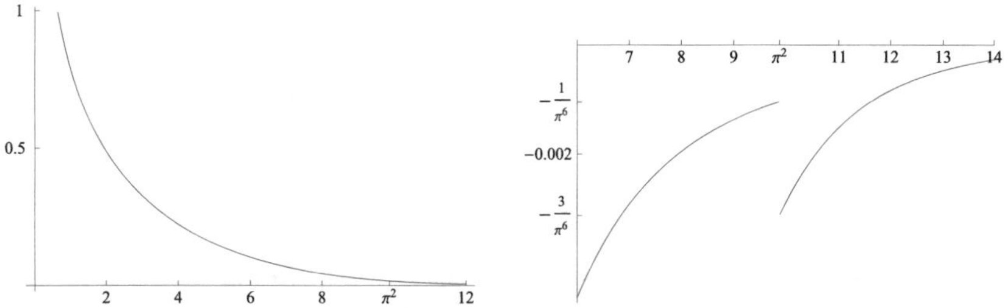

Fig. 8 The graphs of $T \mapsto F(T)$ (on the left) and of $T \mapsto F^{(3)}(T)$ near $T = \pi^2$ (on the right)

3 The Makeenko–Migdal Equations

3.1 First Approach

It is now time that we discuss the equations discovered by Yuri Makeenko and Alexander Migdal and which give their title to these notes. These equations are a powerful tool for the study of the Wilson loop expectations of which we gave a few examples in the previous section. They are related to the approach that we called dynamical, in which an expectation of the form $\mathbb{E}[\mathrm{tr}(H_\ell)]$, where ℓ is some nice loop on a surface M, is seen as a function of the areas of the faces cut by ℓ on the surface M. The Makeenko–Migdal equations give a remarkably elegant expression of the alternated sum of the derivatives of $\mathbb{E}[\mathrm{tr}(H_\ell)]$ with respect to the areas of the four faces that surround a generic point of self-intersection of ℓ. This expression is of the form $\mathbb{E}[\mathrm{tr}(H_{\ell'})\mathrm{tr}(H_{\ell''})]$, where ℓ' and ℓ'' are two loops obtained from ℓ by a very simple operation at this point of self-intersection ℓ. This operation consists in taking the two incoming strands of ℓ at this point and connecting them with the two outgoing strands in the 'other' way, the way that is not realised by ℓ, see Figure 9.

On this figure, we see four faces around the self-intersection point, which need not be pairwise distinct. We denote their areas by t_1, t_2, t_3, t_4 as indicated on Figure 9. The Makeenko–Migdal equation in this case reads

$$\left(\frac{\partial}{\partial t_1} - \frac{\partial}{\partial t_2} + \frac{\partial}{\partial t_3} - \frac{\partial}{\partial t_4}\right)\mathbb{E}[\mathrm{tr}(H_\ell)] = \mathbb{E}[\,\mathrm{tr}(H_{\ell'})\mathrm{tr}(H_{\ell''})] \tag{MM}$$

The relation (30), that we derived earlier in an elementary way, is an instance of this equation.

The relation (MM) would become particularly useful if we could combine it with a result saying that $\mathbb{E}[\mathrm{tr}(H_{\ell'})\mathrm{tr}(H_{\ell''})] = \mathbb{E}[\mathrm{tr}(H_{\ell'})]\mathbb{E}[\mathrm{tr}(H_{\ell''})]$. A crucial fact is that this equality, which is of course false in general, becomes true in the large N limit in all cases where this limit has been studied, that is, on the plane and on the sphere. It corresponds to a concentration phenomenon, namely to the fact that the complex-valued random variable $\mathrm{tr}(H_\ell)$ converges, in the large N limit, to a deterministic

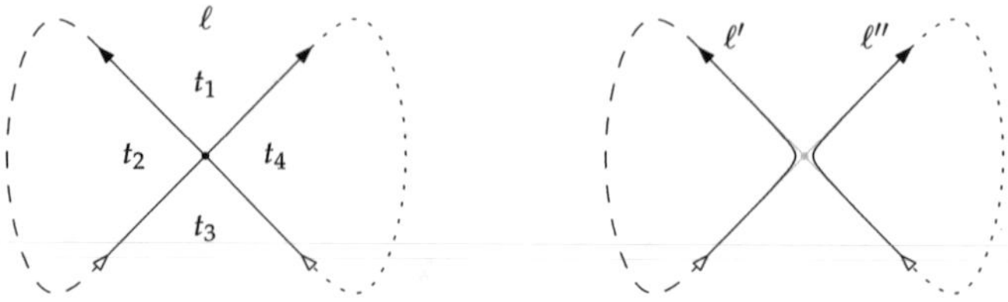

Fig. 9 On the left, we see a loop ℓ around a generic self-intersection point. The dotted and dashed part of ℓ can be arbitrarily complicated, and can meet many times outside the small region of the surface that we are focusing on. It is nevertheless true that after escaping this small region through the North-East corner (resp. North-West corner), the first time ℓ comes back is through the South-East corner (resp. South-West corner). This is why the 'desingularisation' operation illustrated on the right produces exactly two loops, that we call ℓ' and ℓ''

complex, indeed real number $\Phi(\ell)$. This behaviour is expected to occur on any compact surface, and the function $\Phi : \mathscr{L}(M) \to \mathbb{R}$, whose existence has so far been proved when M is the plane or the sphere, is called the *master field*.

In the large N limit, the Makeenko–Migdal equation (MM) becomes a kind of differential equation satisfied by this master field Φ:

$$\left(\frac{\partial}{\partial t_1} - \frac{\partial}{\partial t_2} + \frac{\partial}{\partial t_3} - \frac{\partial}{\partial t_4}\right)\Phi(\ell) = \Phi(\ell')\Phi(\ell'') \qquad (\mathrm{MM}_\infty)$$

On the plane, we will see that this equation, together with the very simple equation (19), essentially characterises the function Φ.

3.2 Makeenko and Migdal's Proof

Makeenko and Migdal discovered the relation (MM), and the extensions that we will describe later, by doing a very clever integration by parts in the functional integral with respect to the Yang–Mills measure (see (3)) that defines a Wilson loop expectation:

$$\mathbb{E}[\mathrm{tr}(H_\ell)] = \frac{1}{Z}\int_{\mathscr{A}} \mathrm{tr}(\mathrm{hol}(\omega, \ell))e^{-\frac{1}{2}S_{\mathrm{YM}}(\omega)}\, \mathrm{d}\omega$$

or instead, as we will explain, in a closely related integral (see [31] and [9]). That this integration by parts performed in an ill-defined integral yields as a final product a perfectly meaningful formula, makes Makeenko and Migdal's original derivation the more intriguing. It is described in mathematical language in the introduction of [24], but this derivation is so beautiful that we reproduce its description here.

The finite-dimensional prototype of the so-called Schwinger–Dyson equations, obtained by integration by parts in functional integrals, is the fact that for all smooth function $f : \mathbb{R}^n \to \mathbb{R}$ with bounded differential, and for all $h \in \mathbb{R}^n$, the equality

$$\int_{\mathbb{R}^n} d_x f(h) e^{-\frac{1}{2}\|x\|^2}\, dx = \int_{\mathbb{R}^n} \langle x, h\rangle f(x) e^{-\frac{1}{2}\|x\|^2}\, dx$$

holds. This equality ultimately relies on the invariance by translation of the Lebesgue measure on $\mathbb{R}^n$ and it can be proved by writing

$$0 = \frac{d}{dt}_{|t=0} \int_{\mathbb{R}^n} f(x+th) e^{-\frac{1}{2}\|x+th\|^2}\, dx$$

In our description of the Yang–Mills measure μ_{YM} (see (3)), we mentioned that the measure $\mathrm{d}\omega$ on the space $\mathscr{A}$ of connections was meant to be a kind of Lebesgue measure, invariant by translations. This is the key to the derivation of the Schwinger–Dyson equations, as we will now explain. In what follows, we will use the differential geometric language introduced in Section 1.2.

Let $\psi : \mathscr{A} \to \mathbb{R}$ be an observable, that is, a function. In general, we are interested in the integral of ψ with respect to the measure μ_{YM}. The tangent space to the affine space $\mathscr{A}$ is the linear space $\Omega^1(M) \otimes \mathrm{Ad}(P)$. To say that the measure $\mathrm{d}\omega$ is translation invariant means that for every element η of this linear space,

$$0 = \frac{d}{dt}_{|t=0} \int_{\mathscr{A}} \psi(\omega + t\eta) e^{-\frac{1}{2} S_{\mathsf{YM}}(\omega+t\eta)}\, \mathrm{d}\omega$$

and the Schwinger–Dyson equations follow in their abstract form

$$\int_{\mathscr{A}} d_\omega \psi(\eta)\, \mathrm{d}\mu_{\mathsf{YM}}(\omega) = \frac{1}{2} \int_{\mathscr{A}} \psi(\omega) d_\omega S_{\mathsf{YM}}(\eta)\, \mathrm{d}\mu_{\mathsf{YM}}(\omega) \tag{43}$$

The directional differential of the Yang–Mills action is well known (see for example [3]) and most easily expressed using the covariant exterior differential $d^\omega : \Omega^0(M) \otimes \mathrm{Ad}(P) \to \Omega^1(M) \otimes \mathrm{Ad}(P)$ defined by $d^\omega \alpha = d\alpha + [\omega \wedge \alpha]$. It is given by

$$d_\omega S_{\mathsf{YM}}(\eta) = 2 \int_M \langle \eta \wedge d^\omega * \Omega \rangle$$

The problem is now to apply this formula to a well-chosen observable ψ and to differentiate in the right direction.

Given a loop ℓ on M, Makeenko and Migdal applied (43) to the observable defined by choosing a skew-Hermitian matrix $X \in \mathfrak{u}(N)$ and setting, for all $\omega \in \mathscr{A}$,

$$\psi_X(\omega) = \mathrm{Tr}(X\, \mathrm{hol}(\omega, \ell)) \tag{44}$$

To make this definition perfectly meaningful, one needs to choose a reference point in the fibre of P over the base point of ℓ: we will assume that such a point has been chosen and fixed, and compute holonomies with respect to this point.

Let us choose a parametrisation $\ell : [0, 1] \to M$ of ℓ. The directional derivative of the observable ψ_X in the direction of a 1-form $\eta \in \Omega^1(M) \otimes \mathrm{Ad}(P)$ is given by

$$d_\omega \psi_X(\eta) = -\int_0^1 \mathrm{Tr}\left(X\,\mathrm{hol}(\omega, \ell_{[s,1]})\eta(\dot{\ell}(s))\mathrm{hol}(\omega, \ell_{[0,s]})\right)\, ds \tag{45}$$

where we denote by $\ell_{[a,b]}$ the restriction of ℓ to the interval $[a, b]$.[28]

One must now choose the direction of differentiation η. Let us assume that ℓ is a nice loop which around each point of self-intersection looks like the left half of Figure 9. Let us assume that for some $s_0 \in (0, 1)$, we have $\ell(s_0) = \ell(0)$ and $\det(\dot{\ell}(0), \dot{\ell}(s_0)) = 1$. Makeenko and Migdal choose for η a distributional 1-form supported at the self-intersection point $\ell(0)$, which one could write as[29]

$$\forall m \in M, \forall v \in T_m M,\ \eta_m(v) = \delta_{m,\ell(0)} \det(\dot{\ell}(0), v) X$$

with $\det(\dot{\ell}(0), v)$ denoting the determinant of the two vectors $\dot{\ell}(0)$ and v. With this choice of η, the directional derivative of ψ_X is given by

$$d_\omega \psi_X(\eta) = -\mathrm{Tr}\left(X\,\mathrm{hol}(\omega, \ell_{[s_0,1]})X\,\mathrm{hol}(\omega, \ell_{[0,s_0]})\right) = -\mathrm{Tr}\left(X\,\mathrm{hol}(\omega, \ell')X\,\mathrm{hol}(\omega, \ell'')\right) \tag{46}$$

where ℓ' and ℓ'' are the loops defined on the right of Figure 9. Recall that $\mathfrak{u}(N)$ is endowed with the invariant scalar product $\langle X, Y\rangle = -N\mathrm{Tr}(XY)$. The directional derivative of the Yang–Mills action is thus given by

$$d_\omega S_{\mathsf{YM}}(\eta) = -2\langle X, (d^\omega * \Omega)(\dot{\ell}(0))\rangle = -2N\mathrm{Tr}\left(X d^\omega * \Omega(\dot{\ell}(0))\right)$$

or so it seems from a naive computation. We shall soon see that this expression needs to be reconsidered. For the time being, our Schwinger–Dyson equation reads

$$\int_{\mathscr{A}} \mathrm{Tr}\left(X\,\mathrm{hol}(\omega, \ell')X\,\mathrm{hol}(\omega, \ell'')\right)\, d\mu_{\mathsf{YM}}(\omega) = N\int_{\mathscr{A}} \mathrm{Tr}(X\,\mathrm{hol}(\omega, \ell))\mathrm{Tr}(X\, d^\omega * \Omega(\dot{\ell}(0)))\, d\mu_{\mathsf{YM}}(\omega) \tag{SD_X}$$

[28]At first glance, (45) may seem to require the choice of a point in $P_{\ell(s)}$ for each s, but in fact it does not, for the way in which the two holonomies and the term $\eta(\dot{\ell}(s))$ would depend on the choice of this point cancel exactly.

[29]It may seem that we are progressively letting go of the intrinsic character of our construction, but the interested reader can check that everything is still geometrically meaningful at this point.

Let us add the equalities (SD_X) obtained by letting X take all the values $X_1, \ldots, X_{N^2}$ of an orthonormal basis of $\mathfrak{u}(N)$. With the scalar product which we chose, the relations[30]

$$\sum_{k=1}^{N^2} \mathrm{Tr}(X_k A X_k B) = -\frac{1}{N}\mathrm{Tr}(A)\mathrm{Tr}(B) \text{ and } \sum_{k=1}^{N^2} \mathrm{Tr}(X_k A)\mathrm{Tr}(X_k B) = -\frac{1}{N}\mathrm{Tr}(AB) \tag{47}$$

hold for any two matrices A and B, so that we find

$$\int_{\mathscr{A}} \mathrm{tr}\left(\mathrm{hol}(\omega,\ell')\right)\mathrm{tr}(\mathrm{hol}(\omega,\ell''))\; \mathrm{d}\mu_{\mathsf{YM}}(\omega) = \int_{\mathscr{A}} \mathrm{tr}\left(\mathrm{hol}(\omega,\ell) d^{\omega} * \Omega(\dot{\ell}(0))\right)\; \mathrm{d}\mu_{\mathsf{YM}}(\omega).$$

The left-hand side of this equation is the right-hand side of (MM). The last and most delicate heuristic step is to interpret the right-hand side of this equation. For this, we must understand the term $d^{\omega} * \Omega(\dot{\ell}(0))$ and we do this by combining two facts: the fact that d^{ω} acts by differentiation in the horizontal direction and the fact that $*\Omega$ computes the holonomy along infinitesimal rectangles. We must also remember that this term comes from the computation of the exterior product of the distributional form η with the form $d^{\omega} * \Omega$. It turns out that, instead of a derivative in the horizontal direction with respect to s at $s = 0$, we should think of the difference between the values at 0^+ and at 0^-, which we denote by $\Delta_{|s=0}$.

With all this preparation and, it must be said, a small leap of faith, the right-hand side of the Schwinger–Dyson equation can finally be drawn as follows:

$$\Delta_{|s=0} \frac{d}{d\epsilon}_{|\epsilon=0} \int_{\mathscr{A}} \mathrm{tr\,hol}\Bigg(\omega,$$
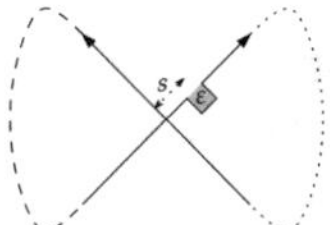
$$\Bigg)\; d\mu_{YM}(\omega)$$

$$= \frac{d}{d\epsilon}_{|\epsilon=0} \int_{\mathscr{A}} \mathrm{tr\,hol}\Bigg(\omega,$$
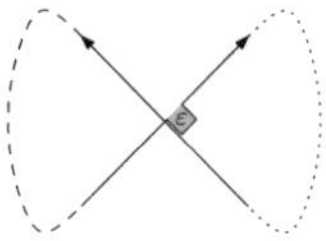
$$\Bigg)\; d\mu_{YM}(\omega)$$

$$-\frac{d}{d\epsilon}_{|\epsilon=0} \int_{\mathscr{A}} \mathrm{tr\,hol}\Bigg(\omega,$$

$$\Bigg)\; d\mu_{YM}(\omega)$$

This is indeed the left-hand side of the Makeenko–Migdal equation (MM).

[30] These relations are strictly equivalent to (16). They are, in one form or the other, *the* fundamental fact of all this story.

3.3 The Equations, Their Merits and Demerits

The strategy of proof described in the previous section can be used, and was used by Makeenko and Migdal, to derive equations slightly more general than (MM). Let us indeed consider a collection $\ell_1, \ldots, \ell_n$ of loops on the surface M. We assume that these loops are nice and in generic position, in the sense that every crossing between two portions of these loops, be they two portions of the same loop or portions of two different loops, is a simple transverse intersection. Around such a crossing, we see, as before, four faces of the graph cut on M by $\ell_1, \ldots, \ell_n$, and we label the areas of these faces t_1, t_2, t_3, t_4 as indicated on Figures 9 and 10. The Makeenko–Migdal equations express the alternated sum of the derivatives with respect to t_1, t_2, t_3, t_4 of $\mathbb{E}[\mathrm{tr}(H_{\ell_1}) \ldots \mathrm{tr}(H_{\ell_n})]$. The equations come in two variants, depending on whether the crossing is between two strands of the same loop (let us call this the case I) or between strands of two distinct loops (the case II). In the case II, where the crossing is between strands of two distinct loops, say ℓ_1 and ℓ_2, the same desingularisation operation explained at the beginning of Section 3.1 gives rise to one new loop ℓ_{12}, as explained in Figure 10.

Calling, in all cases, ℓ_1 the loop containing the South-West – North-East strand, one should replace the observable ψ_X defined in (44) by

$$\psi_X(\omega) = \mathrm{Tr}(X\mathrm{hol}(\omega, \ell_1))\mathrm{Tr}(\mathrm{hol}(\omega, \ell_2)) \ldots \mathrm{Tr}(\mathrm{hol}(\omega, \ell_n))$$

Then the directional derivative of ψ_X is given by

$$d_\omega \psi_X(\eta) = \begin{cases} \mathrm{Tr}\big(X\,\mathrm{hol}(\omega, \ell')X\,\mathrm{hol}(\omega, \ell'')\big)\mathrm{Tr}(\mathrm{hol}(\omega, \ell_2)) \ldots \mathrm{Tr}(\mathrm{hol}(\omega, \ell_n)) & \text{(case I)} \\ \mathrm{Tr}(X\,\mathrm{hol}(\omega, \ell_1))\mathrm{Tr}(X\,\mathrm{hol}(\omega, \ell_2))\mathrm{Tr}(\mathrm{hol}(\omega, \ell_3)) \ldots \mathrm{Tr}(\mathrm{hol}(\omega, \ell_n)) & \text{(case II)} \end{cases}$$

Then, the key to the computation is, as always, given by the equations (47). The final result, with the current notation, is the following.

Theorem 3.1 (Makeenko–Migdal equations) *Let $\ell_1, \ldots, \ell_n$ be nice loops on M in generic position. Consider a crossing point of two strands of ℓ_1 (case I) or of one strand of ℓ_1 and one strand of ℓ_2 (case II). Let t_1, t_2, t_3, t_4 denote the areas of the*

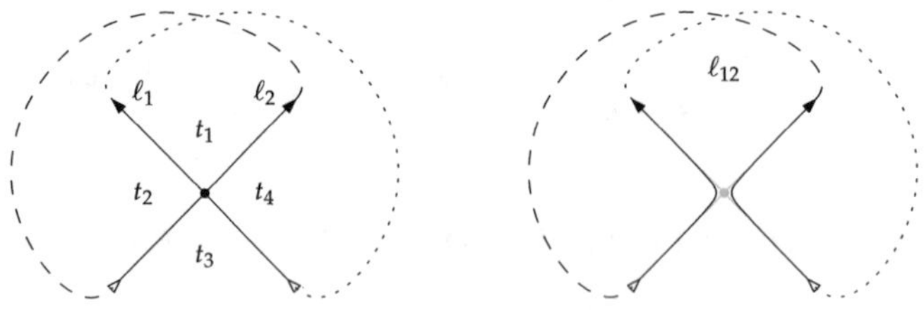

Fig. 10 When performed at a crossing of two distinct loops ℓ_1 and ℓ_2, the operation of reconnecting the incoming and outgoing strands in the other way that is consistent with orientation produces, from ℓ_1 and ℓ_2, one bigger loop that we denote by ℓ_{12}

four faces around this crossing point, as illustrated on Figures 9 and 10. Then, with the notation of these figures,

$$\left(\frac{\partial}{\partial t_1} - \frac{\partial}{\partial t_2} + \frac{\partial}{\partial t_3} - \frac{\partial}{\partial t_4}\right)\mathbb{E}[tr(H_{\ell_1})\dots tr(H_{\ell_n})] = \begin{vmatrix} \mathbb{E}[tr(H_{\ell'})tr(H_{\ell''})tr(H_{\ell_2})\dots\ tr(H_{\ell_n})]\ (I) \\ \frac{1}{N^2}\mathbb{E}[tr(H_{\ell_{12}})tr(H_{\ell_3})\dots\ tr(H_{\ell_n})] \quad (II) \end{vmatrix}$$

It is understood that if two of the four faces around the crossing under consideration are identical, then the corresponding derivative should be taken twice. Moreover, in the case where $M = \mathbb{R}^2$, any term corresponding to the derivative with respect to the area of the unbounded face should be ignored.

Makeenko and Migdal's original paper on this subject is [31]. The first mathematical proof of the equations was given in [24]. It was rather long and convoluted, and restricted to the case where the surface M is the plane $\mathbb{R}^2$. Three very short and elegant proofs of the equations were then given, still for the case of the plane, by Bruce Driver, Brian Hall and Todd Kemp in [8]. Immediately after, the same team joined by Franck Gabriel proved in [7] that the equations hold on any compact surface. There is little point in reproducing here the content of these beautiful papers. Let us simply emphasise that the fundamental computations remain those summarised in (47).

In addition to their simplicity, the Makeenko–Migdal equations have one major quality which is the fact that the collection of loops appearing in the right-hand side has one crossing less compared with the original collection of loops. Indeed, the operation of desingularisation replaces the crossing where it takes place by a tangential contact which, to the price of an arbitrarily small deformation of the loops, can be suppressed. This suggests the possibility of a recursive computation of Wilson loop expectations. We will explain in the next section that it is indeed possible to use the Makeenko–Migdal equations to set up a recursive computation of the large N limit of Wilson loop expectations.

What the Makeenko–Migdal do not do however, is to give a simple formula for the derivative of a Wilson loop expectation with respect to the area of a single face of the graph traced by a given configuration of loops. Only very special linear combinations of these derivatives are accessible. Of course, unless one is working on the plane, the total area of the surface is prescribed and the best one could hope for is a formula describing the variation of the Wilson loop expectations under an arbitrary variation of the areas of the faces that preserves the total area. However, this is, in general, not given by the Makeenko–Migdal equations, see for example Figure 11.

It is, in fact, not too difficult to understand what information is available in the Makeenko–Migdal equations. Let us consider n loops $\ell_1, \dots, \ell_n$ on our surface M. Let $F_1, \dots, F_r$ denote the faces of the graph traced by these loops. Let us identify a vector $(c_1, \dots, c_r)$ of the vector space $\mathbb{R}^r$ with the linear combination of derivatives

$$c_1 \frac{\partial}{\partial |F_1|} + \dots + c_r \frac{\partial}{\partial |F_r|}$$

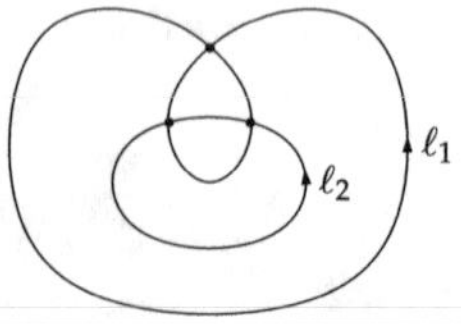

Fig. 11 Consider this configuration of two loops on a sphere. It has five faces and three vertices. Moreover, of the three instances of the Makeenko–Migdal equations, two compute the same linear combination of derivatives. There is no hope that the Makeenko–Migdal equations alone will allow one to compute the corresponding Wilson loop expectation

acting on Wilson loop expectations. Let us define the linear subspace $M \subset \mathbb{R}^r$ generated by the linear combinations given by the Makeenko–Migdal equations applied at each crossing of the loops $\ell_1, \ldots, \ell_n$. This subspace M is of course contained in the hyperplane $\mathbb{R}^r_0$ of equation $c_1 + \ldots + c_r = 0$. Every element of $\mathbb{R}^r$ can naturally be identified with a function on M that is constant on each face of the graph. To each loop ℓ_i, we can associate the unique element n_{ℓ_i} of $\mathbb{R}^r_0$ which, as a function on M, varies by 1 across ℓ_i[31] and is constant across every other loop. This function is a substitute for the winding number of the loop ℓ_i on the surface M.

It is not difficult to check that it is equivalent, for an element of $\mathbb{R}^r$, to be orthogonal, for the simplest scalar product, to the subspace M, or to have a constant jump across every loop, the constant possibly depending on the loop. A more formal statement is the following. We denote by $\mathbf{1}$ the vector $(1, \ldots, 1)$.

Proposition 3.2 *In* $\mathbb{R}^r$, *one has the equality of linear subspaces*

$$M = \mathrm{Vect}(\mathbf{1}, \mathsf{n}_{\ell_1}, \ldots, \mathsf{n}_{\ell_n})^{\perp}$$

In particular, $\dim M = \dim \mathbb{R}^r_0 - n$.

The greater the number of loops, the worse the situation. Even with one single loop, we see that all the information about the Wilson loop expectations is not contained in the Makeenko–Migdal equations.

It is time to turn to a case where things improve drastically, namely the large N limit of the Wilson loop expectations.

3.4 *The Master Field on Compact Surfaces*

We saw in Section 2 that when $G = \mathrm{U}(N)$, Wilson loop expectations tend to take simpler forms in the limit where N tends to infinity (compare for example (20) and (21)). We also observed some instances of a property of factorisation, see for

[31] A convention must be chosen regarding the definition of a positive crossing of ℓ_i.

example (26). The factorisation is due to a phenomenon of concentration, with the effect that, as N tends to infinity, and provided one scales the scalar product on $\mathfrak{u}(N)$ correctly (which we did), the Wilson loop functionals, that is, the normalised traces of the random holonomies, become deterministic. The limit is thus a number depending on a loop, and this function is relatively simple, at least when one is working on the plane, because it satisfies, and is essentially determined, by the Makeenko–Migdal equations.

The main theorem of convergence is the following.

Theorem 3.3 (Master field) *Let M be either the plane $\mathbb{R}^2$ or the sphere S^2. For each $N \geqslant 1$, let $(H_{N,\ell})_{\ell \in \mathscr{L}(M)}$ be the Yang–Mills holonomy process on M with structure group $G = \mathrm{U}(N)$, and with scalar product $\langle X, Y \rangle = N\mathrm{Tr}(X^*Y)$ on $\mathfrak{u}(N)$. Then for every loop $\ell \in \mathscr{L}(M)$, the convergence of complex-valued random variables*

$$\mathrm{tr}(H_{N,\ell}) \xrightarrow[N\to\infty]{P} \Phi(\ell) \tag{48}$$

holds in probability, towards a deterministic real limit.

This theorem was proved in [24] in the case of the plane, and in [5] in the case of the sphere, see also [17]. In the case of the plane, which is simpler, it is also known that the convergence occurs quickly, in the sense that the series $\sum_{N\geqslant 1} \mathrm{Var}(\mathrm{tr}(H_{N,\ell}))$ converges. Thus, the convergence (48) holds almost surely. The conclusion is also known to be true if one replaces the unitary group by the special unitary group, the special orthogonal group, or the symplectic group.

It is expected that Theorem 3.3 is true on any compact surface, but a proof of this fact still has to be given.

In any case, when this theorem holds, the aforementioned asymptotic factorisation takes place, in the sense that for all loops $\ell_1, \ldots, \ell_n$,

$$\lim_{N\to\infty} \mathbb{E}[\mathrm{tr}(H_{\ell_1}) \ldots \mathrm{tr}(H_{\ell_n})] = \lim_{N\to\infty} \mathbb{E}[\mathrm{tr}(H_{\ell_1})] \ldots \lim_{N\to\infty} \mathbb{E}[\mathrm{tr}(H_{\ell_n})] = \Phi(\ell_1) \ldots \Phi(\ell_n)$$

The function $\Phi : \mathscr{L}(M) \to \mathbb{R}$ which appears in (48) is called the *master field*. This is a continuous function with respect to the convergence of loops with fixed endpoints (see the beginning of Section 1.4.4) and it satisfies, crucially, the Makeenko–Migdal equation (MM_∞), which is all that there is left of the full set of equations stated in Theorem 3.1 as N tends to infinity.

Theorem 3.4 *Assume that M is either the plane $\mathbb{R}^2$ or the sphere S^2. The function $\Phi : \mathscr{L}(M) \to \mathbb{R}$ is the unique function that is continuous, invariant under area-preserving diffeomorphisms, satisfying the Makeenko–Migdal equation* (MM_∞) *and such that for every simple loop ℓ enclosing a domain of area t, one has, depending on whether M is the plane or a sphere of total area T,*

$$\Phi(\ell) = e^{-\frac{t}{2}} \qquad (M = \mathbb{R}^2)$$

or

$$\Phi(\ell) = \frac{1}{\pi}\int_{\mathbb{R}} \cosh\left(\frac{x}{2}(T-2t)\right)\sin(\pi\rho_T(x))\,\mathrm{d}x \qquad (M = S^2)$$

3.5 A Value of the Master Field on the Plane

As a conclusion to these notes, we give an example of computation of a value of the master field Φ on the plane, and choose an example that is not listed at the end of [24]. We choose the loop ℓ represented on the left half of Figure 12.

Although we did not include this in our description of the function Φ on the plane $\mathbb{R}^2$, it is not difficult to check that the derivative of Φ of any loop with respect to the area of a face adjacent to the unbounded face is equal to $-\frac{1}{2}$ times the value of Φ on this loop. This factor $-\frac{1}{2}$ comes of course from the stochastic differential equation (15) satisfied by the Brownian motion on $\mathrm{U}(N)$.

Given the value of Φ on simple loops and (29), the Makeenko–Migdal equation applied to the vertex of ℓ_0 that is marked in Figure 12 yields

$$(2\partial_s - \partial_{t_2})\Phi(\ell_0) = (-1 - \partial_{t_2})\Phi(\ell_0) = e^{-\frac{s}{2}-t_1-t_2}(1-t_1)$$

which is solved in

$$\Phi(\ell_0) = e^{-\frac{s}{2}-t_1-t_2}(1-t_1)(1-t_2)$$

If we can determine $\partial_u\Phi(\ell)$ explicitly, we are done, since $\Phi(\ell_0)$ is exactly the value of $\Phi(\ell)$ at $u = 0$. Applying the Makeenko–Migdal equations at the three marked vertices in Figure 12 yields the derivatives $(\partial_{s_1} + \partial_{s_2} - \partial_{t_2})\Phi(\ell)$, $(\partial_{s_1} + \partial_{s_2} - \partial_{t_1})\Phi(\ell)$, and $(\partial_{t_1} + \partial_{t_2} - \partial_{s_2} - \partial_u)\Phi(\ell)$. Adding the three expressions and using the fact that $\partial_{s_1}\Phi(\ell) = \partial_{s_2}\Phi(\ell) = -\frac{1}{2}\Phi(\ell)$, we find

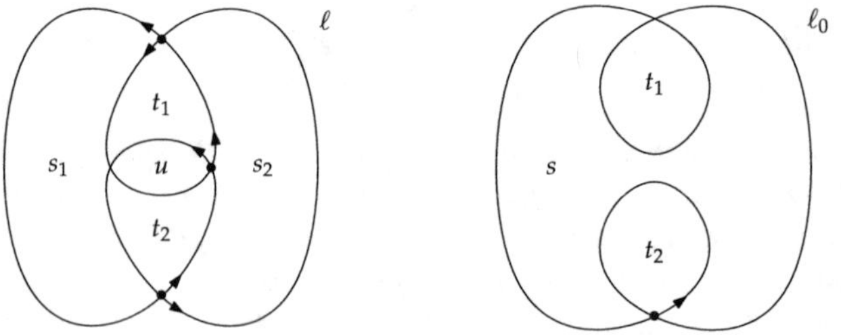

Fig. 12 We are interested in computing $\Phi(\ell)$. The strategy is to use the Makeenko–Migdal equations to compute $\partial_u\Phi(\ell)$. As $u = 0$, the two inner windings of ℓ disentangle, and ℓ becomes identical to ℓ_0. This loop ℓ_0 is similar to the loop that we studied in Section 2.3, and becomes exactly this loop when $t_2 = 0$. Our first task is thus to compute $\partial_{t_2}\Phi(\ell_0)$

$$\Big(-\frac{3}{2}-\partial_u\Big)\Phi(\ell) = e^{-\frac{s_1+s_2}{2}-t_1-t_2-\frac{3u}{2}}(3-t_1-t_2-u)$$

and finally

$$\Phi(\ell) = e^{-\frac{s_1+s_2}{2}-(t_1+t_2)-\frac{3u}{2}}\Big(\frac{u^2}{2}+(t_1+t_2-3)u+(1-t_1)(1-t_2)\Big) \tag{49}$$

Evaluating this expression with $s_1 = s_2 = t_1 = t_2 = 0$ yields the large N limit of the third moment of the unitary Brownian motion at time u, as expressed by (22) with $n = 3$. This is consistent with the fact that shrinking all faces but the face of area u reduces ℓ to a loop winding three times around a simple domain of area u.

Acknowledgments I am grateful to Nalini Anantharaman and Ashkan Nikeghbali for organising the *6th Strasbourg/Zurich - Meeting: Frontiers in Analysis and Probability* and for their invitation to give the talk from which these notes are an expanded version. Part of the content of these notes was also covered in a series of three lectures that I gave in Lyon in June 2018 in a workshop on *Random matrices, maps and gauge theories* organised by Alice Guionnet, Adrien Kassel and Grégory Miermont, whom I also want to thank. I am also indebted to Adrien Kassel for his careful reading of a first version of this manuscript.

References

1. M.F. Atiyah, R. Bott, The Yang-Mills equations over Riemann surfaces. Philos. Trans. R. Soc. Lond. A **308**, 523–615 (1983). https://doi.org/10.1098/rsta.1983.0017
2. P. Biane, Free Brownian motion, free stochastic calculus and random matrices, in *Free Probability Theory (Waterloo, ON, 1995)*. Fields Institute Communications, vol. 12 (American Mathematical Society, Providence, 1997), pp. 1–19
3. D. Bleecker, *Gauge Theory and Variational Principles*. Global Analysis Pure and Applied Series A, vol. 1 (Addison-Wesley, Reading, 1981), xviii+179
4. I. Chevyrev, Yang-mills measure on the two-dimensional torus as a random distribution (2018). http://arxiv.org/abs/arXiv:1808.09196
5. A. Dahlqvist, J.R. Norris, Yang-mills measure and the master field on the sphere (2017). http://arxiv.org/abs/arXiv:1703.10578
6. B.K. Driver, YM_2: continuum expectations, lattice convergence, and lassos. Commun. Math. Phys. **123**(4), 575–616 (1989)
7. B.K. Driver, F. Gabriel, B.C. Hall, T. Kemp, The Makeenko-Migdal equation for Yang-Mills theory on compact surfaces. Commun. Math. Phys. **352**(3), 967–978 (2017). https://doi.org/10.1007/s00220-017-2857-2
8. B.K. Driver, B.C. Hall, T. Kemp, Three proofs of the Makeenko-Migdal equation for Yang-Mills theory on the plane. Commun. Math. Phys. **351**(2), 741–774 (2017). https://doi.org/10.1007/s00220-016-2793-6
9. A. Dubin, Y. Makeenko, Loop equations and nonperturbative QCD, in *At the Frontier of Particle Physics*, ed. by S. Misha. Handbook of QCD. Boris Ioffe Festschrift, vol. 4 (World Scientific, Singapore, 2002), pp. 2479–2525
10. B. Durhuus, On the structure of gauge invariant classical observables in lattice gauge theories. Lett. Math. Phys. **4**(6), 515–522 (1980). https://doi.org/10.1007/BF00943439

11. W.M. Goldman, The symplectic nature of fundamental groups of surfaces. Adv. Math. **54**, 200–225 (1984). https://doi.org/10.1016/0001-8708(84)90040-9
12. W.M. Goldman, The symplectic geometry of affine connections on surfaces. J. Reine Angew. Math. **407**, 126–159 (1990). https://doi.org/10.1515/crll.1990.407.126
13. L. Gross, A Poincaré lemma for connection forms. J. Funct. Anal. **63**(1), 1–46 (1985)
14. L. Gross, The Maxwell equations for Yang-Mills theory, in *Mathematical Quantum Field Theory and Related Topics (Montreal, PQ, 1987)*. CMS Conference Proceedings, vol. 9 (American Mathematical Society, Providence, 1988), pp. 193–203
15. L. Gross, C. King, A. Sengupta, Two-dimensional Yang-Mills theory via stochastic differential equations. Ann. Phys. **194**(1), 65–112 (1989)
16. A. Guionnet, M. Maïda, Character expansion method for the first order asymptotics of a matrix integral. Probab. Theory Related Fields **132**(4), 539–578 (2005). https://doi.org/10.1007/s00440-004-0403-6
17. B.C. Hall, The large-N limit for two-dimensional Yang-Mills theory. Comm. Math. Phys. **363**(3), 789–828 (2018). https://doi.org/10.1007/s00220-018-3262-1
18. N.J. Hitchin, The Yang-Mills equations and the topology of 4-manifolds (after Simon K. Donaldson), in *Bourbaki seminar, Vol. 1982/83*. Astérisque, vol. 105 (Soc. Math. France, Paris, 1983), pp. 167–178
19. N.J. Hitchin, The self-duality equations on a Riemann surface. Proc. Lond. Math. Soc. (3) **55**(1), 59–126 (1987). https://doi.org/10.1112/plms/s3-55.1.59
20. C. King, A. Sengupta, The semiclassical limit of the two-dimensional quantum Yang-Mills model. J. Math. Phys. **35**(10), 5354–5361 (1994). https://doi.org/10.1063/1.530756
21. S. Kobayashi, K. Nomizu, *Foundations of Differential Geometry. Vol. I.* Wiley Classics Library (Wiley, New York, 1996), xii+329. Reprint of the 1963 original, A Wiley-Interscience Publication
22. T. Lévy, Wilson loops in the light of spin networks. J. Geom. Phys. **52**(4), 382–397 (2004). https://doi.org/10.1016/j.geomphys.2004.04.003
23. T. Lévy, Two-dimensional Markovian holonomy fields. Astérisque **329**, 172 (2010)
24. T. Lévy, The master field on the plane. Astérisque **388**, ix+201 (2017)
25. T. Lévy, M. Maïda, On the Douglas-Kazakov phase transition. Weighted potential theory under constraint for probabilists, in *Modélisation Aléatoire et Statistique—Journées MAS 2014*. ESAIM Proceedings Surveys, vol. 51 (EDP Sci., Les Ulis, 2015), pp. 89–121. https://doi.org/10.1051/proc/201551006
26. T. Lévy, A. Sengupta, Four chapters on low-dimensional gauge theories, in *Stochastic Geometric Mechanics. CIB, Lausanne, Switzerland, January–June 2015* (Springer, Cham, 2017), pp. 115–167
27. K. Liechty, D. Wang, Nonintersecting Brownian motions on the unit circle. Ann. Probab. **44**(2), 1134–1211 (2016). https://doi.org/10.1214/14-AOP998
28. K. Liu, Heat kernel and moduli space. Math. Res. Lett. **3**(6), 743–762 (1996). https://doi.org/10.4310/MRL.1996.v3.n6.a3
29. K. Liu, Heat kernel and moduli spaces. II. Math. Res. Lett. **4**(4), 569–588 (1997). https://doi.org/10.4310/MRL.1997.v4.n4.a12
30. R.C. Lyndon, P.E. Schupp, *Combinatorial Group Theory*. Classics in Mathematics (Springer, Berlin, 2001), xiv+339. Reprint of the 1977 edition
31. Y. Makeenko, A.A. Migdal, Exact equation for the loop average in multicolor QCD. Phys. Lett. B **88B**, 135 (1979)
32. A.A. Migdal, Recursion equations in gauge field theories. Sov. Phys. JETP **42**(3), 413–418 (1975)
33. S. Morita, *Geometry of Differential Forms*. Translations of Mathematical Monographs, vol. 201 (American Mathematical Society, Providence, 2001), xxiv+321. Translated from the two-volume Japanese original (1997, 1998) by Teruko Nagase and Katsumi Nomizu, Iwanami Series in Modern Mathematics

34. A. Nica, R. Speicher, *Lectures on the Combinatorics of Free Probability*. London Mathematical Society Lecture Note Series, vol. 335 (Cambridge University Press, Cambridge, 2006), xvi+417. https://doi.org/10.1017/CBO9780511735127
35. D. Puder, O. Parzanchevski, Measure preserving words are primitive. J. Amer. Math. Soc. **28**(1), 63–97 (2015). https://doi.org/10.1090/S0894-0347-2014-00796-7
36. J. Råde, On the Yang-Mills heat equation in two and three dimensions. J. Reine Angew. Math. **431**, 123–163 (1992). https://doi.org/10.1515/crll.1992.431.123
37. E.M. Rains, Combinatorial properties of Brownian motion on the compact classical groups. J. Theoret. Probab. **10**(3), 659–679 (1997). https://doi.org/10.1023/A:1022601711176
38. E.B. Saff, V. Totik, *Logarithmic Potentials with External Fields*. Grundlehren der Mathematischen Wissenschaften [Fundamental Principles of Mathematical Sciences], vol. 316 (Springer, Berlin, 1997), xvi+505. Appendix B by Thomas Bloom. https://doi.org/10.1007/978-3-662-03329-6
39. A. Sengupta, Gauge invariant functions of connections. Proc. Amer. Math. Soc. **121**(3), 897–905 (1994). https://doi.org/10.2307/2160291
40. A.N. Sengupta, Gauge theory on compact surfaces. Mem. Amer. Math. Soc. **126**(600), viii+85 (1997)
41. H. Shen, Stochastic quantization of an abelian gauge theory (2018). http://arxiv.org/abs/arXiv:1801.04596
42. I.M. Singer, On the master field in two dimensions, in *Functional Analysis on the Eve of the 21st Century, Vol. 1 (New Brunswick, NJ, 1993)*. Progress in Mathematics, vol. 131 (Birkhäuser Boston, Boston, 1995), pp. 263–281
43. M. Spivak, *A Comprehensive Introduction to Differential Geometry. Vol. I*, 2nd edn. (Publish or Perish, Inc., Wilmington, 1979), xiv+668.
44. K.K. Uhlenbeck, Connections with L^p bounds on curvature. Commun. Math. Phys. **83**(1), 31–42 (1982)
45. E. Witten, On quantum gauge theories in two dimensions. Commun. Math. Phys. **141**(1), 153–209 (1991). https://doi.org/10.1007/BF02100009
46. E. Witten, Two dimensional gauge theories revisited. J. Geom. Phys. **9**(4), 303–368 (1992). https://doi.org/10.1016/0393-0440(92)90034-X
47. F. Xu, A random matrix model from two-dimensional Yang-Mills theory. Commun. Math. Phys. **190**(2), 287–307 (1997)
48. C.N. Yang, R.L. Mills, Conservation of isotopic spin and isotopic gauge invariance. Phys. Rev. II. Ser. **96**, 191–195 (1954)

Limit Operators for Circular Ensembles

Kenneth Maples, Joseph Najnudel, and Ashkan Nikeghbali

Notation

If $v \in \mathbb{C}^n$ is a vector, then we write $v[m]$ for the image of v under the canonical projection map $\mathbb{C}^n \to \mathbb{C}^m$ onto the first m standard basis vectors and we write v_k or $(v)_k$ for its k-th coordinate. We denote by $\mathbb{C}^\infty$ the set of infinite sequences of complex numbers.

We write $U(n)$ for the unitary group of dimension n, which preserves the standard complex inner product.

We write $\mathbb{U} = U(1)$ for the unit circle in $\mathbb{C}$, i.e. those complex numbers with modulus 1.

Calligraphic characters denote σ-algebras, i.e. $\mathcal{A}$, $\mathcal{B}$, $\mathcal{C}$, etc. If $\mathcal{A}$ and $\mathcal{B}$ are σ-algebras on a common set, then $\mathcal{A} \vee \mathcal{B}$ denotes the smallest σ-algebra containing both $\mathcal{A}$ and $\mathcal{B}$.

We also write $a \vee b$, for $a, b \geq 0$, to mean $\max(a, b)$ and $a \wedge b$ to mean $\min(a, b)$.

We employ asymptotic notation for inequalities where precise constants are not important. In particular, we write $X = O(Y)$ to mean that there exists a constant $C > 0$, possibly random, such that $|X| \leq CY$. We also use (modified) Vinogradov notation, where $X \lesssim Y$ means $X = O(Y)$, for convenience.

If t is a real number, we write $\lfloor t \rfloor$ its integer part.

If H is a Hilbert space with scalar product $\langle ., . \rangle$, and if $F \subset H$, then $F^\perp = \{x \in H;\quad \langle x, y \rangle = 0\ \forall y \in F\}$. If H is a complex Hilbert space, then we will always

K. Maples · A. Nikeghbali (✉)
Institut für Mathematik, Universität Zürich, Winterthurerstrasse 190, CH-8057 Zürich, Switzerland
e-mail: kenneth.maples@math.uzh.ch; ashkan.nikeghbali@math.uzh.ch

J. Najnudel
School of Mathematics, University of Bristol, Woodland Rd, Bristol, BS8 1UG, UK
e-mail: joseph.najnudel@math.univ-toulouse.fr

N. Anantharaman et al. (eds.), *Frontiers in Analysis and Probability*,
https://doi.org/10.1007/978-3-030-56409-4_8

use the scalar product, which is linear in the first variable and conjugate linear in the second, i.e. $\langle ax, by\rangle = a\overline{b}\langle x, y\rangle$.

1 Introduction

It has been observed that for many models of random matrices, the eigenvalues have a limiting short-scale behavior when the dimension goes to infinity, which depends on the global symmetries of the model, but not on its detailed features. For example, the Gaussian Orthogonal Ensemble (GOE), for which the matrices are real symmetric with independent gaussian entries on and above the diagonal, corresponds to a limiting short-scale behavior for the eigenvalues that is also obtained for several other models of random real symmetric matrices. Similarly, the limiting spectral behavior of a large class of random hermitian and unitary ensembles, including the Gaussian Unitary Ensemble (GUE, with independent, complex gaussians above the diagonal), and the Circular Unitary Ensemble (CUE, corresponding to the Haar measure on the unitary group of a given dimension), involves a remarkable random point process, called the *determinantal sine-kernel process*. It is a point process for which the k-point correlation function is given by

$$\rho_k(x_1, \dots, x_k) = \det\left(\frac{\sin(\pi(x_p - x_q))}{\pi(x_p - x_q)}\right)_{1\leq p,q\leq k}.$$

From an observation of Montgomery, it has been conjectured that the limiting short-scale behavior of the imaginary parts of the zeros of the Riemann zeta function is also described by a determinantal sine-kernel process. This similar behavior supports the conjecture of Hilbert and Pólya, who suggested that the non-trivial zeros of the Riemann zeta functions should be interpreted as the spectrum of an operator $\frac{1}{2} + iH$ with H an unbounded Hermitian operator.

In order to understand the kind of randomness that would be involved in such an operator, it is natural to try to construct a random version H_0 of it, for which the spectrum has the conjectured limiting behavior of the zeros of the Riemann zeta function, i.e. is a determinantal sine-kernel point process. Since the spectrum of H_0 should also correspond to the limiting behavior of many ensembles of hermitian and unitary matrices, it is natural to expect that these ensembles can, in one way and another, be related to H_0.

Instead of looking directly for H_0, one can also directly seek the flow of linear operators $(U_0^\alpha)_{\alpha\in\mathbb{R}} := (e^{i\alpha H_0})_{\alpha\in\mathbb{R}}$ generated by exponentiation. This point of view is both consistent with what would be a possible interpretation in quantum mechanics (U_0^α playing the role of the operator of evolution at time α, whereas H_0 corresponds to the Hamiltonian), and with the number theoretic point of view: if the Hilbert–Pólya operator H would exist, the Chebyshev function ψ_0, which is defined as

$$\psi_0(x) = \sum_{p^m < x} \log p + \frac{1}{2} \begin{cases} \log p, & x = p^m \text{ for some prime power } p^m \\ 0, & \text{otherwise} \end{cases}$$

where the summation is through all powers of primes bounded by x, would formally satisfy

$$\psi_0(e^x) = \int_{-\infty}^{x} \left(e^y - e^{y/2} \operatorname{Tr}(e^{iHy}) \right) dy + O(1),$$

which suggests an important role played by the conjectural flow $(e^{iHy})_{y \in \mathbb{R}}$ of unitary operators. To see this, recall the von Mangoldt formula (see e.g. [19]):

$$\psi_0(x) = x - \sum_{\rho} \frac{x^\rho}{\rho} + O(1),$$

where the summation is over all zeros ρ of the Riemann zeta function. Now taking $x = e^y$ in the above, writing $e^y - \sum_\rho \frac{e^{y\rho}}{\rho}$ as the integral of its derivative and then writing the last sum as the trace of the operator yields the formal identity above. On the other hand, as we will see below, considering the flow $(e^{i\alpha H_0})_{\alpha \in \mathbb{R}}$, instead of taking directly H_0, is also consistent with the main construction of the present article, since, in a sense which will be made precise, this flow is an approximation, for large n, of the successive powers of a random matrix following the Haar measure on $U(n)$.

It should be mentioned that the problem of the existence of the operator H_0 is also hinted at in the work by Katz and Sarnak [6]. For instance, the zeta function of algebraic curves is related to the unitary group or some other compact groups (e.g. the symplectic group), and there the question of coupling all different dimensions of the unitary group together in a consistent way in order to prove strong limit theorems (i.e. almost sure convergence) and having an infinite dimension space and operator sitting above is raised.

The main goal of the present paper is the construction of a flow of random operators $(V^\alpha)_{\alpha \in \mathbb{R}}$, whose spectrum, in a sense which can be made precise, is a determinantal sine-kernel process, and which is directly related to the Circular Unitary Ensemble. A similar, but simpler, construction has been made in [15], in which we consider permutation matrices instead of unitary matrices. The construction that is made in the present paper uses the recent construction of virtual isometries given in [2] and needs several further steps.

First, the space on which $(V^\alpha)_{\alpha \in \mathbb{R}}$ acts must be infinite dimensional. In order to relate this space with the Circular Unitary Ensemble, we will construct a coupling between the matrix models of each dimension. That is, we define a sequence $(u_n)_{n \geq 1}$ of random unitary matrices, in such a way that for all $n \geq 1$, each matrix $u_n \in U(n)$ is Haar distributed on the unitary group. Of course, there are many ways to couple the random variables u_n to each other: for example, by taking all the matrices to be independent. However, in order to have sufficient consistency to construct limiting objects, we need to be more subtle.

In fact, we will couple $(u_n)_{n\geq 1}$ in such a way that almost surely, this sequence of random matrices is a *virtual isometry*. The notion of virtual isometry has been introduced in one of our former articles [2], generalizing both the notion of *virtual permutation* studied by Kerov, Olshanski and Vershik [9], and the previous notion of virtual unitary group introduced by Neretin [16].

By definition, a virtual isometry is a sequence $(u_n)_{n\geq 1}$ of unitary matrices u_n of dimension n, such that for all n, u_n is the matrix u such that, for u_{n+1} fixed, the rank of the difference

$$\begin{pmatrix} u & 0 \\ 0 & 1 \end{pmatrix} - u_{n+1}$$

is minimal (which is always 0 or 1). From this definition, one can deduce that for all $n \geq 1$, u_n completely determines the sequence $u_1, u_2, \ldots, u_{n-1}$, and from these matrices, one directly obtains a decomposition of u_n as a product of complex reflections. Moreover, the Haar measures in different dimensions are compatible with respect to the notion of virtual isometries, i.e. it is possible to construct a probability distribution on the space of virtual isometries in such a way that for all $n \geq 1$, the marginal distribution of the n-dimensional component coincides with the Haar measure on $U(n)$. Therefore, the Circular Unitary Ensemble can be coupled in all dimensions by considering a random virtual isometry following a suitable probability distribution.

Note that in the notion of virtual isometry defined here, the vectors of the canonical basis of $\mathbb{C}^n$ play a particular role. One could attempt to generalize the notion of virtual isometries by considering sequences of unitary operators on E_n, $n \geq 1$, where $(E_n)_{n\geq 1}$ is a sequence of complex inner product spaces, E_n being of dimension n. However, this reduces to the particular case $E_n = \mathbb{C}^n$ by a change of basis and so we have chosen to use the standard basis for simplicity.

In [2], it is shown that if $(u_n)_{n\geq 1}$ follows this distribution, then for all k, the kth positive (respectively negative) eigenangle of u_n, multiplied by $n/2\pi$ (i.e. the inverse of the average spacing between eigenangles for any matrix in $U(n)$), converges almost surely to a random variable y_k (respectively y_{1-k}). The random set $(y_k)_{k\in\mathbb{Z}}$ is a determinantal sine-kernel process, and for each k, the convergence holds with a rate dominated by some negative power of n. In [12], we improve our estimate of this rate, and more importantly, we prove that almost sure convergence not only holds for the eigenangles of u_n but also for the components of the corresponding eigenvectors. More precisely, we show that, for all $k, \ell \geq 1$, the ℓth component of the eigenvector of u_n associated with the kth positive (respectively negative) eigenangle converges almost surely to a non-zero limit $t_{k,\ell}$ (respectively $t_{1-k,\ell}$) when n goes to infinity, if the norm of the eigenvector is taken equal to $\sqrt{n}$ and if the phases are suitably chosen. Moreover, the variables $(t_{k,\ell})_{k\in\mathbb{Z},\ell\geq 1}$ are iid complex gaussians. Note that taking a norm equal to $\sqrt{n}$ is natural in this setting: with this normalization, the expectation of the squared modulus of each coordinate of a given eigenvector of u_n is equal to 1, so we can expect a convergence to a non-trivial limit. If the norm of the eigenvectors is taken equal to 1 instead of $\sqrt{n}$, then the coordinates converge to zero when n goes to infinity.

Knowing the joint convergence of the renormalized eigenvalues and the corresponding eigenvectors, it is natural to expect that, in a sense which has to be made precise, the limiting behavior of $(u_n)_{n\geq 1}$ when n goes to infinity can be described by an operator whose eigenvectors are the sequences $(t_{k,\ell})_{\ell\geq 1}$, $k \in \mathbb{Z}$ and for which the corresponding eigenvalues are $(y_k)_{k\in\mathbb{Z}}$.

The precise statement of our main result can in fact be described as follows:

Theorem 1.1 *Almost surely there exists a random vector subspace $\mathcal{F}$ of $\mathbb{C}^\infty$ and a flow of linear maps $(V^\alpha)_{\alpha\in\mathbb{R}}$ on $\mathcal{F}$, such that $V^{\alpha+\beta} = V^\alpha V^\beta$, and satisfying the following properties, for any sequence $w = (w_\ell)_{\ell\geq 1}$ in $\mathcal{F}$:*

(1) *For any fixed $\ell \geq 1$, the ℓth component of the n-dimensional vector of $u_n^{\lfloor \alpha n\rfloor}(w_1, \ldots, w_n)$ tends to the ℓth component of $V^\alpha(w)$ when n goes to infinity.*
(2) *If $w \neq 0$, the L^2 distance between $u_n^{\lfloor \alpha n\rfloor}(w_1, \ldots, w_n)$ and $((V^\alpha w)_1, \ldots, (V^\alpha w)_n)$ is negligible with respect to the norm of $(w_1, \ldots, w_n)$.*

Moreover, the eigenvectors of the flow, i.e. the sequences $w \in \mathcal{F}$ such that there exists $\lambda \in \mathbb{R}$ for which $V^\alpha w = e^{2i\pi\lambda\alpha} w$ for all $\alpha \in \mathbb{R}$, are exactly the sequences proportional to $(t_{k,\ell})_{\ell\geq 1}$ for some $k \in \mathbb{Z}$. The corresponding value of λ is equal to y_k.

Let us emphasize that the last part of this theorem gives the complete family of eigenvectors of the flow $(V^\alpha)_{\alpha\in\mathbb{R}}$. In particular, the eigenvalues form exactly the determinantal sine-kernel process $(y_k)_{k\in\mathbb{Z}}$, with no extra eigenvalue. Note that the notion of eigenvalue and eigenvector is not exactly the same as in the usual situation where a single operator is considered.

Intuitively, the operator V^α can be viewed as a limit, for sequences in the space $\mathcal{F}$, of the iteration $u_n^{\lfloor \alpha n\rfloor}$, when n goes to infinity. The flow $(V^\alpha)_{\alpha\in\mathbb{R}}$ can be compared with the flow constructed in [15] for permutation matrices. The space $\mathcal{F}$ is random and explicitly defined in terms of the virtual isometry $(u_n)_{n\geq 1}$.

Some natural questions that can then be asked are the following:

(1) Is it possible to replace the space $\mathcal{F}$ by another space with a simpler description?
(2) Is it possible to define a version of the flow $(V^\alpha)_{\alpha\in\mathbb{R}}$ of operators, which can be naturally related to ensembles of unitary or Hermitian matrices, which are different from the CUE?

An answer to this last question would give a more generic version of $(V^\alpha)_{\alpha\in\mathbb{R}}$, which may enlighten in a new, more geometric way, the properties of universality enjoyed by the sine-kernel process in random matrix theory.

Such a generalization seems to be very difficult to construct. In particular, we do not know how one could couple the GUE in such a way that the renormalized eigenvalues converge almost surely. Note that the successive minors of an infinite GUE matrix cannot converge almost surely to a non-trivial (i.e. non-constant) limiting distribution. Indeed, if we note $(A_n)_{n\geq 1}$ the successive minors, and $F_n : \mathcal{M}_n(\mathbb{C}) \to \mathbb{R}$, which depends only on the eigenvalues and such that $F_n(A_n)$ converges a.s. to X, then $F_n(A_n)$ and $F_{2n}(A_{2n})$ both converge a.s. to X. In particular $(F_n(A_n), F_{2n}(A_{2n}))$ converges in law to (X, X). Since we work with GUE matrices, it also follows that $(F_n(B_n), F_{2n}(A_{2n}))$ also converges to (X, X), where B_n is the n by n bloc matrix obtained by taking the lower right bloc in A_{2n}. Thus it follows that $F_n(A_n) - F_{2n}(A_{2n})$ converges in law (and also in probability) to $X - X = 0$. Similarly $F_n(B_n) - F_{2n}(A_{2n})$ also converges in probability to 0. Summing up we obtain that $F_n(B_n) - F_n(A_n)$ converges in probability to 0. On the other hand, $F_n(B_n) - F_n(A_n)$ also converges to $X - Y$ where X and Y are independent with the same law. This implies that $|\mathbb{E}[e^{itX}]|^2 = 1$, and hence X is equal to a constant a.s.

It would also be interesting to relate the operators we construct to the Brownian carousel construction, which was introduced by Valkó and Virág in [20] (or alternatively to the work by Killip and Stoiciu [10]), in order to generalize the sine-kernel process to the setting of β-ensembles for any $\beta \in (0, \infty)$ (instead of just $\beta = 2$).

The paper is devoted to providing the details of the construction of the flow of operators given in Theorem 1.1. In Section 2, we state some results proven in [12], which are used in the present article. In Section 3, we construct a flow of operators satisfying Theorem 1.1, on the random vector subspace $\mathcal{E}$ of $\mathbb{C}^\infty$ containing the finite linear combinations of the sequences $(t_{k,\ell})_{\ell\geq 1}$ for $k \in \mathbb{Z}$. In Section 4, we define a scalar product on $\mathcal{E}$, and we use it in order to extend the definition of the flow constructed in Section 3 to a larger space. Another extension of the flow is defined in Section 5, in such a way that its definition does not explicitly involve the limiting eigenvectors of the random virtual isometry $(u_n)_{n\geq 1}$. Some open problems are discussed in Section 6.

The results of the present article, as well as the convergence of eigenvectors proven in [12], are also stated in our preprint [11].

2 Preliminary Results from [12]

We start with a random virtual isometry $(u_n)_{n\geq 1}$, such that u_n is Haar distributed for each $n \geq 1$. As in [12], we may assume that this virtual isometry is constructed as follows:

(1) We consider a sequence $(x_n)_{n\geq 1}$ of independent random vectors, x_n being uniform on the unit sphere of $\mathbb{C}^n$.
(2) Almost surely, for all $n \geq 1$, x_n is different from the last basis vector e_n of $\mathbb{C}^n$, which implies that there exists a unique $r_n \in U(n)$ such that $r_n(e_n) = x_n$ and $r_n - I_n$ has rank one.
(3) We define $(u_n)_{n\geq 1}$ by induction as follows: $u_1 = x_1$ and for all $n \geq 2$,

$$u_n = r_n \begin{pmatrix} u_{n-1} & 0 \\ 0 & 1 \end{pmatrix}.$$

We denote by $\lambda_1^{(n)} = e^{i\theta_1^{(n)}}, \ldots, \lambda_n^{(n)} = e^{i\theta_n^{(n)}}$ the eigenvalues of u_n, almost surely distinct and different from 1, with the ordering $0 < \theta_1^{(n)} < \cdots < \theta_n^{(n)} < 2\pi$. It will be convenient to extend the notation $\lambda_k^{(n)}$ and $\theta_k^{(n)}$ to all $k \in \mathbb{Z}$, in such a way that $\theta_{k+n}^{(n)} = \theta_k^{(n)} + 2\pi$ and $\lambda_{k+n}^{(n)} = \lambda_k^{(n)}$, i.e. the sequence $(\lambda_k^{(n)})_{k\in\mathbb{Z}}$ is n-periodic. Note that with this convention, $(\theta_k^{(n)})_{k\in\mathbb{Z}}$ is the increasing sequence of eigenangles of u_n, taken in the whole real line. The following result is proven in [12]:

Theorem 2.1 *There is a sine-kernel point process $(y_k)_{k\in\mathbb{Z}}$ such that almost surely,*

$$\frac{n}{2\pi}\theta_k^{(n)} = y_k + O((1+k^2)n^{-\frac{1}{3}+\epsilon}),$$

for all $n \geq 1$, $|k| \leq n^{1/4}$ and $\epsilon > 0$, where the implied constant may depend on $(u_m)_{m\geq 1}$ and ϵ, but not on n and k.

We can now construct, for each $n \geq 1$, a basis $(f_k^{(n)})_{1\leq k\leq n}$ of unit eigenvectors of u_n, $f_k^{(n)}$ corresponding to the eigenvalue $\lambda_k^{(n)}$. The construction is done as follows. For $n = 1$, we define $f_1^{(1)} := 1$. If we assume that the basis $(f_k^{(n)})_{1\leq k\leq n}$ is constructed for some $n \geq 1$, then we can expand $r_{n+1}(e_{n+1})$ in this basis:

$$r_{n+1}(e_{n+1}) = \sum_{j=1}^{n} \mu_j^{(n)} f_j^{(n)} + \nu_n e_{n+1}.$$

Notice that here $f_j^{(n)} \in \mathbb{C}^n$ and the other terms are in $\mathbb{C}^{n+1}$: we identify $\mathbb{C}^n$ as a subset of $\mathbb{C}^{n+1}$ by adding a $(n+1)$-th coordinate equal to 0. Then, as written in [12], the eigenvalues of u_{n+1} are precisely the zeros of the equation:

$$\sum_{j=1}^{n} |\mu_j^{(n)}|^2 \frac{\lambda_j^{(n)}}{\lambda_j^{(n)} - z} + \frac{|1-\nu_n|^2}{1-z} = 1 - \overline{\nu}_n$$

Moreover, we can define the unit eigenvectors $(f_k^{(n+1)})_{1\leq k\leq n+1}$ of u_{n+1} as the solutions of the system of equations:

$$C_k f_k^{(n+1)} = \sum_{j=1}^{n} \frac{\mu_j^{(n)}}{\lambda_j^{(n)} - \lambda_k^{(n+1)}} f_j^{(n)} + \frac{\nu_n - 1}{1 - \lambda_k^{(n+1)}} e_{n+1}, \tag{1}$$

where $C_k \in \mathbb{R}^+$ is a suitably chosen constant. Moreover, we can extend the notation $f_k^{(n)}, \mu_j^{(n)}$ to all $j, k \in \mathbb{Z}$, in such a way that the sequences $(f_k^{(n)})_{k\in\mathbb{Z}}$ and $(\mu_j^{(n)})_{j\in\mathbb{Z}}$ are n-periodic.

Comparing the norm of the vectors, we see that $C_k = (h_k^{(n+1)})^{1/2}$ where

$$h_k^{(n+1)} := \sum_{j=1}^{n} \frac{|\mu_j^{(n)}|^2}{|\lambda_j^{(n)} - \lambda_k^{(n+1)}|^2} + \frac{|\nu_n - 1|^2}{|1 - \lambda_k^{(n+1)}|^2}.$$

For $n \geq k \geq 1$, we introduce another normalization of the k-th eigenvector of u_n, namely

$$g_k^{(n)} := D_k^{(n)} f_k^{(n)}$$

where

$$D_k^{(n)} := \prod_{s=k}^{n-1} (h_k^{(s+1)})^{1/2} \frac{\lambda_k^{(s)} - \lambda_k^{(s+1)}}{\mu_k^{(s)}}.$$

The advantage of this normalization is the martingale properties satisfied by the sequence $(g_k^{(n)})_{n\geq k}$, which come from the following equalities, satisfied for all $k, \ell \in \{1, \ldots, n\}$, and deduced from (1):

$$\begin{aligned}
\langle g_k^{(n+1)}, e_\ell \rangle &= D_k^{(n+1)} (h_k^{(n+1)})^{-\frac{1}{2}} \sum_{j=1}^{n} \frac{\mu_j^{(n)}}{\lambda_j^{(n)} - \lambda_k^{(n+1)}} \langle f_j^{(n)}, e_\ell \rangle \\
&= D_k^{(n)} \frac{\lambda_k^{(n)} - \lambda_k^{(n+1)}}{\mu_k^{(n)}} \sum_{j=1}^{n} \frac{\mu_j^{(n)}}{\lambda_j^{(n)} - \lambda_k^{(n+1)}} \langle f_j^{(n)}, e_\ell \rangle \\
&= \langle g_k^{(n)}, e_\ell \rangle + D_k^{(n)} \frac{\lambda_k^{(n)} - \lambda_k^{(n+1)}}{\mu_k^{(n)}} \sum_{\substack{1\leq j\leq n \\ j\neq k}} \frac{\mu_j^{(n)}}{\lambda_j^{(n)} - \lambda_k^{(n+1)}} \langle f_j^{(n)}, e_\ell \rangle.
\end{aligned} \tag{2}$$

More precisely, as in [12], let us define the following σ-algebras:

(1) For $n \geq 1$, the σ-algebra $\mathcal{A}_n$ generated by the eigenvalues of u_m, for $1 \leq m \leq n$. This σ-algebra is equal, up to completion, to the σ-algebra generated by u_1, the variables $|\mu_j^{(m)}|$ and ν_m for $1 \leq m \leq n-1$ and $1 \leq j \leq m$.
(2) The σ-algebra $\mathcal{A} := \vee_{n=1}^{\infty} \mathcal{A}_n$ generated by the eigenvalues of u_m for all $m \geq 1$.

(3) For $n \geq 1$, the σ-algebra $\mathcal{B}_n$ generated by $\mathcal{A}$ and by the phases $\mu_j^{(n)}/|\mu_j^{(n)}|$ for $1 \leq m \leq n-1$, $1 \leq j \leq m$. This σ-algebra is equal, up to completion, to the σ-algebra generated by $\mathcal{A}$ and the eigenvectors $f_j^{(m)}$ for $1 \leq j \leq m \leq n$.
(4) The σ-algebra $\mathcal{B} := \vee_{n=1}^{\infty} \mathcal{B}_n$.

With these definition, it is not difficult to check that $|D_k^{(n)}|^2$ (but not $D_k^{(n)}$) is $\mathcal{A}$-measurable. The following result is proven in [12]:

Proposition 2.2 *For each $k \geq 1$ and $\ell \geq 1$, there exists an increasing sequence $(H_j)_{j\geq 1}$ of events in $\mathcal{A}$, with probability tending to 1, such that for all $j \geq 1$, $(\mathbf{1}_{H_j} \langle g_k^{(n)}, e_\ell \rangle)_{n \geq k \vee \ell}$ is a martingale with respect to the filtration $(\mathcal{B}_n)_{n \geq k\vee\ell}$ and the conditional expectation of $\mathbf{1}_{H_j} \langle g_k^{(n)}, e_\ell \rangle$, given $\mathcal{A}$, is almost surely bounded when n varies.*

Moreover, in [12], we prove that, for $1 \leq k, \ell \leq n$,

$$\mathbb{E}[|\langle g_k^{(n+1)} - g_k^{(n)}, e_\ell \rangle|^2 \,|\mathcal{A}] = |D_k^{(n)}|^2 \frac{|\lambda_k^{(n)} - \lambda_k^{(n+1)}|^2}{|\mu_k^{(n)}|^2} S, \tag{3}$$

where

$$S = \sum_{\substack{1 \leq j \leq n \\ j \neq k}} \frac{|\mu_j^{(n)}|^2}{|\lambda_j^{(n)} - \lambda_k^{(n+1)}|^2} \mathbb{E}[|\langle f_j^{(n)}, e_\ell \rangle|^2 \mid \mathcal{A}]. \tag{4}$$

From Proposition 2.2, the following corollary is deduced:

Corollary 2.3 *Almost surely, for all $k \in \mathbb{Z}$ and $\ell \geq 1$, the scalar product $\langle g_k^{(n)}, e_\ell \rangle$ converges to a limit $g_{k,\ell}$ when n goes to infinity.*

The main theorem of [12] is strongly related to this corollary and is stated as follows:

Theorem 2.4 *Let $(u_n)_{n\geq 1}$ be a virtual isometry, following the Haar measure. For $k \in \mathbb{Z}$ and $n \geq 1$, let $v_k^{(n)}$ be a unit eigenvector corresponding to the kth smallest nonnegative eigenangle of u_n for $k \geq 1$, and the $(1-k)$th largest strictly negative eigenangle of u_n for $k \leq 0$. Then for all $k \in \mathbb{Z}$, there almost surely exist some complex numbers $(\psi_k^{(n)})_{n\geq 1}$ of modulus 1, and a sequence $(t_{k,\ell})_{\ell \geq 1}$, such that for all $\ell \geq 1$,*

$$\sqrt{n}\, \langle \psi_k^{(n)} v_k^{(n)}, e_\ell \rangle \underset{n\to\infty}{\longrightarrow} t_{k,\ell}.$$

Almost surely, for all $k \in \mathbb{Z}$, the sequence $(t_{k,\ell})_{\ell\geq 1}$ depends, up to a multiplicative factor of modulus one, only on the virtual rotation $(u_n)_{n\geq 1}$. Moreover, if $(\psi_k)_{k\in\mathbb{Z}}$ is a sequence of iid, uniform variables on $\mathbb{U}$, independent of $(t_{k,\ell})_{\ell\geq 1}$, *then* $(\psi_k t_{k,\ell})_{k\in\mathbb{Z},\ell\geq 1}$ *is an iid family of standard complex gaussian variables* ($\mathbb{E}[|\psi_k t_{k,\ell}|^2] = 1$).

Remark 2.5 The vectors $v_k^{(n)}$ are equal to $f_k^{(n)}$, up to a multiplicative factor of modulus 1. The independent phases ψ_k introduced in the last part of the theorem are needed in order to get iid complex gaussian variables. This is not the case, for example, if we normalize $(t_{k,\ell})_{\ell\geq 1}$ in such a way that $t_{k,1} \in \mathbb{R}_+$.

The random variables $t_{k,\ell}$ are strongly related to the variables $g_{k,\ell}$. More precisely, the following estimate is proven in [12]:

$$|D_k^{(n)}|^2 = D_k n(1 + O(n^{-\frac{1}{3}+\epsilon})), \tag{5}$$

where D_k is a non-zero random variable that depends only on k. Then, the following equality is deduced:

$$g_{k,\ell} = \sqrt{D_k} t_{k,\ell}. \tag{6}$$

3 A Flow of Operators on the Space Generated by the Limiting Eigenvectors

For each $\alpha \in \mathbb{R}$, let $(\alpha_n)_{n\geq 1}$ be a sequence such that α_n is equivalent to αn when n goes to infinity. For $n \geq 1$, $k \in \mathbb{Z}$, we have

$$u_n^{\alpha_n} f_k^{(n)} = e^{i\theta_k^{(n)}\alpha_n} f_k^{(n)}.$$

Now, $e^{i\theta_k^{(n)}\alpha_n}$ tends to $e^{2i\pi\alpha y_k}$ and after normalization, the coordinates of $f_k^{(n)}$ tend to the corresponding coordinates of the sequence $(t_{k,\ell})_{\ell\geq 1}$. It is then natural to expect that, in a sense which needs to be made precise, $u_n^{\alpha_n}$ tends to some operator U, acting on some infinite sequences, such that

$$U((t_{k,\ell})_{\ell\geq 1}) = e^{2i\pi\alpha y_k}(t_{k,\ell})_{\ell\geq 1}.$$

This motivates the following definition:

Definition 3.1 The space $\mathcal{E}$ is the random vector subspace of $\mathbb{C}^\infty$, consisting of all finite linear combinations of the sequences $(t_{k,\ell})_{\ell\geq 1}$, or equivalently, $(g_{k,\ell})_{\ell\geq 1}$, for $k \in \mathbb{Z}$.
For $\alpha \in \mathbb{R}$, the operator U^α is the unique linear application from $\mathcal{E}$ to $\mathcal{E}$ such that for all $k \in \mathbb{Z}$,

$$U^\alpha((t_{k,\ell})_{\ell\geq 1}) = e^{2i\pi\alpha y_k}(t_{k,\ell})_{\ell\geq 1},$$

or equivalently,

$$U^{\alpha}((g_{k,\ell})_{\ell\geq 1}) = e^{2i\pi\alpha y_k}(g_{k,\ell})_{\ell\geq 1}.$$

Remark 3.2 For each k, the sequence $(t_{k,\ell})_{\ell\geq 1}$ is almost surely well-defined up to a multiplicative constant of modulus 1. Moreover, the sequences $(t_{k,\ell})_{\ell\geq 1}$ for $k \in \mathbb{Z}$ are a.s. linearly independent, since for a suitable normalization, $(t_{k,\ell})_{k\in\mathbb{Z},\ell\geq 1}$ are iid standard complex gaussian. This ensures that the definition of U^{α} given above is meaningful. The notation U^{α} is motivated by the immediate fact that $(U^{\alpha})_{\alpha\in\mathbb{R}}$ is a flow of operators on $\mathcal{E}$, i.e. $U^0 = I_{\mathcal{E}}$ and $U^{\alpha+\beta} = U^{\alpha}U^{\beta}$ for all $\alpha, \beta \in \mathbb{R}$.

As suggested before, we expect that U^{α} is a kind of limit for $u_n^{\alpha_n}$ when n goes to infinity. Of course, these operators do not act on the same space, so we need to be more precise.

Theorem 3.3 *Almost surely, for any sequence $(s_\ell)_{\ell\geq 1}$ in $\mathcal{E}$ and for all integers $m \geq 1$,*

$$\left[u_n^{\alpha_n}((s_\ell)_{1\leq\ell\leq n})\right]_m \underset{n\to\infty}{\longrightarrow} \left[U^{\alpha}((s_\ell)_{\ell\geq 1})\right]_m,$$

where $[\,\cdot\,]_m$ denotes the mth coordinate of a vector or a sequence.

By linearity, it is sufficient to show the theorem for $(s_\ell)_{\ell\geq 1} = (g_{k,\ell})_{\ell\geq 1}$. Hence, Theorem 3.3 can be deduced from the following proposition:

Proposition 3.4 *For all $k \in \mathbb{Z}$, $\ell \geq 1$, one has almost surely:*

$$\langle u_n^{\alpha_n}(g_k[n]), e_\ell\rangle \to e^{2\pi i\alpha y_k} g_{k,\ell}$$

as $n \to \infty$, for $g_k[n] := (g_{k,\ell})_{1\leq\ell\leq n}$.

In the next section, we will give a more intrinsic way to define a flow of operators similar to $(U^{\alpha})_{\alpha\in\mathbb{R}}$. In order to make this construction, we need a more precise and stronger result than Proposition 3.4, which is given by the following two propositions:

Proposition 3.5 *Let $\epsilon > 0$. Almost surely, for all $k \in \mathbb{Z}$, we have the following:*

(1) *The euclidean norm $\|g_k[n]\|$ is equivalent to a strictly positive random variable times $\sqrt{n}$, when n goes to infinity.*
(2) $\|g_k[n] - g_k^{(n)}\| = O(n^{\frac{1}{3}+\epsilon})$.
(3) *For any $T > 0$ and $\delta \in (0, 1/6)$,*

$$\sup_{\alpha\in[-T,T]} \sup_{\alpha_n\in[n(\alpha-n^{-\delta}),n(\alpha+n^{-\delta})]} \|u_n^{\alpha_n} g_k[n] - e^{2\pi i\alpha y_k} g_k[n]\| = O(n^{\frac{1}{2}-\delta}).$$

Proposition 3.6 *Almost surely, for all $k \in \mathbb{Z}$, $\ell \geq 1$, $\alpha, \gamma \in \mathbb{R}$, and for all sequences $(\alpha_n)_{n\geq 1}$ and $(\gamma_n)_{n\geq 1}$ such that $\alpha_n/n = \alpha + o(n^{-\delta})$ and $\gamma_n/n = \gamma + o(n^{-\delta})$ for some $\delta \in [0, 1/6)$,*

$$\langle u_n^{\alpha_n}(g_k[n]) - e^{2\pi i\alpha y_k} g_k[n], u_n^{\gamma_n}(e_\ell)\rangle = o(n^{-\delta}),$$

when n goes to infinity. Moreover, for $\delta \in (0, 1/6)$, we get the uniform estimate:

$$\sup_{\substack{\alpha_n\in[n(\alpha-n^{-\delta}),n(\alpha+n^{-\delta})]\\ \gamma_n\in[n(\gamma-n^{-\delta}),n(\gamma+n^{-\delta})]}} \langle u_n^{\alpha_n}(g_k[n]) - e^{2\pi i\alpha y_k} g_k[n], u_n^{\gamma_n}(e_\ell)\rangle = O(n^{-\delta}).$$

Proof We now prove Propositions 3.5 and 3.6: it is clear that this last proposition implies Proposition 3.4 (by taking $\delta = \gamma_n = 0$).

Using (6), we get the following:

$$\|g_k[n]\|^2 = |D_k| \sum_{\ell=1}^{n} |t_{k,\ell}|^2,$$

where $(t_{k,\ell})_{\ell\geq 1}$ are up to independent and uniform random phases iid standard complex gaussian variables. By the law of large numbers, $\|g_k[n]\|^2$ is a.s. equivalent to $n|D_k|$ when n goes to infinity, which shows the first item of Proposition 3.5.

Now, the third item can be quickly deduced from the second one. Indeed, if we have the estimate $\|g_k[n] - g_k^{(n)}\| = O(n^{\frac{1}{3}+\epsilon})$ for all $\epsilon > 0$, then, since $\delta < 1/6$, we also have $\|g_k[n] - g_k^{(n)}\| = O(n^{\frac{1}{2}-\delta})$, which implies, for all $\alpha \in [-T, T]$ and $\alpha_n \in [n(\alpha - n^{-\delta}), n(\alpha + n^{-\delta})]$,

$$\begin{aligned}
&\|u_n^{\alpha_n} g_k[n] - e^{2\pi i\alpha y_k} g_k[n]\| \\
&\leq \|u_n^{\alpha_n}(g_k[n] - g_k^{(n)})\| + \|u_n^{\alpha_n} g_k^{(n)} - e^{2\pi i\alpha y_k} g_k^{(n)}\| + \|e^{2\pi i\alpha y_k}(g_k[n] - g_k^{(n)})\| \\
&= 2\|g_k[n] - g_k^{(n)}\| + |e^{i\alpha_n\theta_k^{(n)}} - e^{2\pi i\alpha y_k}|\, \|g_k^{(n)}\|.
\end{aligned}$$

Now,

$$\|g_k[n] - g_k^{(n)}\| = O(n^{\frac{1}{2}-\delta}),$$

$$\|g_k^{(n)}\| \leq \|g_k[n]\| + \|g_k[n] - g_k^{(n)}\| = O(\sqrt{n}) + O(n^{\frac{1}{2}-\delta}) = O(\sqrt{n}),$$

and

$$\begin{aligned}
|e^{i\alpha_n\theta_k^{(n)}} - e^{2\pi i\alpha y_k}| &\leq |\alpha_n\theta_k^{(n)} - 2\pi\alpha y_k| \\
&\leq |\theta_k^{(n)}||\alpha_n - \alpha n| + |\alpha||n\theta_k^{(n)} - 2\pi y_k| \\
&= O(1/n)O(n^{1-\delta}) + O(n^{-1/4}) = O(n^{-\delta}),
\end{aligned}$$

where the implied constant does not depend on α and α_n (recall that $|\alpha|$ is assumed to be uniformly bounded by T). Note that we used Theorem 2.1 (for k fixed and $\epsilon = 1/12$) at the last step of the computation. This gives

$$\|u_n^{\alpha_n} g_k[n] - e^{2\pi i \alpha y_k} g_k[n]\| = O(n^{\frac{1}{2}-\delta}),$$

uniformly with respect to α and α_n, i.e. the third item of Proposition 3.5. In order to complete the proof of this proposition, it then remains to show the second item, which needs several intermediate steps. □

Lemma 3.7 *For fixed $k \geq 1$, $\epsilon > 0$, there exists a $\mathcal{A}$-measurable random variable $M > 0$, such that, almost surely, for all $N \geq n \geq k$,*

$$\mathbb{E}[\|g_k^{(n)} - g_k[n]\|^2 \,|\mathcal{A}] \leq M\, n^{\frac{2}{3}+\epsilon},$$

and

$$\mathbb{E}[\|g_k^{(n)} - g_{k,n}^{(N)}\|^2 \,|\mathcal{A}] \leq M\,(N-n)\, n^{-\frac{1}{3}+\epsilon},$$

where $g_{k,n}^{(N)}$ denotes the vector obtained by taking the n first coordinates of $g_k^{(N)}$.

Proof One has, for $n \geq 1$, and $N \in \{n, n+1, n+2, ...\} \cup \{\infty\}$,

$$\mathbb{E}[\|g_k^{(n)} - g_{k,n}^{(N)}\|^2 \,|\mathcal{A}] = \sum_{\ell=1}^{n} \mathbb{E}[|\langle g_k^{(n)}, e_\ell\rangle - \langle g_k^{(N)}, e_\ell\rangle|^2 \,|\mathcal{A}],$$

for and $g_k^{(\infty)} := g_k = (g_{g,\ell})_{\ell \geq 1}$ and $g_{k,n}^{(\infty)} := g_k[n]$. By the martingale property satisfied by the scalar products $\langle g_k^{(m)}, e_\ell\rangle$ for $m \geq k$ (see [12] for more detail),

$$\mathbb{E}[|\langle g_k^{(n)}, e_\ell\rangle - \langle g_k^{(N)}, e_\ell\rangle|^2 \,|\mathcal{A}] = \sum_{n \leq m < N} \Delta_\ell^{(m)}$$

where

$$\Delta_\ell^{(m)} = \mathbb{E}[|\langle g_k^{(m+1)}, e_\ell\rangle - \langle g_k^{(m)}, e_\ell\rangle|^2 \,|\mathcal{A}],$$

and then by (3) and (4),

$$\begin{aligned}
\Delta_\ell^{(m)} &= |D_k^{(m)}|^2 \frac{|\lambda_k^{(m)} - \lambda_k^{(m+1)}|^2}{|\mu_k^{(m)}|^2} \sum_{1 \leq j \leq m,\, j \neq k} \frac{\cdot |\mu_j^{(m)}|^2}{|\lambda_j^{(m)} - \lambda_k^{(m+1)}|^2} \mathbb{E}\left[|\langle f_j^{(m)}, e_\ell\rangle|^2 \,|\mathcal{A}\right] \\
&\leq M_1\, m^{-3+\epsilon} \sum_{j \in J_m \setminus \{k\}} |\lambda_j^{(m)} - \lambda_k^{(m+1)}|^{-2} \mathbb{E}\left[|\langle f_j^{(m)}, e_\ell\rangle|^2 \,|\mathcal{A}\right], \qquad (7)
\end{aligned}$$

where J_m is the random set of m consecutive integers, such that $\theta_k^{(m+1)} - \pi < \theta_j^{(m)} \leq \theta_k^{(m+1)} + \pi$, and where

$$M_1 := \sup_{m \geq k, 1 \leq j \leq m, j \neq k} m^{3-\epsilon} |D_k^{(m)}|^2 \frac{|\lambda_k^{(m)} - \lambda_k^{(m+1)}|^2 |\mu_j^{(m)}|^2}{|\mu_k^{(m)}|^2}.$$

In [12], we prove the following estimate:

$$|\lambda_k^{(m+1)} - \lambda_k^{(m)}| = \theta_k^{(m+1)} |\mu_k^{(m)}|^2 (1 + O(m^{-\frac{1}{3}+\epsilon})). \tag{8}$$

Moreover, from the convergence of the renormalized eigenangles, we have $\theta_k^{(m+1)} = O(1/m)$ almost surely, and by classical tail estimates on beta random variables, $|\mu_j^{(m)}|^2$ and $|\mu_k^{(m)}|^2$ are almost surely dominated by $m^{-1+\frac{\epsilon}{2}}$. Using (8) and (5), we deduce that the $\mathcal{A}$-measurable quantity M_1 is almost surely finite.

Similarly, one has, for all $j \in J_m \backslash \{k\}$,

$$|\lambda_j^{(m)} - \lambda_k^{(m+1)}|^{-2} \leq M_2 \left(m^{2+\epsilon} \wedge \frac{m^{\frac{10}{3}+\epsilon}}{|k-j|^2} \right) \leq M_2 \frac{m^{\frac{8}{3}+\epsilon}}{|k-j|}, \tag{9}$$

where

$$M_2 := \sup_{m \geq k, j \in J_m \backslash \{k\}} |\lambda_j^{(m)} - \lambda_k^{(m+1)}|^{-2} \left(m^{2+\epsilon} \wedge \frac{m^{\frac{10}{3}+\epsilon}}{|k-j|^2} \right)^{-1}.$$

Now, M_2 is $\mathcal{A}$-measurable, and we have proven in [12] that the following estimates almost surely hold, uniformly in $m \geq k$, $j \in J_m$:

$$|\lambda_j^{(m)} - \lambda_k^{(m+1)}| \gtrsim m^{-1-\epsilon},$$

$$|\lambda_j^{(m)} - \lambda_k^{(m+1)}| \gtrsim |j-k| m^{-\frac{5}{3}-\epsilon}.$$

The estimates show that M_2 is almost surely finite. From (7) and (9), we deduce (with the change of variable $p = j - k$):

$$\Delta_\ell^{(m)} \leq M_1 M_2 m^{-\frac{1}{3}+2\epsilon} \sum_{p \in \{-m-1, -m+1, \ldots, -1, 1, \ldots m+1\}} \frac{1}{|p|} \mathbb{E}\left[|\langle f_{k+p}^{(m)}, e_\ell \rangle|^2 \, |\mathcal{A} \right].$$

Therefore,

$$\mathbb{E}[|\langle g_k^{(n)}, e_\ell\rangle - \langle g_k^{(N)}, e_\ell\rangle|^2 \,|\mathcal{A}]$$
$$\leq M_1 M_2 \sum_{n \leq m < N} m^{-\frac{1}{3}+2\epsilon} \sum_{p \in \{-m,-m+1,\dots,-1,1,\dots m\}} \frac{1}{|p|} \mathbb{E}\left[|\langle f_{k+p}^{(m)}, e_\ell\rangle|^2 \,|\mathcal{A}\right],$$

and then, summing for ℓ between 1 and n,

$$\mathbb{E}[\|g_k^{(n)} - g_{k,n}^{(N)}\|^2 \,|\mathcal{A}]$$
$$\leq M_1 M_2 \sum_{n \leq m < N} m^{-\frac{1}{3}+2\epsilon} \sum_{p \in \{-m,-m+1,\dots,-1,1,\dots m\}} \frac{1}{|p|} \mathbb{E}\left[\sum_{\ell=1}^{n} |\langle f_{k+p}^{(m)}, e_\ell\rangle|^2 \,|\mathcal{A}\right].$$

Let us first suppose that N is finite. One has

$$\sum_{\ell=1}^{n} |\langle f_{k+p}^{(m)}, e_\ell\rangle|^2 \leq \sum_{\ell=1}^{m} |\langle f_{k+p}^{(m)}, e_\ell\rangle|^2 = \|f_{k+p}^{(m)}\|^2 = 1,$$

which implies

$$\mathbb{E}[||g_k^{(n)} - g_{k,n}^{(N)}\|^2 \,|\mathcal{A}] \leq M_1\, M_2 \sum_{n \leq m < N} m^{-\frac{1}{3}+2\epsilon} \sum_{p \in \{-m,-m+1,\dots,-1,1,\dots m\}} \frac{1}{|p|}$$
$$\leq M_3 \sum_{n \leq m < N} m^{-\frac{1}{3}+3\epsilon} \leq M_3(N-m)m^{-\frac{1}{3}+3\epsilon},$$

where

$$M_3 := M_1 M_2 \sup_{m \geq 1} \left(m^{-\epsilon} \sum_{p \in \{-m,-m+1,\dots,-1,1,\dots m\}} \frac{1}{|p|} \right) < \infty$$

is $\mathcal{A}$-measurable. Hence, we have the desired bound, after changing ϵ and taking $M = M_3$. In the case where N is infinite, we proceed as follows. Consider the vector $f_j^{(m)}$ for each fixed j and m. By the invariance by conjugation of the Haar measure on $U(m)$, this eigenvector is, up to multiplication by a complex of modulus 1, a uniform vector on the unit sphere of $\mathbb{C}^m$. More precisely, if $\xi \in \mathbb{C}$ is uniform on the unit circle, and independent of $f_j^{(m)}$, then $\xi f_j^{(m)}$ is uniform on the unit sphere. One deduces, from classical estimates on beta and gamma variables, that for all $m, j, \ell, y > 0$,

$$\mathbb{P}(|\langle f_j^{(m)}, e_\ell\rangle|^2 > m^{-1}y) = O(\exp(-\kappa y))$$

for some universal constant $\kappa > 0$. We deduce

$$\begin{aligned}\mathbb{E}[|\langle f_j^{(m)}, e_\ell\rangle|^{8/\epsilon}] &= \int_0^\infty \mathbb{P}[|\langle f_j^{(m)}, e_\ell\rangle|^2 \geq \delta^{\epsilon/4}]d\delta \\ &\lesssim \int_0^\infty e^{-m\kappa\delta^{\epsilon/4}} d\delta \\ &= \int_0^\infty e^{-\kappa z^{\epsilon/4}} d(z/m^{4/\epsilon}) = O(m^{-4/\epsilon}).\end{aligned}$$

We deduce

$$\begin{aligned}\mathbb{P}\left(\mathbb{E}[|\langle f_j^{(m)}, e_\ell\rangle|^{8/\epsilon}|\mathcal{A}] \geq m^{4-\frac{4}{\epsilon}}\right) &\leq m^{\frac{4}{\epsilon}-4}\mathbb{E}\left[\mathbb{E}[|\langle f_j^{(m)}, e_\ell\rangle|^{8/\epsilon}|\mathcal{A}]\right] \\ &= m^{\frac{4}{\epsilon}-4}\mathbb{E}[|\langle f_j^{(m)}, e_\ell\rangle|^{8/\epsilon}] = O(m^{-4}).\end{aligned}$$

By the Borel–Cantelli lemma, almost surely, for all but finitely many $m \geq 1$, $1 \leq j, \ell \leq m$,

$$\mathbb{E}[|\langle f_j^{(m)}, e_\ell\rangle|^{8/\epsilon}|\mathcal{A}] \leq m^{4-\frac{4}{\epsilon}}.$$

By the Hölder inequality applied to the conditional expectation, for ϵ sufficiently small, this implies

$$\mathbb{E}[|\langle f_j^{(m)}, e_\ell\rangle|^2|\mathcal{A}] \leq \left(\mathbb{E}[|\langle f_j^{(m)}, e_\ell\rangle|^{8/\epsilon}|\mathcal{A}]\right)^{\epsilon/4} \leq m^{-1+\epsilon}.$$

Hence, for $p \in \{-m, \ldots, -1, 1, \ldots, m\}$,

$$\mathbb{E}\left[|\langle f_{k+p}^{(m)}, e_\ell\rangle|^2 \,|\mathcal{A}\right] \leq M_4 m^{-1+\epsilon},$$

where

$$M_4 := \sup_{m\geq 1, 1\leq k,\ell\leq m} m^{1-\epsilon}\mathbb{E}\left[|\langle f_k^{(m)}, e_\ell\rangle|^2 \,|\mathcal{A}\right]$$

is $\mathcal{A}$-measurable and almost surely finite. Hence,

$$\begin{aligned}&\mathbb{E}[\|g_k^{(n)} - g_k[n]\|^2 \,|\mathcal{A}] \\ &\leq M_1M_2M_4 \sum_{m\geq n} m^{-\frac{4}{3}+3\epsilon} \sum_{p\in\{-m,-m+1,\ldots,-1,1,\ldots m\}} \frac{n}{|p|},\end{aligned}$$

which is easily dominated by a finite, $\mathcal{A}$-measurable quantity, multiplied by $n^{\frac{2}{3}+4\epsilon}$.

□

We have now an L^2 bound on $\|g_k^{(n)} - g_{k,n}^{(\infty)}\|$, conditionally on $\mathcal{A}$. The next goal is to deduce an almost sure bound, by using Borel–Cantelli lemma. This cannot be made directly, since the corresponding probabilities do not decay sufficiently fast, but one can solve this problem by using subsequences.

Lemma 3.8 *For fixed $k \geq 1$, $\epsilon \in (0, 1)$, let $\nu := 1 + \lfloor 3/\epsilon \rfloor$, and for $r \geq 1$, let $n_r := k + r^{\nu}$. Then, almost surely,*

$$\|g_k^{(n_r)} - g_k[n_r]\|^2 = O(n_r^{\frac{2}{3}+\epsilon}).$$

Proof From Lemma 3.7 (applied to $\epsilon/2$ instead of ϵ), there exists M' almost surely finite and $\mathcal{A}$-measurable such that for all $r \geq 1$,

$$\begin{aligned}\mathbb{P}[\|g_k^{(n_r)} - g_k[n_r]\|^2 \geq n_r^{\frac{2}{3}+\epsilon} | \mathcal{A}] &\leq n_r^{-\frac{2}{3}-\epsilon} \mathbb{E}[\|g_k^{(n_r)} - g_k[n_r]\|^2 | \mathcal{A}] \\ &\leq M' n_r^{-\frac{2}{3}-\epsilon} n_r^{\frac{2}{3}+\frac{\epsilon}{2}} \leq M' n_r^{-\epsilon/2} \\ &= M'(k + r^{\nu})^{-\epsilon/2} \leq M'(r^{3/\epsilon})^{-\epsilon/2} = M' r^{-3/2},\end{aligned}$$

and then

$$\frac{1}{M'+1} \mathbb{E}\left[\sum_{r\geq 1} \mathbf{1}_{\|g_k^{(n_r)} - g_k[n_r]\|^2 \geq n_r^{\frac{2}{3}+\epsilon}} | \mathcal{A}\right] \leq \frac{M'}{M'+1} \sum_{r\geq 1} r^{-3/2} \leq 3.$$

By taking the expectation and using the fact that $1/(M'+1)$ is $\mathcal{A}$-measurable, one deduces

$$\mathbb{E}\left[\frac{1}{M'+1} \sum_{r\geq 1} \mathbf{1}_{\|g_k^{(n_r)} - g_k[n_r]\|^2 \geq n_r^{\frac{2}{3}+\epsilon}}\right] \leq 3,$$

and then

$$\frac{1}{M'+1} \sum_{r\geq 1} \mathbf{1}_{\|g_k^{(n_r)} - g_k[n_r]\|^2 \geq n_r^{\frac{2}{3}+\epsilon}} < \infty$$

almost surely. Hence, almost surely, $\|g_k^{(n_r)} - g_k[n_r]\|^2 \leq n_r^{\frac{2}{3}+\epsilon}$ for all but finitely many $r \geq 1$. □

The next lemma gives a way to go from a given subsequence to a less sparse subsequence. We will say that a value $\delta \in (0, 1]$ is *good* if the conclusion of Lemma 3.8 remains valid after replacing the sequence $(k+r^{\nu})_{r\geq 1}$ by $(k+\lfloor r^{1/\delta} \rfloor)_{r\geq 1}$, the brackets denoting the integer part: from Lemma 3.8, we know that $1/\nu$ is good.

Lemma 3.9 *If $\delta \in [1/\nu, (\nu-1)/\nu]$ is good, then $\delta + 1/\nu$ is good.*

Proof For $r \geq 1$, let $n_r := k + \lfloor r^{1/(\delta+1/\nu)} \rfloor$ and let N_r the smallest element of the sequence $(k + \lfloor s^{1/\delta} \rfloor)_{s \geq 1}$, such that $N_r \geq n_r$. The sequence $(N_r)_{r\geq 1}$, as $(n_r)_{r\geq 1}$, tends to infinity with r: moreover, it is a subsequence of $(k + \lfloor s^{1/\delta} \rfloor)_{s\geq 1}$ (but some terms of this sequence may appear several times in $(N_r)_{r\geq 1}$). By assumption, one has almost surely:

$$\|g_k^{(N_r)} - g_k[N_r]\|^2 = O(N_r^{\frac{2}{3}+\epsilon}).$$

Now, it is easy to check that $N_r = O(n_r)$ (with a constant depending only on δ). Hence, by restricting the vectors to their n_r first coordinates, one deduces, almost surely,

$$\|g_{k,n_r}^{(N_r)} - g_k[n_r]\|^2 \leq \|g_k^{(N_r)} - g_k[N_r]\|^2 = O(n_r^{\frac{2}{3}+\epsilon}). \tag{10}$$

On the other hand, the distance between two consecutive terms of the sequence $(k + \lfloor s^{1/\delta} \rfloor)_{s\geq 1}$ satisfies the following:

$$\begin{aligned}(k + \lfloor (s+1)^{1/\delta} \rfloor) - (k + \lfloor s^{1/\delta} \rfloor) &= (s+1)^{1/\delta} - s^{1/\delta} + O(1) \\ &\lesssim s^{1/\delta - 1} \lesssim (k + \lfloor s^{1/\delta} \rfloor)^{1-\delta},\end{aligned}$$

which implies

$$N_r - n_r = O(n_r^{1-\delta}),$$

where the implied constant depends only on k and δ. One deduces, by Lemma 3.7:

$$\mathbb{E}[\|g_k^{(n_r)} - g_{k,n_r}^{(N_r)}\|^2 \,|\mathcal{A}] \leq M' \,(N_r - n_r)\,(n_r)^{-\frac{1}{3}+\frac{\epsilon}{2}} \lesssim M' n_r^{\frac{2}{3}+\frac{\epsilon}{2}-\delta},$$

and then

$$\mathbb{P}[\|g_k^{(n_r)} - g_{k,n_r}^{(N_r)}\|^2 \geq n_r^{\frac{2}{3}+\epsilon} \,|\mathcal{A}] \lesssim M' n_r^{-\frac{\epsilon}{2}-\delta} \lesssim M' r^{(-\frac{\epsilon}{2}-\delta)/\left(\delta+\frac{1}{\nu}\right)},$$

where the implied constant depends only on k and δ. Since the exponent of r is strictly smaller than -1, one deduces, by using Borel–Cantelli lemma similarly as in the proof of Lemma 3.8, that almost surely,

$$\|g_k^{(n_r)} - g_{k,n_r}^{(N_r)}\|^2 = O(n_r^{\frac{2}{3}+\epsilon}). \tag{11}$$

Combining (10) and (11) gives the desired result. □

By applying Lemma 3.8 and ($\nu - 1$ times) Lemma 3.9, one deduces that 1 is good, which gives the second item of Proposition 3.5 for $k \geq 1$: the situation for $k \leq 0$ is similar.

Let us now prove Proposition 3.6. By the triangle inequality, we get

$$|\langle u_n^{\alpha_n}(g_k[n]) - e^{2\pi i\alpha y_k} g_k[n], u_n^{\gamma_n}(e_\ell)\rangle|$$
$$\leq |\langle u_n^{\alpha_n}(g_k[n] - g_k^{(n)}), u_n^{\gamma_n}(e_\ell)\rangle| + |\langle u_n^{\alpha_n} g_k^{(n)} - e^{2\pi i\alpha y_k} g_k^{(n)}, u_n^{\gamma_n}(e_\ell)\rangle|$$
$$+ |\langle e^{2\pi i\alpha y_k}(g_k^{(n)} - g_k[n]), u_n^{\gamma_n}(e_\ell)\rangle| \leq |\langle g_k^{(n)} - g_k[n], u_n^{\gamma_n - \alpha_n}(e_\ell)\rangle|$$
$$+ |e^{i\alpha_n\theta_k^{(n)}} - e^{2\pi i\alpha y_k}||\langle g_k^{(n)}, u_n^{\gamma_n}(e_\ell)\rangle| + |\langle g_k^{(n)} - g_k[n], u_n^{\gamma_n}(e_\ell)\rangle|.$$

To prove the first part or the proposition, since the sequence $(\alpha_n - \gamma_n)_{n\geq 1}$ satisfies exactly the same assumptions as $(\gamma_n)_{n\geq 1}$ (replacing γ by $\gamma - \alpha$), it is sufficient to prove the almost sure estimates:

$$|e^{i\alpha_n\theta_k^{(n)}} - e^{2\pi i\alpha y_k}||\langle g_k^{(n)}, u_n^{\gamma_n}(e_\ell)\rangle| = o(n^{-\delta}) \tag{12}$$

and

$$|\langle g_k^{(n)} - g_k[n], u_n^{\gamma_n}(e_\ell)\rangle| = o(n^{-\delta}). \tag{13}$$

For the uniform part of the Proposition 3.6, it is sufficient to prove the same estimates, with $o(n^{-\delta})$ replaced by $O(n^{-\delta})$, this bound being uniform with respect to $\alpha_n \in [n(\alpha - n^{-\delta}), n(\alpha + n^{-\delta})]$, and $\gamma_n \in [n(\gamma - 2n^{-\delta}), n(\gamma + 2n^{-\delta})]$ (the factor 2 comes from the term where γ_n plays the role of $\gamma_n - \alpha_n$). Now, under the assumption of the first part of the proposition, (12) is a consequence of the two following estimates:

$$|e^{i\alpha_n\theta_k^{(n)}} - e^{2\pi i\alpha y_k}| \leq |\alpha_n\theta_k^{(n)} - 2\pi\alpha y_k|$$
$$\leq |\theta_k^{(n)}||\alpha_n - \alpha n| + |\alpha||n\theta_k^{(n)} - 2\pi y_k|$$
$$= O(1/n)o(n^{1-\delta}) + O(n^{-1/4}) = o(n^{-\delta}),$$

and

$$|\langle g_k^{(n)}, u_n^{\gamma_n}(e_\ell)\rangle| = |\langle u_n^{-\gamma_n} g_k^{(n)}, e_\ell\rangle| = |\langle e^{-i\gamma_n\theta_k^{(n)}} g_k^{(n)}, e_\ell\rangle|$$
$$= |\langle g_k^{(n)}, e_\ell\rangle| = O(1),$$

the last estimate coming from the almost sure convergence of $\langle g_k^{(n)}, e_\ell\rangle$ towards $g_{k,\ell}$. This computation is also available (after replacing small o by big O) for the second part of the proposition, since all the estimates are easily checked to be uniform with respect to $\alpha_n \in [n(\alpha - n^{-\delta}), n(\alpha + n^{-\delta})]$ and $\gamma_n \in [n(\gamma - 2n^{-\delta}), n(\gamma + 2n^{-\delta})]$.

It remains to prove (13). From the second item of Proposition 3.5 and the fact that $\delta < 1/6$ we have the estimate:

$$\|g_k^{(n)} - g_k[n]\| = o(n^{\frac{1}{2}-\delta'})$$

for some $\delta' > \delta$. Hence, in order to complete the proof of Proposition 3.6, it is sufficient to show the following property of delocalization:

$$\sup_{\gamma_n \in [n(\gamma-2n^{-\delta}), n(\gamma+2n^{-\delta})]} \frac{|\langle g_k^{(n)} - g_k[n], u_n^{\gamma_n}(e_\ell)\rangle|}{\|g_k^{(n)} - g_k[n]\|} = O(n^{-\frac{1}{2}+\delta'-\delta}). \tag{14}$$

This will be a consequence of the following lemma:

Lemma 3.10 *For all $n \geq 1$, the quotient*

$$\frac{|\langle g_k^{(n)} - g_k[n], u_n^{\gamma_n}(e_\ell)\rangle|^2}{\|g_k^{(n)} - g_k[n]\|^2}$$

is a beta random variable of parameters 1 and $n-1$, i.e. it has the same law as $|x_1|^2$, where $(x_1, \ldots, x_n)$ is a uniform vector on the complex sphere S^n.

Proof Let $m \geq 1$, let σ be a random matrix in $U(m)$, independent of $(u_n)_{n\geq 1}$ and following the Haar measure. Let us define $(u'_n)_{n\geq 1}$ as the unique virtual isometry such that for all $n \geq m$,

$$u'_n = \begin{pmatrix} \sigma & 0 \\ 0 & I_{n-m} \end{pmatrix} u_n \begin{pmatrix} \sigma & 0 \\ 0 & I_{n-m} \end{pmatrix}^{-1}.$$

The invariance by conjugation of the Haar measure on the space of virtual isometries implies that $(u'_n)_{n\geq 1}$ has the same law as $(u_n)_{n\geq 1}$. Let $(g_k^{(n)'})_{n\geq 1}$ be the sequence of eigenvectors constructed from $(u'_n)_{n\geq 1}$ in the same way as $(g_k^{(n)})_{n\geq 1}$ is constructed from $(u_n)_{n\geq 1}$. Since $(u'_n)_{n\geq 1}$ has the same law as $(u_n)_{n\geq 1}$, one deduces that almost surely, each coordinate of $g_k^{(n)'}$ with a given index converges to a limit when n goes to infinity: let g'_k be the corresponding limiting sequence, which is the analog of g_k when $(u_n)_{n\geq 1}$ is replaced by $(u'_n)_{n\geq 1}$. One knows that for $n \geq m$,

$$\tilde{g}_k^{(n)} := \begin{pmatrix} \sigma & 0 \\ 0 & I_{n-m} \end{pmatrix} g_k^{(n)}$$

is an eigenvector of $(u'_n)_{n\geq 1}$, corresponding to the eigenvalue $\lambda_k^{(n)}$ (recall that for $n \geq m$, u_n and u'_n have the same eigenvalues). Hence, almost surely, there exists

$\kappa_n \in \mathbb{C}^*$ such that $g_k^{(n)'} = \kappa_n \tilde{g}_k^{(n)}$. Moreover, from (2), and from the fact that $g_k^{(n)}$ is orthogonal to $f_j^{(n)}$ for $j \neq k$, one obtains that

$$\langle g_{k,n}^{(n+1)} - g_k^{(n)}, g_k^{(n)} \rangle = \langle g_{k,n}^{(n+1)'} - g_k^{(n)'}, g_k^{(n)'} \rangle = 0,$$

and then

$$\langle \kappa_{n+1} \tilde{g}_{k,n}^{(n+1)} - \kappa_n \tilde{g}_k^{(n)}, \kappa_n \tilde{g}_k^{(n)} \rangle = 0.$$

Applying $\mathrm{diag}(\sigma^{-1}, I_{n-m})$ to the vectors in the scalar product and dividing by $|\kappa_n|^2$ give

$$\left\langle \frac{\kappa_{n+1}\, g_{k,n}^{(n+1)}}{\kappa_n} - g_k^{(n)}, g_k^{(n)} \right\rangle = 0,$$

and then

$$\begin{aligned}\left(\frac{\kappa_{n+1}}{\kappa_n} - 1\right) & \langle g_{k,n}^{(n+1)}, g_k^{(n)} \rangle \\ &= \left\langle \frac{\kappa_{n+1}\, g_{k,n}^{(n+1)}}{\kappa_n} - g_k^{(n)}, g_k^{(n)} \right\rangle - \langle g_{k,n}^{(n+1)} - g_k^{(n)}, g_k^{(n)} \rangle = 0,\end{aligned}$$

which implies $\kappa_{n+1} = \kappa_n$, since

$$\langle g_{k,n}^{(n+1)}, g_k^{(n)} \rangle = \langle g_{k,n}^{(n+1)} - g_k^{(n)}, g_k^{(n)} \rangle + \|g_k^{(n)}\|^2 = \|g_k^{(n)}\|^2 > 0.$$

Hence, one has $g_k^{(n)'} = \kappa_m \tilde{g}_k^{(n)}$ for all $n \geq m$: by taking the limit for $n \to \infty$, one deduces that $g_k' = \kappa_m \tilde{g}_k$, where $\tilde{g}_k$ is the infinite sequence obtained from g_k by applying σ to the vector formed by the m first coordinates, and by letting the coordinates fixed for the indices strictly larger than m. One deduces the following (with obvious notation):

$$\begin{aligned}\langle g_k^{(m)'} - g_k'[m], (u_m')^{\gamma_m}(e_{\ell_0}) \rangle &= \kappa_m \langle \tilde{g}_k^{(m)} - \tilde{g}_k[m], (u_m')^{\gamma_m}(e_{\ell_0}) \rangle \\ &= \kappa_m \langle \sigma(g_k^{(m)}) - \sigma(g_k[m]), \sigma(u_m)^{\gamma_m} \sigma^{-1}(e_{\ell_0}) \rangle \\ &= \kappa_m \langle g_k^{(m)} - g_k[m], (u_m)^{\gamma_m} \sigma^{-1}(e_{\ell_0}) \rangle \\ &= \kappa_m \langle (u_m)^{-\gamma_m}(g_k^{(m)} - g_k[m]), \sigma^{-1}(e_{\ell_0}) \rangle.\end{aligned}$$

Similarly,

$$\|g_k^{(m)\prime} - g_k'[m]\|^2 = |\kappa_m|^2 \, \|\tilde{g}_k^{(m)} - \tilde{g}_k[m]\|^2 = |\kappa_m|^2 \, \|\sigma(g_k^{(m)}) - \sigma(g_k[m])\|^2$$
$$= |\kappa_m|^2 \, \|g_k^{(m)} - g_k[m]\|^2 = |\kappa_m|^2 \, \|(u_m)^{-\gamma_m}(g_k^{(m)} - g_k[m])\|^2.$$

Hence,

$$\frac{|\langle g_k^{(m)\prime} - g_k'[m], (u_m')^{\gamma_m}(e_{\ell_0})\rangle|^2}{\|g_k^{(m)\prime} - g_k'[m]\|^2} = \frac{|\langle x, y\rangle|^2}{\|x\|^2}, \tag{15}$$

where $x = (u_m)^{-\gamma_m}(g_k^{(m)} - g_k[m])$ and $y = \sigma^{-1}(e_{\ell_0})$ is independent of x (since σ is independent of $(u_n)_{n\geq 1}$) and uniform on the unit sphere of $\mathbb{C}^m$ (since σ is uniform on $U(m)$). One deduces that the left-hand side of (15) is a beta random variable with parameters 1 and $m-1$. Since $(u_n')_{n\geq 1}$ and $(u_n)_{n\geq 1}$ have the same distribution, one can remove the primes from this left-hand side, which gives the announced result. □

Now, applying Lemma 3.10, we get

$$\mathbb{P}\left(\sup_{\gamma_n \in [n(\gamma - 2n^{-\delta}), n(\gamma+2n^{-\delta})]} \frac{|\langle g_k^{(n)} - g_k[n], u_n^{\gamma_n}(e_\ell)\rangle|}{\|g_k^{(n)} - g_k[n]\|} \geq n^{-\frac{1}{2}+\delta'-\delta}\right)$$
$$\leq (1 + 4n^{1-\delta})\mathbb{P}[\beta(1, n-1) \geq n^{-1+2(\delta'-\delta)}].$$

By classical estimates on beta random variables and Borel–Cantelli lemma, we deduce the estimate (14), which completes the proof of Proposition 3.6. □

4 An Inner Product on the Domain of the Operators

We return now to the original space $\mathcal{E}$ consisting of the finite linear combinations of the eigenvectors of the flow. Most of the infinite-dimensional operators that are considered in the literature are defined on Hilbert spaces. Here, the space $\mathcal{E}$ does not have a Hilbert structure, since infinite linear combinations of the basis vectors are not permitted. On the other hand, it is possible to construct an inner product on $\mathcal{E}$. Since eigenspaces of Hermitian and unitary operators are pairwise orthogonal, it is natural to define our scalar product in such a way that the sequences $(t_{k,\ell})_{\ell\geq 1}$, $k \in \mathbb{Z}$ are orthogonal. If we also suppose that these sequences have norm 1, we then define a scalar product on $\mathcal{E}$ as follows:

$$\langle w, w'\rangle = \sum_{k\in\mathbb{Z}} \lambda_k \overline{\lambda_k'}$$

for

$$w_\ell = \sum_{k \in \mathbb{Z}} \lambda_k t_{k,\ell}, \; w'_\ell = \sum_{k \in \mathbb{Z}} \lambda'_k t_{k,\ell},$$

where $(\lambda_k)_{k\in\mathbb{Z}}$ and $(\lambda'_k)_{k\in\mathbb{Z}}$ are sequences containing finitely many non-zero terms. This definition does not depend on the phases of the vectors $(t_{k,\ell})_{\ell\geq 1}$, $k \in \mathbb{Z}$, which are chosen: indeed, if $t_{k,\ell}$ is multiplied by $z_k \in \mathbb{U}$, for all $k \in \mathbb{Z}$, $\ell \geq 1$, then λ_k and λ'_k are both multiplied by z_k^{-1}, and $\lambda_k \overline{\lambda'_k}$ is not changed. Hence, one can choose the phases in such a way that the variables $(t_{k,\ell})_{\ell\geq 1, k\in\mathbb{Z}}$ are iid, complex gaussian. In particular, the vectors $(t_{k,\ell})_{\ell\geq 1}$, $k \in \mathbb{Z}$, are linearly independent, and then the sequences $(\lambda_k)_{k\in\mathbb{Z}}$ and $(\lambda'_k)_{k\in\mathbb{Z}}$ are uniquely determined by $w, w' \in \mathcal{E}$.

In fact, the scalar product $\langle w, w'\rangle$ we have defined can almost surely be written as a function of the coordinates of w and w', without referring to the sequences $(t_{k,\ell})_{\ell\geq 1}$, $k \in \mathbb{Z}$:

Proposition 4.1 *Let $(w_\ell)_{\ell\geq 1}$ and $(w'_\ell)_{\ell\geq 1}$ be two vectors in $\mathcal{E}$. Then*

$$\langle w, w'\rangle = \lim_{n\to\infty} \frac{1}{n} \sum_{\ell=1}^{n} w_\ell \overline{w'_\ell} = \lim_{s\to 1, s<1} (1-s) \sum_{\ell=1}^{\infty} s^{\ell-1} w_\ell \overline{w'_\ell}.$$

Proof By linearity, it is sufficient to show the convergence of the two limits and the equality almost surely for $w = (t_{k,\ell})_{\ell\geq 1}$ and $w' = (t_{k',\ell})_{\ell\geq 1}$ for every $k, k' \in \mathbb{Z}$. The first equality can then be written as follows:

$$\frac{1}{n} \sum_{\ell=1}^{n} X_{k,k',\ell} \underset{n\to\infty}{\longrightarrow} 0,$$

where $X_{k,k',\ell} = t_{k,\ell}\overline{t_{k',\ell}} - \mathbf{1}_{k=k'}$. This is now a consequence of the law of large numbers, since the variables $(X_{k,k',\ell})_{\ell\geq 1}$ are iid, integrable and centered. It remains to prove that

$$(1-s) \sum_{\ell=1}^{\infty} s^{\ell-1} X_{k,k',\ell} \underset{s\to 1, s<1}{\longrightarrow} 0.$$

The sum written here is bounded in L^1, and then a.s. finite for all $s \in (0,1)$: moreover, it is equal to

$$(1-s) \sum_{n=1}^{\infty} (s^{n-1} - s^n) \left(\sum_{\ell=1}^{n} X_{k,k',\ell} \right).$$

For any $\epsilon > 0$, there exists $n_0 \geq 1$ such that for any $n \geq n_0$,

$$\left|\sum_{\ell=1}^{n} X_{k,k'\ell}\right| \leq \epsilon n.$$

Hence,

$$\left|(1-s)\sum_{\ell=1}^{\infty} s^{\ell-1} X_{k,k',\ell}\right| \leq (1-s)\left(\sum_{n=1}^{n_0}(s^{n-1}-s^n)\left|\sum_{\ell=1}^{n} X_{k,k',\ell}\right| + \epsilon \sum_{n=n_0+1}^{\infty} n(s^{n-1}-s^n)\right).$$

The right-hand side of this inequality tends to ϵ, hence

$$\limsup_{s\to 1, s<1}\left|(1-s)\sum_{\ell=1}^{\infty} s^{\ell-1} X_{k,k',\ell}\right| \leq \epsilon,$$

and we are done by sending $\epsilon \to 0$. □

The space $\mathcal{E}$ that we have constructed is not equipped with a complete metric topology. One could attempt to construct a completion $\overline{\mathcal{E}}$ (along with an implicit topology) as the family of formal series:

$$\sum_{k\in\mathbb{Z}} \lambda_k\,(t_{k,\ell})_{\ell\geq 1},$$

where $(\lambda_k)_{k\in\mathbb{Z}}$ is in $\ell^2(\mathbb{Z})$. Unfortunately, it is not true that all such sequences in $\overline{\mathcal{E}}$ would converge to sequences of complex numbers. For example, if λ_k is chosen in such a way that $|\lambda_k| = 1/(1+|k|)$ and $\lambda_k t_{k,1}$ is a nonnegative real number, then the first coordinate would be

$$\sum_{k\in\mathbb{Z}} \lambda_k t_{k,1} = \sum_{k\in\mathbb{Z}} |\lambda_k t_{k,1}| = \sum_{k\in\mathbb{Z}} \frac{|t_{k,1}|}{1+|k|},$$

which is almost surely infinite, since by dominated convergence,

$$\begin{aligned}
\mathbb{E}\left[e^{-\sum_{k\in\mathbb{Z}}\frac{|t_{k,1}|}{1+|k|}}\right] &= \lim_{n\to\infty}\mathbb{E}\left[e^{-\sum_{|k|\leq n}\frac{|t_{k,1}|}{1+|k|}}\right]\\
&= \prod_{k\in\mathbb{Z}}\mathbb{E}\left[e^{-\frac{|t_{k,1}|}{1+|k|}}\right]\\
&\leq \prod_{k\in\mathbb{Z}}\left[\mathbb{P}(|t_{1,1}|\leq 1) + e^{-1/(1+|k|)}\mathbb{P}(|t_{1,1}|>1)\right]\\
&\leq \prod_{k\neq 0}[1-c|k|^{-1}+O(k^{-2})]\\
&= 0.
\end{aligned}$$

We can avoid this problem by restricting, for $\delta > 0$, to the subspace $\mathcal{E}_\delta$ of $\overline{\mathcal{E}}$ given by combinations (λ_k) such that

$$\sum_{k \in \mathbb{Z}} (1 + |k|^{1+\delta}) |\lambda_k|^2 < \infty.$$

Indeed, under this assumption, for all $\ell \geq 1$, by Cauchy–Schwarz

$$\sum_{k \in \mathbb{Z}} |\lambda_k t_{k,\ell}| \leq \left(\sum_{k \in \mathbb{Z}} (1 + |k|^{1+\delta}) |\lambda_k|^2 \right)^{1/2} \left(\sum_{k \in \mathbb{Z}} \frac{|t_{k,\ell}|^2}{1 + |k|^{1+\delta}} \right)^{1/2}. \tag{16}$$

The first factor is finite from the definition of $\mathcal{E}_\delta$, and the second factor is almost surely finite, since

$$\mathbb{E} \left[\sum_{k \in \mathbb{Z}} \frac{|t_{k,\ell}|^2}{1 + |k|^{1+\delta}} \right] = \sum_{k \in \mathbb{Z}} \frac{1}{1 + |k|^{1+\delta}} < \infty.$$

One then has the following:

Proposition 4.2 *Let w and w' be two sequences in $\mathcal{E}_\delta$, such that*

$$w_\ell = \sum_{k \in \mathbb{Z}} \lambda_k t_{k,\ell}, \; w'_\ell = \sum_{k \in \mathbb{Z}} \lambda'_k t_{k,\ell},$$

where

$$\sum_{k \in \mathbb{Z}} (1 + |k|^{1+\delta}) (|\lambda_k|^2 + |\lambda'_k|^2) < \infty. \tag{17}$$

Then, for

$$\langle w, w' \rangle := \sum_{k \in \mathbb{Z}} \lambda_k \overline{\lambda'_k}, \tag{18}$$

the conclusion of Proposition 4.1 is satisfied.

Remark 4.3 For $w, w' \in \mathcal{E}_\delta$, it is a priori not obvious that the coordinates $(\lambda_k)_{k \in \mathbb{Z}}$ and $(\lambda'_k)_{k \in \mathbb{Z}}$ are uniquely determined, and then the definition given by the formula (18) is a priori ambiguous. However, the present and the previous propositions show that for all $w \in \mathcal{E}_\delta$, one has

$$\lambda_k = \langle w, (t_{k,\ell})_{\ell \geq 1} \rangle = \lim_{n \to \infty} \frac{1}{n} \sum_{\ell=1}^{n} w_\ell \overline{t_{k,\ell}},$$

which implies that $(\lambda_k)_{k\in\mathbb{Z}}$ is uniquely determined by w. Notice also that the convergence of the series $\sum_{k\in\mathbb{Z}} \lambda_k \overline{\lambda'_k}$ is an immediate consequence of the assumption (17).

Proof Let us first show that almost surely, for $w_\ell = \sum_{k\in\mathbb{Z}} \lambda_k t_{k,\ell}$ and

$$\|\lambda\|_\delta^2 := \sum_{k\in\mathbb{Z}} (1+|k|^{1+\delta})|\lambda_k|^2 < \infty,$$

one has

$$\limsup_{n\to\infty} \frac{1}{n} \sum_{\ell=1}^{n} |w_\ell|^2 \le C_\delta \|\lambda\|_\delta^2, \tag{19}$$

where $C_\delta > 0$ depends only on δ. Indeed, by (16), one has

$$|w_\ell|^2 \le \|\lambda\|_\delta^2 \mathcal{W}_\ell,$$

where

$$\mathcal{W}_\ell = \sum_{k\in\mathbb{Z}} \frac{|t_{k,\ell}|^2}{1+|k|^{1+\delta}}.$$

Now, the variables $(\mathcal{W}_\ell)_{\ell\ge 1}$ are iid, positive, with expectation:

$$C_\delta = \sum_{k\in\mathbb{Z}} \frac{1}{1+|k|^{1+\delta}} < \infty.$$

By the law of large numbers, one deduces that almost surely, (19) holds for all $(\lambda_k)_{k\in\mathbb{Z}}$ such that $\|\lambda\|_\delta^2 < \infty$.

By Cauchy–Schwarz, for $w_\ell = \sum_{k\in\mathbb{Z}} \lambda_k t_{k,\ell}$ and $w'_\ell = \sum_{k\in\mathbb{Z}} \lambda'_k t_{k,\ell}$, one deduces

$$\limsup_{n\to\infty} \frac{1}{n} \left| \sum_{\ell=1}^{n} w_\ell \overline{w'_\ell} \right| \le C_\delta \|\lambda\|_\delta \|\lambda'\|_\delta.$$

Now, for $K \ge 1$, let use define

$$w_{\ell,K} = \sum_{|k|\le K} \lambda_k t_{k,\ell}, \quad w'_{\ell,K} = \sum_{|k|\le K} \lambda'_k t_{k,\ell},$$

note that the corresponding sequences are in $\mathcal{E}$. One has

$$\frac{1}{n}\sum_{\ell=1}^{n} w_\ell \overline{w'_\ell} = \frac{1}{n}\sum_{\ell=1}^{n} w_{\ell,K}\overline{w'_{\ell,K}} + \frac{1}{n}\sum_{\ell=1}^{n}(w_\ell - w_{\ell,K})\overline{w'_{\ell,K}} + \frac{1}{n}\sum_{\ell=1}^{n} w_\ell(\overline{w'_\ell} - \overline{w'_{\ell,K}}).$$

By Proposition 4.1, the first mean tends to

$$\langle (w_{\ell,K})_{\ell\geq 1}, (w'_{\ell,K})_{\ell\geq 1}\rangle = \sum_{|k|\leq K} \lambda_k \overline{\lambda'_k}$$

when n goes to infinity. The second mean is bounded by

$$C_\delta \left(\sum_{|k|>K} (1+|k|^{1+\delta})|\lambda_k|^2\right)^{1/2} \left(\sum_{|k|\leq K} (1+|k|^{1+\delta})|\lambda'_k|^2\right)^{1/2},$$

and the third mean is bounded by

$$C_\delta \|\lambda\|_\delta \left(\sum_{|k|>K} (1+|k|^{1+\delta})|\lambda'_k|^2\right)^{1/2}.$$

We deduce

$$\limsup_{n\to\infty} \left| \frac{1}{n}\sum_{\ell=1}^{n} w_\ell \overline{w'_\ell} - \sum_{k\in\mathbb{Z}} \lambda_k \overline{\lambda'_k} \right| \leq \sum_{|k|>K} \lambda_k \overline{\lambda'_k} + C_\delta \|\lambda\|_\delta \left(\sum_{|k|>K} (1+|k|^{1+\delta})|\lambda'_k|^2\right)^{1/2}$$
$$+ C_\delta \|\lambda'\|_\delta \left(\sum_{|k|>K} (1+|k|^{1+\delta})|\lambda_k|^2\right)^{1/2}.$$

Letting $K \longrightarrow \infty$ gives the first equality in Proposition 4.1. The second equality is proven if we show that for $s \in (0,1)$, and $X_\ell = w_\ell \overline{w'_\ell} - \langle w, w'\rangle$,

$$(1-s)\sum_{\ell=1}^{\infty} s^{\ell-1} X_\ell \tag{20}$$

is convergent and tends to 0 when s goes to 1. For $N \geq 1$,

$$(1-s)\sum_{\ell=1}^{N} s^{\ell-1} X_\ell = (1-s)\sum_{n=1}^{N}(s^{n-1} - s^n)\left(\sum_{\ell=1}^{n} X_\ell\right) + (1-s)s^N \sum_{\ell=1}^{N} X_\ell.$$

Since we know that

$$\sum_{\ell=1}^{n} X_\ell = o(n),$$

the series (20) converges to the sum of the series:

$$(1-s)\sum_{n=1}^{\infty}(s^{n-1}-s^n)\left(\sum_{\ell=1}^{n} X_\ell\right),$$

which is absolutely convergent. We can then show that this sum tends to zero when s goes to 1, in the same way as in the proof of Proposition 4.1. □

The scalar product we have defined on $\mathcal{E}$, and then in $\mathcal{E}_\delta$, can be compared with the following situation. Let $(B_t^{(k)})_{t\in[0,1]}$, $k \in \mathbb{Z}$, be independent Brownian motions. If $(\alpha_k)_{k\in\mathbb{Z}}$ is a family of real numbers, such that $\alpha_k = 0$ for all but finitely many indices $k \in \mathbb{Z}$, then one can consider the stochastic process:

$$\left(B_t^{(\alpha_k)_{k\in\mathbb{Z}}} := \sum_{k\in\mathbb{Z}} \alpha_k B_t^{(k)}\right)_{t\in[0,1]}.$$

For two sequences $(\alpha_k)_{k\in\mathbb{Z}}$ and $(\beta_k)_{k\in\mathbb{Z}}$ containing finitely non-zero terms, the quadratic covariation of $B^{(\alpha_k)_{k\in\mathbb{Z}}}$ and $B^{(\beta_k)_{k\in\mathbb{Z}}}$ is given by

$$\langle B^{(\alpha_k)_{k\in\mathbb{Z}}}, B^{(\beta_k)_{k\in\mathbb{Z}}}\rangle = \sum_{k\in\mathbb{Z}} \alpha_k \beta_k.$$

Its defines a scalar product on the vector space of stochastic processes of the form $(B_t^{(\alpha_k)_{k\in\mathbb{Z}}})_{t\in[0,1]}$.

Let us now go back to the vector space $\mathcal{E}_\delta$. This space contains some infinite sequences of complex numbers. It is natural to ask if it is possible to embed $\mathcal{E}_\delta$ into a space that has a richer structure. An example is obtained by considering the space of analytic functions on the open unit disc. Indeed, it is possible to identify the sequence $w = (w_\ell)_{\ell\geq 1}$ with the function:

$$F(w) \;:\; z \mapsto \sum_{\ell\geq 1} w_\ell z^{\ell-1}.$$

The series for F converges absolutely on the unit disc, since we have

$$|F(w)| = |\sum_{\ell\geq 1} w_\ell z^{\ell-1}| \leq \left(\sum_{\ell\geq 1} |w_\ell|^2 |z|^{\ell-1}\right)^{1/2} \left(\sum_{\ell\geq 1} |z|^{\ell-1}\right)^{1/2} < \infty$$

by Cauchy–Schwarz inequality and the proof of Proposition 4.2. It is then possible to express the scalar product $\langle w, w' \rangle$ on $\mathcal{E}_\delta$ in terms of integrals involving the holomorphic functions $F(w)$ and $F(w')$, as follows:

Proposition 4.4 *For all* $w, w' \in \mathcal{E}_\delta$,

$$\langle w, w' \rangle = 2 \lim_{s \to 1, s<1} (1-s) \int_0^{2\pi} F(w)(se^{i\theta}) \overline{F(w')(se^{i\theta})} \frac{d\theta}{2\pi}.$$

Proof We expand the integral:

$$\int_0^{2\pi} F(w)(se^{i\theta}) \overline{F(w')(se^{i\theta})} \frac{d\theta}{2\pi} = \int_0^{2\pi} \left(\sum_{\ell=1}^\infty w_\ell s^{\ell-1} e^{i\theta(\ell-1)} \right) \left(\sum_{\ell'=1}^\infty \overline{w_{\ell'}} s^{\ell'-1} e^{-i\theta(\ell'-1)} \right) \frac{d\theta}{2\pi}.$$

Since

$$\sum_{\ell=1}^\infty (|w_\ell| + |w'_\ell|) s^{\ell-1} < \infty,$$

we can apply Fubini's theorem, which gives

$$\int_0^{2\pi} F(w)(se^{i\theta}) \overline{F(w')(se^{i\theta})} \frac{d\theta}{2\pi} = \sum_{\ell=1}^\infty s^{2(\ell-1)} w_\ell \overline{w'_\ell}.$$

By Proposition 4.2, one deduces

$$(1-s^2) \int_0^{2\pi} F(w)(se^{i\theta}) \overline{F(w')(se^{i\theta})} \frac{d\theta}{2\pi} \underset{s \to 1, s<1}{\longrightarrow} \langle w, w' \rangle,$$

which proves the proposition, since $2(1-s)/(1-s^2) = 2/(1+s)$ tends to 1 when s goes to 1. □

The flow $(U^\alpha)_{\alpha \in \mathbb{R}}$ can be naturally extended to the space $\mathcal{E}_\delta$, by setting

$$U^\alpha \left(\sum_{k \in \mathbb{Z}} \lambda_k t_{k,\ell} \right) = \sum_{k \in \mathbb{Z}} e^{2i\pi\alpha y_k} \lambda_k t_{k,\ell}.$$

Note that for all $\alpha \in \mathbb{R}$, the extension of U^α preserves both the norm

$$w \mapsto \langle w, w \rangle^{1/2} = \left(\sum_{k \in \mathbb{Z}} |\lambda_k|^2 \right)^{1/2},$$

and the norm

$$w \mapsto \left(\sum_{k\in\mathbb{Z}}(1+|k|^{1+\delta})|\lambda_k|^2\right)^{1/2}$$

defining the space $\mathcal{E}_\delta$.

5 A More Intrinsic Definition of the Limiting Flow of Operators

The random space $\mathcal{E}$ and the flow $(U^\alpha)_{\alpha\in\mathbb{R}}$ of random operators on $\mathcal{E}$ defined previously have the disadvantage of involving explicitly the limiting eigenvectors of the virtual isometry $(u_n)_{n\geq 1}$. This makes the definition artificial. In this section, we construct a random space of sequences that contains $\mathcal{E}$ and a flow of operators that restricts to U^α on $\mathcal{E}$. This random space is not constructed directly from the eigenvectors but is rather the space of sequences on which the action of the finite matrices $(u_n)_{n=1}^\infty$ converges in a suitable way.

All probabilistic statements in this section are assumed to hold almost surely. For this section, let $\overline{B}(x,r)\subset\mathbb{R}$ denote the closed interval in $\mathbb{R}$ with center x and radius r. For any sequence v, we write $v[n]$ for the vector $(v_\ell)_{1\leq\ell\leq n}\in\mathbb{C}^n$ for all $n\geq 1$.

We define our random space as follows.

Definition 5.1 Let $\mathcal{F}$ denote the space of sequences $(w_\ell)_{\ell\geq 1}$ such that for all $\alpha\in\mathbb{R}$, there exists a sequence $V^\alpha w$, satisfying the following properties:

(1) For all $\ell\geq 1$, $\alpha,\gamma\in\mathbb{R}$ and $0<\delta'<\delta<1/6$, there is a constant $C>0$ such that

$$\sup_{\substack{\alpha_n\in\overline{B}(\alpha n,n^{1-\delta})\\ \gamma_n\in\overline{B}(\gamma n,n^{1-\delta})}} |\langle u_n^{\alpha_n}(w[n])-(V^\alpha w)[n],u_n^{\gamma_n}(e_\ell)\rangle|\leq Cn^{-\delta'}.$$

(2) For all $T>0$ and $0<\delta'<\delta<1/6$, there is a constant $C>0$ such that

$$\sup_{\alpha\in[-T,T]}\ \sup_{\alpha_n\in\overline{B}(\alpha n,n^{1-\delta})} \|u_n^{\alpha_n}w[n]-(V^\alpha w)[n]\|\leq Cn^{\frac{1}{2}-\delta'}.$$

Here the constants may depend on the choices of $\ell,\alpha,\gamma,\delta',\delta$ and T.

Note that the first condition implies, upon taking $\alpha_n=\lfloor\alpha n\rfloor$ and $\gamma_n=0$, that for $w\in\mathcal{F}$, $(V^\alpha w)_\ell$ is the limit of $\langle u_n^{\lfloor\alpha n\rfloor}(w[n]),e_\ell\rangle$ when n goes to infinity. Hence $V^\alpha w$ is uniquely determined.

The above definition gives the random space $\mathcal{F}$ and the family of operators V^α together. It turns out that this definition gives a flow of operators on a vector space, which restricts to $\mathcal{E}$ in the natural way.

Theorem 5.2 *The set $\mathcal{F}$ is a vector space. There is a family $(V^\alpha)_{\alpha\in\mathbb{R}}$ of linear maps, $V^\alpha : \mathcal{F} \to \mathcal{F}$, given by the correspondence $w \mapsto V^\alpha w$. The family satisfies the semigroup properties $V^0 = id$ and $V^\alpha V^\beta = V^{\alpha+\beta}$ for all $\alpha, \beta \in \mathbb{R}$. Moreover, almost surely for all $k \in \mathbb{Z}$ and $\alpha \in \mathbb{R}$, one has $(g_{k,\ell})_{\ell\geq 1} \in \mathcal{F}$ and $V^\alpha((g_{k,\ell})_{\ell\geq 1}) = (e^{2\pi i\alpha y_k} g_{k,\ell})_{\ell\geq 1}$, so that $\mathcal{E}$ is a subspace of $\mathcal{F}$ and that U^α is the restriction of V^α to $\mathcal{E}$.*

Proof We begin by showing that $\mathcal{F}$ is a vector space. Clearly it suffices, for $w_1, w_2 \in \mathcal{F}$ and $\lambda \in \mathbb{C}$, to show that $\lambda w_1 + w_2 \in \mathcal{F}$. Let $w = \lambda w_1 + w_2$ as sequences and define $w_\alpha = \lambda V^\alpha w_1 + V^\alpha w_2$. Now consider all $\ell \geq 1$, $\alpha, \gamma \in \mathbb{R}$ and $0 < \delta < 1/6$. For any $\alpha_n \in \overline{B}(\alpha n, n^{1-\delta})$ and $\gamma_n \in \overline{B}(\gamma n, n^{1-\delta})$, by the triangle inequality:

$$|\langle u_n^{\alpha_n}(w[n]) - w_\alpha[n], u_n^{\gamma_n}(e_\ell)\rangle| \leq |\lambda||\langle u_n^{\alpha_n}(w_1[n]) - (V^\alpha w_1)[n], u_n^{\gamma_n}(e_\ell)\rangle| + |\langle u_n^{\alpha_n}(w_2[n]) - (V^\alpha w_2)[n], u_n^{\gamma_n}(e_\ell)\rangle|.$$

Since each $w_1, w_2 \in \mathcal{F}$, there are constants C_1, C_2, depending on $\ell, \alpha, \gamma, \delta'$ and δ, such that

$$|\langle u_n^{\alpha_n}(w[n]) - w_\alpha[n], u_n^{\gamma_n}(e_\ell)\rangle| \leq |\lambda| C_1 n^{-\delta'} + C_2 n^{-\delta'}$$

so that the first condition is satisfied with constant $|\lambda|C_1 + C_2$. Similarly, we have

$$\begin{aligned}\|u_n^{\alpha_n} w[n] - w_\alpha[n]\| &\leq |\lambda| \|u_n^{\alpha_n} w_1[n] - (V^\alpha w_1)[n]\| + \|u_n^{\alpha_n} w_2[n] - (V^\alpha w_2)[n]\| \\ &\leq |\lambda| C_1' n^{\frac{1}{2}-\delta'} + C_2' n^{\frac{1}{2}-\delta'}\end{aligned}$$

for some constants C_1' and C_2', which implies the second condition with constant $|\lambda|C_1' + C_2'$. We deduce that $w \in \mathcal{F}$ with $V^\alpha w = w_\alpha$ and so $\mathcal{F}$ is a vector space as required.

Now we consider the correspondence $w \mapsto V^\beta w$ for $\beta \in \mathbb{R}$. The linearity of such a map follows from the above argument, so it suffices to show that it maps $\mathcal{F}$ to $\mathcal{F}$ and that it obeys the semigroup properties.

To see that $V^0 = \text{id}$, it suffices to note that $(V^0 w)_\ell$ is the limit of $\langle u_n^{\lfloor 0n \rfloor}(w[n]), e_\ell\rangle = \langle w, e_\ell\rangle$ as n goes to ∞.

Now, let $w \in \mathcal{F}$, $\alpha, \beta \in \mathbb{R}$, and let us show that $V^\beta w \in \mathcal{F}$ and $V^\alpha(V^\beta w) = V^{\alpha+\beta} w$. We thus need to show

(1) For all $\ell \geq 1$, $\alpha, \gamma \in \mathbb{R}$ and $0 < \delta < \delta' < 1/6$,

$$\sup_{\substack{\alpha_n \in \overline{B}(\alpha n, n^{1-\delta}) \\ \gamma_n \in \overline{B}(\gamma n, n^{1-\delta})}} |\langle u_n^{\alpha_n}(V^\beta w)[n]) - (V^{\alpha+\beta} w)[n], u_n^{\gamma_n}(e_\ell)\rangle| = O(n^{-\delta'}).$$

(2) For all $T > 0$ and $0 < \delta < \delta' < 1/6$,

$$\sup_{\alpha\in[-T,T]}\ \sup_{\alpha_n\in\overline{B}(\alpha n,n^{1-\delta})} \|u_n^{\alpha_n}(V^\beta w)[n]-(V^{\alpha+\beta}w)[n]\| = O\left(n^{\frac{1}{2}-\delta'}\right).$$

Note that the semigroup property follows from this because $V^{\alpha+\beta}w$ satisfies the conditions of the definition for $V^\alpha(V^\beta w)$.

Let us therefore fix the parameters $\ell \geq 1$, $\alpha,\gamma \in \mathbb{R}$ and $0 < \delta' < \delta < 1/6$. Consider any choice of $\alpha_n \in \overline{B}(\alpha n, n^{1-\delta})$ and $\gamma_n \in \overline{B}(\gamma n, n^{1-\delta})$. Define the sequence $(\beta_n)_{n\geq 1}$, $\beta_n \in \mathbb{Z}$ by

$$\beta_n = \begin{cases} \lfloor \beta n\rfloor, & \alpha_n \geq \alpha n \\ \lfloor \beta n\rfloor + 1, & \alpha_n < \alpha n. \end{cases}$$

It is easy to check that $\beta_n \in \overline{B}(\beta n, n^{1-\delta})$ and $\alpha_n+\beta_n \in \overline{B}((\alpha+\beta)n, n^{1-\delta})$. Now, by the triangle inequality:

$$\begin{aligned}|\langle u_n^{\alpha_n}(V^\beta w)[n]-(V^{\alpha+\beta}w)[n], u_n^{\gamma_n}(e_\ell)\rangle| &\leq |\langle u_n^{\alpha_n}(V^\beta w)[n]-u_n^{\alpha_n+\beta_n}(w[n]), u_n^{\gamma_n}(e_\ell)\rangle| \\ &\quad + |\langle u_n^{\alpha_n+\beta_n}(w[n])-(V^{\alpha+\beta}w)[n], u_n^{\gamma_n}(e_\ell)\rangle|.\end{aligned}$$

By unitary invariance, we see that

$$|\langle u_n^{\alpha_n}(V^\beta w)[n]-u_n^{\alpha_n+\beta_n}(w[n]), u_n^{\gamma_n}(e_\ell)\rangle| = |\langle (V^\beta w)[n]-u_n^{\beta_n}(w[n]), u_n^{\gamma_n-\alpha_n}(e_\ell)\rangle|$$

and therefore, taking suprema,

$$\begin{aligned}&\sup_{\substack{\alpha_n\in\overline{B}(\alpha n,n^{1-\delta})\\ \gamma_n\in\overline{B}(\gamma n,n^{1-\delta})}} |\langle u_n^{\alpha_n}(V^\beta w)[n]-(V^{\alpha+\beta}w)[n], u_n^{\gamma_n}(e_\ell)\rangle| \\ &\quad\leq \sup_{\substack{\alpha_n\in\overline{B}(\alpha n,n^{1-\delta})\\ \gamma_n\in\overline{B}(\gamma n,n^{1-\delta})}} |\langle (V^\beta w)[n]-u_n^{\beta_n}(w[n]), u_n^{\gamma_n-\alpha_n}(e_\ell)\rangle| \\ &\qquad + \sup_{\substack{\alpha_n\in\overline{B}(\alpha n,n^{1-\delta})\\ \gamma_n\in\overline{B}(\gamma n,n^{1-\delta})}} |\langle u_n^{\alpha_n+\beta_n}(w[n])-(V^{\alpha+\beta}w)[n], u_n^{\gamma_n}(e_\ell)\rangle|.\end{aligned}$$

Denoting $\alpha_n' = \alpha_n+\beta_n$ and $\gamma_n' = \gamma_n-\alpha_n$, we have $|\beta_n-\beta n| \leq 1$, $|\alpha_n'-(\alpha+\beta)n| \leq n^{1-\delta}$ and $|\gamma_n'-(\gamma-\alpha)n| \leq 2n^{1-\delta}$. Hence,

$$\sup_{\substack{\alpha_n\in\overline{B}(\alpha n,n^{1-\delta})\\ \gamma_n\in\overline{B}(\gamma n,n^{1-\delta})}} |\langle u_n^{\alpha_n}(V^\beta w)[n]-(V^{\alpha+\beta}w)[n], u_n^{\gamma_n}(e_\ell)\rangle|$$

$$\leq \sup_{\substack{\beta_n\in\overline{B}(\beta n,1)\\ \gamma_n'\in\overline{B}((\gamma-\alpha)n,2n^{1-\delta})}} |\langle (V^\beta w)[n] - u_n^{\beta_n}(w[n]), u_n^{\gamma_n'}(e_\ell)\rangle|$$

$$+ \sup_{\substack{\alpha_n'\in\overline{B}((\alpha+\beta)n,n^{1-\delta})\\ \gamma_n\in\overline{B}(\gamma n,n^{1-\delta})}} |\langle u_n^{\alpha_n'}(w[n]) - (V^{\alpha+\beta}w)[n], u_n^{\gamma_n}(e_\ell)\rangle|.$$

For $\delta'' \in (\delta', \delta)$, we deduce for n large enough (so that $2n^{-\delta} \leq n^{-\delta''}$),

$$\sup_{\substack{\alpha_n\in\overline{B}(\alpha n,n^{1-\delta})\\ \gamma_n\in\overline{B}(\gamma n,n^{1-\delta})}} |\langle u_n^{\alpha_n}(V^\beta w)[n] - (V^{\alpha+\beta}w)[n], u_n^{\gamma_n}(e_\ell)\rangle|$$

$$\leq \sup_{\substack{\beta_n\in\overline{B}(\beta n,n^{1-\delta''})\\ \gamma_n'\in\overline{B}((\gamma-\alpha)n,n^{1-\delta''})}} |\langle (V^\beta w)[n] - u_n^{\beta_n}(w[n]), u_n^{\gamma_n'}(e_\ell)\rangle|$$

$$+ \sup_{\substack{\alpha_n'\in\overline{B}((\alpha+\beta)n,n^{1-\delta})\\ \gamma_n\in\overline{B}(\gamma n,n^{1-\delta})}} |\langle u_n^{\alpha_n'}(w[n]) - (V^{\alpha+\beta}w)[n], u_n^{\gamma_n}(e_\ell)\rangle|.$$

Now, this last expression is dominated by $n^{-\delta'}$, by definition of the space $\mathcal{F}$ and the maps V^β and $V^{\alpha+\beta}$. This verifies the first condition.

The second condition follows from a similar computation. Indeed, we have

$$\|u_n^{\alpha_n}(V^\beta w)[n] - (V^{\alpha+\beta}w)[n]\| \leq \|u_n^{\alpha_n}(V^\beta w)[n] - u_n^{\alpha_n+\beta_n}w[n]\| + \|u_n^{\alpha_n+\beta_n}w[n] - (V^{\alpha+\beta}w)[n]\|,$$

which, again by unitary invariance, reduces to

$$\|u_n^{\alpha_n}(V^\beta w)[n]-(V^{\alpha+\beta}w)[n]\| \leq \|(V^\beta w)[n]-u_n^{\beta_n}w[n]\|+\|u_n^{\alpha_n+\beta_n}w[n]-(V^{\alpha+\beta}w)[n]\|.$$

Defining $\alpha' = \alpha + \beta$, $\alpha_n' = \alpha_n + \beta_n$, $T' = T + |\beta|$ and taking suprema for $\alpha \in [-T, T]$ and $\alpha_n \in \overline{B}(\alpha n, n^{1-\delta})$, we get

$$\sup_{\alpha\in[-T,T]} \sup_{\alpha_n\in\overline{B}(\alpha n,n^{1-\delta})} \|u_n^{\alpha_n}(V^\beta w)[n] - (V^{\alpha+\beta}w)[n]\|$$

$$\leq \sup_{\beta_n\in\overline{B}(\beta n,1)} \|(V^\beta w)[n] - u_n^{\beta_n}w[n]\|$$

$$+ \sup_{\alpha'\in[-T',T']} \sup_{\alpha_n'\in\overline{B}(\alpha' n,n^{1-\delta})} \|u_n^{\alpha_n'}w[n] - (V^{\alpha'}w)[n]\|.$$

Both of the quantities on the right-hand side are bounded by $O(n^{\frac{1}{2}-\delta'})$. We have therefore proven the stability of $\mathcal{F}$ by the family of maps $(V^\alpha)_{\alpha\in\mathbb{R}}$ and the semigroup property.

It remains to show that the eigenvectors are contained in $\mathcal{F}$ and V^α restricts to U^α there, but this is an immediate consequence of Propositions 3.5 and 3.6. □

We have now defined a random space $\mathcal{F}$ containing $\mathcal{E}$, on which the flow $(U^\alpha)_{\alpha\in\mathbb{R}}$ is extended to a flow of operators $(V^\alpha)_{\alpha\in\mathbb{R}}$. The definition of $\mathcal{F}$ is quite complicated, but it has the strong advantage, compared with the case of $\mathcal{E}$, to be given intrinsically in terms of $(u_n)_{n\geq 1}$, without referring explicitly to eigenvectors. A natural question that can now be asked is the following: by extending the space $\mathcal{E}$ to $\mathcal{F}$, do there exist eigenvectors of the flow that are not contained in $\mathcal{E}$? The answer is negative, in the following precise sense.

Definition 5.3 For $w \in \mathcal{F}$ different from zero, we say that w is an eigenvector of the flow $(V^\alpha)_{\alpha\in\mathbb{R}}$, if and only if there exists $\chi \in \mathbb{R}$ such that $V^\alpha(w) = e^{2i\pi\alpha\chi}w$ for all $\alpha \in \mathbb{R}$.

Using this definition, we see that for all $k \in \mathbb{Z}$, the sequences $(t_{k,\ell})_{\ell\geq 1}$ are eigenvectors of $(V^\alpha)_{\alpha\in\mathbb{R}}$, corresponding to $\chi = y_k$. The following result shows that these sequences are the only eigenvectors of the flow:

Theorem 5.4 *The only eigenvectors of $(V^\alpha)_{\alpha\in\mathbb{R}}$ are the non-zero sequences that are proportional to $(t_{k,\ell})_{\ell\geq 1}$ (or $(g_{k,\ell})_{\ell\geq 1}$) for some $k \in \mathbb{Z}$.*

From this theorem, we deduce in particular that the set of parameters χ associated with the flow $(V^\alpha)_{\alpha\in\mathbb{R}}$ is a determinantal sine-kernel process.

Before we begin the proof of this theorem, it is convenient to separately establish two partial results that will aid in the proof. For convenience, we will define

$$M_p(\lambda) := \frac{1}{p}\sum_{j=0}^{p-1}\lambda^j = \frac{1-\lambda^p}{p(1-\lambda)}$$

for $p \geq 1$ and $\lambda \in \mathbb{C}$, $|\lambda| = 1$.

First, we show that a non-zero element of $\mathcal{F}$ cannot be too small in a precise sense. This in particular shows that $\mathcal{F} \cap \ell^2 = \{0\}$.

Proposition 5.5 *Let w be an element of $\mathcal{F}$. Suppose there exists a $\delta > 0$ and a strictly increasing sequence $(n_q)_{q\geq 1}$ of integers such that $\|w[n_q]\| = O(n_q^{\frac{1}{2}-\delta})$. Then, w is identically equal to zero.*

Proof In the first part of the definition of $\mathcal{F}$, let us take $(\delta/3) \wedge (1/7)$ instead of δ, and $\alpha = \gamma_n = 0$. Since $V^0 w = w$, we deduce, for all $\ell \geq 1$,

$$\sup_{|\alpha_n|\leq n^{1-\frac{\delta}{3}}} |\langle u_n^{\alpha_n}(w[n]) - w[n], e_\ell\rangle| = O(n^{-\delta'}),$$

for (say) $\delta' = (\delta/4) \wedge (1/8) > 0$. Hence, if we decompose $w[n]$ in terms of eigenvectors of u_n:

$$w[n] = \sum_{k=1}^{n} \eta_k^{(n)} f_k^{(n)},$$

we get

$$\sup_{|\alpha_n| \leq n^{1-\frac{\delta}{3}}} \left| w_\ell - \sum_{k=1}^{n} (\lambda_k^{(n)})^{\alpha_n} \eta_k^{(n)} \langle f_k^{(n)}, e_\ell \rangle \right| = O(n^{-\delta'}),$$

Taking $p = \lfloor n^{1-\frac{\delta}{3}} \rfloor + 1$ and averaging for $\alpha_n \in \{0, 1, \dots, p-1\}$ give

$$w_\ell = \lim_{n \to \infty} \sum_{k=1}^{n} M_p(\lambda_k^{(n)}) \eta_k^{(n)} \langle f_k^{(n)}, e_\ell \rangle.$$

Moreover, for all $n \geq 1$ and $k \in \{1, \dots, n\}$, $|\langle f_k^{(n)}, e_\ell \rangle|^2$ is a beta random variable of parameters 1 and $n-1$. From the Borel–Cantelli lemma, we deduce that almost surely,

$$\langle f_k^{(n)}, e_\ell \rangle = O\left(n^{-\frac{1}{2}+\frac{\delta}{3}}\right),$$

and then, using Cauchy–Schwarz inequality, we get

$$\begin{aligned} |w_\ell| &\lesssim n^{-\frac{1}{2}+\frac{\delta}{3}} \left(\sum_{k=1}^{n} |M_p(\lambda_k^{(n)})|^2 \right)^{1/2} \left(\sum_{k=1}^{n} |\eta_k^{(n)}|^2 \right)^{1/2} \\ &= n^{-\frac{1}{2}+\frac{\delta}{3}} \|w[n]\| \left(\sum_{k=1}^{n} |M_p(\lambda_k^{(n)})|^2 \right)^{1/2}. \end{aligned} \tag{21}$$

Let us now assume the following bound:

$$\sum_{k=1}^{n} |M_p(\lambda_k^{(n)})|^2 = O(n^{\delta}). \tag{22}$$

Then, for $n = n_q$ ($q \geq 1$), the right-hand side of (21) is dominated by $n_q^{-\delta/6}$, which is only possible if $w_\ell = 0$. Hence, Proposition 5.5 is proven if we are able to show (22). Now, we have

$$|M_p(\lambda)|^2 = \frac{1}{p^2} \sum_{j=-p+1}^{p-1} (p - |j|)\lambda^j,$$

which implies

$$\sum_{k=1}^{n} |M_p(\lambda_k^{(n)})|^2 = \frac{1}{p^2} \sum_{j=-p+1}^{p-1} (p - |j|) \operatorname{Tr}(u_n^j).$$

By [5], it is known, that for all integers $j_1, j_2, \dots, j_r$ such that $|j_1| + |j_2| + \cdots + |j_r| \leq n$, one has

$$\mathbb{E}\left[\prod_{s=1}^{r} \operatorname{Tr}(u_n^{j_s})\right] = \mathbb{E}\left[\prod_{s=1}^{r} Z_{j_s}\right],$$

where $Z_0 = n$, $(Z_j/\sqrt{j})_{j\geq 1}$ are iid standard complex gaussian variables ($\mathbb{E}[|Z_j|^2] = j$) and $Z_{-j} = \overline{Z_j}$ for $j \geq 1$. For a fixed integer $A > 0$, we deduce, from the fact that $p = o(n)$, that for n large enough:

$$\mathbb{E}\left[\left(\sum_{k=1}^{n} |M_p(\lambda_k^{(n)})|^2\right)^A\right] = \mathbb{E}[W^A],$$

where

$$W := \frac{1}{p^2} \sum_{j=-p+1}^{p-1} (p - |j|) Z_j.$$

The variable W is a real-valued gaussian variable, with expectation n/p (coming from the term $j = 0$), and variance

$$\frac{1}{p^4} \sum_{1\leq|j|\leq p-1} |j|(p - |j|)^2 \leq \frac{1}{p^4} \sum_{1\leq|j|\leq p-1} p^3 \leq 1.$$

Hence,

$$\|W\|_{L^A} \leq (n/p) + \|W - (n/p)\|_{L^A} \leq (n/p) + \|N(0, 1)\|_{L^A} \leq c(A) + n/p,$$

where $c(A) > 0$ depends only on A. We deduce (for A fixed)

$$\mathbb{E}\left[\left(\sum_{k=1}^{n}|M_p(\lambda_k^{(n)})|^2\right)^A\right] \lesssim (n/p)^A \lesssim n^{A\delta/3}.$$

Using Markov's inequality, we get

$$\mathbb{P}\left[\sum_{k=1}^{n}|M_p(\lambda_k^{(n)})|^2 \geq n^\delta\right] \leq n^{-A\delta}\mathbb{E}\left[\left(\sum_{k=1}^{n}|M_p(\lambda_k^{(n)})|^2\right)^A\right] \lesssim n^{-2A\delta/3}.$$

Taking any A strictly larger that $3/2\delta$ and using the Borel–Cantelli lemma give the estimate (22). □

We also require the following easy technical lemma.

Lemma 5.6 *There exists a universal constant $c > 0$, such that for all $\lambda \in \mathbb{C}$, $|\lambda| = 1$, and $n \geq 1$,*

$$|M_n(\lambda)|^2 \leq 1 - c((n|\lambda - 1|) \wedge 1)^2.$$

Proof If $|\lambda - 1| \geq 3/n$, we have

$$|M_n(\lambda)|^2 = \frac{|1-\lambda^n|^2}{n^2|1-\lambda|^2} \leq 4/9,$$

and the inequality is true for $c = 5/9$. If $|\lambda - 1| \leq 3/n$, we can write $\lambda = e^{i\theta}$ where

$$|\lambda - 1| \leq |\theta| \leq \pi|\lambda - 1|/2 \leq 3\pi/2n.$$

Then,

$$\begin{aligned} 1 - |M_n(\lambda)|^2 &= \frac{1}{n^2}\sum_{j=-n+1}^{n-1}(n-|j|)(1-\lambda^j) \\ &= \frac{1}{n^2}\sum_{j=-n+1}^{n-1}(n-|j|)(1-\cos(j\theta)), \end{aligned}$$

where $|j\theta| \leq 3\pi/2$, which implies that $1 - \cos(j\theta) \gtrsim j^2\theta^2$. Hence,

$$\begin{aligned} 1 - |M_n(\lambda)|^2 &\gtrsim \frac{1}{n^2}\sum_{j=-n+1}^{n-1}(n-|j|)j^2\theta^2 \\ &\gtrsim n^2\theta^2 \geq |\lambda - 1|^2 n^2. \end{aligned}$$

□

Proof of Theorem 5.4 Let w be an eigenvector of $(V^{\alpha})_{\alpha\in\mathbb{R}}$, and let χ be the corresponding eigenvalue. Taking, in the second part of the definition of $\mathcal{F}$, $T = 1$, $\alpha = j/n$ for $j \in \{0, 1, \ldots, n-1\}$, $\alpha_n = \alpha n = j$, we obtain, for any $\delta \in (0, 1/6)$,

$$\sup_{0\le j\le n-1} \|u_n^j w[n] - e^{2i\pi\chi j/n} w[n]\| = O\left(n^{\frac{1}{2}-\delta}\right),$$

or equivalently,

$$w[n] = u_n^j e^{-2i\pi\chi j/n} w[n] + v_{n,j},$$

where

$$\sup_{0\le j\le n-1} \|v_{n,j}\| = O\left(n^{\frac{1}{2}-\delta}\right).$$

Decomposing in the eigenvector basis of u_n gives, with the notation of the proof of Proposition 5.5:

$$w[n] = v_{n,j} + \sum_{k=1}^{n} (\lambda_k^{(n)} e^{-2i\pi\chi/n})^j \eta_k^{(n)} f_k^{(n)}.$$

Averaging with respect to $j \in \{0, \ldots, n-1\}$ gives

$$w[n] = v_n + \sum_{k=1}^{n} M_n(\lambda_k^{(n)} e^{-2i\pi\chi/n}) \eta_k^{(n)} f_k^{(n)},$$

where

$$\|v_n\| = O\left(n^{\frac{1}{2}-\delta}\right).$$

Hence,

$$\|w[n]\| = \left(\sum_{k=1}^{n} |M_n(\lambda_k^{(n)} e^{-2i\pi\chi/n})|^2 |\eta_k^{(n)}|^2\right)^{1/2} + O\left(n^{\frac{1}{2}-\delta}\right). \tag{23}$$

First let us assume that χ is not one of the values of y_k for $k \in \mathbb{Z}$. Then, there exists $k \in \mathbb{Z}$, $\chi_1, \chi_2 \in \mathbb{R}$, such that

$$y_k < \chi_1 < \chi < \chi_2 < y_{k+1}.$$

Hence, for n large enough,

$$\frac{n\theta_k^{(n)}}{2\pi} < \chi_1 < \chi < \chi_2 < \frac{n\theta_{k+1}^{(n)}}{2\pi},$$

and there are no eigenangles of u_n between $2\pi\chi_1/n$ and $2\pi\chi_2/n$. One deduces that for all $k \in \{1, \ldots, n\}$,

$$|\lambda_k^{(n)} e^{-2i\pi\chi/n} - 1| \gtrsim 1/n,$$

and by Lemma 5.6, there exists $d < 1$ such that for all n large enough, and all $k \in \{1, \ldots, n\}$,

$$|M_n(\lambda_k^{(n)} e^{-2i\pi\chi/n})| \leq d.$$

Therefore, (23) gives, for n large enough,

$$\|w[n]\| \leq d \left(\sum_{k=1}^{n} |\eta_k^{(n)}|^2 \right)^{1/2} + O\left(n^{\frac{1}{2}-\delta}\right) = d\|w[n]\| + O\left(n^{\frac{1}{2}-\delta}\right),$$

and, since $1 - d > 0$ is independent of n,

$$\|w[n]\| \lesssim (1-d)\|w[n]\| = O\left(n^{\frac{1}{2}-\delta}\right).$$

By Proposition 5.5, w is identically zero, which implies that $(V^\alpha)_{\alpha\in\mathbb{R}}$ has no eigenvectors for $\chi \notin \{y_k, k \in \mathbb{Z}\}$.

Now, let us assume that $\chi = y_k$ for $k \geq 1$ (the case $k \leq 0$ is similar), and let w_n be the projection of $w[n]$ on the orthogonal of $f_k^{(n)}$. We have

$$w[n] = w_n + \eta_k^{(n)} f_k^{(n)},$$

and then

$$u_n^j w[n] - e^{2i\pi\chi j/n} w[n] = [u_n^j(w_n) - e^{2i\pi\chi j/n} w_n] + \eta_k^{(n)} (e^{2i\pi j\theta_k^{(n)}} - e^{2i\pi j\chi/n}) f_k^{(n)},$$

where the two terms in the last sum are orthogonal vectors. Hence,

$$\sup_{0\leq j\leq n-1} \|u_n^j(w_n) - e^{2i\pi\chi j/n} w_n\| \leq \sup_{0\leq j\leq n-1} \|u_n^j w[n] - e^{2i\pi\chi j/n} w[n]\| \lesssim n^{\frac{1}{2}-\delta}.$$

Now, decomposing w_n in the eigenvector basis of u_n and performing the same computation as for $w[n]$, we obtain an estimate which is similar to (23):

$$\|w_n\| = \left(\sum_{1 \le k' \le n, k' \neq k} |M_n(\lambda_{k'}^{(n)} e^{-2i\pi\chi/n})|^2 |\eta_k^{(n)}|^2 \right)^{1/2} + O\left(n^{\frac{1}{2}-\delta}\right). \tag{24}$$

Notice that the term $k' = k$ is not in the sum, since w_n is obtained from $w[n]$ by removing the component proportional to $f_k^{(n)}$. Now, it is easy to check that for $k' \in \{1, \ldots n\} \backslash \{k\}$, one has the estimate:

$$|\lambda_k^{(n)} e^{-2i\pi\chi/n} - 1| \gtrsim 1/n$$

(note that this estimate is not true for $k' = k$, since $n\theta_k^{(n)}/2\pi$ tends to $\chi = y_k$ when n goes to infinity). Hence, for n large enough, $|M_n(\lambda_{k'}^{(n)} e^{-2i\pi\chi/n})|$ is uniformly bounded by a quantity that is strictly smaller than 1. As above for $w[n]$, this implies the estimate:

$$\|w_n\| = O\left(n^{\frac{1}{2}-\delta}\right).$$

If we set $\kappa_n := \eta_k^{(n)}/D_k^{(n)}$, we deduce

$$\|w[n] - \kappa_n g_k^{(n)}\| = O\left(n^{\frac{1}{2}-\delta}\right),$$

and then

$$\|w[n] - \kappa_n g_k[n]\| = O\left((1 + |\kappa_n|)n^{\frac{1}{2}-\delta}\right), \tag{25}$$

since

$$\|g_k[n] - g_k^{(n)}\| = O\left(n^{\frac{1}{2}-\delta}\right).$$

Now, for any integer m such that $n \le m \le 2n$, we have

$$\|w[m] - \kappa_m g_k[m]\| \lesssim (1 + |\kappa_m|)m^{\frac{1}{2}-\delta} = O\left((1 + |\kappa_m|)n^{\frac{1}{2}-\delta}\right),$$

and taking the n first components, we obtain

$$\|w[n] - \kappa_m g_k[n]\| = O\left((1 + |\kappa_m|)n^{\frac{1}{2}-\delta}\right). \tag{26}$$

Comparing (25) and (26) gives the following:

$$|\kappa_m - \kappa_n| \|g_k[n]\| = O\left((1 + |\kappa_n| + |\kappa_m|)n^{\frac{1}{2}-\delta}\right),$$

and then

$$|\kappa_m - \kappa_n| = O\left((1 + |\kappa_n| + |\kappa_m|)n^{-\delta}\right), \tag{27}$$

since $\|g_k[n]\|$ is equivalent to a strictly positive constant times $\sqrt{n}$. In particular, for n large enough and $n \le m \le 2n$,

$$|\kappa_m| - |\kappa_n| \le |\kappa_m - \kappa_n| \le \frac{1}{2}(1 + |\kappa_n| + |\kappa_m|),$$

which implies

$$|\kappa_m| \le 1 + 3|\kappa_n| = O(1 + |\kappa_n|),$$

and (27) can be replaced by

$$|\kappa_m - \kappa_n| = O\left((1 + |\kappa_n|)n^{-\delta}\right). \tag{28}$$

Hence,

$$|\kappa_m| = |\kappa_n| + O\left((1 + |\kappa_n|)n^{-\delta}\right),$$

$$\sup_{n \le m \le 2n} (1 + |\kappa_m|) = (1 + |\kappa_n|)(1 + O(n^{-\delta})),$$

and

$$S_{q+1} \le S_q(1 + O(2^{-\delta q})),$$

where S_q denotes the supremum of $1 + |\kappa_m|$ for $2^q \le m \le 2^{q+1}$. We deduce that the sequence $(S_q)_{q \ge 0}$ and then the sequence $(\kappa_n)_{n \ge 1}$ are bounded. The estimate (28) becomes

$$|\kappa_m - \kappa_n| = O\left(n^{-\delta}\right),$$

for $n \le m \le 2n$. If for $q \ge 1$, $2^q n \le m \le 2^{q+1} n$, we also get

$$|\kappa_m - \kappa_n| \le |\kappa_m - \kappa_{2^q n}| + \sum_{r=0}^{q-1} |\kappa_{2^{r+1} n} - \kappa_{2^r n}| \lesssim \sum_{r=0}^{q} (2^r n)^{-\delta} = O(n^{-\delta}).$$

Hence $(\kappa_n)_{n \ge 1}$ is a Cauchy sequence, and if κ denotes its limit, one has

$$|\kappa - \kappa_n| = O(n^{-\delta}).$$

Using (25), we deduce

$$\begin{aligned}\|(w-\kappa g_k)[n]\| &\leq \|w[n]-\kappa_n g_k[n]\| + |\kappa-\kappa_n|\|g_k[n]\| \\ &\lesssim (1+|\kappa_n|)n^{\frac{1}{2}-\delta} + n^{-\delta}\sqrt{n} = O\left(n^{\frac{1}{2}-\delta}\right).\end{aligned}$$

Now, $w - \kappa g_k$ is a sequence in $\mathcal{F}$: by Proposition 5.5, $w - \kappa g_k = 0$, which implies that w is proportional to g_k. □

6 Further Questions

A natural question that can be asked is whether one could construct a similar flow of operators for other ensembles of random matrices. In [14], convergence of the renormalized eigenvalues and eigenvectors is proven for a large class of infinite-dimensional Hermitian random matrices that are invariant in law by unitary conjugation. The coordinates of limiting eigenvectors can still be taken as iid complex Gaussian variables. A flow of operators can then be constructed as in Section 3, on a space similar to $\mathcal{E}$. However, it is still an open problem to relate this flow to the infinite random Hermitian matrix considered at the beginning of the construction, and to define and extend it without explicit reference to the limiting eigenvectors.

On the other hand, many unitary invariant ensembles of random matrices have a determinantal structure, in such a way that the eigenvalues tend to a determinantal sine-kernel process on the microscopic scale, and the components of the suitably renormalized eigenvectors converge in law to iid complex Gaussian variables. One can then ask if there is a universal way to couple these ensembles in all dimensions, in order to get almost sure convergence for renormalized eigenvalues and eigenvectors.

Another problem consists to see if there exists a natural version of the flow $(V^{\alpha})_{\alpha\in\mathbb{R}}$, which is defined on a random space with a Hilbert structure. In this case, V^{α} would be a "true" unitary operator on this space. Moreover, it would be possible to define a self-adjoint operator H, whose spectrum is the determinantal sine-kernel process $(y_m)_{m\in\mathbb{Z}}$, and which is equal to $1/2i\pi$ times the infinitesimal generator of the flow $(V^{\alpha})_{\alpha\in\mathbb{R}}$. Note that H would be an unbounded operator, and then its domain would not be the whole Hilbert space where $(V^{\alpha})_{\alpha\in\mathbb{R}}$ is defined. However, the operator H^{-1} would be a bounded, and even compact operator.

References

1. P. Bourgade, C.-P. Hughes, A. Nikeghbali, M. Yor, The characteristic polynomial of a random unitary matrix: a probabilistic approach. Duke Math. J. **145**(1), 45–69 (2008)
2. P. Bourgade, J. Najnudel, A. Nikeghbali, A unitary extension of virtual permutations. Int. Math. Res. Not. **2013**(18), 4101–4134 (2013)

3. P. Bourgade, A. Nikeghbali, A. Rouault, Circular Jacobi ensembles and deformed Verblunsky coefficients. Int. Math. Res. Not. **23**, 4357–4394 (2009)
4. P. Diaconis, M. Shahshahani, The subgroup algorithm for generating random variables. Prob. Eng. Inf. Sc. **1**, 15–32 (1987)
5. P. Diaconis, M. Shahshahani, On the eigenvalues of random matrices. J. Appl. Probab. **31A**, 49–62 (1994). Studies in applied probability
6. N.M. Katz, P. Sarnak, *Random Matrices, Frobenius Eigenvalues, and Monodromy*. American Mathematical Society Colloquium Publications, vol. 45 (American Mathematical Society, Philadelphia, 1999)
7. J.P. Keating, N.C. Snaith, *Random matrix theory and* $\zeta(1/2+it)$. Commun. Math. Phys. **214**, 57–89 (2000)
8. J.P. Keating, N.C. Snaith, Random matrix theory and L-functions at $s = 1/2$. Commun. Math. Phys. **214**, 91–110 (2000)
9. S.-V. Kerov, G.-I. Olshanski, A.-M. Vershik. Harmonic analysis on the infinite symmetric group. C. R. Acad. Sci. Paris **316**, 773–778 (1993)
10. R. Killip, M. Stoiciu, Eigenvalue statistics for CMV matrices: from Poisson to clock via random matrix ensembles. Duke Math. J. **146**(3), 361–399 (2009)
11. K. Maples, J. Najnudel, A. Nikeghbali, Limit operators for circular ensembles (2014). Preprint. arXiv:1304.3757
12. K. Maples, J. Najnudel, A. Nikeghbali, Strong convergence of eigenangles and eigenvectors for the circular unitary ensemble. Ann. Probab. **47**(4), 2417–2458 (2019)
13. F. Mezzadri, How to generate random matrices from the classical compact groups. Not. AMS **54**, 592–604 (2007)
14. J. Najnudel, Eigenvector convergence for minors of unitarily invariant infinite random matrices. Int. Math. Res. Not. (2020). https://doi.org/10.1093/irmn/rnz330
15. J. Najnudel, A. Nikeghbali, *On a flow of Operators Associated to Virtual Permutations*. Séminaire de Probabilités XLVI (Springer, Cham, 2014), pp. 481–512
16. Y.-A. Neretin, Hua type integrals over unitary groups and over projective limits of unitary groups. Duke Math. J. **114**, 239–266 (2002)
17. J. Pitman, *Combinatorial stochastic processes*. Lecture Notes in Mathematics, vol. 1875 (Springer, Berlin, 2006). Lectures from the 32nd Summer School on Probability Theory held in Saint-Flour, July 7–24, 2002, With a foreword by Jean Picard
18. A. Soshnikov, Determinantal random point fields. Uspekhi Mat. Nauk. **55**(5(335)), 107–160 (2000)
19. E.C. Titchmarsh, *The Theory of the Riemann Zeta Function*, 2nd edn. (Oxford University Press, Oxford, 1986)
20. B. Valkó, B. Virág, Continuum limits of random matrices and the Brownian carousel. Invent. Math. **177**(3), 463–508 (2009)

Gibbs Measures of Nonlinear Schrödinger Equations as Limits of Quantum Many-Body States in Dimension $d \leq 3$

Vedran Sohinger

1 Derivation of Gibbs Measures for NLS from Many-Body Quantum Systems

We consider the spatial domain $\Lambda = \mathbb{R}^d$ or $\mathbb{T}^d$ for $d = 1, 2, 3$. Given a one-body potential $V \geq 0$ and an even *positive (defocusing)* interaction potential w on Λ, we consider the *nonlinear Schrödinger equation (NLS)*:

$$\mathrm{i}\partial_t \phi(x) \;=\; -\Delta\phi(x) + V(x)\phi(x) + \int \mathrm{d}y\, |\phi(y)|^2\, w(x-y)\, \phi(x)\,. \tag{1}$$

We consider either *nonlocal (Hartree)* interactions $w \in L^\infty(\Lambda)$ or *local* interactions $w = \delta$. The NLS (1) is formally given as the Hamiltonian equation of motion on the space of fields $\phi : \Lambda \to \mathbb{C}$ with Hamilton function:

$$H(\phi) \;:=\; \int \mathrm{d}x\, |\nabla\phi(x)|^2 + V(x)|\phi(x)|^2 + \frac{1}{2}\int \mathrm{d}x\, \mathrm{d}y\, |\phi(x)|^2\, w(x-y)\, |\phi(y)|^2 \tag{2}$$

and Poisson bracket satisfying

$$\{\phi(x), \bar{\phi}(y)\} \;=\; \mathrm{i}\delta(x-y), \qquad \{\phi(x), \phi(y)\} \;=\; \{\bar{\phi}(x), \bar{\phi}(y)\} \;=\; 0\,. \tag{3}$$

V. Sohinger (✉)
Mathematics Institute, University of Warwick, Coventry, UK
e-mail: V.Sohinger@warwick.ac.uk

N. Anantharaman et al. (eds.), *Frontiers in Analysis and Probability*,
https://doi.org/10.1007/978-3-030-56409-4_9

The *Gibbs measure* $d\mathbb{P}$ associated with H is formally defined as the probability measure on the space of fields $\phi : \Lambda \to \mathbb{C}$ given by

$$\mathbb{P}(d\phi) := \frac{1}{Z} e^{-H(\phi)} \, d\phi \,, \tag{4}$$

for a normalisation constant Z and the (formal) Lebesgue measure $d\phi$ on the space of fields. By formal arguments, $d\mathbb{P}$ is invariant under the flow of (1). The first rigorous result of invariance of Gibbs measures for the NLS was obtained by Bourgain [2]. In proving this, a significant challenge is the *infinite dimensionality of the Hamiltonian system*, which is overcome by an approximation argument. The latter requires a suitable stability analysis of (1). Gibbs measures have since been extensively studied as tools to construct global solutions of time-dependent NLS equations with rough initial data, see [3–12, 18, 28, 29, 31, 38–41] and references therein. Note that the problem of constructing the Gibbs measure (without the invariance) was first addressed in the constructive quantum field literature in the 1970s [20, 33], and it was introduced to the PDE community in [24].

The main question that we study is *whether it is possible to derive* (4) *as a classical limit of many-body quantum objects*. The analogous question for the equation (1) was first studied by Hepp [22] and Ginibre and Velo [19]. We refer the reader to [13–15, 36] and the references therein, as well as to the expository work [32] for a more detailed discussion on this topic. In our work, we analyse both stationary and dynamic aspects of the problem. Let us address each of these aspects separately.

1.1 The Stationary Problem

One starts from a many-body quantum system of n particles with Hamiltonian of the form:

$$H^{(n)} := \sum_{i=1}^{n} \big(-\Delta_{x_i} + V(x_i) \big) + \lambda \sum_{1 \leq i < j \leq n} w(x_i - x_j) \,, \tag{5}$$

for some coupling constant $\lambda > 0$ acting on $L^2_{\mathrm{sym}}(\Lambda^n)$ (the symmetric subspace of $L^2(\Lambda^n)$). At temperature $\tau > 0$, the equilibrium state of $H^{(n)}$ is governed by the *canonical ensemble* $P^{(n)}_\tau := e^{-H^{(n)}/\tau}$. The goal is to appropriately combine the canonical ensembles and obtain the Gibbs measure $d\mathbb{P}$ when $\tau \to \infty$.

This problem was first considered in $1D$ and in higher dimensions with nontranslation-invariant interactions by Lewin, Nam, and Rougerie in [25] (see also the more recent results [26, 27] by the same group of authors). In [16], we consider $d = 1, 2, 3$ and $\Lambda = \mathbb{T}^d$ or $\mathbb{R}^d$. On the *one-particle space* $\mathfrak{H} := L^2(\Lambda; \mathbb{C})$, we work with the densely defined *one-body Hamiltonian*:

$$h := -\Delta + \kappa + v, \tag{6}$$

for $\kappa > 0$ and $v : \Lambda \to [0, \infty)$. We assume that h has compact resolvent and that for some $s < 1$ we have

$$\operatorname{Tr} h^{s-1} < \infty. \tag{7}$$

The two main cases that we study are when (7) holds for $s = -1$ and $s = 0$. When $d \leq 3$, the former holds for $\Lambda = \mathbb{T}^d$ and $v = 0$ or $\Lambda = \mathbb{R}^d$ and v sufficiently confining, e.g. $v = |x|^r$ for large enough r. The latter holds for $\Lambda = \mathbb{T}^1$ and $v = 0$.

Write $h = \sum_{k \in \mathbb{N}} \lambda_k u_k u_k^*$ with eigenvalues λ_k and L^2-normalised eigenvectors u_k. Let us consider an infinite sequence of independent standard complex Gaussians $\mu_k := \pi^{-1} \mathrm{e}^{-|z|^2/2} \mathrm{d}z$, where $\mathrm{d}z$ is Lebesgue measure on $\mathbb{C}$. We work on the probability space $(\mathbb{C}^{\mathbb{N}}, \mathscr{G}, \mu)$, where $\mathscr{G}$ is taken to be the product sigma-algebra and $\mu := \bigotimes_{k \in \mathbb{N}} \mu_k$. The points of the probability space $\mathbb{C}^{\mathbb{N}}$ are denoted by $\omega = (\omega_k)_{k \in \mathbb{N}}$. Given $\omega = (\omega_k)_k \in \mathbb{C}^{\mathbb{N}}$, $K \in \mathbb{N}$, one defines the *truncated classical free field* as

$$\phi_{[K]} := \sum_{k=0}^{K} \frac{\omega_k}{\sqrt{\lambda_k}} u_k .$$

Then $\phi_{[K]} \to \phi$ as $K \to \infty$ in $\mathfrak{H}_s$, where $\langle f, g \rangle_{\mathfrak{H}_s} := \langle f, h^s g \rangle_{\mathfrak{H}}$. The obtained limit ϕ is called the *classical free field*. The *classical interaction* is given by

$$W := \frac{1}{2} \int \mathrm{d}x\, \mathrm{d}y\, |\phi(x)|^2\, w(x-y)\, |\phi(y)|^2 .$$

Note that, μ-almost surely, $W \geq 0$ under appropriate positivity assumptions on w. Moreover W is μ-almost surely finite if $w \in L^\infty(\Lambda)$ and if (7) holds with $s = 0$. In this case the *classical Gibbs state* $\rho(\cdot)$ associated with h and w is defined by

$$\rho(X) := \frac{\int X \mathrm{e}^{-W} \mathrm{d}\mu}{\int \mathrm{e}^{-W} \mathrm{d}\mu}, \tag{8}$$

for X a random variable. On $\mathfrak{H}^{(p)}$ (by which we henceforth denote the *symmetric subspace* of $\mathfrak{H}^{\otimes p}$), the *classical p-particle correlation function* γ_p is defined by

$$\gamma_p(x_1, \dots, x_p; y_1, \dots, y_p) := \rho\big(\bar{\phi}(y_1) \cdots \bar{\phi}(y_p) \phi(x_1) \cdots \phi(x_p)\big) . \tag{9}$$

When (7) holds with $s = -1$, but not with $s = 0$ (which occurs when $d = 2, 3$), it is necessary to modify this construction by using the procedure of *Wick ordering*, as in [3] for instance. Assuming that this is the case, we note that $\phi \notin \mathfrak{H}$ almost surely and so for general $w \in L^\infty(\Lambda)$ the classical interaction W *does not make sense*. For $K \in \mathbb{N}$, the *truncated Wick-ordered classical interaction* is defined as

$$W_{[K]} := \frac{1}{2}\int \mathrm{d}x\,\mathrm{d}y\,\big(|\phi_{[K]}(x)|^2 - \varrho_{[K]}(x)\big)\,w(x-y)\,\big(|\phi_{[K]}(y)|^2 - \varrho_{[K]}(y)\big)\,, \tag{10}$$

where the *truncated classical density at* x is given by $\varrho_{[K]}(x) := \int \mathrm{d}\mu\,|\phi_{[K]}(x)|^2$. The latter quantity diverges to infinity almost surely as $K \to \infty$. One then has $W_{[K]} \to W$ in $\bigcap_{m\geq 1} L^m(\mu)$. The limit W is the *Wick-ordered classical interaction*. Hence for $d = 2, 3$, it is necessary to modify the definitions (8) and (9) by replacing the interaction W by its Wick-ordered version.

In the quantum setting one works on the *Bosonic Fock space* $\mathscr{F} \equiv \mathscr{F}(\mathfrak{H}) := \bigoplus_{n\in\mathbb{N}} \mathfrak{H}^{(n)}$. At temperature $\tau > 0$, the *many-body quantum Hamiltonian* on the sector $\mathfrak{H}^{(n)}$ is given by (5) with $\lambda = \frac{1}{\tau}$. The latter choice of coupling constant λ is necessary in order to ensure a well-defined limit as $\tau \to \infty$, see [16, (1.25)]. On $\mathscr{F}$, the *many-body quantum Hamiltonian* is given by

$$H_\tau := \frac{1}{\tau}\bigoplus_{n\in\mathbb{N}} H^{(n)}\,, \tag{11}$$

and the *grand canonical ensemble* defined by

$$P_\tau := \bigoplus_{n\in\mathbb{N}} P_\tau^{(n)} = \mathrm{e}^{-H_\tau}\,. \tag{12}$$

Furthermore, we define the *quantum state* $\rho_\tau(\cdot)$ by

$$\rho_\tau(\mathscr{A}) := \frac{\mathrm{Tr}\,(\mathscr{A}\,P_\tau)}{\mathrm{Tr}\,(P_\tau)}\,, \tag{13}$$

for $\mathscr{A}$ a closed operator on $\mathscr{F}$.

When $d = 2, 3$, it is necessary to modify (11)–(13) by means of Wick ordering. More precisely, given $f \in \mathfrak{H}$, consider the *bosonic creation and annihilation operators* $b(f)$ and $b^*(f)$ on the Fock space $\mathscr{F}$ satisfying the canonical commutation relations (i.e. $[b(f), b^*(g)] = \langle f, g\rangle$ and $[b(f), b(g)] = [b^*(f), b^*(g)] = 0$). By rescaling, the *quantum field* is defined as $\phi_\tau(f) := \tau^{-1/2}\,b(f), \phi_\tau^*(f) := \tau^{-1/2}\,b^*(f)$. The associated distribution kernels are denoted by $\phi_\tau(x)$ and $\phi_\tau^*(x)$. One defines the *free Hamiltonian* $H_{\tau,0}$, the associated *free quantum state* $\rho_{\tau,0}(\cdot)$ and *quantum density at* x, denoted by $\varrho_\tau(x)$ according to

$$H_{\tau,0} := \int \mathrm{d}x\,\mathrm{d}y\,\phi_\tau^*(x)\,h(x;y)\,\phi_\tau(y)\,,$$

$$\rho_{\tau,0}(\mathscr{A}) := \frac{\mathrm{Tr}\,(\mathscr{A}\,\mathrm{e}^{-H_{\tau,0}})}{\mathrm{Tr}\,(\mathrm{e}^{-H_{\tau,0}})}\,,$$

$$\varrho_\tau(x) := \rho_{\tau,0}\big(\phi_\tau^*(x)\phi_\tau(x)\big)\,.$$

With these definitions, we can rewrite (11) as

$$H_\tau = H_{\tau,0} + \frac{1}{2}\int \mathrm{d}x\,\mathrm{d}y\, \phi_\tau^*(x)\phi_\tau^*(y)w(x-y)\phi_\tau(x)\phi_\tau(y)\,. \tag{14}$$

This is the definition that we use when $d = 1$. When $d = 2, 3$, instead of H_τ given by (11) and (14), we work with the *Wick-ordered many-body Hamiltonian*, which is given by $H_\tau := H_{\tau,0} + W_\tau$ where the *Wick-ordered quantum interaction* is

$$W_\tau := \frac{1}{2}\int \mathrm{d}x\,\mathrm{d}y\, \big(\phi_\tau^*(x)\phi_\tau(x) - \varrho_\tau(x)\big)\, w(x-y)\, \big(\phi_\tau^*(y)\phi_\tau(y) - \varrho_\tau(y)\big)\,. \tag{15}$$

In [16], we need to modify (12) when $d = 2, 3$. This is needed for technical reasons concerning the analysis of the asymptotic series (18), which will be explained in more detail in the discussion below. The modification of (12) in [16] consists in considering a *modified grand canonical density operator*:

$$P_\tau^\eta := \mathrm{e}^{-\eta H_{\tau,0}}\mathrm{e}^{-(1-2\eta)H_{\tau,0}-W_\tau}\mathrm{e}^{-\eta H_{\tau,0}}\,, \tag{16}$$

for fixed $\eta \in [0, 1/4]$. In accordance with (16), we consider a modified quantum state $\rho_\tau^\eta(\cdot)$. For $p \in \mathbb{N}$, we define the *quantum p-particle correlation function* $\gamma_{\tau,p}^\eta$ by

$$\gamma_{\tau,p}^\eta(x_1,\ldots,x_p;y_1,\ldots,y_p) := \rho_\tau^\eta\big(\phi_\tau^*(y_1)\cdots\phi_\tau^*(y_p)\phi_\tau(x_1)\cdots\phi_\tau(x_p)\big)\,.$$

In [16] the main result we prove is the following.

Theorem 1 *Let $\Lambda = \mathbb{T}^d$ or $\Lambda = \mathbb{R}^d$. Let $\kappa > 0$ and $v : \Lambda \to [0,\infty)$. Suppose that h given by* (6) *has compact resolvent and satisfies* (7) *with $s = -1$. Let $w \in L^\infty(\Lambda)$ be an even function, which is pointwise nonnegative when $d = 1$ or of positive type (i.e. $\widehat{w} \geq 0$) when $d = 2, 3$. Then for every $\eta \in (0, 1/4]$ and $p \in \mathbb{N}$, we have*

$$\lim_{\tau\to\infty} \|\gamma_{\tau,p}^\eta - \gamma_p\|_{\mathfrak{S}^2(\mathfrak{H}^{(p)})} = 0\,. \tag{17}$$

Here $\mathfrak{S}^2$ denotes the class of Hilbert–Schmidt operators.

One interprets (17) as a convergence result of the quantum Gibbs states $\rho_\tau^\eta(\cdot)$ to the classical Gibbs state $\rho(\cdot)$ *in the sense of p-particle correlation functions*. In [16, Theorem 1.8] we also consider the $1D$ problem where it is possible to take $\eta = 0$ and no Wick ordering is necessary. In particular we give an alternative proof of the $1D$ result given in [25, Theorem 5.3]. In this case, we assume that (7) holds with $s = 0$, w is taken to be pointwise nonnegative and (17) holds in the trace class. Furthermore, in [16, Theorem 1.9] we consider a local nonlinearity on $\Lambda = \mathbb{T}^1$ with quantum interactions w_τ being approximations of the delta function instead of $w_\tau = \delta$ as in [25, Theorem 5.3].

Remark 1 We note that, after the completion of our manuscript, a derivation of the Gibbs measure for a 2D nonlocal NLS with sufficiently regular interaction potential w was obtained with unmodified grand canonical ensemble (12) in [27] by using different methods.

Remark 2 For $\Lambda = \mathbb{T}^d$, the author has recently extended the result of Theorem 1 to the optimal regime of w in the sense of L^p-integrability [34]. In particular, it is shown in [34, Theorem 1.9] that an appropriate variant of (17) holds for even $w \in L^p(\Lambda)$ with the same positivity properties in Theorem 1, where

$$p \in \begin{cases} [1, \infty] & \text{if} \quad d = 1 \\ (1, \infty] & \text{if} \quad d = 2 \\ (3, \infty] & \text{if} \quad d = 3 \,. \end{cases}$$

Moreover, in [34, Theorem 1.10], it is shown that when $d = 2$ and $w \in L^1$, an analogous result holds if $|\hat{w}(k)| \leq C/(1 + |k|)^\epsilon$ for some $\epsilon > 0$. This is in accordance with results obtained in the classical setting by Bourgain [4].

The strategy of our proof is based on a perturbative expansion in the interaction and is best illustrated in the example of the (relative) partition function. The general result follows by the same principle after adding an appropriate observable and concluding by a duality argument. In particular, if we consider the classical and quantum partition functions defined, respectively, by

$$A(z) := \int \mathrm{e}^{-zW} \, \mathrm{d}\mu \,, \quad A_\tau(z) = \frac{\mathrm{Tr}\,(P_\tau^\eta(z))}{\mathrm{Tr}\,(\mathrm{e}^{-H_{\tau,0}})} \,,$$

our goal is to prove that

$$\lim_{\tau \to \infty} A_\tau(z) = A(z) \quad \text{for } \mathrm{Re}\, z > 0 \,.$$

For $M \in \mathbb{N}$, we expand

$$A(z) = \sum_{m=0}^{M-1} a_m z^m + R_M(z) \,, \quad A_\tau(z) = \sum_{m=0}^{M-1} a_{\tau,m} z^m + R_{\tau,M}(z) \tag{18}$$

into asymptotic series by means of Duhamel's formula. A fundamental difficulty is that both series have radius of convergence zero. This is remedied by using the method of *Borel summation* as formulated in the work of Sokal [35] (see also the earlier known results [21, Theorem 136] and [30]). Given a formal power series $\mathscr{A}(z) = \sum_{m \geq 0} \alpha_m z^m$, its *Borel transform* is defined as $\mathscr{B}(z) := \sum_{m \geq 0} \frac{\alpha_m}{m!} z^m$. Formally one can recover $\mathscr{A}$ from $\mathscr{B}$ by a rescaled Laplace transform

$$\mathscr{A}(z) = \int_0^\infty \mathrm{d}t\, \mathrm{e}^{-t}\, \mathscr{B}(tz)\,.$$

Using the result of [35], this method can be applied to A and A_τ provided that we have

$$\begin{cases} |a_m| + |a_{\tau,m}| \leq C^m m! \\ |R_M(z)| + |R_{\tau,M}(z)| \leq C^M M! |z|^M \text{ for } \mathrm{Re} z \geq 0\,. \end{cases} \tag{19}$$

In proving the estimate on the explicit terms a_m and $a_{\tau,m}$, we use the *classical Wick theorem* for Gaussian measures and the *quantum Wick theorem* for quasi-free states, respectively. In the quantum setting, this result informally states that an expression of the form:

$$\frac{1}{\mathrm{Tr}\,(\mathrm{e}^{-H_{\tau,0}})} \mathrm{Tr}\left(\phi_\tau^*(x_1)\cdots\phi_\tau^*(x_m)\phi_\tau(y_1)\cdots\phi_\tau(y_m)\,\mathrm{e}^{-H_{\tau,0}}\right) \tag{20}$$

is given by a sum over all *pairings* of the labels $x_1, \ldots, x_m, y_1, \ldots, y_m$, where each pairing contributes a product over pairs of two-point functions of the form:

$$\frac{1}{\mathrm{Tr}\,(\mathrm{e}^{-H_{\tau,0}})} \mathrm{Tr}\left(\phi_\tau^*(x)\phi_\tau(y)\,\mathrm{e}^{-H_{\tau,0}}\right) = G_\tau(x;y)\,,$$

which is the *quantum Green function*. Expressions analogous to those in (20), but with a different number of creation and annihilation operators, vanish by gauge invariance of the Hamiltonian.

Due to the Duhamel expansion, we really consider products of *time-evolved quantum Green functions*:

$$G_{\tau,t} := \frac{\mathrm{e}^{-th/\tau}}{\tau(\mathrm{e}^{h/\tau}-1)} \quad \text{for } t \geq -1 \tag{21}$$

and time-evolved delta functions:

$$S_{\tau,t} := \mathrm{e}^{-th/\tau} \quad \text{for } t \geq 0\,. \tag{22}$$

Each contribution from the quantum Wick theorem is then encoded in terms of a graph and estimated by using an inductive integration algorithm [16, Sections 2.4–2.6]. A subtle point in the analysis arises from the time evolutions applied to the operators in (21) and (22). Namely, these operators are bounded only in operator norm. Part of the algorithm involves cancelling the time evolution [16, Section 2.5]. Finally, the modification of the grand canonical ensemble with $\eta \neq 0$ (16) is required to prove the bound in (19) on the remainder term $R_{\tau,M}(z)$.

1.2 The Time-Dependent Problem

Consider first a general Hamiltonian system $(\Gamma, H, \{\cdot, \cdot\})$ and denote by $\mathbb{P}(\mathrm{d}\phi) := \frac{1}{Z}\mathrm{e}^{-H(\phi)}\,\mathrm{d}\phi$ the associated Gibbs measure, which is assumed to be well-defined. Let S_t denote the flow map of H. Given $m \in \mathbb{N}$, $X^1, \dots, X^m \in C^\infty(\Gamma)$ (which are henceforth referred to as *observables*), and times $t_1, \dots, t_m \in \mathbb{R}$, the *m-particle time-dependent correlation function* is defined as

$$\mathscr{Q}_{\mathbb{P}}(X^1, \dots, X^m; t_1, \dots, t_m) := \int X^1(S_{t_1}\phi) \cdots X^m(S_{t_m}\phi)\,\mathrm{d}\mathbb{P}\,.$$

A time-dependent variant of the problem studied in the previous section is to obtain a derivation of $\mathscr{Q}_{\mathbb{P}}$ from many-body quantum expectation values in the setting where S_t is the Hamiltonian flow map generated by (2) and (3). In [17], we consider this problem when $\Lambda = \mathbb{T}^1$ or $\mathbb{R}$, $V = \kappa > 0$ and $w \in L^\infty(\Lambda)$. The fact that the flow is well-defined in the periodic problem is nontrivial and was first shown in [1]. Before recalling the main result of [17], it is necessary to set up some notation. For ξ a bounded operator on $\mathfrak{H}^{(p)}$, we define

$$\Theta_\tau(\xi) := \int \mathrm{d}x_1 \cdots \mathrm{d}x_p\,\mathrm{d}y_1 \cdots \mathrm{d}y_p\,\xi(x_1, \dots, x_p; y_1, \dots, y_p) \times \phi_\tau^*(x_1) \cdots \phi_\tau^*(x_p)\phi_\tau(y_1) \cdots \phi_\tau(y_p) \tag{23}$$

and

$$\Theta(\xi) := \int \mathrm{d}x_1 \cdots \mathrm{d}x_p\,\mathrm{d}y_1 \cdots \mathrm{d}y_p\,\xi(x_1, \dots, x_p; y_1, \dots, y_p) \times \bar{\phi}(x_1) \cdots \bar{\phi}(x_p)\phi(y_1) \cdots \phi(y_p)\,. \tag{24}$$

Given an operator $\mathscr{A}$ on the Fock space $\mathscr{F} \equiv \mathscr{F}(\mathfrak{H})$ we define

$$\Psi_{\tau,t}\,\mathscr{A} := \mathrm{e}^{\mathrm{i}t\tau H_\tau}\,\mathscr{A}\,\mathrm{e}^{-\mathrm{i}t\tau H_\tau}\,. \tag{25}$$

Furthermore, given an operator $\mathscr{B}$ on $\mathfrak{H}$, we define

$$(\Psi_t\,\mathscr{B})(f) := \mathscr{B}(f(t))\,, \tag{26}$$

where $f(t)$ is the solution of (1) with initial data $f \in \mathfrak{H}$. If we set $\mathscr{A} = \Theta_\tau(\xi)$, then (25) is the canonical time evolution of the observable given by (23). Likewise, if $\mathscr{B} = \Theta(\xi)$, then (26) is the canonical time evolution of (24). We now state the main result of [17].

Theorem 2 *Consider $w \in L^\infty(\Lambda)$ pointwise nonnegative and suppose that* (7) *holds with $s = 0$. Let $m \in \mathbb{N}$, $p_1, \ldots, p_m \in \mathbb{N}$, observables $\xi_j \in \mathscr{L}(\mathfrak{H}^{(p_j)})$ and times $t_1, \ldots, t_m \in \mathbb{R}$ be given. We have*

$$\rho_\tau\big(\Psi_{\tau,t_1}\Theta_\tau(\xi_1)\cdots\Psi_{\tau,t_m}\Theta_\tau(\xi_m)\big) \to \rho\big(\Psi_{t_1}\Theta(\xi_1)\cdots\Psi_{t_m}\Theta(\xi_m)\big) \quad \textit{as} \quad \tau \to \infty\,. \tag{27}$$

Let us note that when $m = 1$ and $t_1 = 0$, (27) follows directly from the corresponding result in [16]. Moreover, if $m = 1$, one can use (27) and cyclicity of the trace to give an alternative proof of the invariance of the Gibbs measure in this setting. A version of (27) had earlier been proved in the setting when the domain is a lattice in [23, Section 3.4].

The strategy of proof in [17] is based on the one in [23, Section 3.4]. One first divides the problem into two parts by adding a smooth cut-off in *particle number*:

$$\mathscr{N} := \int \mathrm{d}x\,|\phi(x)|^2\,,$$

and the *rescaled particle number*:

$$\mathscr{N}_\tau := \int \mathrm{d}x\,\phi_\tau^*(x)\,\phi_\tau(x)\,.$$

The large-particle number contribution can be made arbitrarily small by applying a version of Markov's inequality. For the small-particle number contribution, one applies a Schwinger–Dyson expansion of $\Psi_{\tau,t_j}\Theta(\xi_j)$ and $\Psi_{t_j}\Theta_\tau(\xi_j)$ in t_j in the spirit of [23, Section 4.2]. The main technical step that one needs to prove is that for a function $f \in C_c^\infty(\mathbb{R})$ with $f \geq 0$ we have

$$\rho_\tau\big(\Theta_\tau(\xi) f(\mathscr{N}_\tau)\big) \to \rho\big(\Theta(\xi) f(\mathscr{N})\big) \quad \text{as} \quad \tau \to \infty\,. \tag{28}$$

Note that, in [16], the analogous result is obtained with $f = 1$. The reason why (28) for compactly supported f is more challenging is that the presence of f destroys the symmetries that allow us to apply Wick's theorem. More precisely, it destroys the Gaussianity of the free measure in the classical setting and the quasi-freeness of the free measure in the quantum setting. Moreover, the cut-off in the number of particles coming from f is necessary to ensure the convergence of the Schwinger–Dyson series [17, Sections 3.2–3.3]. The strategy to prove (28) is to reduce it to the analysis of [16] by expanding the expressions according to the *Helffer–Sjöstrand formula* and then applying the Wick theorem for a Hamiltonian translated by a real chemical potential.

In [17, Section 5], the local version of (1) is also studied for $\Lambda = \mathbb{T}^1$ and $v = 0$. In particular, a partial result is shown by using an approximation argument from the nonlocal problem. More precisely, let w be a continuous compactly supported nonnegative function satisfying $\int \mathrm{d}x\, w(x) = 1$. For $\epsilon > 0$, define the two-body potential as

$$w^{\epsilon}(x) := \frac{1}{\epsilon}\, w\left(\frac{[x]}{\epsilon}\right), \tag{29}$$

where $[x]$ denotes the unique element of the set $(x+\mathbb{Z})\cap[-1/2, 1/2)$. We state the result [17, Theorem 1.5].

Theorem 3 *There exists a sequence* (ϵ_τ) *of positive numbers satisfying* $\lim_{\tau\to\infty}\epsilon_\tau = 0$, *such that, for arbitrary* $m \in \mathbb{N}$, $p_1, \ldots, p_m \in \mathbb{N}$, *observables* $\xi^j \in \mathscr{L}(\mathfrak{H}^{(p_j)})$ *and times* $t_1 \in \mathbb{R}, \ldots, t_m \in \mathbb{R}$, *we have*

$$\lim_{\tau\to\infty} \rho_\tau^{\epsilon_\tau}\Big(\Psi_\tau^{t_1,\epsilon_\tau}\,\Theta_\tau(\xi^1)\cdots\Psi_\tau^{t_m,\epsilon_\tau}\,\Theta_\tau(\xi^m)\Big) = \rho\Big(\Psi^{t_1}\,\Theta(\xi^1)\cdots\Psi^{t_m}\,\Theta(\xi^m)\Big).$$

Here, all the objects on the left-hand side are defined with interaction w^{ϵ_τ} *given as in (29), and all the objects on the right-hand side are defined with interaction* $w=\delta$.

The approximation argument used to deduce Theorem 3 from Theorem 2 reduces to a PDE problem. Namely, for $\phi_0 \in H^{s_0}(\mathbb{T}^1)$ with appropriately chosen s_0, one compares the Cauchy problems:

$$\begin{cases} \mathrm{i}\partial_t u + \Delta u = |u|^2 u \\ u|_{t=0} = \phi_0 \end{cases} \quad \text{and} \quad \begin{cases} \mathrm{i}\partial_t u^\epsilon + \Delta u^\epsilon = (w^\epsilon * |u^\epsilon|^2)u^\epsilon \\ u^\epsilon|_{t=0} = \phi_0\,, \end{cases}$$

and shows that for all $T>0$ the solutions satisfy

$$\lim_{\epsilon\to 0}\|u^\epsilon(\cdot,t)-u(\cdot,t)\|_{\mathfrak{H}} = 0 \quad \text{uniformly in } t\in[-T,T]. \tag{30}$$

The proof of (30) is based on dispersive properties of the equation through the use of $X^{s,b}$ *spaces* (also known as *dispersive Sobolev spaces*), which are given by the norm:

$$\|f\|_{X^{s,b}} := \|\mathrm{e}^{-\mathrm{i}t\Delta} f\|_{H_t^b H_x^s} \sim \left\|\left(1+|2\pi k|\right)^s\left(1+|\eta+2\pi k^2|\right)^b \tilde{f}\right\|_{L_\eta^2 l_k^2},$$

where

$$\tilde{f}(k,\eta) := \int_{-\infty}^{\infty}\mathrm{d}t \int_\Lambda \mathrm{d}x\; f(x,t)\,\mathrm{e}^{-2\pi\mathrm{i}kx-2\pi\mathrm{i}\eta t}$$

denotes the *spacetime Fourier transform*. These spaces are a fundamental tool to study nonlinear dispersive PDEs (see [37, Chapter 2.6]).

References

1. J. Bourgain, Fourier transform restriction phenomena for certain lattice subsets and applications to nonlinear evolution equations, I: Schrödinger equations. Geom. Funct. Anal. **3**, 107–156 (1993)
2. J. Bourgain, Periodic nonlinear Schrödinger equation and invariant measures. Commun. Math. Phys. **166**(1), 1–26 (1994)
3. J. Bourgain, Invariant measures for the 2D-defocusing nonlinear Schrödinger equation. Commun. Math. Phys. **176**(2), 421–445 (1996)
4. J. Bourgain, Invariant measures for the Gross-Pitaevskii equation. J. Math. Pures et Appl. **T76**, F. 8, 649–702 (1997)
5. J. Bourgain, Invariant measures for NLS in infinite volume. Commun. Math. Phys. **210**, 605–620 (2000)
6. J. Bourgain, A. Bulut, Gibbs measure evolution in radial nonlinear wave and Schrödinger equations on the ball. C. R. Math. Acad. Sci. Paris **350**, 571–575 (2012)
7. J. Bourgain, A. Bulut, Almost sure global well posedness for the radial nonlinear Schrödinger equation on the unit ball II: the 3D case. J. Eur. Math. Soc. **16**, 1289–1325 (2014)
8. J. Bourgain, A. Bulut, Almost sure global well posedness for the radial nonlinear Schrödinger equation on the unit ball I: the 2D case, Ann. Inst. H. Poincaré Anal. Non Linéaire, **31**, 1267–1288 (2014)
9. N. Burq, L. Thomann, N. Tzvetkov, Long time dynamics for the one dimensional non linear Schrödinger equation. Ann. Inst. Fourier (Grenoble) **63**, 2137–2198 (2013)
10. F. Cacciafesta, A.-S. de Suzzoni, Invariant measure for the Schrödinger equation on the real line. J. Funct. Anal. **269**, 271–324 (2015)
11. F. Cacciafesta, A.-S. de Suzzoni, On Gibbs measure and weak flow for the cubic NLS with non-localised initial data. Preprint. arXiv:1507.03820
12. Y. Deng, Two-dimensional nonlinear Schrödinger equation with random initial data. Anal. PDE **5**, 913–960 (2012)
13. A. Elgart, L. Erdős, B. Schlein, H.-T. Yau, Gross-Pitaevskii equation as the mean field limit of weakly coupled Bosons. Arch. Ration. Mech. Anal. **179**, 265–283 (2006)
14. L. Erdős, H.-T. Yau, Derivation of the nonlinear Schrödinger equation from a many body Coulomb system. Adv. Theor. Math. Phys. **5**(6), 1169–1205 (2001)
15. L. Erdős, B. Schlein, H.-T. Yau, Derivation of the cubic non-linear Schrödinger equation from quantum dynamics of many-body systems, Invent. Math. **167**(3), 515–614 (2007)
16. J. Fröhlich, A. Knowles, B. Schlein, V. Sohinger, Gibbs measures of nonlinear Schrödinger equations as limits of many-body quantum states in dimensions $d \leq 3$. Commun. Math. Phys. **356**(3), 883–980 (2017)
17. J. Fröhlich, A. Knowles, B. Schlein, V. Sohinger, A microscopic derivation of time-dependent correlation functions of the $1D$ cubic nonlinear Schrödinger equation. Adv. Math. **353**, 67–115 (2019)
18. G. Genovese, R. Lucà, D. Valeri, Gibbs measures associated to the integrals of motion of the periodic dNLS. Sel. Math. New Ser. **22**(3), 1663–1702 (2016)
19. J. Ginibre, G. Velo, The classical field limit of scattering theory for nonrelativistic many-Boson systems. I and II. Commun. Math. Phys. **66**(1), 37–76 (1979). And **68**(1), 45–68 (1979)
20. J. Glimm, A. Jaffe, *Quantum Physics. A Functional Integral Point of View*, 2nd edn. (Springer, Berlin, Heidelberg, 1987)
21. G.H. Hardy, *Divergent Series* (Clarendon Press, Oxford, 1949)
22. K. Hepp, The classical limit for quantum mechanical correlation functions. Commun. Math. Phys. **35**, 265–277 (1974)
23. A. Knowles, Limiting dynamics in large quantum systems, Ph. D. Thesis, ETH, June 2009
24. J. Lebowitz, H. Rose, E. Speer, Statistical mechanics of the nonlinear Schrödinger equation. J. Stat. Phys. **50**, 657–687 (1988)

25. M. Lewin, P.T. Nam, N. Rougerie, Derivation of nonlinear Gibbs measures from many-body quantum mechanics. J. Éc. Polytech. Math. **2**, 65–115 (2015)
26. M. Lewin, P.T. Nam, N. Rougerie, Gibbs measures based on 1D (an)harmonic oscillators as mean-field limits. J. Math. Phys. **59**(4), 041901 (2018)
27. M. Lewin, P.T. Nam, N. Rougerie, Classical field theory limit of 2D many-body quantum Gibbs states. Preprint, arXiv: 1805.08370
28. A. Nahmod, L. Rey-Bellet, S. Sheffield, G. Staffilani, Absolute continuity of Brownian bridges under certain gauge transformations. Math. Res. Lett. **18**, 875–887 (2011)
29. A. Nahmod, T. Oh, L. Rey-Bellet, G. Staffilani, Invariant weighted Wiener measures and almost sure global well-posedness for the periodic derivative NLS. J. Eur. Math. Soc. **14**, 1275–1330 (2012)
30. F. Nevanlinna, Zur Theorie der asymptotischen Potenzreihen. Ann. Acad. Sci. Fen. Ser. A **12**(3), 1918–1919
31. T. Oh, J. Quastel, On invariant Gibbs measures conditioned on mass and momentum. J. Math. Soc. Jpn. **65**, 13–35 (2013)
32. D. Ellwood, I. Rodniansky, G. Staffilani, J. Wunsch (eds.), *Evolution equations: Clay Mathematics Institute Summer School, Evolution Equations, Eidgenössische Technische Hochschule, Zürich, Switzerland, June 23–July 18, 2008*. Clay Mathematics Proceedings, vol. 17 (2008)
33. B. Simon, *The* $P(\Phi)_2$ *Euclidean (Quantum) Field Theory* (Princeton University Press, Princeton, 1974)
34. V. Sohinger, A microscopic derivation of Gibbs measures for nonlinear Schrödinger equations with unbounded interaction potentials, Preprint, arXiv:1904.08137
35. A.D. Sokal, An improvement of Watson's theorem on Borel summability. J. Math. Phys. **21**(2), 261–263 (1980)
36. H. Spohn, Kinetic equations from Hamiltonian Dynamics. Rev. Mod. Phys. **52**(3), 569–615 (1980)
37. T. Tao, *Nonlinear Dispersive Equations: Local and Global Analysis*. CBMS Regional Conference Series in Mathematics, vol. 106 (AMS, Providence, RI, 2006)
38. L. Thomann, N. Tzvetkov, Gibbs measure for the periodic derivative nonlinear Schrödinger equation. Nonlinearity **23**, 2771 (2010)
39. N. Tzvetkov, Invariant measures for the Nonlinear Schrödinger equation on the disc. Dynam. PDE **2**, 111–160 (2006)
40. N. Tzvetkov, Invariant measures for the defocusing nonlinear Schrödinger equation, Ann. Inst. Fourier (Grenoble) **58**, 2543–2604 (2008)
41. P.E. Zhidkov, An invariant measure for the nonlinear Schrödinger equation (Russian). Dokl. Akad. Nauk SSSR **317** (1991) 543–546; translation in Soviet Math. Dokl. **43**, 431–434

Interfaces in Spectral Asymptotics and Nodal Sets

Steve Zelditch

1 Introduction

This is mainly an expository article on interfaces in spectral asymptotics. Interfaces are studied in many fields of mathematics and physics but seem to be a novel area of spectral asymptotics. Spectral asymptotics refers to the behavior of spectral projections and nodal sets for a quantum Hamiltonian $\hat{H}_\hbar$, which might be a Schrödinger operator on $L^2(\mathbb{R}^d)$ or on a Riemannian manifold (M, g), with or without boundary, or a Toeplitz Hamiltonian acting on holomorphic sections $H^0(M, L^k)$ of line bundles over a Kähler manifold. Interface asymptotics refers to the change in behavior of the spectral projections or nodal sets as a hypersurface is crossed, either in physical space (configuration space) or in phase space. Interfaces exist in diverse settings and indeed the purpose of this article is to compare interface behavior in different settings and to consider possible future settings that have yet to be explored.

What is meant by an "interface" in the sense of this article? The general idea is that there is a hypersurface in the phase space separating two regions in which the asymptotic behavior of a spectral projections kernel has different types of behavior: In the first, that we will term the "allowed" region, the asymptotics are constant and, after normalization, equal to 1, so that one has a plateau over the region; in the second "forbidden" region the asymptotics are rapidly decaying, so that one has a rather flat 0 region. The interface is the shape of the graph of the spectral kernel connecting 1 and 0 in a thin region separating the allowed and forbidden

Research partially supported by NSF grant DMS-1810747.

S. Zelditch (✉)
Department of Mathematics, Northwestern University, 60208 Evanston, IL, USA
e-mail: zelditch@math.northwestern.edu

N. Anantharaman et al. (eds.), *Frontiers in Analysis and Probability*,
https://doi.org/10.1007/978-3-030-56409-4_10

region. One expects that when scaled properly, the limit shape is universal. More precisely, universality holds in each type of model (e.g. Schrödinger or Kähler) but is model-dependent: one expects "Airy interfaces" in the Schrödinger setting and Erf interfaces in the Kähler setting. The separation into different regions for the spectral projections kernel often coincides with the separation of other spectral behavior, such as nodal sets of the eigenfunctions.

The terminology (classically) "allowed" and (classically) "forbidden" is standard in quantum mechanics for regions inside, resp. outside, of an energy surface in phase space, or more commonly, the projection of these regions to configuration space. This will indeed be the meaning of "interface" for most of this article. We will describe results of B. Hanin, P. Zhou and the author [HZZ15,HZZ16] on the different behavior of nodal sets of Schrödinger eigenfunctions in allowed resp. forbidden regions for the simplest Schrödinger Hamiltonian $\hat{H}_\hbar$, namely the isotropic Harmonic oscillator on $\mathbb{R}^d$. We then consider phase space interfaces of Wigner distributions for the same model, following [HZ19,HZ19b]. We then turn to phase space interfaces in the Kähler (complex holomorphic) setting, and discuss results of Pokorny-Singer [33], Ross-Singer [35], P. Zhou and the author [ZZ16,ZZ17] on interfaces for *partial Bergman kernel* asymptotics. In Section 8 we explain that the exact analogue of the results on Wigner distributions for the isotropic harmonic oscillator in the complex setting is a series of results on interfaces for disc bundles in the Bargmann-Fock space of a line bundle. This Bargmann-Fock space and the interface results constitute the new results of the article.

Roughly speaking, interfaces in spectral asymptotics involve two types of localization: (i) spectral, i.e. quantum, localization where the eigenvalues are constrained to lie in an interval I, (ii) classical, i.e. phase space, localization where a phase space point is constrained to lie in an open set U of phase space. It has long been understood that spectral localization $E_j(\hbar) \in I$ implies phase space localization in the sense that quantum objects decay in the complement of the allowed region $H^{-1}(I)$. But the study of interfaces is devoted to the precise behavior of quantum objects as one crosses the interface between allowed and forbidden regions, and more generally, considers all possible combinations of spectral localization $E_j(\hbar) \in I$ and phase space localization $\zeta \in U$, where U may have any position relative to $H^{-1}(I)$.

Often, the interface corresponds to a sharp cutoff in a spectral parameter and signals something discontinuous. In fact, the earliest studies of interface asymptotics are classical analysis studies of Bernstein polynomials of discontinuous functions with jump discontinuities [13, 28–31]. These studies were intended to be analogues of Gibbs phenomena for Fourier series of discontinuous functions, which have been generalized to wave equations on Riemannian manifolds in [32].

In this article we review the following results on interface asymptotics:

- Interface behavior for spectral projections and for nodal sets of random eigenfunctions of energy $E_N(\hbar) = \hbar(N + \frac{d}{2}) = E$ of the isotropic harmonic oscillator on $\mathbb{R}^d$ across the *caustic set* in physical space, where the potential $V(x) = |x|^2/2 = E$.

- Interface behavior for Wigner distributions of the same eigenspace projections, and more generally for various types of Wigner–Weyl sums across an energy surface in phase space.
- Interface behavior for the holomorphic analogues of such Wigner distributions, namely for *partial Bergman kernels* for general Berezin-Toeplitz Hamiltonians on general Kähler phase spaces.
- Interface results for partial Bergman kernels corresponding to the canonical S^1 action on the total space L^* of the dual line bundle of an ample line bundle $L \to M$ over a Kähler manifold.

In the case of Schrödinger operators, the results are only proved in the special case of the isotropic harmonic oscillator. It is plausible that some of the results should be universal among Schrödinger operators, but at the present time the generalizations have not been formulated or proved. See Section 9.1 for further problems. Among other gaps in the theory, Wigner distributions per se are only defined when the Riemannian manifold is $\mathbb{R}^d$ and are closely connected to the representation theory of the Heisenberg and metaplectic groups. Wigner distributions of eigenfunctions are special types of "microlocal lifts" of eigenfunctions; there is no generally accepted canonical microlocal lift on a general Riemannian manifold. Despite the restrictive setting, Wigner distributions are important in mathematical physics, in particular in quantum optics. The results in the complex holomorphic (Kähler) setting are much more complete, due to the fact that the theory of Bergman kernels is technically simpler and more complete than the corresponding theory of Wigner distributions for Schrödinger operators. The results are proved for any Toeplitz Hamiltonian on any projective Kähler manifold. In fact, the exact analogue of the Wigner result is proved in Section 8, where a new construction is introduced in this article: the Bargmann-Fock space of a holomorphic line bundle. It is a Gaussian space of holomorphic functions on the total space L^* of the dual of a holomorphic Hermitian line bundle $L \to M$ over a Kähler manifold. This total space carries a natural S^1 action[1] and this S^1 action plays the role of the propagator of the isotropic Harmonic Oscillator. Thus, the interfaces are the boundaries of the co-disc bundles $D_E^* \subset L^*$ of different energy levels (i.e. radii). The interface results in Section 8 are a "new result" of this article, but the proofs are similar to, and simpler than, those in [ZZ17,ZZ18].

This survey is organized as follows:

(1) In Section 2, we review the basic linear models: the Harmonic oscillator in the Schrödinger representation on $L^2(\mathbb{R}^d)$ and in the Bargmann-Fock (holomorphic) representation on entire holomorphic functions on $\mathbb{C}^d$. We also present a list of analogies between the real Schrödinger setting and the complex holomorphic quantization. Section 3 is devoted to the Bergman kernel on Bargmann-Fock space, and the Bargmann-Fock representations of the Heisenberg and Symplectic groups on Bargmann-Fock space.

[1] S^1 always denotes the unit circle.

(2) In Section 4, we review the interface results in physical space for spectral projections for the isotropic Harmonic Oscillator. These imply interface results for nodal sets of random eigenfunctions in a fixed eigenspace.
(3) In Section 5, we change the setting to phase space $T^*\mathbb{R}^d$ and review the interface results in physical space for Wigner distributions of spectral projections for the isotropic Harmonic Oscillator.
(4) In Section 6, we switch to the complex holomorphic setting and review interface results for partial Bergman kernels on general compact Kähler manifolds.
(5) In Section 7 we specialize to the isotropic harmonic oscillator on the standard Bargmann-Fock space and describe its interfaces.
(6) In Section 8 we introduce a new model: the Bargmann-Fock space of a holomorphic line bundle. We then consider interfaces with respect to a natural S^1 action on this space, generalizing the previous result on the Bargmann-Fock isotropic Harmonic oscillator.
(7) In Section 9.1 we list some further problems on interfaces.
(8) In Section 10 we give some background to the holomorphic setting.

1.1 Results Surveyed in This Article

The articles surveyed in this article are the following:

References

[HZZ15] Boris Hanin, Steve Zelditch, Peng Zhou Nodal Sets of Random Eigenfunctions for the Isotropic Harmonic Oscillator, International Mathematics Research Notices, Vol. 2015, No. 13, pp. 4813–4839, (2015) (arXiv:1310.4532)

[HZZ16] Boris Hanin, Steve Zelditch and Peng Zhou, Scaling of harmonic oscillator eigenfunctions and their nodal sets around the caustic. Comm. Math. Phys. 350 (2017), no. 3, 1147-1183 (arXiv:1602.06848).

[HZ19] B. Hanin and S. Zelditch, Interface Asymptotics of Eigenspace Wigner distributions for the Harmonic Oscillator, to appear in Comm. PDE (arXiv:1901.06438).

[HZ19b] B. Hanin and S. Zelditch, Interface Asymptotics of Wigner-Weyl Distributions for the Harmonic Oscillator, arXiv:1903.12524.

[ZZ16] S. Zelditch and P. Zhou, Interface asymptotics of partial Bergman kernels on S^1-symmetric Kaehler manifolds, J. Symplectic Geom. 17 no. 3, (2019), 793-856. (arXiv:1604.06655).

[ZZ17] S. Zelditch and P. Zhou, Central Limit theorem for spectral Partial Bergman kernels, Geom. Topl. 23 (2019), 1961-2004 (arXiv:1708.09267.)

[ZZ18] S. Zelditch and P. Zhou, Interface asymptotics of Partial Bergman kernels around a critical level, Ark. Mat. 57 (2019), 471-492 (arXiv:1805.01804).

[ZZ18b] S. Zelditch and P. Zhou, Pointwise Weyl law for Partial Bergman kernels, *Algebraic and Analytic Microlocal Analysis* pp. 589–634. M. Hitrik, D. Tamarkin, B. Tsygan, S. Zelditch (eds). Springer Proceedings in Mathematics and Statistics, Springer-Verlag (2018).

2 The Basic Linear Models

As mentioned above, our aim in this survey is not only to describe interface results in various settings but also to compare the results in the real Schrödinger setting and the complex holomorphic Bargmann-Fock or Berezin-Toeplitz setting. The real setting is self-explanatory to mathematical physicists but the complex holomorphic setting is probably less familiar. In this section, we give some background on the basic linear models (isotropic Harmonic Oscillator in both settings) to make the relations between the real and complex settings more familiar. We then give a list of analogies between the two settings. In addition, we present a list of open problems on interfaces to amplify the scope of spectral interface problems. It would be laborious to present all of the background for the geometric setting before getting to the main results and phenomena, so we have put that background into an Appendix Section 10.

A preliminary remark: Since the early days of quantum mechanics, it was understood that there are many equivalent representations (or "pictures") of quantum mechanics. In the case of $\mathbb{R}^d$ they correspond to different but unitarily equivalent representations of the Heisenberg and metaplectic groups (see [16] for background). The most common are the Schrödinger representation on $L^2(\mathbb{R}^d)$ and the Bargmann-Fock representation on $H^2(\mathbb{C}^d, e^{-|Z|^2} dL(Z))$, the Bargmann-Fock space of entire holomorphic functions on $\mathbb{C}^d$ which are in L^2 with respect to Gaussian measure; here dL is Lebesgue measure. One refers to $\mathbb{R}^d$ as "configuration space" or "physical space" and to $T^*\mathbb{R}^d$ as phase space. Of course, $T^*\mathbb{R}^d \simeq \mathbb{C}^d$, so that Bargmann-Fock space employs a complex structure on phase space. A natural unitary intertwining operator is the Bargmann transform (see (26) below). We refer to [16] and to [25] for background on Bargmann-Fock space and metaplectic operators.

The first item is to give background on the isotropic Harmonic oscillator in both the Schrödinger representation and the Bargmann-Fock representation.

2.1 Schrödinger Representation of the Isotropic Harmonic Oscillator

The Schrödinger representation of quantum mechanics is too familiar to need a detailed review here. The isotropic Harmonic Oscillator on $L^2(\mathbb{R}^d, dx)$ is the operator,

$$\widehat{H}_\hbar = \sum_{j=1}^{d} \left(-\frac{\hbar^2}{2} \frac{\partial^2}{\partial x_j^2} + \frac{x_j^2}{2} \right). \tag{1}$$

It has a discrete spectrum of eigenvalues

$$E_N(\hbar) = \hbar(N + d/2), \qquad (N = 0, 1, 2, \dots) \tag{2}$$

with multiplicities given by the composition function $p(N, d)$ of N and d (i.e. the number of ways to write N as an ordered sum of d non-negative integers). That is, the eigenspaces

$$V_{\hbar, E_N(\hbar)} := \{\psi \in L^2(\mathbb{R}^d) : \widehat{H}_\hbar \psi = E_N(\hbar)\psi\} \tag{3}$$

have dimensions given by

$$\dim V_{\hbar_N, E} = p(N, d) = \frac{1}{(d-1)!} N^{d-1}(1 + O(N^{-1})). \tag{4}$$

When $E_N(\hbar) = E$ we also write

$$\hbar = \hbar_N(E) := \frac{E}{N + \frac{d}{2}}. \tag{5}$$

An orthonormal basis of its eigenfunctions is given by the product Hermite functions,

$$\phi_{\alpha,h}(x) = h^{-d/4} p_\alpha\left(x \cdot h^{-1/2}\right) e^{-x^2/2h}, \tag{6}$$

where $\alpha = (\alpha_1, \dots, \alpha_d) \geq (0, \dots, 0)$ is a $d-$dimensional multi-index and $p_\alpha(x)$ is the product $\prod_{j=1}^{d} p_{\alpha_j}(x_j)$ of the Hermite polynomials p_k (of degree k) in one variable.

The eigenspace projections are the orthogonal projections

$$\Pi_{\hbar, E_N(\hbar)} : L^2(\mathbb{R}^d) \to V_{\hbar, E_N(\hbar)}. \tag{7}$$

When $E_N(\hbar) = E$ (5), their Schwartz kernels are given in terms of an orthonormal basis by,

$$\Pi_{h_N,E}(x, y) = \sum_{|\alpha|=N} \phi_{\alpha,h_N}(x)\phi_{\alpha,h_N}(y). \tag{8}$$

The high multiplicities are due to the $U(d)$-invariance of the isotropic Harmonic Oscillator. Due to extreme degeneracy of the spectrum of (1) when $d \geq 2$, the eigenspace projections have very special semi-classical asymptotic properties, reflecting the periodicity of the classical Hamiltonian flow and of the Schrödinger propagator $\exp[-\frac{it}{\hbar}\widehat{H}_\hbar]$. In particular, the eigenspace projections (7) are semi-classical Fourier integral operators (see, e.g., [19, 20], [HZ19]). We exploit this very rare property to obtain scaling asymptotics across the caustic. This explains why the results to date are only available for isotropic oscillators. For general Harmonic Oscillators with incommensurate frequencies the eigenvalues have multiplicity one and the eigenspace projections are of a very different type. For general Schrödinger operator, one would need to take appropriate combinations of eigenspace projections with eigenvalues in an interval.

As with any 1-parameter metaplectic unitary group [16, 25], one has an explicit Mehler formula for the Schwartz kernel $U_h(t, x, y)$ of the propagator, $e^{-\frac{i}{\hbar}tH_h}$. The Mehler formula [16] reads

$$U_h(t, x, y)=e^{-\frac{i}{\hbar}tH_h}(x, y)=\frac{1}{(2\pi ih\sin t)^{d/2}}\exp\left(\frac{i}{h}\left(\frac{|x|^2+|y|^2}{2}\frac{\cos t}{\sin t}-\frac{x\cdot y}{\sin t}\right)\right), \tag{9}$$

where $t \in \mathbb{R}$ and $x, y \in \mathbb{R}^d$. The right-hand side is singular at $t = 0$. It is well-defined as a distribution, however, with t understood as $t - i0$. Indeed, since H_h has a positive spectrum the propagator U_h is holomorphic in the lower half-plane and $U_h(t, x, y)$ is the boundary value of a holomorphic function in $\{\mathrm{Im} t < 0\}$.

One may express the Nth spectral projection as a Fourier coefficient of the propagator. It is somewhat simpler to work with the number operator $\mathcal{N}$, i.e. the Schrödinger operator with the same eigenfunctions as H_h and eigenvalues $h|\alpha|$. If we replace $U_h(t)$ by $e^{-\frac{it}{\hbar}\mathcal{N}}$, then the spectral projections $\Pi_{h,E}$ are simply the Fourier coefficients of $e^{-\frac{it}{\hbar}\mathcal{N}}$. In [HZZ15, HZZ16] it is shown that

$$\Pi_{h_N,E}(x, y) = \int_{-\pi}^{\pi} U_h(t - i\epsilon, x, y)e^{\frac{i}{\hbar}(t-i\epsilon)E}\frac{dt}{2\pi}. \tag{10}$$

The integral is independent of ϵ. Combining (10) with the Mehler formula (9), one has an explicit integral representation of (8).

2.1.1 Wigner Distributions

For any Schwartz kernel $K_\hbar \in L^2(\mathbb{R}^d \times \mathbb{R}^d)$ one may define the Wigner distribution of $K_\hbar$ by

$$W_{K,\hbar}(x,\xi) := \int_{\mathbb{R}^d} K_\hbar\left(x+\frac{v}{2}, x-\frac{v}{2}\right) e^{-\frac{i}{\hbar} v\xi} \frac{dv}{(2\pi h)^d}. \tag{11}$$

The map from $K_\hbar \to W_{K,\hbar}$ defines the unitary "Wigner transform,"

$$\mathcal{W}_\hbar : L^2(\mathbb{R}^d \times \mathbb{R}^d) \to L^2(T^*\mathbb{R}^d).$$

The inverse Wigner transform is given by (see page 79 of [16])

$$f \otimes g^*(x,y) = \int W_{f,g}(\frac{x+y}{2}, \xi) e^{i\langle x-y,\xi\rangle} d\xi. \tag{12}$$

Here, $W_{f,g} := W_{f\otimes g^*}$ is the Wigner transform of the rank one operator $f \otimes g^*$.

The unitary group $U(d)$ acts on $L^2(\mathbb{R}^d \times \mathbb{R}^d)$ by conjugation,$U(g) \cdot K = gKg^*$, where we identify $K(x,y) \in L^2(\mathbb{R}^d \times \mathbb{R}^d)$ with the associated Hilbert–Schmidt operator. Metaplectic covariance implies that,

$$\mathcal{W}_\hbar U(g) = T_g \mathcal{W}_\hbar.$$

Definition 2.1 The Wigner distributions $W_{\hbar,E_N(\hbar)}(x,p) \in L^2(T^*\mathbb{R}^d)$ of the eigenspace projections $\Pi_{\hbar,E_N(\hbar)}$ are defined by,

$$W_{\hbar,E_N(\hbar)}(x,\xi) = \int_{\mathbb{R}^d} \Pi_{\hbar,E_N(\hbar)}\left(x+\frac{v}{2}, x-\frac{v}{2}\right) e^{-\frac{i}{\hbar} v\cdot\xi} \frac{dv}{(2\pi h)^d}. \tag{13}$$

When $E_N(\hbar) = E$, the Wigner distribution $W_{\hbar,E_N(\hbar)}$ of a single eigenspace projection (13) is the "quantization" of the energy surface of energy E and should therefore be localized at the classical energy level $H(x,\xi) = E$, where $H(x,\xi) = \frac{1}{2}\sum_{j=1}^d (\xi_j^2 + x_j^2)$. We denote the (energy) level sets by,

$$\Sigma_E = \{(x,\xi) \in T^*\mathbb{R}^d : H(x,\xi) := \frac{1}{2}(||x||^2 + ||\xi||^2) = E\}. \tag{14}$$

The Hamiltonian flow of H is 2π periodic, and its orbits form the complex projective space $\mathbb{CP}^{d-1} \simeq \Sigma_E / \sim$ where $\sim$ is the equivalence relation of belonging to the same Hamilton orbit. Due to this periodicity, the projections (7) are semi-classical Fourier integral operators (see [19, 20], [HZZ15]). This is also true for the Wigner distributions (13). Their properties are basically unique to the isotropic oscillator (1). These properties are visible in Figure 1 depicting the graph of $W_{\hbar,1/2}$.

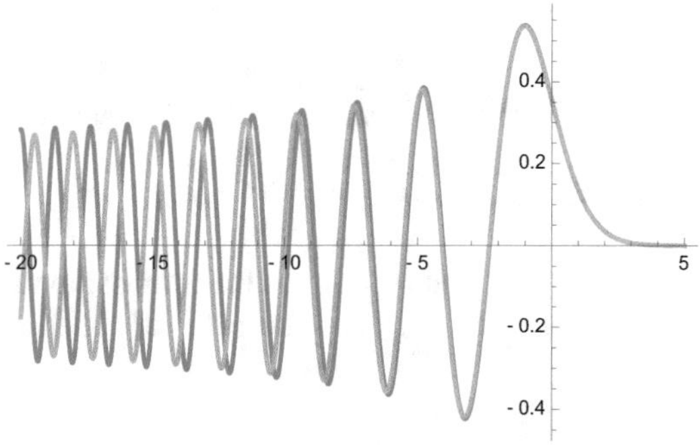

Fig. 1 The Wigner function $W_{\hbar,E_N(\hbar)}$ of the eigenspace projection $\Pi_{\hbar,E_N(\hbar)}$ is always radial (see Proposition 5.1). Displayed above is the graph of the Airy function (orange) and of $W_{\hbar,E_N(\hbar)}$ with $N = 500$ (blue) as a function of the rescaled radial variable ρ in a $\hbar^{2/3}$ tube around the energy surface $H(x,\xi) = E_N(\hbar) = 1/2$. Theorem 5.3 predicts that, when properly scaled, $W_{\hbar,E_N(\hbar)}$ should converge to the Airy function (with the rate of convergence being slower farther from the energy surface, which is defined here by $\rho = 0$).

2.1.2 Weyl Pseudo-Differential Operators, Metaplectic Covariance

A semi-classical Weyl pseudo-differential operator is defined by the formula,

$$Op_h^w(a)u(x) = \int_{\mathbb{R}^d}\int_{\mathbb{R}^d} a_\hbar(\frac{1}{2}(x+y),\xi)e^{\frac{i}{\hbar}\langle x-y,\xi\rangle}u(y)dyd\xi.$$

See [16, 47] for background. By using the identity

$$\langle Op^w(a)f, f\rangle = \int_{T^*\mathbb{R}^d} a(x,\xi)W_{f,f}(x,\xi)dxd\xi,$$

of [16, Proposition 2.5] for orthonormal basis elements $f = \phi_{\alpha,\hbar_N}$ of $V_{\hbar,E_N(\hbar)}$ and summing over α, one obtains the (well-known) identity,

$$\text{Tr } Op_h^w(a)\Pi_{\hbar,E_N(\hbar)} = \int_{T^*\mathbb{R}^d} a(x,\xi)W_{\hbar,E_N(\hbar)}(x,\xi)dxd\xi. \tag{15}$$

This formula is one of the key properties of Wigner distributions and Weyl quantization.

The Wigner transform (40) taking kernels to Wigner functions is therefore an isometry from Hilbert–Schmidt kernels $K(x,y)$ on $\mathbb{R}^d \times \mathbb{R}^d$ to their Wigner distributions on $T^*\mathbb{R}^d$ [16]. From (15) and this isometry, it is straightforward to check that,

$$\begin{cases} (i) \ \int_{T^*\mathbb{R}^d} W_{\hbar,E_N(\hbar)}(x,\xi)dxd\xi = \mathrm{Tr}\Pi_{\hbar,E_N(\hbar)} = \dim V_{\hbar,E_N(\hbar)} = \binom{N+d-1}{d-1} \\ (ii) \ \int_{T^*\mathbb{R}^d} \left|W_{\hbar,E_N(\hbar)}(x,\xi)\right|^2 dxd\xi = \mathrm{Tr}\Pi^2_{\hbar,E_N(\hbar)} = \dim V_{\hbar,E_N(\hbar)} = \binom{N+d-1}{d-1} \\ (iii) \ \int_{T^*\mathbb{R}^d} W_{\hbar,E_N(\hbar)}(x,\xi)\overline{W_{\hbar,E_M(\hbar)}(x,\xi)}dxd\xi = \mathrm{Tr}\Pi_{\hbar,E_N(\hbar)}\Pi_{\hbar,E_M(\hbar)} = 0, \text{ for } M \neq N. \end{cases} \tag{16}$$

In these equations, $N = \frac{E}{\hbar} - \frac{d}{2}$, and $\binom{N+d-1}{d-1}$ is the composition function of (N, d) (i.e. the number of ways to write N as an ordered us of d non-negative integers). Thus, the sequence

$$\{\frac{1}{\sqrt{\dim V_{\hbar,E_N(\hbar)}}} W_{\hbar,E_N(\hbar)}\}_{N=1}^{\infty} \subset L^2(\mathbb{R}^{2n})$$

is orthonormal.

In comparing (15), (16)(i)–(ii) one should keep in mind that $W_{\hbar,E_N(\hbar)}$ is rapidly oscillating in $\{H \leq E\}$ with slowly decaying tails in the interior of $\{H \leq E\}$, with a large "bump" near Σ_E and with maximum given by Proposition 5.7. Integrals (e.g. of $a \equiv 1$) against $W_{\hbar,E_N(\hbar)}$ involve a lot of cancellation due to the oscillations. The square integrals in (ii) enhance the "bump" and decrease the tails and of course are positive.

Another key property of Weyl quantization is its metaplectic covariance (see Section 3.2 for background). Let $Sp(2d,\mathbb{R}) = Sp(T^*\mathbb{R}^d, \sigma)$ denote the symplectic group and let $\mu(g)$ denote the metaplectic representation of its double cover. Then, $\mu(g)Op_h^w(a)\mu(g) = Op_h^w(a \circ T_g)$, where $T_g : T^*\mathbb{R}^d \to T^*\mathbb{R}^d$ denotes translation by g. See [16] and Section 3.2 for background. In particular, $U \in U(d)$ acts on $L^2(T^*\mathbb{R}^d)$ by translation T_U of functions, using the identification $T^*\mathbb{R}^d \simeq \mathbb{C}^d$ defined by the standard complex structure J. $U(d) \subset Sp(2d,\mathbb{R})$ is a subgroup of the symplectic group and the complete symbol $H(x,\xi)$ of (1) is $U(d)$ invariant, so by metaplectic covariance, $\hat{H}_\hbar$ commutes with the metaplectic representation of $U(d)$.

3 Bargmann-Fock Space and the Toeplitz Representation of the Isotropic Oscillator

Bargmann-Fock space of degree k on $\mathbb{C}^{m+1}$ is defined by

$$\mathcal{H}_k = \{f(z) \text{ holomorphic function on } \mathbb{C}^{m+1}, \quad \int_{\mathbb{C}^{m+1}} |f|^2 e^{-k|z|^2} dVol_{\mathbb{C}^{m+1}} < \infty\}.$$

The volume form on $\mathbb{C}^{m+1}$ is $d\,\mathrm{Vol}_{\mathbb{C}^{m+1}} = \omega^{m+1}/(m+1)!$, and $dL(z)$ denotes Lebesgue measure. We note that

$$\int_{\mathbb{C}^{m+1}} e^{-k|z|^2} dL(z) = \omega_{m+1} \int_0^\infty e^{-k\rho^2} \rho^{2m+1} d\rho = \omega_{m+1} \int_0^\infty e^{-kx} x^m dx$$

and that

$$\int_0^\infty e^{-kx}x^m dx = k^{-(m+1)}\Gamma(m+1) = m!k^{-(m+1)},$$

where we use polar coordinates (θ, ρ) on $\mathbb{C}^{m+1}$ and where $\omega_{m+1} = |S^{2m+1}|$ is the surface measure of the unit sphere in $\mathbb{C}^{m+1}$. We normalize the Gaussian measure to have mass 1 and denote it by,

$$d\Gamma_{m+1,k} := \frac{k^{(m+1)}}{m!\omega_{m+1}} e^{-k|z|^2} dL(z). \tag{17}$$

Let us fix $k = 1$. An orthonormal basis is given by the holomorphic monomials,

$$\{\frac{z^\alpha}{\sqrt{\alpha!}}\}|_{\alpha\in\mathbb{N}^{m+1}},$$

where $\alpha = (\alpha_1, \ldots, \alpha_{m+1})$ is a lattice point in the orthant $\alpha_j \in \mathbb{N}$ and $z^\alpha = \prod_{j=1}^{m+1} z_j^{\alpha_j}$, $\alpha! := \prod_{j=1}^{m+1} \alpha_j!$. If we fix the degree $|\alpha| = \sum_{j=1}^{m+1} \alpha_j$ we get the subspaces

$$\mathcal{H}_N = \text{Span}\ \{\mathrm{z}^\alpha : |\alpha| = \mathrm{N}\},$$

and one has the orthogonal decomposition,

$$L^2_{\text{hol}}(\mathbb{C}^{m+1}, d\Gamma_{m+1,k}) = \bigoplus_{N=0}^\infty \mathcal{H}_N.$$

Further, there is a canonical isomorphism

$$\mathcal{H}_N \simeq H^0(\mathbb{CP}^m, \mathcal{O}(N))$$

between $\mathcal{H}_N$ and the space of holomorphic sections of the Nth power of the standard line bundle $\mathcal{O}(1) \to \mathbb{CP}^m$ over projective space. The isomorphism is essentially by the lift

$$\hat{s}(z, \lambda) = \lambda^{\otimes N}(s(z))$$

of a section $s \in H^0(M, \mathcal{O}(N))$ to the total space $\mathcal{O}(-1) \to \mathbb{CP}^m$ of the line bundle dual to $\mathcal{O}(1)$, as an equivariant holomorphic function $\hat{s}$ of degree N. The lifted function vanishes at the zero section. If one blows down the zero section to a point, then $\mathcal{O}(-1) \simeq \mathbb{C}^{m+1}$ and the lifted sections are, again, homogeneous holomorphic polynomials of degree N. This implies that Bargmann-Fock space is, as a vector space, isomorphic to $\bigoplus_{N=0}^\infty H^0(\mathbb{CP}^m, \mathcal{O}(N))$. The direct sum is endowed with the

Bargmann-Fock Hilbert space inner product and, up to a scalar, this inner product on $\mathcal{H}_N$ is the same as the Fubini–Study inner product on $H^0(M, \mathcal{O}(N))$.

The degree k *Bargmann-Fock Bergman kernel* is the orthogonal projection from $L^2(\mathbb{C}^{m+1}, d\Gamma_{m+1,k}) \to \mathcal{H}_k$. Its Schwartz kernel relative to Gaussian measure $d\Gamma_{m+1,k}$ is given by

$$\Pi_k(z, w) = \left(\frac{k}{2\pi}\right)^{m+1} e^{kz\bar{w}},$$

i.e. for any function $f \in L^2(\mathbb{C}^{m+1}, d\Gamma_{m+1,k})$, its orthogonal projection to Bargmann-Fock space is given by

$$(\Pi_k f)(z) = \int_{\mathbb{C}^m} \Pi_k(z, w) f(w) d\Gamma_{m+1,k}(dw)).$$

More generally, fix (V, ω) be a real $2m$ dimensional symplectic vector space. Let $J : V \to V$ be a ω compatible linear complex structure, that is, $g(v, w) := \omega(v, Jw)$ is a positive-definite bilinear form and $\omega(v, w) = \omega(Jv, Jw)$. There exists a canonical identification of $V \cong \mathbb{C}^m$ up to $U(m)$ action, identifying ω and J. We denote the BF space for (V, ω, J) by $\mathcal{H}_{k,J}$.

To put Bargmann-Fock space into the general framework of holomorphic line bundles over Kähler manifolds, we let $M = \mathbb{C}^m$ with coordinate $z_i = x_i + \sqrt{-1} y_i$, $L \to M$ be the trivial line bundle, let $L \cong \mathbb{C}^m \times \mathbb{C}$, and let $\omega = i \sum_i dz_i \wedge d\bar{z}_i$ be the Kähler form, whose potential is $\varphi(z) = |z|^2 := \sum_i |z_i|^2$.

3.1 Lifting to the Heisenberg Group

It is useful to lift holomorphic sections of line bundles to equivariant functions on the dual L^* of the total space of the line bundle. Since they are equivariant with respect to the natural S^1 action, one often restricts them to the unit circle bundle $X = X_h$ defined by a Hermitian metric h on L^*.

In the case of Bargmann-Fock space, X is the Heisenberg group $\mathbb{H}^m_{red} = \mathbb{C}^m \times S^1$, with group multiplication

$$(z, \theta) \circ (z', \theta') = (z + z', \theta + \theta' + \mathrm{Im}(z\bar{z}')).$$

The circle bundle $\pi : X \to M$ can be trivialized as $X \cong \mathbb{C}^m \times S^1$. The contact form on X is

$$\alpha = d\theta + (i/2) \sum_j (z_j d\bar{z}_j - \bar{z}_j dz_j).$$

The contact form $\alpha = d\theta + \frac{i}{2}\sum_j (z_j d\bar{z}_j - \bar{z}_j dz_j)$ on $\mathbb{H}^m_{red}$ is invariant under the left multiplication

$$L_{(z_0,\theta_0)} : (z,\theta) \mapsto (z_0,\theta_0) \circ (z,\theta) = (z + z_0, \theta + \theta_0 + \frac{z_0\bar{z} - \bar{z}_0 z}{2i}).$$

The volume form on $X = \mathbb{C}^m \times S^1$ is $d\,\mathrm{Vol}_X = (d\theta/2\pi) \wedge \omega^m/m!$.

The action of the Heisenberg group is by *Heisenberg translations* on phase space. As seen in the next Lemma, Heisenberg translations are Euclidean translations in the $\mathbb{C}^m$ component but also have a non-trivial change in the angular component. The infinitesimal Heisenberg group action on X can be identified with the contact vector field generated by a linear Hamiltonian function $H : \mathbb{C}^m \to \mathbb{R}$.

Lemma 3.1 ([ZZ17, Section 3.2]) *For any $\beta \in \mathbb{C}^m$, we define a linear Hamiltonian function on $\mathbb{C}^m$ by*

$$H(z) = z\bar{\beta} + \beta\bar{z}.$$

The Hamiltonian vector field on $\mathbb{C}^m$ is

$$\xi_H = -i\beta\partial_z + i\bar{\beta}\partial_{\bar{z}},$$

and its contact lift is

$$\hat{\xi}_H = -i\beta\partial_z + i\bar{\beta}\partial_{\bar{z}} - \frac{1}{2}(z\bar{\beta} + \beta\bar{z})\partial_\theta.$$

The time t flow $\hat{g}^t$ on X is given by left multiplication

$$\hat{g}^t(z,\theta) = (-i\beta t, 0) \circ (z,\theta) = (z - i\beta t, \theta - tRe(\beta\bar{z})).$$

The lift of a holomorphic section of $L^k \to \mathbb{C}^m$ is the CR-holomorphic function defined by,

$$\hat{s}(z,\theta) = e^{k(i\theta - \frac{1}{2}|z|^2)} s(z).$$

Indeed, the horizontal lift of $\partial_{\bar{z}_j}$ is $\partial^h_{\bar{z}_j} = \partial_{\bar{z}_j} - \frac{i}{2} z_j \partial_\theta$, and $\partial^h_{\bar{z}_j}\hat{s}(z,\theta) = 0$.

The corresponding lift of the degree k Bergman (or, Szegö) kernel $\hat{\Pi}_k(\hat{z},\hat{w})$ to $X = \mathbb{C}^m \times S^1$ is given by

$$\hat{\Pi}_k(\hat{z},\hat{w}) = \left(\frac{k}{2\pi}\right)^m e^{k\hat{\psi}(\hat{z},\hat{w})}, \tag{18}$$

where $\hat{z} = (z,\theta_z)$, $\hat{w} = (w,\theta_w)$ and the phase function is

$$\hat{\psi}(\hat{z},\hat{w}) = i(\theta_z - \theta_w) + z\bar{w} - \frac{1}{2}|z|^2 - \frac{1}{2}|w|^2. \tag{19}$$

3.2 Metaplectic Representation

The Harmonic oscillator is a quadratic operator. Such operators form the symplectic Lie algebra. Their representations on Bargmann-Fock space are a unitary representation of the Lie algebra. The integration of this representation gives the metaplectic representation. There exist exact formulae for the Schwartz kernels of metaplectic propagators, generalizing the Mehler formula. We need these formulae later on. A thorough treatment can be found in [16, 25].

Let $\mathbb{R}^{2m}, \omega = 2\sum_{j=1}^{m} dx_j \wedge dy_j$ be a symplectic vector space. The space $Sp(m, \mathbb{R})$ consists of linear transformation $S : \mathbb{R}^{2m} \to \mathbb{R}^{2m}$, such that $S^*\omega = \omega$. In coordinates, we write

$$\begin{pmatrix} x' \\ y' \end{pmatrix} = S \begin{pmatrix} x \\ y \end{pmatrix} = \begin{pmatrix} A & B \\ C & D \end{pmatrix} \begin{pmatrix} x \\ y \end{pmatrix}.$$

The semi-direct product of the symplectic group and Heisenberg group (sometimes called the Jacobi group) thus consists of linear transformations fixing 0 together with Heisenberg translations moving 0 to any point.

In complex coordinates $z_i = x_i + iy_i$, we have then

$$\begin{pmatrix} z' \\ \bar{z}' \end{pmatrix} = \begin{pmatrix} P & Q \\ \bar{Q} & \bar{P} \end{pmatrix} \begin{pmatrix} z \\ \bar{z} \end{pmatrix} =: \mathcal{A} \begin{pmatrix} z \\ \bar{z} \end{pmatrix},$$

where

$$\begin{pmatrix} P & Q \\ \bar{Q} & \bar{P} \end{pmatrix} = \mathcal{W}^{-1} \begin{pmatrix} A & B \\ C & D \end{pmatrix} \mathcal{W}, \quad \mathcal{W} = \frac{1}{\sqrt{2}} \begin{pmatrix} I & I \\ -iI & iI \end{pmatrix}. \tag{20}$$

The choice of normalization of $\mathcal{W}$ is such that $W^{-1} = W^*$.Thus,

$$P = \frac{1}{2}(A + D + i(C - B)).$$

We say such $\mathcal{A} \in Sp_c(m, \mathbb{R}) \subset M(2n, \mathbb{C})$. The following identities are often useful.

Proposition 3.2 ([16] Prop 4.17) *Let* $\mathcal{A} = \begin{pmatrix} P & Q \\ \bar{Q} & \bar{P} \end{pmatrix} \in Sp_c$, *then*

(1) $\begin{pmatrix} P & Q \\ \bar{Q} & \bar{P} \end{pmatrix}^{-1} = \begin{pmatrix} P^* & -Q^t \\ -Q^* & P^t \end{pmatrix} = K\mathcal{A}^*K$, *where* $K = \begin{pmatrix} I & 0 \\ 0 & -I \end{pmatrix}$.

(2) $PP^* - QQ^* = I$ *and* $PQ^t = QP^t$.

(3) $P^*P - Q^t\bar{Q} = I$ *and* $P^t\bar{Q} = Q^*P$.

The (double cover) of $Sp(m, \mathbb{R})$ acts on the Bargmann-Fock space $\mathcal{H}_k$ of $\mathbb{C}^m$ by integral operators with the following kernels: given $M = \begin{pmatrix} P & Q \\ \bar{Q} & \bar{P} \end{pmatrix} \in Sp_c$, we define

$$\mathcal{K}_{k,M}(z, w) = \left(\frac{k}{2\pi}\right)^m (\det P)^{-1/2} \exp\left\{k\frac{1}{2}\left(z\bar{Q}P^{-1}z + 2\bar{w}P^{-1}z - \bar{w}P^{-1}Q\bar{w}\right)\right\},$$

where the ambiguity of the sign of the square root $(\det P)^{-1/2}$ is determined by the lift to the double cover. When $\mathcal{A} = Id$, then $\mathcal{K}_{k,\mathcal{A}}(z, \bar{w}) = \Pi_k(z, \bar{w})$. The lifted kernel upstairs on the reduced Heisenberg group X is given by,

$$\hat{\mathcal{K}}_{k,\mathcal{A}}(\hat{z}, \hat{w}) = \mathcal{K}_{k,M}(z, \bar{w}) e^{k(i\theta_z - |z|^2/2) + k(-i\theta_w - |w|^2/2)}. \tag{21}$$

3.3 Toeplitz Construction of the Metaplectic Representation

The analogue of Weyl pseudo-differential operators on $L^2(\mathbb{R}^m)$ is (Berezin-) Toeplitz operators on Bargmann-Fock space. Given the semi-classical parameter k, the Berezin-Toeplitz quantization of a multiplication operator by a semi-classical symbol $\sigma_k(Z, \bar{Z})$ on $\mathbb{C}^m$ is defined by

$$\Pi_k \sigma_k(Z, \bar{Z}) \Pi_k. \tag{22}$$

It operates on Bargmann-Fock space by multiplying a holomorphic function by σ_k and then projecting back onto Bargmann-Fock space. More generally, one could let σ_k be a semi-classical pseudo-differential operator.

The isotropic Harmonic oscillator is represented on $\mathcal{H}_k(\mathbb{C}^d)$ as

$$\hat{H}_k = \Pi_k |Z|^2 \Pi_k.$$

It is equally well represented by $\sum_{j=1}^m a_j^* a_j + \frac{d}{2} = \sum_{j=1}^m z_j \frac{\partial}{\partial z_j} + \frac{d}{2}$, where $a_j = \frac{\partial}{\partial z_j}$ and $a_j^* = z_j$ are the annihilation/creation operators. The operator $\sum_{j=1}^m a_j^* a_j$ is called the degree or number operator since its action on a holomorphic polynomial is to give its degree. In a similar way, the infinitesimal metaplectic representation of quadratic polynomials $Q = Q(z, \bar{z})$ is by Toeplitz operators $\Pi_k Q \Pi_k$.

The Toeplitz construction of the metaplectic representation is due to Daubechies [14]. The integrated metaplectic representation $W_J(S)$ of $S \in Mp(n, \mathbb{R})$ on $\mathcal{H}_J$ is defined as follows: Let $S \in Sp(n, \mathbb{R})$ and let U_S be the unitary translation operator on $L^2(\mathbb{R}^{2n}, dL)$ defined by $U_S F(x, \xi) := F(S^{-1}(x, \xi))$. The metaplectic representation of S on $\mathcal{H}_J$ is given by ([14], (5.5) and (6.3 b))

$$W_J(S) = \eta_{J,S} \Pi_J U_S \Pi_J, \tag{23}$$

where (see [14] (6.1) and (6.3a)),

$$\eta_{J,S} = 2^{-n} \det(I - iJ) + S(I + iJ)^{\frac{1}{2}} \tag{24}$$

and Π_J is the Bargmann-Fock Szegö projector.

In the notation of the previous section, a quadratic Hamiltonian function $H : \mathbb{C}^m \to \mathbb{R}$ generates a one-parameter family of symplectic linear transformations $\mathcal{A}_t = g^t : \mathbb{C}^m \to \mathbb{C}^m$, which in general is only $\mathbb{R}$-linear and not $\mathbb{C}$-linear, i.e. M_t does not preserve the complex structure of $\mathbb{C}^m$. Hence, one needs to orthogonal project back to holomorphic sections. To compensate for the loss of norm due to the projection, one needs to multiply a factor $\eta_{\mathcal{A}_t}$.

Proposition 3.3 *Let $\mathcal{A} : \mathbb{C}^m \to \mathbb{C}^m$ be a linear symplectic map, $\mathcal{A} = \begin{pmatrix} P & Q \\ \bar{Q} & \bar{P} \end{pmatrix}$, and let $\hat{\mathcal{A}} : X \to X$ be the contact lift that fixes the fiber over 0, then*

$$\hat{\mathcal{K}}_{k,\mathcal{A}}(\hat{z}, \hat{w}) = (\det P^*)^{1/2} \int_X \hat{\Pi}_k(\hat{z}, \hat{\mathcal{A}}\hat{u})\hat{\Pi}_k(\hat{u}, \hat{w}) d\,\mathrm{Vol}_X(\hat{u}).$$

Proof The contact lift $\hat{\mathcal{A}} : \mathbb{C}^m \times S^1 \to \mathbb{C}^m \times S^1$ is given by $\mathcal{A}$ acting on the first factor:

$$\hat{\mathcal{A}} : (z, \theta) \mapsto (Pz + Q\bar{z}, \theta),$$

one can check that $\hat{\mathcal{A}}^*\alpha = \alpha$. The integral over X is a standard complex Gaussian integral, analogous to [16, Prop 4.31], and with determinant Hessian $1/|\det P|$, hence we have $(\det P^*)^{1/2}/|\det P| = (\det P)^{-1/2}$. □

3.4 Toeplitz Quantization of Hamiltonian Flows

The Toeplitz construction of the metaplectic representation generalizes to the construction of a Toeplitz quantization of any symplectic map on any Kähler manifold as a Toeplitz operator on the quantizing line bundles [45]. In this section we briefly review the construction of a Toeplitz parametrix for the propagator $U_k(t)$ of the quantum Hamiltonian (58). We refer to Section 10 and to [ZZ17,ZZ18] for the details.

Let (M, ω, L, h) be a polarized Kähler manifold, and $\pi : X \to M$ the unit circle bundle in the dual bundle (L^*, h^*). X is a contact manifold, equipped with the Chern connection contact one-form α, whose associated Reeb flow R is the rotation ∂_θ in the fiber direction of X. Any Hamiltonian vector field ξ_H on M generated by a smooth function $H : M \to R$ can be lifted to a contact Hamiltonian vector field $\hat{\xi}_H$ on X, which generates a contact flow $\hat{g}^t$. The following Proposition from [45] expresses the lift of (76) to $\mathcal{H}(X) = \bigoplus_{k \geq 0} \mathcal{H}_k(X)$.

Proposition 3.4 *There exists a semi-classical symbol $\sigma_k(t)$ so that the unitary group* (76) *has the form*

$$\hat{U}_k(t) = \hat{\Pi}_k(\hat{g}^{-t})^* \sigma_k(t) \hat{\Pi}_k \tag{25}$$

modulo smooth kernels of order $k^{-\infty}$.

3.5 Bargmann Intertwining Operator Between Schrödinger and Bargmann-Fock

The standard unitary intertwining operator between the Schrodinger representation and the Bargmann-Fock representation is the (Segal-)Bargmann transform,

$$Bf(Z) = \int_{\mathbb{R}^n} \exp\left(-(Z \cdot Z - 2\sqrt{2} Z \cdot X + X \cdot X)/2\right) f(X) dX. \tag{26}$$

Its inverse is its adjoint,

$$B^* F(x) = \int_{\mathbb{C}^n} \exp\left(-(\bar{Z} \cdot \bar{Z} - 2\sqrt{2}\bar{Z} \cdot X + X \cdot X)/2\right) F(Z) e^{-|Z|^2} L(dZ).$$

Another inversion formula is

$$f(x) = \pi^{-n/4} (2\pi)^{-n/2} e^{-|x|^2} \int_{\mathbb{R}^n} (Bf)(x + iy) e^{-|y|^2/2} dy.$$

The Bargmann transform is obtained from the Euclidean heat kernel by analytic continuation in the first variable. It might be surprising that this transform is useful in studying the Harmonic oscillator. One could just as well analytically continue the propagator (9), which also defines a unitary intertwining operator. However, that operator would simply analytically continue Hermite functions, which does not simply the analysis. The Bargmann transform maps Hermite functions to holomorphic polynomials, and the Hermite operator to the degree operator (up to a constant) and this is a significant simplification.

One may also use the Bargmann transform to convert Wigner distributions associated with spectral projections of the Harmonic oscillator to the much simpler orthogonal projections onto spaces of holomorphic polynomials of fixed degree. The density of states (diagonal of a Bergman kernel) is known as a Husimi distribution in physics. An interesting historical fact is that Cahill–Glauber studied the relation between Wigner distributions $W_{\Pi_{\hbar, E_M}}(x, \xi)$ and the Bargmann-conjugate Bergman Husimi distributions

$$B \Pi_{\hbar, E_N} B^*(Z, \bar{Z})$$

in [7, 8]. The Bargmann transform is the same as the spectral projections of the Bargmann-Fock quantization $\Pi_{BF,k}|Z|^2\Pi_{BF,k}$ of $|Z|^2$. They showed that

$$B_x \otimes B_y \int W_{\Pi_{\hbar,E_N}}(\tfrac{x+y}{2}, \xi)e^{i\langle x-y,\xi\rangle}d\xi = \int_{\mathbb{R}^n}\int_{\mathbb{R}^n}\int_{\mathbb{R}^n} B(x,Z)B(y,Z) \\ W_{\Pi_{\hbar,E_N}}(\tfrac{x+y}{2}, \xi)e^{i\langle x-y,\xi\rangle}d\xi dx dy$$

is convolution of $W_{\Pi_{\hbar,E_M}}(x, \xi)$ with a complex Gaussian.

3.6 Analogies and Correspondences Between the Real and Complex Settings

We now list some important analogies to help navigate the results of this article, and to compare the results in the real and complex settings. The undefined notation and terminology will be provided in the relevant section of this article. The reader is encouraged to consult this list as the article proceeds; it is probably not possible to understand much of it from the start.

Microlocal analysis provides a generalization of this equivalence to general manifolds. The generalization of the Bargmann transform (see Section 26) is called an FBI transform. It is well-recognized that the setting of holomorphic sections of high powers $L^k \to M$ of ample line bundles over Kähler manifolds is quite analogous to the setting of Schrödinger operators on Riemannian manifolds, to the extent that one may expect parallel results in both domains. The role of the Planck constant $\hbar$ in semi-classical analysis is analogous to k^{-1} in the line bundle setting. In fact, the relation between Wigner distributions and "Husimi distributions" (or partial Bergman density of states) was first given by Cahill–Glauber in 1969 [7, 8] for applications in quantum optics. We refer to [34, 47] for background in semi-classical analysis and to [5] for background on Toeplitz operators.

Here is a list of analogies which are relevant to the present survey.

- The cotangent bundle $(T^*\mathbb{R}^d, \sigma)$ equipped with its canonical symplectic structure is analogous to a Kähler manifold (M, ω). One may equip $T^*\mathbb{R}^d$ with a complex structure J so that it becomes the Kähler manifold $\mathbb{C}^d$.
- The total space of the dual line bundle L^* of a holomorphic line bundle $L \to M$ is analogous to $\mathbb{C}^d$. Indeed, if $M = \mathbb{CP}^{d-1}$ (complex projective space), then $\mathbb{C}^d = L^*$ where $L^* = \mathcal{O}(-1)$ is the tautological line bundle over $\mathbb{CP}^{d-1}$ (More precisely, $\mathbb{C}^d = \mathcal{O}(-1)$ with the zero section "blown down.").
- When L is an "ample" line bundle, sections $s_k \in H^0(M, L^k)$ in the space of holomorphic sections of the kth power of L lift in a canonical way to equivariant holomorphic functions $\hat{s}_k$ on L^*. In the case $(M, L) = (\mathbb{CP}^{d-1}, \mathcal{O}(-1))$, lifts of sections of L^k are the holomorphic homogeneous polynomials on $\mathbb{C}^d$ of degree k.
- The total space L carries an S^1 (circle) action, namely rotation in the fibers L_z of $\pi : L \to M$. The generator D_θ of this circle action is analogous to the isotropic

harmonic oscillator and to the degree operator. Namely, if $D_\theta \hat{s}_k = k\hat{s}_k$. The isotropic harmonic oscillator $\hat{H}_\hbar$ on $L^2(\mathbb{R}^d)$ is unitarily equivalent to the degree operator on $\mathbb{C}^d$ under the Bargmann transform.

- In the case $(M, L) = (\mathbb{CP}^{d-1}, \mathcal{O}(-1))$, $H^0(\mathbb{CP}^{d-1}, \mathcal{O}(k))$ is canonically isomorphic to the eigenspace of eigenvalue $k + \frac{d}{2}$ of the isotropic harmonic oscillator.
- Eigenspace spectral projection kernels $\Pi_{\hbar, E_N(\hbar)}(x, y)$ for eigenspaces V_N of isotropic harmonic oscillators are analogous to Bergman kernels $\Pi_{h^k}(z, w)$ for spaces $H^0(M, L^k)$ of holomorphic sections of powers of a positive Hermitian line bundle (L, h) over a Kähler manifold (M, ω).
- The Wigner distribution $W_{\hbar, E_N(\hbar)}(x, \xi)$ of an eigenspace projection is analogous to the density of states $\Pi_{h^k}(z, z)$ where Π_{h^k} is the Bergman kernel for $H^0(M, L^k)$. The density of states is the contraction of the diagonal of the Bergman kernel.
- Airy scaling asymptotics of scaled Wigner distributions of eigenspace projections of the isotropic harmonic oscillator around an energy surface $\Sigma_E \subset T^*\mathbb{R}^d$ are analogous to Gaussian error function asymptotics of scaled Bergman kernels around an energy surface. Both live on "phase space." The eigenspace projections of the oscillator live on configuration (or, physical) space and have no simple analogue in the Kähler setting.
- The unitary Bargmann transform $\mathcal{B} : L^2(\mathbb{R}^d) \to H^2(\mathbb{C}^d, e^{-|Z|^2} dL(Z))$ intertwines the real Schrödinger and holomorphic Bargmann-Fock representations of quantum mechanics on $\mathbb{R}^d$. There is no simple analogue for general Kähler manifolds. It would be a unitary intertwining operator between the Bargmann-Fock spaces of L^* and $L^2(N)$ where $N \subset M$ would be a totally real Lagrangian submanifold. See Section 26 for background.

There is an important difference between the results on Wigner distributions and the results on partial Bergman kernels, which indicates that there is much more to be done on interfaces in spectral asymptotics. Namely, in the Kähler setting we have two Hamiltonians: (i) A Toeplitz Hamiltonian $\hat{H}_k := \Pi_{h^k} H \Pi_{h^k}$ (where $H : M \to \mathbb{R}$ is a smooth function) and (ii) the operator D_θ on L^* defining the degree k of a lifted section. The latter is analogous to the isotropic oscillator. The interfaces for D_θ are interfaces across "disc bundles" $D_R^* \subset L^*$ defined by a Hermitian metric h on L. The analogue of Airy scaling asymptotics of Wigner distributions is Gaussian error function asymptotics for lifts of Bergman kernels to L^*. A Toeplitz Hamiltonian $\hat{H}_k$ lifts to a Hamiltonian on L^* which commutes with D_θ, and our results on partial Bergman kernels pertain to the pair. So far, we have not considered the analogous problem on $L^2(\mathbb{R}^d)$ defined by a second Schrödinger operator which commutes with the isotropic harmonic oscillator. As this brief discussion indicates, there are many types of interface phenomena that remain to be explored.

4 Interface Problems for Schrödinger Equations

In this section we consider the simplest Schrödinger operator, namely the isotropic Harmonic Oscillator on $\mathbb{R}^d$. We review three types of interface scaling results:

- Scaling of the spectral projections kernel for a single eigenspace around the caustic. At the same time, we consider scaling of nodal sets of random eigenfunctions around the caustic.
- Scaling asymptotics of the Wigner distributions of the spectral projections kernel around an energy level in phase space.
- Scaling asymptotics of the Wigner distributions of Weyl sums of spectral projections kernels over an interval of energies at the boundary of the interval.

4.1 Allowed and Forbidden Regions and the Caustic

Consider a general Schrödinger operator $\hat{H}_\hbar := -\hbar^2\Delta + V$ on $L^2(\mathbb{R}^d)$ with $V(x) \to \infty$ as $|x| \to \infty$. Then $\hat{H}_\hbar$ has a discrete spectrum of eigenfunctions $E_j(\hbar)$,

$$\hat{H}_\hbar \psi_{\hbar,j} = E_j(\hbar)\psi_{\hbar,j}. \tag{27}$$

In the semi-classical limit

$$\hbar \to 0,\ j \to \infty,\ E_j(\hbar) = E, \tag{28}$$

the eigenfunctions of $\hat{H}_\hbar$ are rapidly oscillating in the classically allowed region

$$\mathcal{A}_E := \{V(x) \le E\},$$

and exponentially decaying in the classically forbidden region

$$\mathcal{F}_E := \mathcal{A}_E^c = \{V(x) > E\}.$$

This reflects the fact that a classical particle of energy E is confined to $\mathcal{A}_E = \{V(x) \le E\}$. We define the *caustic* to be

$$\mathcal{C}_E := \partial\mathcal{A}_E = \{V(x) = E\}. \tag{29}$$

The exponential decay rate of eigenfunctions in the forbidden region as $\hbar \to 0$ is measured by the Agmon distance to the caustic. We refer to [1, 24] for background.

In the first series of results we are interested in the transition between the oscillatory and exponential decay behavior of eigenfunctions in a zone around the caustic (29). We review two types of results: (i) Airy scaling asymptotics of spectral projections kernels and (ii) interface asymptotics of nodal (i.e. zero) sets of "random eigenfunctions" in a spectral eigenspace. At this time, results are only proved in the special case of the isotropic harmonic oscillator, but one may expect that suitably generalized results hold rather universally.

In the case of the isotropic Harmonic Oscillator, the allowed region $\mathcal{A}_E$, resp. the forbidden region $\mathcal{F}_E$ are given, respectively, by,

$$\mathcal{A}_E = \{x : |x|^2 < 2E\}, \quad \mathcal{F}_E = \{x : |x|^2 > 2E\}. \tag{30}$$

Thus, $\mathcal{A}_E$ is the projection to $\mathbb{R}^d$ of the energy surface $\{H = E\} \subset T^*\mathbb{R}^d$, $\mathcal{F}_E$ is its complement, and the caustic set is given by,

$$\mathcal{C}_E = \{|x| = 2E\}.$$

The semi-classical limit at the energy level $E > 0$ is the limit as $\hbar \to 0$, $N \to \infty$ with fixed E, so that $\hbar$ only takes the values (5).

4.2 Scaling Asymptotics Around the Caustic in Physical Space

Due to the homogeneity of the isotropic oscillator, it suffices to consider one value of E. We fix $E = \frac{1}{2}$ and consider $E_N(\hbar) = \frac{1}{2}$. For this choice of E, (7) is $\Pi_{\hbar,\frac{1}{2}}$.

When $d = 1$, the eigenspaces $V_{\hbar_N,E}$ have dimension 1 and it is a classical fact (based on WKB or ODE techniques) that Hermite functions and more general Schrödinger eigenfunctions exhibit Airy asympotics at the caustic (turning points). See, for instance, [17, 31, 39]. It is not true for $d > 1$ that individual eigenfunctions exhibit analogous Airy scaling asymptotics around the caustic. Indeed, due to the high multiplicity of eigenvalues, there is a good theory of Gaussian random eigenfunctions of the isotropic oscillator, and random eigenfunctions do not exhibit Airy scaling asymptotics. The proper generalization of the $d = 1$ result is to consider the scaling asymptotics of the eigenspace projection kernels (7) with x, y in an $\hbar^{2/3}$-tube around $\mathcal{C}_E$.

The first result states that *individual* eigenspace projection kernels (7) exhibit Airy scaling asymptotics around a point $x_0 \in \mathcal{C}_E$ of the caustic. Let x_0 be a point on the caustic $|x_0|^2 = 1$ for $E = 1/2$. Points in an $\hbar^{2/3}$ neighborhood of x_0 may be expressed as $x_0 + \hbar^{2/3}u$ with $u \in \mathbb{R}^d$. The caustic is a $(d-1)$-sphere whose normal direction at x_0 is x_0, so the normal component of u is $u_1 x_0$ when $|x_0| = 1$, where $u_1 := \langle x_0, u\rangle$. We also put $u' := u - u_1 x_0$ for the tangential component, and identify $T_{x_0}\mathcal{C}_E \cong T^*_{x_0}\mathcal{C}_E \cong \mathbb{R}^{d-1}$. By rotational symmetry, we may assume $x_0 = (1, 0, \cdots, 0)$, so that $u = (u_1, u_2, \cdots, u_d) =: (u_1; u')$.

Theorem 4.1 *Let x_0 be a point on the caustic $|x_0|^2 = 1$ for $E = 1/2$. Then for $u, v \in \mathbb{R}^d$,*

$$\Pi_{\hbar,1/2}(x_0 + \hbar^{2/3}u, x_0 + \hbar^{2/3}v) = \hbar^{-2d/3+1/3}\Pi_0(u, v)(1 + O(\hbar^{1/3})), \tag{31}$$

where

$$\Pi_0(u_1, u'; v_1, v') := 2^{2/3}(2\pi)^{-d+1} \int_{\mathbb{R}^{d-1}} e^{i\langle u'-v', p\rangle} \operatorname{Ai}(2^{1/3}(u_1 + p^2/2)) \operatorname{Ai}(2^{1/3}(v_1 + p^2/2)) dp, \tag{32}$$

and $u_1 := \langle x_0, u\rangle$, $u' := u - u_1 x_0$ *(similarly for* v_1*.) On the diagonal, let* $|x|^2 = \left|x_0 + \hbar^{2/3}u\right|^2 = 1 + \hbar^{2/3}s + O(\hbar^{4/3})$ *with* $s = 2\langle x_0, u\rangle \in \mathbb{R}$*. Then,*

$$\Pi_\hbar(x, x) = 2^{-d+1}\pi^{-d/2}\hbar^{(1-2d)/3} \operatorname{Ai}_{-d/2}(s)(1 + O(\hbar^{1/3})). \tag{33}$$

The error terms in (31) *and* (33) *are uniform when* u, v, s *vary over a compact set.*

Above, Ai is the Airy function, and $\operatorname{Ai}_{-d/2}$ is a *weighted Airy function*, defined for $k \in \mathbb{R}$ by

$$\operatorname{Ai}_k(s) := \int_C T^k \exp\left(\frac{T^3}{3} - Ts\right) \frac{dT}{2\pi i}, \qquad u \in \mathbb{R}, \tag{34}$$

where C is the usual contour for Airy function, running from $e^{-i\pi/3}\infty$ to $e^{i\pi/3}\infty$ on the right half of the complex plane (see Section 11.1 for a brief review of the Airy function).

Remark 4.2 When $d = 3$, the kernel (32) with $u' = v'$, i.e. $\Pi_0(u_1, u'; v_1, u')$, coincides modulo the factor of $\sqrt{\lambda}$ with the Airy kernel $K(x, y)$ of the Tracy–Widom distribution. The "allowed region" of this article is analogous to the "bulk" in random matrix theory, and the "caustic" of this article is analogous to the "edge of the spectrum."

For results on scaling asymptotics on Riemannian manifolds, see [10].

4.3 Nodal Sets of Random Hermite Eigenfunctions

Theorem 4.1 can be used to determine the interface behavior of nodal (zero) sets of random eigenfunctions of the isotropic oscillator of a fixed eigenvalue. In many ways, the isotropic oscillator is the analogue among Schrödinger operators on $L^2(\mathbb{R}^d)$ of the Laplacian on a standard sphere $\mathbb{S}^d$, and the study of random Hermite eigenfunctions is somewhat analogous to the study of random spherical harmonics. However, there are no forbidden regions in the case of $\mathbb{S}^d$, and the interface behavior of random Hermite eigenfunctions has no parallel for random spherical harmonics.

Definition 4.3 A Gaussian random eigenfunction for H_h with eigenvalue E is the random series

$$\Phi_N(x) := \sum_{|\alpha|=N} a_\alpha \phi_{\alpha,h_N}(x),$$

for $a_\alpha \sim N(0,1)_{\mathbb{R}}$ i.i.d. Equivalently, it is the Gaussian measure γ_N on V_N which is given by $e^{-\sum_\alpha |a_\alpha|^2/2} \prod da_\alpha$.

We denote by

$$Z_{\Phi_N} = \{x : \Phi_N(x) = 0\}$$

the nodal set of Φ_N and by $|Z_{\Phi_{\hbar,E}}|$ the random measure of integration over Z_{Φ_N} with respect to the Euclidean surface measure (the Hausdorff measure) of the nodal set. Thus for any ball $B \subset \mathbb{R}^d$,

$$|Z_{\Phi_{\hbar,E}}|(B) = \mathcal{H}^{d-1}(B \cap Z_{\Phi_N}).$$

Thus $\mathbb{E}|Z_{\Phi_{\hbar,E}}|$ is a measure on $\mathbb{R}^n$ given by

$$\mathbb{E}|Z_{\Phi_{\hbar,E}}|(B) = \int_{V_N} \mathcal{H}^{d-1}(B \cap Z_{\Phi_N}) d\gamma_N.$$

The first result gives semi-classical asymptotics of the hypersurface volumes of the nodal sets of random Hermite eigenfunctions of fixed eigenvalue in the allowed, resp. forbidden region.

Theorem 4.4 *Let $x \in \mathbb{R}^d$ such that $0 < |x| \neq \sqrt{2E}$. Then the measure $\mathbb{E}|Z_{\Phi_{\hbar,E}}|$ has a density $F_N(x)$ with respect to Lebesgue measure given by*

$$\begin{cases} \text{If } x \in \mathcal{A}_E \backslash \{0\}, \; F_N(x) \simeq h^{-1} \cdot c_d \sqrt{2E - |x|^2}(1 + O(h)) \\ \\ \text{If } x \in \mathcal{F}_E, \qquad F_N(x) \simeq h^{-1/2} \cdot C_d \frac{E^{1/2}}{|x|^{1/2}(|x|^2 - 2E)^{1/4}}(1 + O(h)) \end{cases},$$

where the implied constants in the "O" symbols are uniform on compact subsets of the interiors of $\mathcal{A}_E \backslash \{0\}$ and $\mathcal{F}_E$, and where

$$c_d = \frac{\Gamma\left(\frac{d+1}{2}\right)}{\sqrt{d\pi}\,\Gamma\left(\frac{d}{2}\right)} \qquad \text{and} \qquad C_d = \frac{\Gamma\left(\frac{d+1}{2}\right)}{\sqrt{\pi}\,\Gamma\left(\frac{d}{2}\right)}.$$

The key point is the different growth rates in h for the density of zeros in the allowed and forbidden region. In dimension one, eigenfunctions have no zeros in the forbidden region, but in dimensions $d \geq 2$ they do. In the allowed region, nodal sets of eigenfunctions behave in a similar way to nodal sets on Riemannian manifolds [11], but in the forbidden region they are sparser (see [15]).

The next result on nodal sets (Theorem 4.5) gives scaling asymptotics for the average nodal density that "interpolate" between (4.3) and (4.3). Fix $x \in \mathcal{C}_E$, where $E = 1/2$, and study the rescaled ensemble

$$\Phi_{\hbar,E}^{x,\alpha}(u) := \Phi_{\hbar,E}(x + \hbar^\alpha u)$$

and the associated hypersurface measure

$$\left|Z^{x,\alpha}_{\hbar,E}\right|(B) = \mathcal{H}^{d-1}\Big(\{\Phi^{x,\alpha}_{\hbar,E}(v) = 0\} \cap B\Big), \qquad B \subset \mathbb{R}^d.$$

The next result gives the asymptotics of $\mathbb{E}\left|Z^{x,\alpha}_{\hbar,E}\right|$ when $\alpha = 2/3$ is in terms of the weighted Airy functions Ai_k (see (34)).

Theorem 4.5 (Nodal Set in a Shrinking Ball Around a Caustic Point) *Fix* $E = 1/2$ *and* $x \in \mathcal{C}_E$, *i.e.* $|x| = 1$. *For any bounded measurable* $B \subseteq \mathbb{R}^d$,

$$\mathbb{E}\left|Z^{x,2/3}_{\hbar,E}\right|(B) = \int_B \mathcal{F}(u)du,$$

where

$$\mathcal{F}(u) = (2\pi)^{-\frac{d+1}{2}} \int_{\mathbb{R}^d} |\Omega(u)^{1/2}\xi|e^{-|\xi|^2/2}d\xi\ (1 + O(\hbar^{1/3})) \tag{35}$$

and $\Omega = \left(\Omega_{ij}\right)_{1\le i,j\le n}$ *is the symmetric matrix*

$$\Omega_{ij}(u) = x_i x_j \left(\frac{\mathrm{Ai}_{2-d/2}(s)}{\mathrm{Ai}_{-d/2}(s)} - \frac{\mathrm{Ai}^2_{1-d/2}(s)}{\mathrm{Ai}^2_{-d/2}(s)}\right) + \frac{\delta_{ij}}{2}\frac{\mathrm{Ai}_{-1-d/2}(s)}{\mathrm{Ai}_{-d/2}(s)}. \tag{36}$$

where $s = 2\langle u, x\rangle$. *The implied constant in the error estimate from* (35) *is uniform when* u *varies in compact subsets of* $\mathbb{R}^d$.

Remark 4.6 The leading term in $\mathcal{F}$ is $\hbar$-independent and positive everywhere since the matrix $\Omega_{ij}(u)$ as a linear operator has non-trivial range. The matrix $\left(x_i x_j\right)_{i,j}$ in (36) is a rank 1 projection onto the $x-$direction; since the dimension $d \ge 2$, it cannot cancel out the second term. We refer to [HZZ15,HZZ16] for details.

Remark 4.7 Theorem 4.5 says that if $x \in \mathcal{C}_E$ and $\widetilde{B}_\hbar = x + \hbar^{2/3}B$ for some bounded measurable B, then

$$\mathbb{E}\left|Z_{\Phi_{\hbar,E}}\right|(\widetilde{B}_\hbar) = \hbar^{2/3(d-1)}\mathbb{E}\left|Z^{x,\alpha}_{\hbar,E}\right|(B) = \hbar^{-2/3}\int_{\widetilde{B}_\hbar} \mathcal{F}(\hbar^{-2/3}(y-x))dy,$$

which shows that the average (unscaled) density of zeros in a $\hbar^{2/3}-$tube around $\mathcal{C}_E$ grows like $\hbar^{-2/3}$ as $\hbar \to 0$.

Remark 4.8 The scaling asymptotics of zeros around the caustic, especially in the radial (normal) direction, is analogous to the scaling asymptotics of eigenvalues of random Hermitian matrices around the edge of the spectrum.

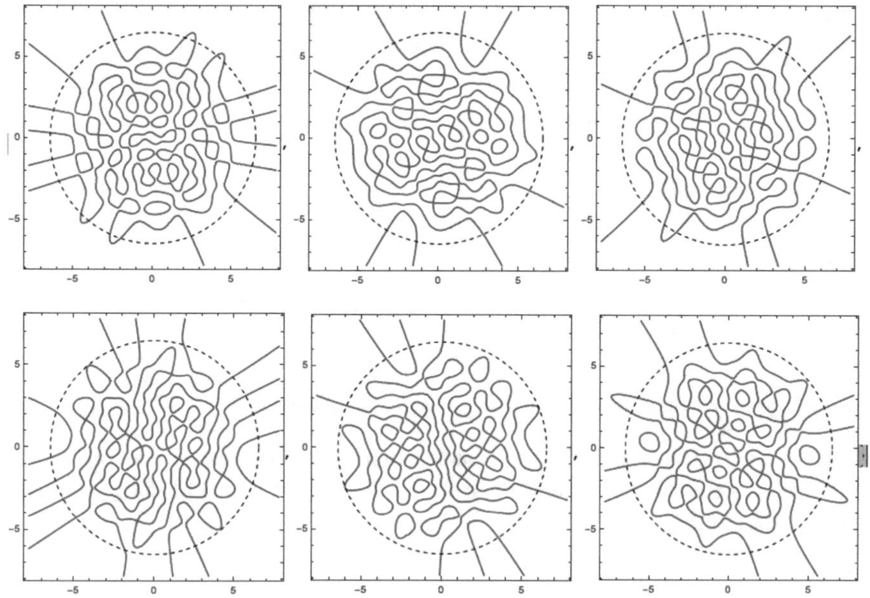

The nodal set is very dense and busy in $\mathcal{A}_E$ and rather sparse and 'non-oscillating' in $\mathcal{F}_E$.

4.4 Discussion of the Nodal Results

Computer graphics of Bies–Heller [3] (reprinted as Figure 4.3 in [HZZ15]) and the displayed graphics of Peng Zhou show that the nodal set in $\mathcal{A}_E$ near the caustic $\partial\mathcal{A}_E$ consists of a large number of highly curved nodal components apparently touching the caustic, while the nodal set in $\mathcal{F}_E$ near $\partial\mathcal{A}_E$ consists of fewer and less curved nodal components all of which touch the caustic. This is because, if $\psi \in V_{\hbar,E}$ is non-zero, $\Delta\psi = (V - E)\psi$ forces ψ and $\Delta\psi$ to have the same sign in $\mathcal{F}_E$. In a nodal domain $\mathcal{D}$ we may assume $\psi > 0$, but then ψ is a positive subharmonic function in $\mathcal{D}$ and cannot be zero on $\partial\mathcal{D}$ without vanishing identically. Hence, every nodal component which intersects $\mathcal{F}_E$ must also intersect $\mathcal{A}_E$ and therefore $\mathcal{C}_E$.

The scaling limit of the density of zeros in a shrinking neighborhood of the caustic, or in annular subdomains of $\mathcal{A}_E$ and $\mathcal{F}_E$ at shrinking distances from the caustic, is given in Theorem 4.5.

4.5 The Kac–Rice Formula

The proof of Theorem 4.5 is based on the Kac–Rice formula for the average density of zeros.

Lemma 4.9 (Kac–Rice for Gaussian Fields) *Let $\Phi_{\hbar,E}$ be the random Hermite eigenfunction of $\widehat{H}_\hbar$ with eigenvalue E. Then the density of zeros of $\Phi_{\hbar,E}$ is given by*

$$F_{\hbar,E}(x) = (2\pi)^{-\frac{d+1}{2}} \int_{\mathbb{R}^d} |\Omega^{1/2}(x)\xi| \; e^{-|\xi|^2/2} \; d\xi, \tag{37}$$

where $\Omega(x)$ is the $d \times d$ matrix

$$\begin{aligned}\Omega_{ij}(x) &= (\partial_{x_i}\partial_{y_j} \log \Pi_{\hbar,E})(x,x)\\ &= \frac{(\Pi_{\hbar,E} \cdot \partial_{x_i}\partial_{y_j}\Pi_{\hbar,E})(x,x) - (\partial_{x_i}\Pi_{\hbar,E} \cdot \partial_{y_j}\Pi_{\hbar,E})(x,x)}{\Pi_{\hbar,E}(x,x)^2}\end{aligned} \tag{38}$$

and $\Pi_{\hbar,E}(x,y)$ is the kernel of eigenspace projection (8).

We refer to [HZZ15, HZZ16] for background. The main task in proving results on zeros near the caustic is therefore to work out the asymptotics of $\Pi_{\hbar,E}(x,x)$ and its derivatives there.

5 Interfaces in Phase Space for Schrödinger Operators: Wigner Distributions

We now turn to phase space interfaces. Instead of studying the scaling asymptotics of the spectral projections (7)

$$\Pi_{\hbar,E_N(\hbar)} : L^2(\mathbb{R}^d) \to V_{\hbar,E_N(\hbar)} \tag{39}$$

we study the scaling asymptotics of their semi-classical Wigner distributions

$$W_{\hbar,E_N(\hbar)}(x,\xi) := \int_{\mathbb{R}^d} \Pi_{\hbar,E_N(\hbar)}\left(x + \frac{v}{2}, x - \frac{v}{2}\right) e^{-\frac{i}{\hbar}v\cdot\xi} \frac{dv}{(2\pi h)^d} \tag{40}$$

across the phase space energy surface (14).

When $E_N(\hbar) = E + o(1)$ as $\hbar \to 0$, $W_{\hbar,E_N(\hbar)}$ is thought of as the "quantization" of the energy surface, and (40) is thought of as an approximate δ-function on (14). This is true in the weak* sense, but the pointwise behavior is quite a bit more complicated and is studied in [HZ19].

Wigner distributions were introduced in [43] as phase space densities. Heuristically, the Wigner distribution (7) is a kind of probability density in phase space of finding a particle of energy $E_N(\hbar)$ at the point $(x,\xi) \in T^*\mathbb{R}^d$. This is not literally true, since $W_{\hbar,E_N(\hbar)}(x,\xi)$ is not positive: it oscillates with heavy tails inside the energy surface (14), has a kind of transition across Σ_E, and then decays rapidly outside the energy surface. The purpose of this paper is to give detailed results on the concentration and oscillation properties of these Wigner distributions in three phase space regimes, depending on the position of (x, ξ) with respect to Σ_E.

There is an exact formula for the Wigner distributions (13) of the eigenspace projections for the isotropic Harmonic oscillator in terms of Laguerre functions (see Appendix 11.2 and [39] for background on Laguerre functions).

Proposition 5.1 *The Wigner distribution of Definition 2.1 is given by,*

$$W_{\hbar, E_N(\hbar)}(x, \xi) = \frac{(-1)^N}{(\pi\hbar)^d} e^{-2H/\hbar} L_N^{(d-1)}(4H/\hbar), \qquad H = H(x, \xi) = \frac{|x|^2 + |\xi|^2}{2}, \tag{41}$$

where $L_N^{(d-1)}$ is the associated Laguerre polynomial of degree N and type $d-1$.

See [26, 31] for $d = 1$ and [39, Theorem 1.3.5] and [HZ19] for general dimensions. The second result is a weak* limit result for normalized Wigner distributions.

Proposition 5.2 *Let a_0 be a semi-classical symbol of order zero and let $Op_h^w(a)$ be its Weyl quantization. Then, as $\hbar \to 0$, with $E_N(\hbar) \to E$,*

$$\frac{1}{\dim V_{\hbar, E_N(\hbar)}} \int_{T^*\mathbb{R}^d} a_0(x, \xi) W_{\hbar, E_N(\hbar)}(x, \xi) dx d\xi \to \fint_{\Sigma_E} a_0 d\mu_E,$$

where $d\mu_E$ is Liouville measure on Σ_E and $\fint_{\Sigma_E} a_0 d\mu_E = \frac{1}{\mu_E(\Sigma_E)} \int_{\Sigma_E} a_0 d\mu_E$.

Thus, $W_{\hbar, E_N(\hbar)}(x, \xi) \to \delta_{\Sigma_E}$ in the sense of weak* convergence. But this limit is due to the oscillations inside the energy ball; the pointwise asymptotics are far more complicated.

5.1 Interface Asymptotics for Wigner Distributions of Individual Eigenspace Projections

Our first main result gives the scaling asymptotics for the Wigner function $W_{\hbar, E_N(\hbar)}(x, \xi)$ of the projection onto the E-eigenspace of $\widehat{H}_\hbar$ when (x, ξ) lies in an $\hbar^{2/3}$ neighborhood of the energy surface Σ_E.

Theorem 5.3 *Fix $E > 0, d \geq 1$. Assume $E_N(\hbar) = E$ and let $\hbar = \hbar_N(E)$ (5). Suppose $(x, \xi) \in T^*\mathbb{R}^d$ satisfies*

$$H(x, \xi) = E + u\left(\frac{\hbar}{2E}\right)^{2/3}, \qquad u \in \mathbb{R}, \ H(x, \xi) = \frac{\|x\|^2 + \|\xi\|^2}{2} \tag{42}$$

with $|u| < \hbar^{-1/3}$.[2] *Then,*

$$W_{\hbar,E_N(\hbar)}(x,\xi) = \begin{cases} \frac{2}{(2\pi\hbar)^d}\left(\frac{\hbar}{2E}\right)^{1/3}\left(\mathrm{Ai}(u/E) + O\left((1+|u|)^{1/4}u^2\hbar^{2/3}\right)\right), & u < 0 \\ \frac{2}{(2\pi\hbar)^d}\left(\frac{\hbar}{2E}\right)^{1/3}\mathrm{Ai}(u/E)\left(1 + O\left((1+|u|)^{3/2}u\hbar^{2/3}\right)\right), & u > 0. \end{cases} \tag{43}$$

Here, $\mathrm{Ai}(x)$ is the Airy function. The Airy scaling of $W_{\hbar,E_N(\hbar)}$ is illustrated in Figure 1. The assumption (42) may be stated more invariantly that (x, ξ) lies in the tube of radius $O(\hbar^{2/3})$ around Σ_E defined by the gradient flow of H with respect to the Euclidean metric on $T^*\mathbb{R}^d$. The asymptotics are illustrated in Figure 1. Due to the behavior of the Airy function $\mathrm{Ai}(s)$, these formulae show that in the semi-classical limit $\hbar \to 0$, $E_N(\hbar) \to E$, $W_{\hbar,E_N(\hbar)}(x,\xi)$ concentrates on the energy surface Σ_E, is oscillatory inside the energy ball $\{H \le E\}$, and is exponentially decaying outside the ball.

5.2 *Interior Bessel Asymptotics*

In addition to the Airy asymptotics in an $\hbar^{2/3}$-tube around Σ_E, $W_{\hbar,E_N(\hbar)}$ exhibits Bessel asymptotics in the interior of Σ_E. There are two (or three, depending on taste) uniform asymptotic regimes for the Laguerre polynomial $L_n^{(\alpha)}(x)$: Bessel, Trigonometric, Airy.

For $t \in [0, 1)$, define

$$A(t) = \frac{1}{2}[\sqrt{t - t^2} + \sin^{-1}\sqrt{t}], \ t \in [0, 1].$$

For $t < 0$ the $\sin^{-1}$ is replaced by $\sinh^{-1}$ and the $\frac{1}{2}$ by $i/2$ (see [17, (2.7)]). Also, let J_{d-1} be the Bessel function (of the first kind) of index $d-1$.

Theorem 5.4 *Fix* $E > 0$ *and suppose* $E_N(\hbar) = E$. *For each* $(x, \xi) \in T^*\mathbb{R}^d$ *write*

$$H_E := \frac{H(x,\xi)}{E} = \frac{\|x\|^2 + \|\xi\|^2}{2E}, \qquad \nu_E := \frac{4E}{\hbar}.$$

Fix $0 < a < 1/2$. *Uniformly over* $a \le H_E \le 1 - a$, *there is an asymptotic expansion,*

$$W_{\hbar,E_N(\hbar)}(x,\xi) = \frac{2}{(2\pi\hbar)^d}\left[\frac{J_{d-1}(\nu_E A(H_E))}{A(H_E)^{d-1}}\alpha_0(H_E) + O\left(\nu_E^{-1}\left|\frac{J_d(\nu_E A(H_E))}{A(H_E)^d}\right|\right)\right].$$

[2] The errors blow up when $u = \hbar^{-1/3}$.

In particular, uniformly over H_E in a compact subset of $(0, 1)$, *we find*

$$W_{\hbar,E_N(\hbar)}(x,\xi) = (2\pi\hbar)^{-d+1/2} P_{H,E} \cos\left(\xi_{\hbar,E,H}\right) + O\left(\hbar^{-d+3/2}\right), \tag{44}$$

where we have set

$$\xi_{\hbar,E,H} = -\frac{\pi}{4} - \frac{2H}{\hbar}\left(H_E^{-1} - 1\right)^{1/2} + \frac{2E}{\hbar}\cos^{-1}\left(H_E^{1/2}\right)$$

and

$$P_{E,H} := \left(\pi E^{1/2}\left(H_E^{-1} - 1\right)^{1/4} (H_E)^{d/2}\right)^{-1}.$$

5.3 Small Ball Integrals

The interior Bessel asymptotics do not encompass the behavior of $W_{\hbar,E_N(\hbar)}$ in shrinking balls around $\rho = 0$. In that case, we have,

Proposition 5.5 *For* $\epsilon > 0$ *sufficiently small and for any* $a(x,\xi) \in C_b(T^*\mathbb{R}^d)$,

$$\int_{T^*\mathbb{R}^d} a(x,\xi) W_{\hbar,E_N(\hbar)}(x,\xi)\psi_{\epsilon,\hbar}(x,\xi)dxd\xi = O(\hbar^{\frac{1-d}{2}-2d\epsilon}\, \|a\|_{L^\infty(B_0(\hbar^{1/2-\epsilon}))}), \tag{45}$$

where $\psi_{\epsilon,\hbar}$ *is a smooth radial cutoff that is identically* 1 *on the ball of radius* $\hbar^{1/2-\epsilon}$ *and is identically* 0 *outside the ball of radius* $2\hbar^{1/2-\epsilon}$.

5.4 Exterior Asymptotics

If $E_N(\hbar) \to E$, then $W_{\hbar,E_N(\hbar)}(x,\xi)$ concentrates on Σ_E and is exponentially decaying in the complement $H = H(x,\xi) > E$. The precise statement is,

Proposition 5.6 *Suppose that* $H_E = H(x,\xi)/E > 1$ *and let* $E_N(\hbar) = E$. *Then, there exists* $C_1 > 0$ *so that*

$$|W_{\hbar,E_N(\hbar)}(x,\xi)| \le C_1 \hbar^{-d+\frac{1}{2}} e^{-\frac{2E}{\hbar}[\sqrt{H_E^2 - H_E} - \cosh^{-1}\sqrt{H_E}]}.$$

Moreover, as $H(x,\xi) \to \infty$, *there exists* $C_2 > 0$ *so that*

$$|W_{\hbar,E_N(\hbar)}(x,\xi)| \le C_2 \hbar^{-d+\frac{1}{2}} e^{-\frac{2H(x,\xi)}{\hbar}}.$$

5.5 Supremum at $\rho = 0$

The reader may notice the "spike" at the origin $\rho = 0$; it is the point at which $W_{\hbar,E_N(\hbar)}$ has its global maximum (see Figure 2). The height is given by

$$W_{\hbar,E_N(\hbar)}(0,0) = \frac{(-1)^N}{(\pi\hbar)^d} L_N^{d-1}(0) = \frac{(-1)^N}{(\pi\hbar)^d} \frac{\Gamma(N+d)}{\Gamma(N+1)\Gamma(d)} \simeq \frac{(-1)^N}{\pi^d} C_d \hbar^{-d} N^{d-1}. \tag{46}$$

The last statement follows from the explicit formula $L_N^{(d-1)}(0) = \frac{\Gamma(N+d)}{\Gamma(N+1)\Gamma(d)} = \frac{(N+d-1)!}{N!(d-1)!}$ (see, e.g., [39, (1.1.39)]).

On the complement of the ball $B(0, \hbar^{\frac{1}{2}-\epsilon})$, the Wigner distribution is much smaller than at its maximum. The following is proved by combining the estimates of Theorem 5.3 , Theorem 5.4, and Proposition 5.6.

Proposition 5.7 *For any* $\epsilon > 0$,

$$\sup_{(x,\xi):H(x,\xi)\geq\epsilon} |W_{\hbar,E_N(\hbar)}(x,\xi)| \leq C\hbar^{-d+\frac{1}{3}}.$$

The supremum in this region is achieved in $\{H \leq E\}$ *at* (x,ξ) *satisfying* (42) *where* u *is the global maximum of* Ai(x).

Why the spike at $\rho = 0$? It is observed in [HZ19] that $W_{\hbar,E_N(\hbar)}$ is an eigenfunction of the (essentially isotropic) Schrödinger operator

$$\left(-\frac{\hbar^2}{8}(\Delta_\xi + \Delta_x) + H(x,\xi)\right) W_{\hbar,E_N(\hbar)} = E_N(\hbar) W_{\hbar,E_N(\hbar)}, \tag{47}$$

on $T^*\mathbb{R}^d$. By [HZZ15, Lemma 10], the eigenspace spectral projections for the isotropic harmonic oscillator in dimension d satisfies,

$$\Pi_{h,E}(x,x) = (2\pi h)^{-(d-1)} \left(2E - |x|^2\right)^{\frac{d}{2}-1} \omega_{d-1}(1 + O(h)),$$

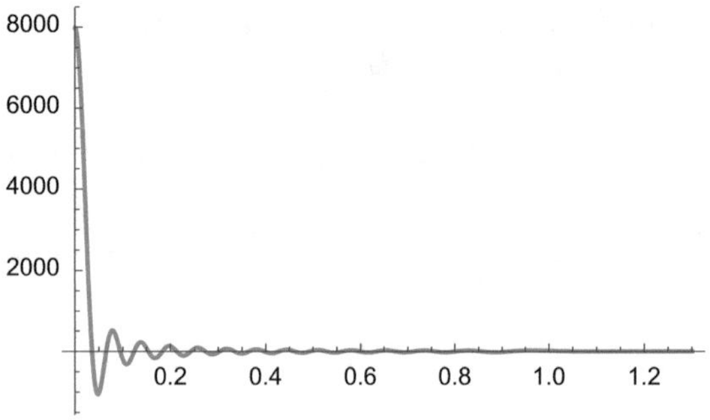

Fig. 2 The Wigner function $W_{\hbar,E_N(\hbar)}$ of the eigenspace projection $\Pi_{\hbar,E_N(\hbar)}$ is always radial (see Proposition 5.1). Displayed above is the blow-up of the Wigner function at (0, 0).

for a dimensional constant ω_d. We apply this result to the eigenspace projections for (47) in dimension $2d$ and find that at the point $(0, 0)$ its diagonal value is of order $\hbar^{-2d+1}$. We then express this eigenspace projection in terms of an orthonormal basis for the eigenspace. From the inner product formulae (16), it is seen that one of the orthonormal basis elements is $\frac{1}{\sqrt{\dim V_{\hbar, E_N(\hbar)}}} W_{\hbar, E_N(\hbar)}$. Note that $\dim V_{\hbar, E_N(\hbar)} \simeq \hbar^{-2d+1}$ in dimension $2d$. Due to the normalization and (46),

$$\frac{1}{\sqrt{\dim V_{\hbar, E_N(\hbar)}}} W_{\hbar, E_N(\hbar)}(0, 0) \simeq \hbar^{-2d+1+d-\frac{1}{2}} = \hbar^{-d+\frac{1}{2}}.$$

There exists a simple spectral geometric explanation for the order of magnitude at the origin: All eigenfunctions of (47) with the exception of the radial eigenfunction $W_{\hbar, E_N(\hbar)}(0, 0)$ vanish at the origin $(0, 0)$ since they transform by non-trivial characters of $U(d)$ and $(0, 0)$ is a fixed point of the action. Consequently, the value of the eigenspace projection on the diagonal at $(0, 0)$ is the square of $W_{\hbar, E_N(\hbar)}(0, 0)$ and that accounts precisely for the order of growth.

5.6 *Sums of Eigenspace Projections*

Let us begin by introducing the three types of spectral localization we are studying and the interfaces in each type.

- (i) $\hbar$-localized Weyl sums over eigenvalues in an $\hbar$-window $E_N(\hbar) \in [E - a\hbar, E + b\hbar]$ of width $O(\hbar)$. More generally we consider smoothed Weyl sums $W_{\hbar, E, f}$ with weights $f(\hbar^{-1}(E_N(\hbar) - E))$; see (49) for such $\hbar$-energy localization. This is the scale of individual spectral projections but is substantially more general than the results of [HZ19]. The scaling and asymptotics are in Theorem 5.9. For general Schrödinger operators, $\hbar$- localization around a single energy level leads to expansions in terms of periodic orbits. Since all orbits of the classical isotropic oscillator are periodic, the asymptotics may be stated without reference to them. The generalization to all Schrödinger operators will be studied in a future article.
- (ii) Airy-type $\hbar^{2/3}$-spectrally localized Weyl sums $W_{\hbar, f, 2/3}(x, \xi)$ over eigenvalues in a window $[E - a\hbar^{2/3}, E + a\hbar^{2/3}]$ of width $O(\hbar^{2/3})$. See Definition 5.10 for the precise definition. The level set Σ_E is viewed as the interface. The scaling asympotics of its Wigner distribution across the interface are given in Theorems 5.11 and 5.12. To our knowledge, this scaling has not previously been considered in spectral asymptotics.
- (iii) Bulk Weyl sums $\sum_{N: \hbar(N+\frac{d}{2}) \in [E_1, E_2]} W_{\hbar, E_N(\hbar)}(x, \xi)$ over energies in an $\hbar$-independent "window" $[E_1, E_2]$ of eigenvalues; this "bulk" Weyl sum runs over $\simeq \hbar^{-1}$ distinct eigenvalues; See Definition 5.13. We are mainly interested in its scaling asymptotics around the interface Σ_{E_2} (see Theorem 5.16). However, we

also prove that the Wigner distribution approximates the indicator function of the shell $\{E_1 \leq H \leq E_2\} \subset T^*\mathbb{R}^d$ (see Proposition 5.15). As far as we know, this is also a new result and many details are rather subtle because of oscillations inside the energy shell. Indeed, the results of [HZ19] show that the individual terms in the sum grow like $W_{\hbar, E_N(\hbar)}(x, \xi) \simeq \hbar^{-d+1/2}$ when $H(x, \xi) \in (E_1, E_2)$. Proposition 5.15, in contrast, shows although the bulk Weyl sums have $\simeq \hbar^{-1}$ such terms, their sum has size $\hbar^{-d}$, implying significant cancellation.

We are particularly interested in "interface asymptotics" of the bulk Wigner–Weyl distributions $W_{\hbar, f, \delta(\hbar)}$ around the edge (i.e. boundary) of the spectral interval when (x, ξ) is near the corresponding classical energy surface Σ_E. Such edges occur when f is discontinuous, e.g. the indicator function of an interval. In other words, we integrate the empirical measures (48) below over an interval rather than against a Schwartz test function. At the interface, there is an abrupt change in the asymptotics with a conjecturally universal shape. Theorem 5.9 gives the shape of the interface for $\hbar$-localized sums, Theorem 5.11 gives the shape for $\hbar^{2/3}$ localized sums, and Theorem 5.16 gives results on the bulk sums.

Our results concern asymptotics of integrals of various types of test functions against the weighted empirical measures,

$$d\mu_{\hbar}^{(x,\xi)}(\tau) := \sum_{N=0}^{\infty} W_{\hbar, E_N(\hbar)}(x, \xi) \delta_{E_N(\hbar)}(\tau), \tag{48}$$

and of recentered and rescaled versions of these measures (see (53) below). A key property of Wigner distributions of eigenspace projections (40) is that the measures (48) are signed, reflecting the fact that Wigner distributions take both positive and negative values, and are of infinite mass:

Proposition 5.8 *The signed measures* (48) *are of infinite mass (total variation norm). On the other hand, the mass of* (48) *is finite on any one-sided interval of the form,* $[-\infty, \tau]$. *Also,* $\int_{\mathbb{R}} d\mu_{\hbar}^{(x,\xi)} = 1$ *for all* (x, ξ).

Moreover, the L^2 norms of the terms $W_{\hbar, E_N(\hbar)}$ grow in N like $N^{\frac{d-1}{2}}$. Hence, the measures (48) are highly oscillatory and the summands can be very large.

5.7 *Interior Asymptotics for ħ-Localized Weyl Sums*

The first result we present pertains to the $\hbar$-spectrally localized Weyl sums of type (i), defined by

$$W_{\hbar, E, f}(x, \xi) := \sum_{N} f(\hbar^{-1}(E - E_N(\hbar))) W_{\hbar, E_N(\hbar)}(x, \xi), \qquad f \in \mathcal{S}(\mathbb{R}). \tag{49}$$

Theorem 5.9 *Fix $E > 0$, and let $W_{\hbar,E,f}$ be the Wigner distribution as in* (49) *with f an even Schwartz function. If $H(x,\xi) > E$, then $W_{\hbar,E,f}(x,\xi) = O(\hbar^\infty)$. In contrast, when $0 < H(x,\xi) < E$, set $H_E := H(x,\xi)/E$ and define*

$$t_{+,\pm,k} := 4\pi k \pm 2\cos^{-1}\left(H_E^{1/2}\right), \quad t_{-,\pm,k} := 4\pi\left(k+\frac{1}{2}\right) \pm 2\cos^{-1}\left(H_E^{1/2}\right),$$
$$k \in \mathbb{Z}.$$

Fix any $\delta > 0$. Then

$$W_{\hbar,f,E}(x,\xi) = \frac{\hbar^{-d+1}(1+O_\delta(\hbar^{1-\delta}))}{(2E)^{1/2}(2\pi)^d H_E^{d/2}(H_E^{-1}-1)^{1/4}} \sum_{\pm_1,\pm_2\in\{+,-\}} \frac{e^{\pm_2 i\left(\frac{\pi}{4}-\frac{4E}{\hbar}\right)}}{(\pm_1)^d}$$
$$\sum_{k\in\mathbb{Z}} \hat{f}(t_{\pm_1,\pm_2,k}) e^{\frac{iE}{\hbar}t_{\pm_1,\pm_2,k}},$$

where the notation O_δ means the implicit constant depends on δ.

Note that there are potentially an infinite number of "critical points" in the support of $\hat{f}$.

5.8 Interface Asymptotics for Smooth $\hbar^{2/3}$-Localized Weyl Sums

We now consider spectrally localized Wigner distributions that are both spectrally localized and phase space localized on the scale $\delta(\hbar) = \hbar^{2/3}$. They are mainly relevant when we study interface behavior around Σ_E of Weyl sums.

Definition 5.10 Let $H(x,\xi) = (\|x\|^2 + \|\xi\|^2)/2$, and assume that (x,ξ) satisfy

$$H(x,\xi) = E + u(\hbar/2E)^{2/3}. \tag{50}$$

Let $\delta(h) = \hbar^{2/3}$ and define the interface-localized Wigner distributions by

$$W_{\hbar,f,2/3}(x,\xi) := \sum_N f(h^{-2/3}(E - E_N(\hbar)))W_{\hbar,E_N}(x,\xi).$$

Theorem 5.11 *Assume that (x,ξ) satisfies* (50) *with $|u| < \hbar^{-2/3}$. Fix a Schwartz function $f \in \mathcal{S}(\mathbb{R})$ with compactly supported Fourier transform. Then*

$$W_{\hbar,f,2/3}(x,\xi) = (2\pi\hbar)^{-d} I_0(u; f, E) + O((1+|u|)\hbar^{-d+2/3}),$$

where

$$I_0(u; f, E) = \int_{\mathbb{R}} f(-\lambda/C_E)\mathrm{Ai}\Big(\lambda + \frac{u}{E}\Big) d\lambda, \qquad C_E = (E/4)^{1/3}.$$

More generally, there is an asymptotic expansion

$$W_{\hbar,f,2/3}(x,\xi) \simeq (2\pi\hbar)^d \sum_{m\geq 0} \hbar^{2m/3} I_m(u; f, E)$$

in ascending powers of $\hbar^{2/3}$ *where* $I_m(u; f, E)$ *are uniformly bounded when* u *stays in a compact subset of* $\mathbb{R}$.

The calculations show that the results are valid with far less stringent conditions on f than $f \in \mathcal{S}(\mathbb{R})$ and $\widehat{f} \in C_0^\infty$. To obtain a finite expansion and remainder it is sufficient that $\int_{\mathbb{R}} |\widehat{f}(t)||t|^k dt < \infty$ for all k. It is not necessary that $\hat{f} \in C^k$ for any $k > 0$.

5.9 Sharp $\hbar^{2/3}$-Localized Weyl Sums

Next we consider the sums of Definition 5.10 when f is the indicator function of a spectral interval,

$$f = \mathbf{1}_{[\lambda_-,\lambda_+]}.$$

Equivalently, we fix integers $0 < n_\pm$ such that

$$\lambda_\pm = \hbar^{1/3} n_\pm \text{ are bounded,}$$

and consider the corresponding Wigner–Weyl sums $W_{\hbar,f,2/3}(x,\xi)$ of Definition 5.10:

$$W_{2/3,E,\lambda_\pm}(x,\xi) :$$
$$= \sum_{N:\lambda_-\hbar^{2/3} \leq E_N(\hbar)-E < \lambda_+\hbar^{2/3}} W_{\hbar,E_N(\hbar)}(x,\xi) = \sum_{N=N(E,\hbar)+n_-}^{N(E,\hbar)+n_+-1} W_{\hbar,E_N(\hbar)}(x,\xi), \quad (51)$$

where $N(E,\hbar) = E/\hbar - d/2$. Thus, the sums run over spectral intervals of size $\simeq \hbar^{2/3}$ centered at a fix $E > 0$ and consist of sum of $\simeq \hbar^{-1/3}$ Wigner functions for spectral projections of individual eigenspaces. The following extends Theorem 5.11 to sharp Weyl sums at the cost of only giving a 1-term expansion plus remainder.

Theorem 5.12 *Assume that* (x,ξ) *satisfies* $\big(\|x\|^2 + \|\xi\|^2\big)/2 = E + u\Big(\frac{\hbar}{2E}\Big)^{2/3}$ *with* $|u| < \hbar^{-2/3}$. *Then,*

$$W_{2/3,E,\lambda_\pm}(x,\xi) = (2\pi\hbar)^{-d} C_E \int_{-\lambda_+}^{-\lambda_-} \mathrm{Ai}\left(\tfrac{u}{E} + \lambda C_E\right) d\lambda$$
$$+ O\left(\hbar^{-d+1/3-\delta} + (1+|u|)\hbar^{-d+2/3-\delta}\right), \tag{52}$$

where $C_E = (E/4)^{1/3}$.

Theorem 5.12 can be rephrased in terms of weighted empirical measures

$$d\mu_\hbar^{u,E,\frac{2}{3}} := \hbar^d \sum_N W_{\hbar,E_N(\hbar)}\left(E + u(\hbar/2E)^{2/3}\right)\delta_{[\hbar^{-2/3}(E-E_N(\hbar))]}, \tag{53}$$

obtained by centering and scaling the family (48). Thus, for (x,ξ) satisfying $\left(\|x\|^2 + \|\xi\|^2\right)/2 = E + u\left(\frac{\hbar}{2E}\right)^{2/3}$, and for $f \in \mathcal{S}(\mathbb{R})$,

$$W_{\hbar,f,2/3}(x,\xi) := \hbar^{-d} \int_{\mathbb{R}} f(\tau) d\mu_\hbar^{u,E,\frac{2}{3}}(\tau), \quad W_{2/3,E,\lambda_\pm}(x,\xi) = \hbar^{-d} \int_{\lambda_-}^{\lambda^+} d\mu_\hbar^{u,E,\frac{2}{3}}(\tau).$$

5.10 Bulk Sums

We next consider Weyl sums of eigenspace projections corresponding to an energy shell (or window) $[E_1, E_2]$. We consider both sharp and smoothed sums.

Definition 5.13 Define the "bulk" Wigner distributions for an $\hbar$-independent energy window $[E_1, E_2]$ by

$$W_{\hbar,[E_1,E_2]}(x,\xi) : \sum_{N: E_N(\hbar) \in [E_1,E_2]} W_{\hbar,E_N(\hbar)}(x,\xi). \tag{54}$$

More generally for $f \in C_b(\mathbb{R})$ define

$$W_{\hbar,f}(x,\xi) := \sum_{N=1}^{\infty} f(\hbar(N + d/2))\, W_{\hbar,E_N(\hbar)}(x,\xi). \tag{55}$$

Our first result about the bulk Weyl sums concerns the smoothed Weyl sums $W_{\hbar,f}$.

Proposition 5.14 *For* $f \in \mathcal{S}(\mathbb{R})$ *with* $\hat{f} \in C_0^\infty$, $W_{\hbar,f}(x,\xi)$ *admits a complete asymptotic expansion as* $\hbar \to 0$ *of the form,*

$$\begin{cases} W_{\hbar,f}(x,\xi) \simeq (\pi\hbar)^{-d} \sum_{j=0}^{\infty} c_{j,f,H}(x,\xi)\hbar^j, \text{ with} \\ c_{0,f,H}(x,\xi) = f(H(x,\xi)) = \int_{\mathbb{R}} \hat{f}(t)e^{itH(x,\xi)}dt. \end{cases}$$

In general $c_{k,f,H}(x,\xi)$ *is a distribution of finite order on* f *supported at the point* (x,ξ).

The proof merely involves Taylor expansion of the phase.

5.11 Interior/Exterior Asymptotics for Bulk Weyl Sums of Definition 5.13

From Proposition 5.14, it is evident that the behavior of $W_{\hbar,[E_1,E_2]}(x,\xi)$ depends on whether $H(x,\xi) \in (E_1, E_2)$ or $H(x,\xi) \notin [E_1, E_2]$. Some of this dependence is captured in the following result.

Proposition 5.15 *We have,*

$$W_{\hbar,[E_1,E_2]}(x,\xi) = \begin{cases} (i)\ (2\pi\hbar)^{-d}(1+O(\hbar^{1/2})), & H(x,\xi) \in (E_1,E_2), \\ (ii)\ O(\hbar^{-d+1/2}), & H(x,\xi) < E_1, \\ (iii)\ O(\hbar^{\infty}), & H(x,\xi) > E_2. \end{cases}$$

The two "sides" $0 < H(x,\xi) < E_1$ and $H(x,\xi) > E_2$ also behave differently because the Wigner distributions have slowly decaying tails inside an energy ball but are exponentially decaying outside of it. If we write $W_{\hbar,[E_1,E_2]}(x,\xi) = W_{\hbar,[0,E_2]}(x,\xi) - W_{\hbar,[0,E_1]}(x,\xi)$, we see that the two cases with $H(x,\xi) > E_1$ are covered by results for $W_{\hbar,[0,E]}$ with $E = E_1$ or $E = E_2$. When $H(x,\xi) < E_1$, then both terms of $W_{\hbar,[0,E_2]}(x,\xi) - W_{\hbar,[0,E_1]}(x,\xi)$ have the order of magnitude $\hbar^{-d}$ and the asymptotics reflect the cancellation between the terms. The boundary case where $H(x,\xi) = E_1$, or $H(x,\xi) = E_2$ is special and is given in Theorem 5.11.

5.12 Interface Asymptotics for Bulk Weyl Sums of Definition 5.13

Our final result concerns the asymptotics of $W_{\hbar,[E_1,E_2]}(x,\xi)$ in $\hbar^{\frac{2}{3}}$-tubes around the "interface" $H(x,\xi) = E_2$. Again, it is sufficient to consider intervals $[0, E]$. It is at least intuitively clear that the interface asymptotics will depend only on the individual eigenspace projections with eigenvalues in an $\hbar^{2/3}$-interval around

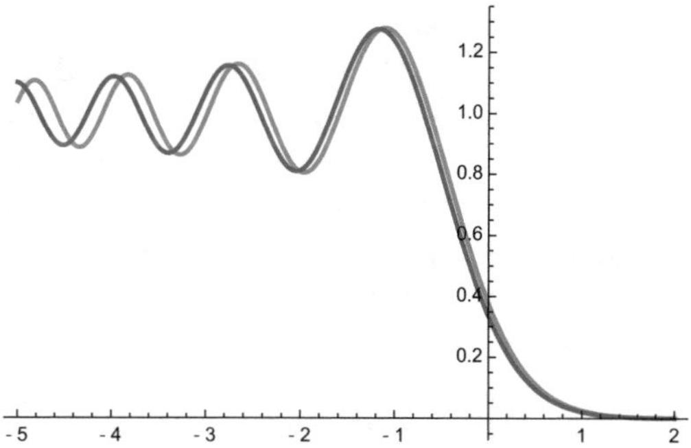

Fig. 3 Plot with $\hbar \approx 0.02,\ E = 1/2$ of scaled bulk Wigner–Weyl sum $(2\pi\hbar)^d W_{\hbar,[0,E]}(x,\xi)$ when $H(x,\xi) = E + u(\hbar/2E)^{2/3}$ as a function of u (blue) against its integrated Airy limit $\int_0^\infty \mathrm{Ai}(\lambda + u/E)d\lambda$ (red) from Theorem 5.16.

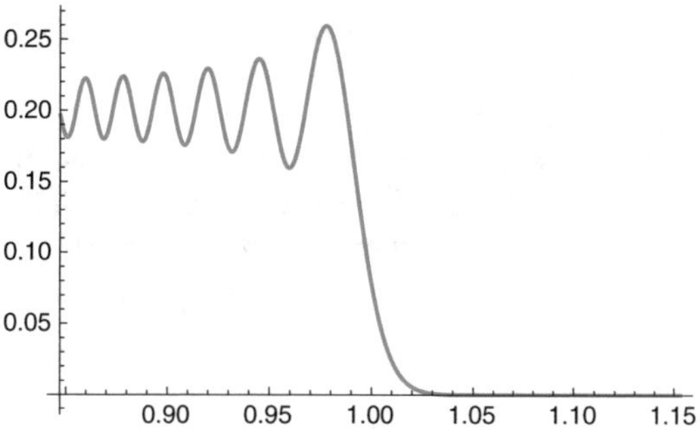

Fig. 4 Scaling at energy surface of Wigner function of projection onto energy interval [0, 1/2].

the energy level E, and since they add to 1 away from the boundary point, one may expect the asymptotics to be similar to the interface asymptotics for individual eigenspace projections in [HZ19].

Theorem 5.16 *Assume that* (x,ξ) *satisfies* $\frac{|x|^2+|\xi|^2}{2} - E = u\left(\frac{\hbar}{2E}\right)^{2/3}$ *with* $|u| < \hbar^{-2/3}$. *Then, for any* $\epsilon > 0$

$$W_{\hbar,[0,E]}(x,\xi) = (2\pi\hbar)^{-d}\left[\int_0^\infty \mathrm{Ai}\left(\frac{u}{E}+\tau\right)d\tau + O(\hbar^{1/3-\epsilon}\,|u|^{1/2}) + O(|u|^{5/2}\,\hbar^{2/3-\epsilon})\right],$$

where the implicit constant depends only on d, ϵ.

The Airy scaling the Wigner function is illustrated in Figures 3 and 4.

5.13 Heuristics

Wigner distributions are normalized so that the Wigner distribution of an L^2 normalized eigenfunction has L^2 norm 1 in $T^*\mathbb{R}^d$. Due to the multiplicity N^{d-1} of eigenspaces (3), the L^2 norm of $W_{\hbar,E_N(\hbar)}$ is of order $N^{\frac{d-1}{2}}$.

In the main results, we sum over windows of eigenvalues, e.g. $\lambda_- \hbar^{2/3} \leq E - E_N(\hbar) < \lambda_+ \hbar^{2/3}$ (51), resp. $E_N(\hbar) \in [0, E]$ in (5.13). Inevitably, the asymptotics are joint in $(\hbar, N)$. As $\hbar \downarrow 0$, the number of N contributing to the sum grows at the rate $\hbar^{-\frac{1}{3}}$, resp. $\hbar^{-1}$. Due to the N-dependence of the L^2 norm, terms with higher N have norms of higher weight in N than those of small N but the precise size of the contribution depends on the position of (x, ξ) relative to the interface $\{H = E\}$ and of course the relation (2).

$W_{\hbar, E_N(\hbar)}(x, \xi)$ peaks when $H(x, \xi) = E_N(\hbar)$, exponentially decays in $\hbar$ when $H(x, \xi) > E_N(\hbar)$ and has slowly decaying tails inside the energy ball $\{H < E_N(\hbar)\}$, which fall into three regimes: (i) Bessel near 0, (ii) oscillatory or trigonometric in the bulk, and (iii) Airy near $\{H = E\}$. In terms of N, when (2) holds, and $H(x, \xi) < E_N(\hbar)$, then $W_{\hbar, E_N(\hbar)}(x, \xi) \simeq \hbar^{-d+1/2} \simeq N^{d-1/2}$. Near the peak point, when $H(x, \xi) - E_N(\hbar) \approx \hbar^{2/3}$, we have in contrast $W_{\hbar, E_N(\hbar)}(x, \xi) \simeq \hbar^{-d+1/3} \simeq_E N^{d-1/3}$.

It follows that the terms with a high value of N and with $E_N(\hbar) \geq H(x, \xi)$ in (48) contribute high weights. There are an infinite number of such terms, and so (48) is a signed measure of infinite mass (as stated in Proposition 5.8). This is why we mainly consider the restriction of the measures (48) to compact intervals.

5.14 Remark on Nodal Sets in Phase Space

In Section 4 we discussed nodal sets of random eigenfunctions of the isotropic Harmonic oscillator. It would also make sense to consider nodal sets in phase space $T^*\mathbb{R}^d$ for Wigner distributions $W_{\Phi_{\hbar,E}}$ of random eigenfunctions of the isotropic Harmonic oscillator. This is of interest because Wigner distributions are signed, i.e. not positive, and their nodal sets and domains signal the extent of this "defect" in their interpretation as phase space densities. But so far, this has not been done. However, the covariance function is simply the Wigner distribution of the spectral projection kernels, so the analysis of Wigner distributions and of their interfaces across energy surfaces provides the necessary techniques.

In the next section we consider interfaces for partial Bergman kernels. The analogue in the complex domain of random nodal sets of isotropic oscillator eigenfunctions is zero sets of random homogeneous holomorphic polynomials of fixed degree in $\mathbb{C}^d$. This is essentially the same as studying such zero sets on complex projective space $\mathbb{CP}^{d-1}$, and to that extent the theory has already been developed. But interface phenomena for complex zero sets have not so far been studied.

6 Interfaces in Phase Space: Partial Bergman Kernels

In this section, we continue to study phase space distributions of orthogonal projections, but change from the Schrödinger quantization to the holomorphic quantization. The holomorphic setting consists of Berezin-Toeplitz operators acting on holomorphic sections of line bundles over Kähler manifolds, and is analytically simpler than the real Schrödinger setting. Hence we are able to present much more general results. Instead of fixing a model Schrödinger operator like the isotropic Harmonic Oscillator, we consider all possible Toeplitz Hamiltonians on holomorphic sections of Hermitian line bundles $(L, h) \to (M, \omega)$ over all possible projective Kähler manifolds. Here it is assumed that $i\partial\bar{\partial} \log h = \omega$, i.e. (L, h) is a positive, ample line bundle. For background on Bargmann-Fock space, and on line bundles over general Kähler manifolds, we refer to Section 10.

Motivation to study partial Bergman kernels comes from two sources. On the one hand, they arise in many problems of complex geometry (see [2, 22, 23, 33, 35] besides the articles surveyed here). On the other hand, they arise in the IQHE (integer quantum Hall effect). The author's interest was stimulated by conversations with A. Abanov, S. Klevtsov, and P. Wiegmann during a Simons' Center program on complex geometry and the IQHE. We refer to [9, 41, 42] for some physics articles where interfaces in the density of states of the IQHE are studied. It should be emphasized that there are many types of partial Bergman kernels, and the ones most interesting in physics are still out of reach of the rigorous techniques described here. What we study here are *spectral partial Bergman kernels*, i.e. orthogonal projection kernels onto spectral subspaces for Toeplitz Hamiltonians. By no means do all pBK's (partial Bergman kernels) arise from spectral problems, but the spectral pBK's are the only types for which there exist general results (or almost any results) and sometimes the pBK's of interest in the IQHE are spectral pBK's.

We do not review the basic definitions here (see Section 10) but head straight for the interface results. In place of the spectral projections of the previous sections, we consider *partial Bergman kernels* on "polarized" Kähler manifolds $(L, h) \to (M^m, \omega, J)$, i.e. Kähler manifolds of (complex) dimension m equipped with a Hermitian holomorphic line bundle whose curvature form F_∇ for the Chern connection ∇ satisfies $\omega = iF_\nabla$. Partial Bergman kernels

$$\Pi_{k,\mathcal{S}_k} : L^2(M, L^k) \to \mathcal{S}_k \subset H^0(M, L^k) \tag{56}$$

are Schwarz kernels for orthogonal projections onto proper subspaces $\mathcal{S}_k$ of the holomorphic sections of L^k.

For general subspaces, there is little one can say about the asymptotics of the partial density of states $\Pi_{k,\mathcal{S}_k}(z)$, i.e. the contraction of the diagonal of the kernel. But for certain sequences $\mathcal{S}_k$ of subspaces, the partial density of states $\Pi_{k,\mathcal{S}_k}(z)$ has an asymptotic expansion as $k \to \infty$ which roughly gives the probability density that a quantum state from $\mathcal{S}_k$ is at the point z. More concretely, in terms of an orthonormal basis $\{s_i\}_{i=1}^{N_k}$ of $\mathcal{S}_k$, the partial Bergman densities are defined by

$$\Pi_{k,\mathcal{S}_k}(z) = \sum_{i=1}^{N_k} \|s_i(z)\|^2_{h^k}. \tag{57}$$

When $\mathcal{S}_k = H^0(M, L^k)$, $\Pi_{k,\mathcal{S}_k} = \Pi_k : L^2(M, L^k) \to H^0(M, L^k)$ is the orthogonal (Szegö or Bergman) projection. We also call the ratio $\frac{\Pi_{k,\mathcal{S}_k}(z)}{\Pi_k(z)}$ the partial density of states.

Corresponding to $\mathcal{S}_k$ there is an allowed region $\mathcal{A}$ where the relative partial density of states $\Pi_{k,\mathcal{S}_k}(z)/\Pi_k(z)$ is one, indicating that the states in $\mathcal{S}_k$ "fill up" $\mathcal{A}$, and a forbidden region $\mathcal{F}$ where the relative density of states is $O(k^{-\infty})$, indicating that the states in $\mathcal{S}_k$ are almost zero in $\mathcal{F}$. On the boundary $\mathcal{C} := \partial\mathcal{A}$ between the two regions there is a shell of thickness $O(k^{-\frac{1}{2}})$ in which the density of states decays from 1 to 0. The $\sqrt{k}$-scaled relative partial density of states is asymptotically Gaussian along this interface, in a way reminiscent of the central limit theorem. This was proved in [35] for certain Hamiltonian holomorphic S^1 actions, then in greater generality in [ZZ17]. In fact, it is a universal property of partial Bergman kernels defined by C^∞ Hamiltonians. The first universal scaling results for full Bergman kernel were obtained in [4].

To begin with, we define the subspaces $\mathcal{S}_k$. They are defined as spectral subspaces for the quantization of a smooth function $H : M \to \mathbb{R}$. By the standard (Kostant) method of geometric quantization, one can quantize H as the self-adjoint zeroth order Toeplitz operator

$$H_k := \Pi_k(\frac{i}{k}\nabla_{\xi_H} + H)\Pi_k : H^0(M, L^k) \to H^0(M, L^k) \tag{58}$$

acting on the space $H^0(M, L^k)$ of holomorphic sections. Here, ξ_H is the Hamiltonian vector field of H, ∇_{ξ_H} is the Chern covariant derivative on sections, and H acts by multiplication. We denote the eigenvalues (repeated with multiplicity) of $\hat{H}_k$ (58) by

$$\mu_{k,1} \le \mu_{k,2} \le \cdots \le \mu_{k,N_k}, \tag{59}$$

where $N_k = \dim H^0(M, L^k)$, and the corresponding orthonormal eigensections in $H^0(M, L^k)$ by $s_{k,j}$.

Let E be a regular value of H. We denote the partial Bergman kernels for the corresponding spectral subspaces by

$$\Pi_{k,E} : H^0(M, L^k) \to \mathcal{H}_{k,E}, \tag{60}$$

where

$$\mathcal{S}_k := \mathcal{H}_{k,E} := \bigoplus_{\mu_{k,j} < E} V_{\mu_{k,j}}, \tag{61}$$

$\mu_{k,j}$ being the eigenvalues of H_k and

$$V_{\mu_{k,j}} := \{s \in H^0(M, L^k) : H_k s = \mu_{k,j} s\}. \tag{62}$$

We denote by $\Pi_{k,j} : H^0(M, L^k) \to V_{\mu_{k,j}}$ the orthogonal projection to $V_{\mu_{k,j}}$. The associated allowed region $\mathcal{A}$ is the classical counterpart to (61), and the forbidden region $\mathcal{F}$ and the interface $\mathcal{C}$ are

$$\mathcal{A} := \{z : H(z) < E\}, \quad \mathcal{F} = \{z : H(z) > E\}, \quad \mathcal{C} = \{z : H(z) = E\}. \tag{63}$$

More generally, for any spectral interval $I \subset \mathbb{R}$ we define the partial Bergman kernels to be the orthogonal projections,

$$\Pi_{k,I} : H^0(M, L^k) \to \mathcal{H}_{k,I}, \tag{64}$$

onto the spectral subspace,

$$\mathcal{H}_{k,I} := \text{span}\{s_{k,j} : \mu_{k,j} \in I\}. \tag{65}$$

Its (Schwartz) kernel is defined by

$$\Pi_{k,I}(z, w) = \sum_{\mu_{k,j} \in I} s_{k,j}(z)\overline{s_{k,j}(w)} \tag{66}$$

and the metric contraction of (66) on the diagonal with respect to h^k is the partial density of states,

$$\Pi_{k,I}(z) = \sum_{\mu_{k,j} \in I} \|s_{k,j,\alpha}(z)\|^2.$$

The classical allowed region $\mathcal{A}$ and forbidden region $\mathcal{F}$ are the open subsets

$$\mathcal{A} := \text{Int}(H^{-1}(I)), \quad \mathcal{F} = \text{Int}(M \backslash \mathcal{A}),$$

and the interface as

$$\mathcal{C} = \partial\mathcal{A} = \partial\mathcal{F}.$$

In [ZZ17] it is proved that

$$\frac{\Pi_{k,I}(z)}{\Pi_k(z)} = \begin{cases} 1 & \text{if } z \in \mathcal{A} \\ 0 & \text{if } z \in \mathcal{F} \end{cases} \mod O(k^{-\infty}).$$

We denote by $\Pi_k(z, w)$ and $\Pi_k(z)$ the (full) Bergman kernel and density function. Here and throughout, we use the notation $K(z)$ for the metric contraction of the diagonal values $K(z, z)$ of a kernel.

For each $z \in \mathcal{C}$, let ν_z be the unit normal vector to $\mathcal{C}$ pointing towards $\mathcal{A}$. And let $\gamma_z(t)$ be the geodesic curve with respect to the Riemannian metric $g(X, Y) = \omega(X, JY)$ defined by the Kähler form ω, such that $\gamma_z(0) = z$, $\dot{\gamma}_z(0) = \nu_z$. For small enough $\delta > 0$, the map

$$\Phi : \mathcal{C} \times (-\delta, \delta) \to M, \quad (z, t) \mapsto \gamma_z(t) \tag{67}$$

is a diffeomorphism onto its image.

Main Theorem *Let $(L, h) \to (M, \omega, J)$ be a polarized Kähler manifold. Let $H : M \to \mathbb{R}$ be a smooth function and E a regular value of H. Let $\mathcal{S}_k \subset H^0(X, L^k)$ be defined as in* (61). *Then we have the following asymptotics on partial Bergman densities $\Pi_{k,\mathcal{S}_k}(z)$:*

$$\left(\frac{\Pi_{k,\mathcal{S}_k}}{\Pi_k}\right)(z) = \begin{cases} 1 & \text{if } z \in \mathcal{A} \\ 0 & \text{if } z \in \mathcal{F} \end{cases} \mod O(k^{-\infty}).$$

For small enough $\delta > 0$, let $\Phi : \mathcal{C} \times (-\delta, \delta) \to M$ be given by (67). *Then for any $z \in \mathcal{C}$ and $t \in \mathbb{R}$, we have*

$$\left(\frac{\Pi_{k,\mathcal{S}_k}}{\Pi_k}\right)(\Phi(z, t/\sqrt{k})) = \operatorname{Erf}(2\sqrt{\pi} t) + O(k^{-1/2}), \tag{68}$$

where $\operatorname{Erf}(x) = \int_{-\infty}^{x} e^{-s^2/2} \frac{ds}{\sqrt{2\pi}}$ *is the cumulative distribution function of the Gaussian, i.e.* $\mathbb{P}_{X\sim N(0,1)}(X < x)$.[3]

Remark 6.1 The analogous result for critical levels is proved in [ZZ18b]. We could also choose an interval (E_1, E_2) with E_i regular values of H[4], and define $\mathcal{S}_k$ as the span of eigensections with eigenvalue within (E_1, E_2). However, the interval case can be deduced from the half-ray case $(-\infty, E)$ by taking difference of the corresponding partial Bergman kernel, hence we only consider allowed region of the type in (63).

Example 6.2 As a quick illustration, holomorphic sections of the trivial line bundle over $\mathbb{C}$ are holomorphic functions on $\mathbb{C}$. We equip the bundle with the Hermitian metric where 1 has the norm-square $e^{-|z|^2}$. The kth power has metric $e^{-k|z|^2}$. Fix

[3]The usual Gaussian error function $\mathrm{erf}(x) = (2\pi)^{-1/2} \int_{-x}^{x} e^{-s^2/2} ds$ is related to Erf by $\operatorname{Erf}(x) = \frac{1}{2}(1 + \mathrm{erf}(\frac{x}{\sqrt{2}}))$.

[4]It does not matter whether the endpoints are included in the interval, since contribution from the eigenspaces $V_{k,\mu}$ with $\mu = E_i$ is of lower order than k^m.

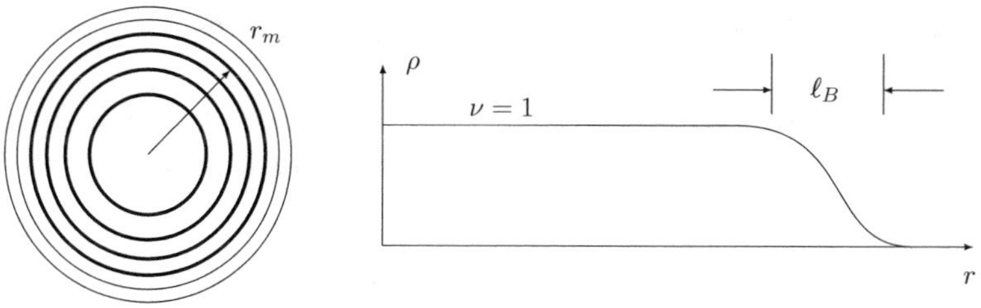

Fig. 5 "The density profile of the $\nu = 1$ droplet, where the first m levels (represented by the thick lines) are filled." From Fig 7.11 in [41].

$\epsilon > 0$ and define the subspaces $\mathcal{S}_k = \bigoplus_{j \leq \epsilon k} z^j$ of sections vanishing to order at most ϵk at 0, or sections with eigenvalues $\mu < \epsilon$ for operator $H_k = \frac{1}{ik}\partial_\theta$ quantizing $H = |z|^2$. The full and partial Bergman densities are

$$\Pi_k(z) = \frac{k}{2\pi}, \quad \Pi_{k,\epsilon}(z) = \left(\frac{k}{2\pi}\right) \sum_{j \leq \epsilon k} \frac{k^j}{j!} |z^j|^2 e^{-k|z|^2}.$$

As $k \to \infty$, we have

$$\lim_{k\to\infty} k^{-1} \Pi_{k,\epsilon}(z) = \begin{cases} 1 & |z|^2 < \epsilon \\ 0 & |z|^2 > \epsilon. \end{cases}$$

For the boundary behavior, one can consider sequence z_k, such that $|z_k|^2 = \epsilon(1 + k^{-1/2}u)$,

$$\lim_{k\to\infty} k^{-1} \Pi_{k,\epsilon}(z_k) = \mathrm{Erf}(u).$$

This example is often used to illustrate the notion of "filling domains" in the IQH (integer Quantum Hall) effect (see Figure 5). In IQH, one considers a free electron gas confined in plane $\mathbb{R}^2 \simeq \mathbb{C}$, with a uniform magnetic field in the perpendicular direction. A one-particle electron state is said to be in the lowest Landau level (LLL) if it has the form $\Psi(z) = e^{-|z|^2/2} f(z)$, where $f(z)$ is holomorphic as in Example 6.2. The following image of the density profile is copied from [41], where the picture on the right illustrates how the states $\frac{(\sqrt{k}\, z)^j}{\sqrt{j!}} e^{-k|z|^2/2}$ with $j \leq \epsilon k$ fill the disc of radius $\sqrt{\epsilon}$, so that the density profile drops from 1 to 0.

The example is S^1 symmetric and therefore the simpler results of [ZZ16] apply. For more general domains $D \subset \mathbb{C}$, it is not obvious how to fill D with LLL states. The Main Theorem answers the question when $D = \{H \leq E\}$ for some H. For a physics discussion of Erf asymptotics and their (as yet unknown) generalization to the fractional QH effect, see [9, 42].

6.1 Three Families of Measures at Different Scales

The rationale for viewing the Erf asymptotics of scaled partial Bergman kernels along the interface $\mathcal{C}$ is explained by considering three different scalings of the spectral problem.

$$\begin{cases} (i) \quad d\mu_k^z(x) \;\; = \sum_j \Pi_{k,j}(z)\delta_{\mu_{k,j}}(x), \\ \\ (ii) \;\; d\mu_k^{z,\frac{1}{2}}(x) = \sum_j \Pi_{k,j}(z)\delta_{\sqrt{k}(\mu_{k,j}-H(z))}(x), \\ \\ (iii) \; d\mu_k^{z,1,\tau}(x) = \sum_j \Pi_{k,j}(z)\delta_{k(\mu_{k,j}-H(z))+\sqrt{k}\tau}(x), \end{cases} \tag{69}$$

where as usual, δ_y is the Dirac point mass at $y \in \mathbb{R}$. We use $\mu(x) = \int_{-\infty}^x d\mu(y)$ to denote the cumulative distribution function.

We view these scalings as analogous to three scalings of the convolution powers μ^{*k} of a probability measure μ supported on $[-1, 1]$ (say). The third scaling (iii) corresponds to μ^{*k}, which is supported on $[-k, k]$. The first scaling (i) corresponds to the Law of Large Numbers, which rescales μ^{*k} back to $[-1, 1]$. The second scaling (ii) corresponds to the CLT (central limit theorem) which rescales the measure to $[-\sqrt{k}, \sqrt{k}]$.

Our main results give asymptotic formulae for integrals of test functions and characteristic functions against these measures. To obtain the remainder estimate (68), we need to apply semi-classical Tauberian theorems to $\mu_k^{z,\frac{1}{2}}$ and that forces us to find asymptotics for $\mu_k^{z,1,\tau}$.

6.2 Unrescaled Bulk Results on $d\mu_k^z$

The first result is that the behavior of the partial density of states in the allowed region $\{z : H(z) < E\}$ is essentially the same as for the full density of states, while it is rapidly decaying outside this region.

We begin with a simple and general result about partial Bergman kernels for smooth metrics and Hamiltonians.

Theorem 6.3 *Let ω be a C^∞ metric on M and let $H \in C^\infty(M)$. Fix a regular value E of H and let $\mathcal{A}, \mathcal{F}, \mathcal{C}$ be given by* (63). *Then for any $f \in C^\infty(\mathbb{R})$, we have*

$$\Pi_k(z)^{-1} \int_{-\infty}^{E} f(\lambda) d\mu_k^z(\lambda) \to \begin{cases} f(H(z)) & \text{if } z \in \mathcal{A} \\ 0 & \text{if } z \in \mathcal{F}. \end{cases} \tag{70}$$

In particular, the density of states of the partial Bergman kernel is given by the asymptotic formula:

$$\Pi_k(z)^{-1}\Pi_{k,E}(z) \sim \begin{cases} 1 \mod O(k^{-\infty}) & \text{if } z \in \mathcal{A} \\ 0 \mod O(k^{-\infty}) & \text{if } z \in \mathcal{F}, \end{cases} \tag{71}$$

where the asymptotics are uniform on compact sets of $\mathcal{A}$ or $\mathcal{F}$.

In effect, the leading order asymptotics says that the normalized measure $\Pi_k(z)^{-1}d\mu_k^z \to \delta_{H(z)}$. This is a kind of Law of Large Numbers for the sequence $d\mu_k^z$. The theorem does not specify the behavior of $\mu_k^z(-\infty, E)$ when $H(z) = E$. The next result pertains to the edge behavior.

6.3 $\sqrt{k}$-Scaling Results on $d\mu_k^{z,1/2}$

The most interesting behavior occurs in $k^{-\frac{1}{2}}$-tubes around the interface $\mathcal{C}$ between the allowed region $\mathcal{A}$ and the forbidden region $\mathcal{F}$. For any $T > 0$, the tube of 'radius' $Tk^{-\frac{1}{2}}$ around $\mathcal{C} = \{H = E\}$ is the flowout of $\mathcal{C}$ under the gradient flow of H

$$F^t := \exp(t\nabla H) : M \to M,$$

for $|t| < Tk^{-1/2}$. Thus it suffices to study the partial density of states $\Pi_{k,E}(z_k)$ at points $z_k = F^{\beta/\sqrt{k}}(z_0)$ with $z_0 \in H^{-1}(E)$. The interface result for any smooth Hamiltonian is the same as if the Hamiltonian flow generates a holomorphic S^1-actions, and thus our result shows that it is a universal scaling asymptotics around $\mathcal{C}$.

Theorem 6.4 *Let ω be a C^∞ metric on M and let $H \in C^\infty(M)$. Fix a regular value E of H and let $\mathcal{A}, \mathcal{F}, \mathcal{C}$ be given by* (63). *Let $F^t : M \to M$ denote the gradient flow of H by time t. We have the following results:*

(1) *For any point $z \in \mathcal{C}$, any $\beta \in \mathbb{R}$, and any smooth function $f \in C^\infty(\mathbb{R})$, there exists a complete asymptotic expansion,*

$$\sum_j f(\sqrt{k}(\mu_{k,j} - E))\Pi_{k,j}(F^{\beta/\sqrt{k}}(z)) \simeq \left(\frac{k}{2\pi}\right)^m (I_0 + k^{-\frac{1}{2}} I_1 + \cdots), \tag{72}$$

in descending powers of $k^{\frac{1}{2}}$, with the leading coefficient as

$$I_0(f, z, \beta) = \int_{-\infty}^{\infty} f(x)e^{-\left(\frac{x}{|\nabla H|(z)|} - \beta|\nabla H(z)|\right)^2} \frac{dx}{\sqrt{\pi}|\nabla H(z)|}.$$

(2) *For any point* $z \in \mathcal{C}$, *and any* $\alpha \in \mathbb{R}$, *the cumulative distribution function* $\mu_k^{z,1/2}(\alpha) = \int_{-\infty}^{\alpha} d\mu_k^{z,1/2}$ *is given by*

$$\mu_k^{z,1/2}(\alpha) = \sum_{\mu_{k,j} < E + \frac{\alpha}{\sqrt{k}}} \Pi_{k,j}(z) = \left(\frac{k}{2\pi}\right)^m \operatorname{Erf}\left(\frac{\sqrt{2}\alpha}{|\nabla H(z)|}\right) + O(k^{m-1/2}). \tag{73}$$

(3) *For any point* $z \in \mathcal{C}$, *and any* $\beta \in \mathbb{R}$, *the Bergman kernel density near the interface is given by*

$$\begin{aligned} \Pi_{k,E}(F^{\beta/\sqrt{k}}(z)) &= \textstyle\sum_{\mu_{j,k} < E} \Pi_{k,j}(F^{\beta/\sqrt{k}}(z)) \\ &= \left(\tfrac{k}{2\pi}\right)^m \operatorname{Erf}\left(-\sqrt{2}\beta|\nabla H(z)|\right) + O(k^{m-1/2}). \end{aligned} \tag{74}$$

Remark 6.5 The leading power $\left(\frac{k}{2\pi}\right)^m$ is the same as in Theorem 6.3, despite the fact that we sum over a packet of eigenvalues of width (and cardinality) $k^{-\frac{1}{2}}$ times the width (and cardinality) in Theorem 6.3. This is because the summands $\Pi_{k,j}(z)$ already localize the sum to $\mu_{k,j}$ satisfying $|\mu_{k,j} - H(z)| < Ck^{-\frac{1}{2}}$.

6.4 Energy Level Localization and $d\mu_k^{z,1,\alpha}$

To obtain the remainder estimate for the $\sqrt{k}$ rescaled measure $d\mu_k^{z,1/2}$ in (73) and (74) , we apply the Tauberian theorem. Roughly speaking, one approximates $d\mu_k^{z,1/2}$ by convoluting the measure with a smooth function W_h of width h, and the difference of the two is proportional to h. The smoothed measure $d\mu_k^{z,1/2} * W_h$ has a density function, the value of which can be estimated by an integral of the propagator $U_k(t, z, z)$ for $|t| \sim k^{-1}/(hk^{-1/2})$. Thus if we choose $h = k^{-1/2}$, and W_h to have Fourier transform supported in $(-\epsilon, +\epsilon)/h$, we only need to evaluate $U_k(t, z, z)$ for $|t| < \epsilon$, where ϵ can be taken to be arbitrarily small.

Theorem 6.6 *Let E be a regular value of H and $z \in H^{-1}(E)$. If ϵ is small enough, such that the Hamiltonian flow trajectory starting at z does not loop back to z for time $|t| < 2\pi\epsilon$, then for any Schwarz function $f \in \mathcal{S}(\mathbb{R})$ with $\hat{f}$ supported in $(-\epsilon, \epsilon)$ and $\hat{f}(0) = \int f(x)dx = 1$, and for any $\alpha \in \mathbb{R}$ we have*

$$\int_{\mathbb{R}} f(x) d\mu_k^{z,1,\alpha}(x) = \left(\frac{k}{2\pi}\right)^{m-1/2} e^{-\frac{\alpha^2}{\|\xi_H(z)\|^2}} \frac{\sqrt{2}}{2\pi\,\|\xi_H(z)\|}(1 + O(k^{-1/2})).$$

6.5 Critical Levels

In this section we consider interfaces at critical levels. Let $H : M \to \mathbb{R}$ be a smooth function with Morse critical points. Henceforth, to simplify notation, we use Kähler local coordinates u centered at z_0 to write points in the $k^{-\epsilon}$ tube around $\mathcal{C}$ by

$$z = z_0 + k^{-\epsilon} u := \exp_{z_0}(k^{-\epsilon} u), \quad u \in T_{z_0}.\mathcal{C}.$$

The abuse of notation in dropping the higher order terms of the normal exponential map is harmless since we are working so close to $\mathcal{C}$. At regular points z_0 we may use the exponential map along $N_{z_0}\mathcal{C}$ but we also want to consider critical points. More generally we write $z_0 + u$ for the point with Kähler normal coordinate u. In these coordinates,

$$\omega(z_0 + u) = i \sum_{j=1}^{m} du_j \wedge d\bar{u}_j + O(|u|).$$

We also choose a local frame e_L of L near z, such that the corresponding $\varphi = -\log h(e_L, e_L)$ is given by

$$\varphi(z_0 + u) = |u|^2 + O(|u|^3).$$

See [36] also for more on such adapted frames and Heisenberg coordinates.

Clearly, the formula (72) breaks down at critical points and near such points on critical levels. Our main goal in this paper is to generalize the interface asymptotics to the case when the Hamiltonian is a Morse function and the interface $\mathcal{C} = \{H = E\}$ is a critical level, so that $\mathcal{C}$ contains a non-degenerate critical point z_c of H. To allow for non-standard scaling asymptotics, we study the smoothed partial Bergman density near the critical value $E = H(z_c)$,

$$\Pi_{k,E,f,\delta}(z) := \sum_j \|s_{k,j}(z)\|^2 \cdot f(k^\delta(\mu_{k,j} - E)),$$

where $f \in \mathcal{S}(\mathbb{R})$ with Fourier transform $\hat{f} \in C_c^\infty(\mathbb{R})$, and $0 \le \delta \le 1$. This is the smooth analog of summing over eigenvalues within $[E - k^{-\delta}, E + k^{-\delta}]$.

The behavior of the scaled density of states is encoded in the following measures,

$$\begin{cases} d\mu_k^z(x) = \sum_j \|s_{k,j}(z)\|^2 \, \delta_{\mu_{k,j}}(x), \\ d\mu_k^{z,\delta}(x) = \sum_j \|s_{k,j}(z)\|^2 \, \delta_{k^\delta(\mu_{k,j} - H(z))}(x), \\ d\mu_k^{(z,u,\epsilon),\delta}(x) = \sum_j \|s_{k,j}(z + k^{-\epsilon}u)\|^2 \, \delta_{k^\delta(\mu_{k,j} - H(z))}(x). \end{cases} \tag{75}$$

For each measure μ we denote by $d\hat{\mu}$ the normalized probability measure

$$d\hat{\mu}(x) = \mu(\mathbb{R})^{-1} d\mu(x).$$

For all $z \in M$, we have the following weak limit, reminiscent of the law of large numbers;

$$\hat{\mu}_k^z(x) \rightharpoonup \delta_{H(z)}(x).$$

For $z \in M$ with $dH(z) \neq 0$, (72) shows that

$$\hat{\mu}_k^{z,1/2}(x) \rightharpoonup e^{-\frac{x^2}{|dH(z)|^2}} \frac{dx}{\sqrt{\pi}|dH(z)|}.$$

6.6 *Interface Asymptotics at Critical Levels*

The next result generalizes the ERF scaling asymptotics to the critical point case. We use the following setup: Let z_c be a non-degenerate Morse critical point of H, then for small enough $u \in \mathbb{C}^m$, we denote the Taylor expansion components by

$$H(z_c + u) = E + H_2(u) + O(|u|^3),$$

where

$$E = H(z_c), \quad H_2(u) = \frac{1}{2}\mathrm{Hess}_{z_c} H(u, u).$$

Theorem 6.7 *For any $f \in \mathcal{S}(\mathbb{R})$ with $\hat{f} \in C_c^\infty(\mathbb{R})$, we have*

$$\begin{aligned}\Pi_{k,E,f,1/2}(z_c + k^{-1/4}u) &:= \textstyle\sum_j \|s_{k,j}(z_c + k^{-1/4}u)\|^2 \cdot f(k^{1/2}(\mu_{k,j} - E)) \\ &= \left(\tfrac{k}{2\pi}\right)^m f(H_2(u)) + O_f(k^{m-1/4}).\end{aligned}$$

Moreover, the normalized rescaled pointwise spectral measure

$$d\hat{\mu}_k^{(z_c,u,1/4),1/2}(x) := \frac{\sum_j \|s_{k,j}(z_c + k^{-1/4}u)\|^2 \, \delta_{k^{1/2}(\mu_{k,j}-E)}(x)}{\sum_j \|s_{k,j}(z_c + k^{-1/4}u)\|^2}$$

converges weakly

$$\hat{\mu}_k^{(z_c,u,1/4),1/2}(x) \rightharpoonup \delta_{H_2(u)}(x).$$

We notice that the scaling width has changed from $k^{-\frac{1}{2}}$ to $k^{-1/4}$ due to the critical point. The difference in scalings raises the question of what happens if we scale by $k^{-\frac{1}{2}}$ around a critical point. The result is stated in terms of the metaplectic representation on the osculating Bargmann-Fock space at z_c.

Theorem 6.8 *Let $1 \gg T > 0$ be small enough, such that there is no non-constant periodic orbit with periods less than T. Then for any $f \in \mathcal{S}(\mathbb{R})$ with $\hat{f} \in C_c^\infty((-T, T))$, we have*

$$\Pi_{k,E,f,1}(z_c + k^{-1/2}u) = \left(\frac{k}{2\pi}\right)^m \int_{\mathbb{R}} \hat{f}(t)\mathcal{U}(t,u)\frac{dt}{2\pi} + O(k^{m-1/2}),$$

where $\mathcal{U}(t,u)$ is the metaplectic quantization of the Hamiltonian flow of $H_2(u)$ defined as

$$\mathcal{U}(t,u) = (\det P)^{-1/2} \exp(\bar{u}(P^{-1} - 1)u + u\bar{Q}P^{-1}u/2 - \bar{u}P^{-1}Q\bar{u}/2).$$

Here $P = P(t)$, $Q = Q(t)$ be complex $m \times m$ matrices such that if $u(t) = \exp(t\xi_{H_2})u$, then

$$\begin{pmatrix} u(t) \\ \bar{u}(t) \end{pmatrix} = \begin{pmatrix} P(t) & Q(t) \\ \bar{Q}(t) & \bar{P}(t) \end{pmatrix} \begin{pmatrix} u \\ \bar{u} \end{pmatrix}.$$

Remark 6.9 Unlike the universal Erf decay profile in the $1/\sqrt{k}$-tube around the smooth part of $\mathcal{C}$, we cannot give the decay profile of $\Pi_{k,I}(z)$ near the critical point z_c. The reason is that there are eigensections that highly peak near z_c and with eigenvalues clustering around $H(z_c)$. Hence it even matters whether we use $[E_1, E_2]$ or (E_1, E_2). See the following case where the Hamiltonian action is holomorphic, where the peak section at z_c is an eigensection, and all other eigensections vanish at z_c.

The next result pertains to Hamiltonians generating $\mathbb{R}$ actions, as studied in [35], [ZZ16]. The Hamiltonian flow always extends to a holomorphic $\mathbb{C}$ action.

Proposition 6.10 *Assume H generate a holomorphic Hamiltonian $\mathbb{R}$ action. The pointwise spectral measure $d\mu_k^{z_c}(x)$ is always a delta-function*

$$\mu_k^{z_c} = \delta_{H(z_c)}(x), \quad \forall k = 1, 2 \cdots$$

Equivalently, for any spectral interval I,

$$\lim_{k\to\infty} \Pi_{k,I}(z_c) = \begin{cases} 1 & E \in I \\ 0 & E \notin I \end{cases}.$$

The above result follows immediately from:

Proposition 6.11 *Let z_c be a Morse critical point of H, $E = H(z_c)$. Then*

(1) *The L^2-normalized peak section $s_{k,z_c}(z) = C(z_c)\Pi_k(z, z_c)$ is an eigensection of $\hat{H}_k$ with eigenvalue $H(z_c)$. And all other eigensections orthogonal to s_{k,z_c} vanish at z_c.*
(2) *If $s_{k,j} \in H^0(M, L^k)$ is an eigensection of $\hat{H}_k$ with eigenvalue $\mu_{k,j} < E$, then $s_{k,j}$ vanishes on $W^+(z_c)$.*
(3) *If $s_{k,j} \in H^0(M, L^k)$ is an eigensection of $\hat{H}_k$ with eigenvalue $\mu_{k,j} > E$, then $s_{k,j}$ vanishes on $W^-(z_c)$.*

In particular, this shows the concentration of eigensection near z_c. Depending on whether the spectral interval I includes boundary point $H(z_c)$ or not, the partial Bergman density will differ by a large Gaussian bump of height $\sim k^m$.

6.7 Sketch of Proof

As in [ZZ17,ZZ18] the proofs involve rescaling parametrices for the propagator

$$U_k(t) = \exp itk\hat{H}_k \tag{76}$$

of the Hamiltonian (58). The parametrix construction is reviewed in Section 3.4. We begin by observing that for all $z \in M$, the time-scaled propagator has pointwise scaling asymptotics with the $k^{-\frac{1}{2}}$ scaling:

Proposition 6.12 ([ZZ17] Proposition 5.3) *If $z \in M$, then for any $\tau \in \mathbb{R}$,*

$$\hat{U}_k(t/\sqrt{k}, \hat{z}, \hat{z}) = \left(\frac{k}{2\pi}\right)^m e^{it\sqrt{k}H(z)} e^{-t^2 \frac{\|dH(z)\|^2}{4}} (1 + O(|t|^3 k^{-1/2})),$$

where the constant in the error term is uniform as t varies over compact subset of $\mathbb{R}$.

The condition $dH(z) \neq 0$ in the original statement in [ZZ17] is never used in the proof, hence both statement and proof carry over to the critical point case. We therefore omit the proof of this Proposition.

We also give asymptotics for the trace of the scaled propagator $U_k(t/\sqrt{k})$. It is based on stationary phase asymptotics and therefore also reflects the structure of the critical points.

Theorem 6.13 *If $t \neq 0$, the trace of the scaled propagator $U_k(t/\sqrt{k}) = e^{i\sqrt{k}t\hat{H}_k}$ admits the following asymptotic expansion*

$$\int_{z\in M} U_k(t/\sqrt{k}, z) d\,\mathrm{Vol}_M(z) =$$

$$\left(\frac{k}{2\pi}\right)^m \left(\frac{t\sqrt{k}}{4\pi}\right)^{-m} \sum_{z_c\in crit(H)} \frac{e^{it\sqrt{k}H(z_c)} e^{(i\pi/4)sgn(\mathrm{Hess}_{z_c}(H))}}{\sqrt{|\det(\mathrm{Hess}_{z_c}(H))|}}$$

$$\cdot(1 + O(|t|^3 k^{-1/2})),$$

where $sgn(\mathrm{Hess}_{z_c}(H))$ *is the signature of the Hessian, i.e. the number of its positive eigenvalues minus the number of its negative eigenvalues.*

7 Interfaces for the Bargmann-Fock Isotropic Harmonic Oscillator

We continue the discussion of Bargmann-Fock space from Section 3 by considering partial Bargmann-Fock Bergman kernels. In this section, we tie together the results on Wigner distributions of spectral projections for the isotropic Harmonic oscillator, and on density of states for partial Bergman kernels associated with the natural S^1 action on Bargmann-Fock space. This is the most direct analogue of the Schrödinger results.

The classical Bargmann-Fock isotropic Harmonic oscillator corresponds to the degree operator on $H^0(\mathbb{CP}^m, \mathcal{O}(N))$. The total space of the associated line bundle is $\mathbb{C}^{m+1}$. The harmonic operator generates the standard diagonal S^1 action on $\mathbb{C}^{m+1}$,

$$e^{i\theta} \cdot (z_1, \dots, z_{m+1}) = (e^{in_1\theta} z_1, \dots, e^{in_m\theta} z_{m+1}).$$

Its Hamiltonian is $H_{\vec{n}}(Z) = \sum_{j=1}^{m+1} n_j |z_j|^2$. The critical point set of $H_{\vec{n}}$ is its minimum set.

The eigenspaces $\mathcal{H}_{k,m,N}$ consist of monomials z^α with $|\alpha| = N$. Given the Planck constant k, the eigenspace projection is given by

$$\Pi_{h^k_{BF}, N}(Z, W) = \sum_{|\alpha|=N} \frac{(kZ)^\alpha)(k\bar{W})^\alpha}{\alpha!}, \tag{77}$$

as a kernel relative to the Bargmann-Fock Gaussian volume form. The partial Bergman kernels arising from spectral projections of the isotropic oscillator thus have the form,

$$\Pi_{h^k_{BF}, E} = \sum_{N: \frac{N}{k} \geq E} \Pi_{h^k_{BF}, N}(Z, W).$$

We claim that the eigenspace projector (77) satisfies,

$$\Pi_{h^k_{BF}, N}(Z, Z) = C_{N,k,m} ||Z||^{2N}, \tag{78}$$

where

$$C_{N,k,m} = \frac{p(N, m+1)}{\omega_m} \frac{k^N}{\Gamma(N+m+1)}.$$

Here, $\omega_m = \text{Vol}(\text{S}^{2\text{m}+1})$ is the surface measure of the unit sphere in $\mathbb{C}^{m+1}$. Also, $\dim \mathcal{H}_{k,m,N} = p(m+1, N)$, the partition function which counts the number of ways to express N as a sum of $m+1$ positive integers. To prove this, we first observe that the $U(m+1)$-invariance of the Harmonic oscillator Hamiltonian $H = ||Z||^2$ implies that $U^* \Pi_{h^k_{BF}, N} U = \Pi_{h^k_{BF}, N}$ and therefore $\Pi_{h^k_{BF}, N}(UZ, UZ) = \Pi_{h^k_{BF}, N}(Z, Z)$. It follows that $\Pi_{h^k_{BF}, N}(Z, Z) = F(||Z||^2)$ is radial. It is also homogeneous of degree $2N$, hence is a constant multiple $C_{N,k,m}||Z||^{2N}$ as claimed in (78). The constant is calculated from the fact that

$$\begin{aligned} p(m, N) = \dim \mathcal{H}_{k,m,N} &= \tfrac{k^{m+1}}{(m+1)!} \int_{\mathbb{C}^{m+1}} \Pi_{h^k_{BF}, N}(Z, Z) e^{-k||Z||^2} dL(Z) \\ &= \omega_m C_m k^{m+1} \int_0^\infty e^{-k\rho^2} \rho^{2N} \rho^{2m+1} d\rho \\ &= \tfrac{1}{2} \omega_m C_m k^{m+1} \int_0^\infty e^{-k\rho} \rho^N \rho^m d\rho \\ &= \tfrac{1}{2} \tfrac{k^{m+1}}{(m+1)!} k^{-(N+m+1)} \omega_m C_{m,k,N} \Gamma(N+m+1). \end{aligned}$$

Solving for $C_{m,k,N}$ establishes the formula. It also follows that the density of states is given by,

$$\begin{aligned} \textstyle\sum_{N \geq \epsilon k} \Pi_{h^k_{BF}, N}(Z, Z) &= \tfrac{k^{m+1}}{(m-1)! \omega_m} e^{-k||Z||^2} \textstyle\sum_{N \geq \epsilon k} \tfrac{(k||Z||^2)^N p(m,N)}{\Gamma(N+m+1)} \\ &\simeq \tfrac{k^{m+1}}{(m-1)! \omega_m} e^{-k||Z||^2} \textstyle\sum_{N \geq \epsilon k} \tfrac{(k||Z||^2)^N}{N!}, \end{aligned} \tag{79}$$

since $p(m+1, N) \simeq \frac{1}{(m+1)!} N^m (1 + O(N^{-1}))$ (4); also, $\Gamma(N+m+1) = (N+m)! \simeq (N+m) \cdots (N+1) N! \simeq N^m N!$.

8 Bargmann-Fock Space of a Line Bundle and Interface Asymptotics

In this section, we introduce a new model, the Bargmann-Fock space of an ample line bundle $\pi : L \to M$ over a Kähler manifold, and generalize the results of the preceding section to density of states for partial Bergman kernels associated with the natural S^1 action on the total space L^* of the dual line bundle. We let $X_h = \partial D_h^* \subset L^*$ be the unit S^1-bundle given by the boundary of the unit co-disc bundle, $D_h^* = \{(z, \lambda) \in L^* : |\lambda|_z < 1\}$. We sketch the proof that "interfaces" for the Hamiltonian generating the standard S^1 action on the Bargmann-Fock space of L satisfy the central limit theorem or cumulative Gaussian Erf interfaces as in the compact case of [ZZ16]. The Hamiltonian is simply the norm-square function $N(z, \lambda) := |\lambda|_{h_z}^2$, so the energy balls are simply the co-disc bundles

$$D_E^* = \{(z, \lambda) \in L^* : |\lambda|_{h_z} \le E^2\}.$$

As usual, we equivariantly lift sections $s_k \in H^0(M, L^k)$ to $\hat{s}_k \in \mathcal{H}_k(L^*)$, which are homogeneous of degree k in the sense that

$$\hat{s}_k(rx) = r^k \hat{s}_k(x).$$

8.1 Volume Forms

X_h is a contact manifold with contact volume form $dV = \alpha \wedge (\pi^*\omega)^m$. This contact volume form induces a volume form $dVol_{L^*}$ on L^*, generalizing the Lebesgue volume form $dVol_{\mathbb{C}^m}$ in the standard Bargmann-Fock space. Namely, the Kähler metric ω_h of the Hermitian metric h on L lifts to the partial Kähler metric $\pi^*\omega_h$. Then,

$$\omega_{L^*} = \pi^*\omega_h + d\lambda \wedge d\bar{\lambda}$$

is a Kähler metric on L^* with potential $|\lambda|^2 e^{-\phi}$ where $\phi = \log |e_L|_{h_z}^2$ is the local Kähler potential on M. Since $L^* \simeq X_h \times \mathbb{R}_+$ we may use polar coordinates (x, ρ) on L^*, which correspond to coordinates $(z, \lambda) \in M \times \mathbb{C}$ in a local trivialization by $\rho = |\lambda|_{h_z}$ and $x = (z, e^{i\theta})$. Since $\dim_{\mathbb{R}} X = 2m + 1$ when $\dim_{\mathbb{C}} M = m$, the volume form on L^* is given by

$$dVol_{L^*}(x, \rho) = \rho^{2m+1} dV(x) d\rho.$$

We then endow L^* with the (normalized) Gaussian measure analogous to (17),

$$d\Gamma_{m+1,\hbar} := \frac{\hbar^{-(m+1)}}{\mathrm{Vol}(X_h)\Gamma(m+1)} e^{-||Z||^2/\hbar} dVol_{L^*}(Z). \tag{80}$$

To check that the measure has mass 1, we note that

$$\int_{L^*} e^{-||Z||^2/\hbar} dVol_{L^*}(Z) = \mathrm{Vol}(\mathrm{X_h}) \int_0^\infty \mathrm{e}^{-\rho^2/\hbar} \rho^{2\mathrm{m}+1} \mathrm{d}\rho = \mathrm{Vol}(\mathrm{X_h}) \hbar^{\mathrm{m}+1} \Gamma(\mathrm{m}+1).$$

Here, we denote a general point of L^* by $Z = \rho x$ with $\rho \in \mathbb{R}_+, x \in X_h$. In the future we put

$$C_m(h) = \frac{1}{\text{Vol}(X_h)\Gamma(m+1)},$$

so that we do not have to keep track of this constant.

Definition 8.1 The Bargmann-Fock space of (L, h) is the Hilbert space

$$\mathcal{H}^2_{BF,\hbar}(L^*) := \bigoplus_{N=0}^{\infty} \mathcal{H}_N(L^*)$$

of entire square integrable holomorphic functions on L^* with respect to the inner product

$$||f||^2_{\hbar,BF} =: \frac{\hbar^{-(m+1)}}{\text{Vol}(X_h)\Gamma(m+1)} \int_{L^*} |f(Z)|^2 e^{-||Z||^2/\hbar} dVol_{L^*}(Z). \tag{81}$$

8.2 Orthonormal Basis

If $s \in H^0(M, L^k)$, then

$$||\hat{s}_k||_{L^2(X_h)} = \frac{1}{m!} \int_{X_h} |\hat{s}(x)|^2 dV(x) = \int_M ||s(z)||^2_{h^k} dV_\omega, \tag{82}$$

where the right side is the inner product on $H^0(M, L^k)$, where $dV_\omega = \omega^m/m!$. Let $N_k = \dim H^0(M, L^k)$ and let $\{\hat{s}_{k,j}\}_{j=1}^{N_k}$ be any orthonormal basis of $\mathcal{H}_k(L^*)$, corresponding to an orthonormal basis $\{s_{k,j}\}$ of $H^0(M, L^k)$. We let $\hbar = k^{-1}$. We also change the notation for powers of a bundle $k \to N$ to agree with the notation for the real Harmonic oscillator but retain the notation $\hbar = k^{-1}$. Thus, in effect, there are two semi-classical parameters: N and k, parallel to the parameters N and $\hbar^{-1}$ for the Schrödinger representation of the harmonic oscillator. The lifts $\hat{s}_{N,j}$ of an orthonormal basis $s_{N,j}$ of $H^0(M, L^N)$ are orthogonal but no longer normalized.

Lemma 8.2 *There exists a constant* $c_m = (\text{Vol}(X_h)\Gamma(m+1))^{-\frac{1}{2}}$ *so that* $\{c_m \hbar^{-N/2} \frac{\hat{s}_{N,j}(Z)}{\sqrt{(N+m+1)!}}\}$ *is an orthonormal basis of* $\mathcal{H}^2_{BF}$.

Proof We have,

$$\begin{aligned} ||\hat{s}_N||^2_{BF,\hbar} &= ||\hat{s}||^2_{L^2(X_h)} C_m \hbar^{-(m+1)} \int_0^\infty e^{-\rho^2/\hbar} \rho^{2N+2m+1} d\rho, \\ &= C_m ||\hat{s}||^2_{L^2(X_h)} \hbar^N \Gamma(N+m+1) = C_m \hbar^N \, (N+m)! ||\hat{s}||^2_{L^2(X_h)}, \end{aligned}$$

since $\hbar^{-(m+1)} \int_0^\infty e^{-\rho^2/\hbar} \rho^{2N+2m+1} d\rho = \hbar^N \Gamma(N+m+1)$. Putting $c_m = C_m^{-\frac{1}{2}}$ completes the proof. □

Corollary 8.3 *In the notation above, an orthonormal basis of $\mathcal{H}^2_{BF,\hbar}(L^*)$ is given by $\{c_m \hbar^{-\frac{N}{2}} \frac{\hat{s}_{N,j}}{\sqrt{(N+m)!}}\}$.*

8.3 Bargmann-Fock Bergman Kernel of a Line Bundle

We now define the Bargmann-Fock Bergman kernel:

Definition 8.4 The Bargmann-Fock Bergman kernel is the kernel of the orthogonal projection,

$$\hat{\Pi}_{BF,\hbar} : L^2(L^*) \to \mathcal{H}_{BF}(L^*),$$

with respect to the Gaussian measure $\Gamma_{m+1,\hbar}$ of the inner product (81). The density of states is the positive measure,

$$\hat{\Pi}_{BF,\hbar}(Z, Z) d\Gamma_{m+1,\hbar}(Z).$$

Let $\Pi_{h^N} : L^2(M, L^N) \to H^0(M, L^N)$ be the orthogonal projection with respect to the inner product (82). It lifts to the orthogonal projection $\hat{\Pi}_N : L^2(X_h) \to \mathcal{H}_N(X_h)$ with respect to the inner product on $L^2(X_h)$ defined by (82). Again by (81), $\hat{\Pi}_N$ is equal up to the constant C_N to the orthogonal projection $\mathcal{H}^2_{BF}(L^*) \to \mathcal{H}_N$. The next Lemma is an immediate consequence of Corollary 8.3.

Lemma 8.5 *The Bargmann-Fock Bergman kernel on $\mathcal{H}^2_{BF}(L^*)$ is given for $Z = (z, \lambda)$, $W = (w, \mu) \in L^*$ by*

$$\hat{\Pi}_{BF,\hbar}(Z, W) := c_m \sum_{N=0}^\infty \frac{\hbar^{-N}}{(N+m)!} \hat{\Pi}_N(Z, W) = C_m \sum_{N=0}^\infty \hbar^{-N} \frac{(\lambda \bar{\mu})^N}{(N+m)!} \hat{\Pi}_N(z, 1, w, 1),$$

where the equivariant kernel $\hat{\Pi}_N$ on X_h is extended by homogeneity to L^*. The density of states is given by

$$\begin{aligned}\hat{\Pi}_{BF,\hbar}(Z, Z) e^{-||Z||^2/\hbar} &:= c_m \hbar^{-(m+1)} e^{-||Z||^2/\hbar} \sum_{N=0}^\infty \frac{\hbar^{-N}}{(N+m)!} \hat{\Pi}_N(Z, Z) \\ &= c_m \hbar^{-(m+1)} e^{-||Z||^2/\hbar} \sum_{N=0}^\infty \hbar^{-N} \frac{|\lambda|^{2N}}{(N+m)!} \Pi_{h^N}(z),\end{aligned}$$

where $\Pi_{h^N}(z)$ is the metric contraction of $\Pi_N(z, z)$ on M.

The following is the main result of this section:

Proposition 8.6 *Let $\hbar = k^{-1}$. For $Z = (z, \lambda)$, the density of states equals*

$$\hat{\Pi}_{BF,k}(Z) := c_m k^{m+1} e^{-k||Z||^2} \sum_{N=0}^{\infty} \frac{|\lambda|^{2N}}{(N+m)!} k^N N^m [1 + O(\frac{1}{N})] dVol_{L^*}(Z).$$

Proof We recall that the density of states admits an asymptotic expansion,

$$\Pi_{h^N}(z) \simeq \frac{N^m}{m!}[1 + \frac{a_1(z)}{N} + \cdots],$$

so by Lemma 8.5, the density of states equals

$$\hat{\Pi}_{BF,\hbar}(Z, Z) d\Gamma_{m+1,\hbar} := c_m \hbar^{-(m+1)} e^{-||Z||^2/\hbar} \sum_{N=1}^{\infty} \hbar^{-N} \frac{|\lambda|^{2N}}{(N+m)!} N^m [1 + \frac{a_1(z)}{N} + \cdots] dVol_{L^*}(Z),$$

where C_m is a dimensional constant. Substituting $\hbar = k^{-1}$ completes the proof. □

We note that $\frac{N^m}{(N+m)!} \simeq \frac{1}{N!}$, so that the asymptotics of Proposition 8.6 agree with the Bargmann-Fock case (79).

8.4 *Interface Asymptotics*

The Hamiltonian is the norm square of the Hermitian metric itself, i.e.

$$H(z, \lambda) = |\lambda|^2_{h_z}.$$

The sublevel set $\{H \leq E\}$ is the disc bundle of radius E^2. We denote its boundary by Σ_E. The normal direction to Σ_E is the gradient ∇H direction, is given by the radial vector on L^* generated by the natural $\mathbb{R}_+$ action in the fibers dual to the S^1 action generated by H. Together, the $\mathbb{R}_+$ and S^1 actions define the standard $\mathbb{C}^*$ action on L^* and $\nabla H = J\xi_H$ where $\xi_H = \frac{\partial}{\partial \theta}$ is the Hamilton vector field of H. Thus, the asymptotics of such partial Bergman kernels falls into the $\mathbb{C}^*$ equivariant setting of [ZZ16].

We fix E and consider the partial Bargmann-Fock Bergman kernel of L^* with the energy interval $[0, E]$. Then as in the standard case, the exterior interface asymptotics pertain to the sums,

$$\sum_{N \geq \epsilon k} \Pi_{h^k_{BF}, N}(Z, Z) = \frac{k^{m+1}}{\omega_m m!} e^{-k||Z||^2} \sum_{N \geq \epsilon k} \frac{(k||Z||^2)^N N^m}{(N+m)!} [1 + \frac{a_1(z)}{N} + \cdots], \tag{83}$$

or to the complementary sums. Comparison with the standard Bargmann-Fock case of (79) shows that they agree to leading order, due to the Bergman kernel asymptotics of the summands $\Pi_N(z, 1, z, 1)$. The interface asymptotics are therefore the same as on Bargmann-Fock space for the Toeplitz isotropic Harmonic oscillator, and are also essentially the same as in Theorem 6, with $H(z, \lambda) = |\lambda|$ and $|\nabla H(z, \lambda)| = |\frac{\partial}{\partial \theta}| = \lambda$. We refer to orbits of the $\mathbb{R}_+$ action as radial orbits.

Theorem 8.7 *Let $\Pi_{h^k_{BF},(E,\infty]}(Z, Z) = \sum_{N \geq Ek} \Pi_{h^k_{BF},N}(Z, Z)$. Let $Z = (z, \lambda) \in L^*$ and let $Z_E = (z, \lambda_E) \in \Sigma_E$ with $|\lambda_E|_{h_z} = E$. Let $Z_k = e^{\frac{\beta}{\sqrt{k}}} \cdot Z_E = (z, e^{\frac{\beta}{\sqrt{k}}} \lambda_E)$ be sequence of points approaching (z, λ_E) along a radial $\mathbb{R}_+$ orbit, where $\beta \in \mathbb{R}$. Then, as $k \to \infty$,*

$$\Pi_{h^k_{BF},(E,\infty]}(Z_k) = k^m \operatorname{Erf}\left(\sqrt{k}\frac{E - e^{\frac{\beta}{\sqrt{k}}} E}{E}\right)(1 + O(k^{-1/2})) = k^m \operatorname{Erf}(-\beta)(1 + O(k^{-1/2})). \tag{84}$$

The proof of Theorem 8.7 is essentially the same as for Theorem 6, or better the same as in [ZZ16] for the $\mathbb{C}^*$ equivariant case. The only difference is that L^* is of infinite volume, but this does not affect pointwise asymptotics. However, there is a more elementary proof in this case.

Let $x = |Z_k|^2 = |\lambda|^2_{h_z} = e^{2\frac{\beta}{\sqrt{k}}} Z_E$ with $|Z_E| = E$. It is well-known that, as $k \to \infty$,

$$e^{-kx} \sum_{N \leq kE^2} \frac{(kx)^N N^m}{(N+m)!} \sim \frac{1}{\sqrt{2\pi x}} \int_{-\infty}^{\sqrt{k}\frac{E^2-x}{\sqrt{x}}} e^{-\frac{t^2}{2x}} dt.$$

Indeed, Lemma 1 of [44] asserts that

$$e^{-kx} \sum_{N=1}^{xk+y\sqrt{k}} \frac{(kx)^N}{N!} \sim \frac{1}{\sqrt{2\pi}} \int_{-\infty}^{\frac{y}{\sqrt{x}}} e^{-\frac{t^2}{2}} dt + O\left(\frac{Ax\sqrt{3x+1}}{\sqrt{k}((\sqrt{x}+y)^3}\right). \tag{85}$$

We have,

$$\sqrt{x} = e^{\frac{\beta}{\sqrt{k}}} E \simeq E + \frac{\beta}{\sqrt{k}} \implies \frac{E^2 - x}{\sqrt{x}} = \frac{E^2 - e^{2\frac{\beta}{\sqrt{k}}} E^2}{e^{\frac{\beta}{\sqrt{k}}} E} = -2E\frac{\beta}{\sqrt{k}}(1+O(\frac{1}{\sqrt{k}})).$$

Then let $kx + y\sqrt{k} = kE^2$, i.e. $\frac{y}{\sqrt{k}} = E^2 - x \simeq 2E\frac{\beta}{\sqrt{k}}$, thus $y = 2\beta E$, and use $\frac{N^m}{(N+m)!} \simeq \frac{1}{N!}$ to obtain the desired asymptotic.

To see this asymptotic implies Theorem 8.7, we let $\sqrt{k}\frac{E-x}{\sqrt{x}} = \beta$ or $\frac{E-x}{\sqrt{x}} = \frac{\beta}{\sqrt{k}}$. Then we get

$$\Pi_{h^k_{BF},(E,\infty]}(Z_k) \simeq k^m e^{-ke^{\frac{\beta}{\sqrt{k}}}E} \sum_{N \leq kE^2} \frac{(ke^{\frac{\beta}{\sqrt{k}}}E)^N N^m}{(N+m)!}$$
$$\sim k^m \frac{1}{\sqrt{2\pi x}} \int_{-\infty}^{\beta} e^{-\frac{t^2}{2E}} dt \, (1 + O(\frac{1}{\sqrt{k}}))$$

Remark 8.8 In [38], Szasz introduces the "Szasz operator"

$$P_f(u, x) := e^{-xu} \sum_{n=1}^{\infty} \frac{(ux)^n}{n!} f(\frac{n}{u}),$$

and shows that, for $f \in C_b(\mathbb{R})$, $\lim_{u\to\infty} P_f(u, x) = f(x)$. If we let $f(v) = \mathbf{1}_{[E,\infty]}(v)$, then $f(\frac{n}{u}) = \mathbf{1}_{u \leq nE}$. Szasz's asymptotic does not apply at the point of discontinuity. Later, Mirakyan introduced the "Szasz- Mirakyan operator" [30]

$$P_{f,N}(u, x) := e^{-xu} \sum_{n=1}^{N} \frac{(ux)^n}{n!} f(\frac{n}{u}),$$

and Omey [31] proved that if $N = N(n, x)$ with $\lim_{n\to\infty} \frac{N-nx}{\sqrt{n}} = C < \infty$, then $\lim_{n\to\infty} P_{f,N}(n, x) = \frac{f(x)}{\sqrt{2\pi}} \int_{-\infty}^{C} e^{-\frac{1}{2}u^2} du$. [44, Lemma 1] is a refinement of this limit formula.

This asymptotic formula arises in the analysis of Bernstein polynomials of discontinuous functions with a jump, and we refer to [13, 28, 31, 38, 44, 46] for the analysis.

9 Further Types of Interface Problems

9.1 Further Types of Interface Problems

Here are some further types of interface asymptotics:

- Entanglement entropy: Sharp spectral cutoffs involve indicator functions $\mathbf{1}_{E_1,E_2}(\hat{H}_\hbar)$ of a quantum Hamiltonian. On the other hand, one might quantize the indicator function $\mathbf{1}_{E_1,E_2}(H)$ of a classical Hamiltonian. This is obviously related but different, since the first is a projection and the second is not. Entanglement entropy is a measure of how the second fails to be a projection and has been studied by Charles-Estienne [12] and by the author (unpublished).
- On a manifold M with boundary ∂M one may study the spectral projections kernel $E^D_{[0,\lambda]}(x, x)$ of the Laplacian with Dirichlet boundary conditions. Away from ∂M, $\lambda^{-n} E^D_{[0,\lambda]}(x, x) \simeq 1$ where $n = \dim M$. Yet $E^D_{[0,\lambda]}(x, x) = 0$ on ∂M. What is the shape of the drop-off from 1 to 0 n a boundary zone of width λ^{-1}?

- For the hydrogen atom Hamiltonian $\hat{H}_\hbar$, there is a phase space interface $\Sigma_0 \subset T^*\mathbb{R}^d$ separating the bound states from the scattering states. The Hamiltonian flow is periodic on the one side of Σ_0 and unbounded on the other side and parabolic on Σ_0. The quantization of the bound state region is the discrete spectral projection $\Pi_{\mathrm{disc},\hbar}(x, y)$. How does its Wigner distribution behave along Σ_0?
- Interfaces arise in the quantum Hall effect, a point process defined by a weight ϕ and a Laughlin state which gives probabilities of N electrons to occur in a given configuration. The Laughlin states concentrate as $N \to \infty$ inside a "droplet." The interface asymptotics across the droplet in dimension one have been studied in [9, 42] and others, and from a mathematical point of view by Hedenmalm and Wennman [22, 23]. In the next section, we discuss higher dimensional droplets.
- Interfaces are studied for nonlinear equations such as the Allen–Cahn equation, and are related to phase transition problems; see, e.g., [18] for references to the literature.

9.2 *Droplets in Phase Space*

Let us describe droplets in more detail. Droplets in phase space arise as coincidence sets in envelope problems for plurisubharmonic functions. The boundary of such coincidence sets is the interface. In special cases, it is the same interface that we have described for spectral interfaces. But in general, the interface is a free boundary that must be determined from the envelope, and even its regularity is a problem. We refer to [2] for the origins of the theory of dimensions > 1.

The definition involves the inner products $\mathrm{Hilb}_N(h, \nu)$ induced by the data (h, ν) on the spaces $H^0(M, L^N)$ of holomorphic sections of powers $L^N \to M$ by

$$||s||^2_{\mathrm{Hilb}_N(h,\nu)} := \int_M |s(z)|^2_{h^N} d\nu(z). \tag{86}$$

We let h be a general C^2 Hermitian metric on L, and denote its positivity set by

$$M(0) = \{x \in M : \omega_\phi|_{T_x M} \text{ has only positive eigenvalues}\}, \tag{87}$$

i.e. the set where ω_ϕ is a positive $(1, 1)$ form. For a compact set $K \subset M$, also define the *equilibrium potential* $\phi_{eq} = V^*_{h,K}$[5]

$$V^*_{h,K}(z) = \phi_{eq}(z) := \sup\{u(z) : u \in PSH(M, \omega_0), u \le \phi \text{ on } K\}, \tag{88}$$

[5]Both notations ϕ_{eq} and $V^*_{h,K}$, and also $P_K(\phi)$, are standard and we use them interchangeably. $V^*_{h,K}$ is called the pluri-complex Green's function.

where ω_0 is a reference Kähler metric on M and $PSH(M, \omega_0)$ are the psh functions u relative to ω_0,

$$PSH(M, \omega_0) = \{u \in L^1(M, \mathbb{R} \cup \infty) : dd^c u + \omega_0 \geq 0, \quad \text{and } u \text{ is } \omega_0 - u.s.c.\}. \tag{89}$$

Further define the coincidence set,

$$D := \{z \in M : \phi(z) = \phi_e(z)\}. \tag{90}$$

The boundary ∂D is the "interface" and the problem is to determine its regularity and other properties. It carries an *equilibrium measure* defined by

$$d\mu_\phi = (dd^c \phi_{eq})^m / m! = \mathbf{1}_{D \cap M(0)} (dd^c \phi)^m / m!. \tag{91}$$

Here, $d^c = \frac{1}{i}(\partial - \bar{\partial})$.

Some droplets are classically forbidden regions for spectrally defined subspaces. The extent to which one may construct a spectral problem with this property is unknown. Since the interface is usually only $C^{1,1}$, it cannot be the level set (even a critical level) for a smooth (Morse-Bott) Hamiltonian in general.

10 Appendix on Kähler Analysis

In this Appendix, we give a quick review of the basic notations of Kähler analysis. First we introduce co-circle bundle $X \subset L^*$ for a positive Hermitian line bundle (L, h), so that holomorphic sections of L^k for different k can all be represented in the same space of CR-holomorphic functions on X, $\mathcal{H}(X) = \oplus_k \mathcal{H}_k(X)$. The Hamiltonian flow g^t generated by ξ_H on (M, ω) lifts to a contact flow $\hat{g}^t$ generated by $\hat{\xi}_H$ on X.

10.1 Holomorphic Sections in L^k and CR-Holomorphic Functions on X

Let $(L, h) \to (M, \omega)$ be a positive Hermitian line bundle, L^* the dual line bundle. Let

$$X := \{p \in L^* \mid \|p\|_h = 1\}, \quad \pi : X \to M$$

be the unit circle bundle over M.

Let $e_L \in \Gamma(U, L)$ be a non-vanishing holomorphic section of L over U, $\varphi = -\log \|e_L\|^2$ and $\omega = i\partial\bar{\partial}\varphi$. We also have the following trivialization of X:

$$U \times S^1 \cong X|_U, (z;\theta) \mapsto e^{i\theta} \frac{e_L^*|_z}{\|e_L^*|_z\|}. \tag{92}$$

X has a structure of a contact manifold. Let ρ be a smooth function in a neighborhood of X in L^*, such that $\rho > 0$ in the open unit disk bundle, $\rho|_X = 0$ and $d\rho|_X \neq 0$. Then we have a contact one-form on X

$$\alpha = -\mathrm{Re}(i\bar{\partial}\rho)|_X, \tag{93}$$

well-defined up to multiplication by a positive smooth function. We fix a choice of ρ by

$$\rho(x) = -\log \|x\|_h^2, \quad x \in L^*,$$

then in local trivialization of X (92), we have

$$\alpha = d\theta - \frac{1}{2} d^c\varphi(z). \tag{94}$$

X is also a strictly pseudoconvex CR manifold. The *CR structure* on X is defined as follows: The kernel of α defines a horizontal hyperplane bundle

$$HX := \ker\alpha \subset TX, \tag{95}$$

invariant under J since $\ker\alpha = \ker d\rho \cap \ker d^c\rho$. Thus we have a splitting

$$TX \otimes \mathbb{C} \cong H^{1,0}X \oplus H^{0,1}X \oplus \mathbb{C}R.$$

A function $f : X \to \mathbb{C}$ is CR-holomorphic, if $df|_{H^{0,1}X} = 0$.

A holomorphic section s_k of L^k determines a CR-function $\hat{s}_k$ on X by

$$\hat{s}_k(x) := \langle x^{\otimes k}, s_k\rangle, \quad x \in X \subset L^*.$$

Furthermore $\hat{s}_k$ is of degree k under the canonical S^1 action r_θ on X, $\hat{s}_k(r_\theta x) = e^{ik\theta}\hat{s}_k(x)$. The inner product on $L^2(M, L^k)$ is given by

$$\langle s_1, s_2\rangle := \int_M h^k(s_1(z), s_2(z)) d\,\mathrm{Vol}_M(z), \quad d\,\mathrm{Vol}_M = \frac{\omega^m}{m!},$$

and inner product on $L^2(X)$ is given by

$$\langle f_1, f_2\rangle := \int_X f_1(x)\overline{f_2(x)} d\,\mathrm{Vol}_X(x), \quad d\,\mathrm{Vol}_X = \frac{\alpha}{2\pi} \wedge \frac{(d\alpha)^m}{m!}.$$

Thus, sending $s_k \mapsto \hat{s}_k$ is an isometry.

10.2 Szegö Kernel on X

On the circle bundle X over M, we define the orthogonal projection from $L^2(X)$ to the CR-holomorphic subspace $\mathcal{H}(X) = \hat{\oplus}_{k\geq 0}\mathcal{H}_k(X)$, and degree-$k$ subspace $\mathcal{H}_k(X)$:

$$\hat{\Pi} : L^2(X) \to \mathcal{H}(X), \quad \hat{\Pi}_k : L^2(X) \to \mathcal{H}_k(X), \quad \hat{\Pi} = \sum_{k\geq 0} \hat{\Pi}_k.$$

The Schwarz kernels $\hat{\Pi}_k(x, y)$ of $\hat{\Pi}_k$ is called the degree-k Szegö kernel, i.e.

$$(\hat{\Pi}_k F)(x) = \int_X \hat{\Pi}_k(x, y) F(y) d\,\mathrm{Vol}_X(y), \quad \forall F \in L^2(X).$$

If we have an orthonormal basis $\{\hat{s}_{k,j}\}_j$ of $\mathcal{H}_k(X)$, then

$$\hat{\Pi}_k(x, y) = \sum_j \hat{s}_{k,j}(x)\overline{\hat{s}_{k,j}(y)}.$$

The degree-k kernel can be extracted as the Fourier coefficient of $\hat{\Pi}(x, y)$

$$\hat{\Pi}_k(x, y) = \frac{1}{2\pi} \int_0^{2\pi} \hat{\Pi}(r_\theta x, y) e^{-ik\theta} d\theta. \tag{96}$$

We refer to (96) as the *semi-classical Bergman kernels.*

10.3 Boutet de Monvel-Sjöstrand Parametrix for the Szegö Kernel

Near the diagonal in $X \times X$, there exists a parametrix due to Boutet de Monvel-Sjöstrand [6] for the Szegö kernel of the form,

$$\hat{\Pi}(x, y) = \int_{\mathbb{R}^+} e^{\sigma \hat{\psi}(x,y)} s(x, y, \sigma) d\sigma + \hat{R}(x, y), \tag{97}$$

where $\hat{\psi}(x, y)$ is the almost-CR-analytic extension of $\hat{\psi}(x, x) = -\rho(x) = \log \|x\|^2$, and $s(x, y, \sigma) = \sigma^m s_m(x, y) + \sigma^{m-1} s_{m-1}(x, y) + \cdots$ has a complete asymptotic expansion. In local trivialization (92),

$$\hat{\psi}(x, y) = i(\theta_x - \theta_y) + \psi(z, w) - \frac{1}{2}\varphi(z) - \frac{1}{2}\varphi(w),$$

where $\psi(z, w)$ is the almost analytic extension of $\varphi(z)$.

10.4 Lifting the Hamiltonian Flow to a Contact Flow on X_h

In this section we review the definition of the lifting of a Hamiltonian flow to a contact flow, following [ZZ17, Section 3.1]. Let $H : M \to \mathbb{R}$ be a Hamiltonian function on (M, ω). Let ξ_H be the Hamiltonian vector field associated with H, such that $dH = \iota_{\xi_H}\omega$. The purpose of this section is to lift ξ_H to a contact vector field $\hat{\xi}_H$ on X. Let α denote the contact 1-form (94) on X, and R the corresponding Reeb vector field determined by $\langle \alpha, R\rangle = 1$ and $\iota_R d\alpha = 0$. One can check that $R = \partial_\theta$.

Definition 10.1

(1) The horizontal lift of ξ_H is a vector field on X denoted by ξ_H^h. It is determined by

$$\pi_* \xi_H^h = \xi_H, \quad \langle \alpha, \xi_H^h\rangle = 0.$$

(2) The contact lift of ξ_H is a vector field on X denoted by $\hat{\xi}_H$. It is determined by

$$\pi_* \hat{\xi}_H = \xi_H, \quad \mathcal{L}_{\hat{\xi}_H}\alpha = 0.$$

Lemma 10.2 *The contact lift $\hat{\xi}_H$ is given by*

$$\hat{\xi}_H = \xi_H^h - HR.$$

The Hamiltonian flow on M generated by ξ_H is denoted by g^t

$$g^t : M \to M, \quad g^t = \exp(t\xi_H).$$

The contact flow on X generated by $\hat{\xi}_H$ is denoted by $\hat{g}^t$

$$\hat{g}^t : X \to X, \quad \hat{g}^t = \exp(t\hat{\xi}_H).$$

Lemma 10.3 *In local trivialization* (92)*, we have a useful formula for the flow, $\hat{g}^t$ has the form (see* [ZZ17, Lemma 3.2]*):*

$$\hat{g}^t(z, \theta) = (g^t(z), \;\; \theta + \int_0^t \frac{1}{2}\langle d^c\varphi, \xi_H\rangle(g^s(z))ds - tH(z)).$$

Since $\hat{g}^t$ preserves α it preserves the horizontal distribution $H(X_h) = \ker\alpha$, i.e.

$$D\hat{g}^t : H(X)_x \to H(X)_{\hat{g}^t(x)}. \tag{98}$$

It also preserves the vertical (fiber) direction and therefore preserves the splitting $V \oplus H$ of TX. Its action in the vertical direction is determined by Lemma 10.3. When g^t is non-holomorphic, $\hat{g}^t$ is not CR-holomorphic, i.e. does not preserve the horizontal complex structure J or the splitting of $H(X) \otimes \mathbb{C}$ into its $\pm i$ eigenspaces.

11 Appendix

11.1 Appendix on the Airy Function

The Airy function is defined by,

$$Ai(z) = \frac{1}{2\pi i} \int_L e^{v^3/3 - zv} dv,$$

where L is any contour that beings at a point at infinity in the sector $-\pi/2 \leq \arg(v) \leq -\pi/6$ and ends at infinity in the sector $\pi/6 \leq \arg(v) \leq \pi/2$. In the region $|\arg z| \leq (1-\delta)\pi$ in $\mathbb{C} - \{\mathbb{R}_-\}$ write $v = z^{\frac{1}{2}} + it^{\frac{1}{2}}$ on the upper half of L and $v = z^{\frac{1}{2}} - it^{\frac{1}{2}}$ in the lower half. Then

$$\mathrm{Ai}(z) = \Psi(z) e^{-\frac{2}{3} z^{3/2}}, \text{ with } \Psi(z) \sim z^{-1/4} \sum_{j=0}^{\infty} a_j z^{-3j/2}, \quad a_0 = \frac{1}{4}\pi^{-3/2}. \tag{99}$$

11.2 Appendix on Laguerre Functions

The Laguerre polynomials $L_k^{\alpha}(x)$ of degree k and of type α on $[0, \infty)$ are defined by

$$e^{-x} x^{\alpha} L_k^{\alpha}(x) = \frac{1}{k!} \frac{d^k}{dx^k} (e^{-x} x^{k+\alpha}). \tag{100}$$

They are solutions of the Laguerre equation(s),

$$xy'' + (\alpha + 1 - x) y(x)' + ky(x) = 0.$$

For fixed α they are orthogonal polynomials of $L^2(\mathbb{R}_+, e^{-x} x^{\alpha} dx)$. An orthonormal basis is given by

$$\mathcal{L}_k^{\alpha}(x) = \left(\frac{\Gamma(k+1)}{\Gamma(k+\alpha+1)} \right)^{\frac{1}{2}} L_k^{\alpha}(x).$$

We will have occasion to use the following generating function:

$$\sum_{k=0}^{\infty} L_k^{\alpha}(x) w^k = (1-w)^{-\alpha-1} e^{-\frac{w}{1-w} x}$$

.

The most useful integral representation for the Laguerre functions is

$$e^{-x/2} L_n^{(\alpha)}(x) = (-1)^n \oint \frac{e^{-\frac{x}{2}\cdot\frac{1-z}{1+z}}}{z^n (1+z)^{\alpha+1}} \frac{dz}{2\pi i z}, \tag{101}$$

where the contour encircles the origin once counterclockwise. Equivalently,

$$e^{-x/2} L_n^{(\alpha)}(x) = \frac{(-1)^n}{2^\alpha} \frac{1}{2\pi i} \int^{1+} e^{-xz/2} \left(\frac{1+z}{1-z}\right)^{\nu/4} (1-z^2)^{\frac{\alpha-1}{2}} dz, \tag{102}$$

where $\nu = 4n + \alpha + 2$ and the contour encircles $z = 1$ in the positive direction and closes at $\mathrm{Re} z = \infty$, $|\mathrm{Im} z| =$ constant. In (5.9) of [17] the Laguerre functions are represented as the oscillatory integrals,

$$e^{-\nu t/2} L_n^{\alpha}(\nu t) = \frac{(-1)^n}{2^\alpha} \frac{1}{2\pi i} \int_{\mathcal{L}} [1 - z^2(u)]^{\frac{\alpha-1}{2}} \exp\{\nu \left(\frac{u^3}{3} - B^2(t) u\right)\} du, \tag{103}$$

where $\nu = 4n + 2\alpha + 2$ and $B(t)$ is defined in (5.5) of [17] and $\mathcal{L}$ is a branch of the hyperbolic curve in the right half-plane.

References

1. S. Agmon, *Lectures on Exponential Decay of Solutions of Second-Order Elliptic Equations: Bounds on Eigenfunctions of N-body Schrödinger Operators*. Mathematical Notes, vol. 29 (Princeton University Press, Princeton, NJ; University of Tokyo Press, Tokyo, 1982)
2. R. Berman, Bergman kernels and equilibrium measures for line bundles over projective manifolds. Am. J. Math. **131**(5), 1485–1524 (2009)
3. W.E. Bies, E.J. Heller, Nodal structure of chaotic eigenfunctions. J. Phys. A **35**(27), 5673–5685 (2002)
4. P. Bleher, B. Shiffman, S. Zelditch, Universality and scaling of correlations between zeros on complex manifolds. Invent. Math. **142**(2), 351–395 (2000)
5. L. Boutet de Monvel, V. Guillemin, *The Spectral Theory of Toeplitz Operators*. Annals of Mathematics Studies, vol. 99 (Princeton University Press, Princeton, 1981)
6. L. Boutet de Monvel, J. Sjöstrand, Sur la singularité des noyaux de Bergman et de Szegö. Asterisque **34–35**, 123–164 (1976)
7. K.E. Cahill, R.J. Glauber, Ordered expansions in boson amplitude operators. Phys. Rev. **177**, 1857–1881 (1969)
8. K.E. Cahill, R.J. Glauber, Density operators and quasiprobability distributions. Phys. Rev. **177**(5), 1882–1902 (1969)
9. T. Can, P.J. Forrester, G. Tellez, P. Wiegmann, Singular behavior at the edge of Laughlin states. Phys. Rev. B **89**, 235137 (2014). arXiv:1307.3334
10. Y. Canzani, B. Hanin, Scaling limit for the kernel of the spectral projector and remainder estimates in the pointwise Weyl law. Anal. PDE **8**(7), 1707–1731 (2015)
11. Y. Canzani, J.A. Toth, Nodal sets of Schroedinger eigenfunctions in forbidden regions. Ann. Henri Poincare **17**(11), 3063–3087 (2016) (arXiv:1502.00732)
12. L. Charles, B. Estienne, Entanglement entropy and Berezin-Toeplitz operators. Comm. Math. Phys. **376**(1), 521–554 (2020). (arXiv:1803.03149)

13. J. Chazarain, Spectre d'un hamiltonien quantique et mécanique classique. Commun. Partial Differ. Equ. **5**(6), 595–644 (1980)
14. I. Daubechies, Coherent states and projective representation of the linear canonical transformations. J. Math. Phys. **21**(6), 1377–1389 (1980)
15. H. Donnelly, C. Fefferman, Nodal sets of eigenfunctions on Riemannian manifolds. Invent. Math. **93**, 161–183 (1988)
16. G.B. Folland, *Harmonic Analysis in Phase Space*. Annals of Mathematics Studies, vol. 122 (Princeton University Press, Princeton, NJ, 1989)
17. C.L. Frenzen, R. Wong, Uniform asymptotic expansions of Laguerre polynomials. SIAM J. Math. Anal. **19**(5), 1232–1248 (1988)
18. P. Gaspar, M. Guaraco, The Allen-Cahn equation on closed manifolds. Calc. Var. Partial Differ. Equ. **57**(4), 101 (2018)
19. V. Guillemin, A. Uribe, Z. Wang, Band invariants for perturbations of the harmonic oscillator. J. Funct. Anal. **263**(5), 1435–1467 (2012)
20. V. Guillemin, A. Uribe, Z. Wang, Canonical forms for perturbations of the harmonic oscillator. New York J. Math. **21**, 163–180 (2015)
21. B. Hanin, S. Zelditch, Universality of Schrodinger scaling asymptotics around the caustic (in preparation)
22. H. Hedenmalm, A. Wennman, Planar orthogonal polynomials and boundary universality in the random normal matrix model, arXiv:1710.06493
23. H. Hedenmalm, A. Wennman, Off-spectral analysis of Bergman kernels, Comm. Math. Phys. **373**(3), 1049–1083 (2020) (arXiv 1805.00854)
24. P.D. Hislop, I.M. Sigal, *Introduction to Spectral Theory. With Applications to Schrödinger Operators*. Applied Mathematical Sciences, vol. 113 (Springer, New York, 1996)
25. M. Hitrik, J. Sjoestrand, Two Minicourses on analytic microlocal analysis, algebraic and analytic microlocal analysis, 483–540, springer Proc. Math, Stat. vol. **269**, (Springer, Cham, 2018)
26. A.J.E.M. Janssen, S. Zelditch, Szegö limit theorems for the harmonic oscillator. Trans. Am. Math. Soc. **280**(2), 563–587 (1983)
27. L. Jin, Semiclassical Cauchy estimates and applications. Trans. Amer. Math. Soc. **369**(2), 975–995 (2017). (arXiv:1302.5363)
28. B. Levikson, On the behavior of a certain class of approximation operators for discontinuous functions. Acta Math. Acad. Sci. Hungar. **33**(3–4), 299–306 (1979)
29. G.G. Lorentz, *Bernstein Polynomials*, 2nd edn. (Chelsea Publishing Co., New York, 1986)
30. G. Mirakyan, Approximation des fonctions continues au moyen de polynomes de la forme $e^{-nx} \sum_{k=0}^{m} C_{k,n} x^j$. C. R. (Doklady) Acad. Sci. URSS (N.S.) **31**, 201–205 (1941)
31. E. Omey, Note on operators of Szasz-Mirakyan type. J. Approx. Theory **47**(3), 246–254 (1986)
32. M.A. Pinsky, M. Taylor, Pointwise Fourier inversion: a wave equation approach. J. Fourier Anal. Appl. **3**(6), 647–703 (1997). MR1481629
33. F. Pokorny, M. Singer, Toric partial density functions and stability of toric varieties. Math. Ann. **358**(3–4), 879–923 (2014)
34. D. Robert, *Autour de l'approximation semi-classique*. Progress in Mathematics, vol. 68 (Birkhäuser Boston, Inc., Boston, MA, 1987)
35. J. Ross, M. Singer, Asymptotics of partial density functions for divisors. J. Geom. Anal. **27**(3), 1803–1854 (2017) (arXiv:1312.1145)
36. S. Bernard, S. Zelditch, Asymptotics of almost holomorphic sections of ample line bundles on symplectic manifolds. J. Reine Angew. Math. **544**, 181–222 (2002)
37. B. Shiffman, S. Zelditch, Random polynomials with prescribed Newton polytope. J. Am. Math. Soc. **17**(1), 49–108
38. O. Szasz, Generalization of S. Bernstein's polynomials to the infinite interval. J. Res. Nat. Bur. Stand. **45**, 239–245 (1950)
39. S. Thangavelu, *Lectures on Hermite and Laguerre expansions*. With a preface by Robert S. Strichartz. Mathematical Notes, vol. 42 (Princeton University Press, Princeton, NJ, 1993)

40. S. Thangavelu, Hermite and Laguerre semigroups: some recent developments. Orthogonal families and semigroups in analysis and probability, 251–284, Semin. Congr., 25, Soc. Math. France, Paris, 2012
41. X.-G. Wen, *Quantum Field Theory of Many-Body Systems*. Oxford Graduate Texts (Oxford University Press, Oxford, 2004)
42. P. Wiegmann, Nonlinear hydrodynamics and fractionally quantized solitons at the fractional quantum Hall edge. Phys. Rev. Lett. **108**, 206810 (2012)
43. E.P. Wigner, On the quantum correction for thermodynamic equilibrium. Phys. Rev. **40**, 749–759 (1932)
44. L. Xie, T. Xie, Approximation theorems for localized Szasz-Mirakjan operators. J. Approx. Theory **152**(2), 125–134 (2008)
45. S. Zelditch, Index and dynamics of quantized contact transformations. Ann. Inst. Fourier **47**, 305–363 (1997). MR1437187, Zbl 0865.47018
46. S. Zelditch, Bernstein polynomials, Bergman kernels and toric Khler varieties. J. Symplectic Geom. **7**, 1–26 (2009)
47. M. Zworski, *Semiclassical Analysis*. Graduate Studies in Mathematics, vol. 138 (American Mathematical Society, Providence, RI, 2012). MR2952218

Printed in the United States
by Baker & Taylor Publisher Services